AQA
A-Level

Further Mathematics
Core Year 2

2

Authors

Ben Sparks

Claire Baldwin

Series editors

Roger Porkess

Catherine Berry

Contributing editors

Heather Davis

Nicola Trubridge

Mark Heslop

Owen Toller

Dr Jane Lawson

Approval message from AQA

This textbook has been approved by AQA for use with our qualification. This means that we have checked that it broadly covers the specification and we are satisfied with the overall quality. Full details of our approval process can be found on our website.

We approve textbooks because we know how important it is for teachers and students to have the right resources to support their teaching and learning. However, the publisher is ultimately responsible for the editorial control and quality of this book.

Please note that when teaching the *AQA A-Level Further Mathematics* course, you must refer to AQA's specification as your definitive source of information. While this book has been written to match the specification, it cannot provide complete coverage of every aspect of the course.

A wide range of other useful resources can be found on the relevant subject pages of our website: www.aqa.org.uk.

DYNAMIC

HODDER EDUCATION
AN HACHETTE UK COMPANY

The Publishers would like to thank the following for permission to reproduce copyright material.

Photo credits

p.1 © Elena Elisseeva - Shutterstock; **p.45** © Iryna Tiumentseva – Shutterstock; **p.77** © Hans / stock.adobe.com; **p.90** © Imagestate Media (John Foxx) / Electrique V3047; **p.105** © Imagestate Media (John Foxx) / Global Trave; SS83; **p.128** © Digoarpi - iStock via Thinkstock; **p.147** © D1min / stock.adobe.com; **p.156** *tl*, *tr* and *bl* © Geograph.com, *br* © David Burrows – Shutterstock; **p.176** By Halibutt / Wikimedia Commons / CC-BY-SA 3.0 (https://creativecommons.org/licenses/by-sa/3.0/deed.en); **p.197** © Snehit / stock.adobe.com; **p.217** © marinkin / stock.adobe.com; **p.247** © Brian Kinney – Fotolia; **p.277** © Ozgur Coskun / stock.adobe.com; **p.297** © Overflightstock Ltd/Alamy Stock Photo; **p.329** © PaulVinten – iStock via Thinkstock; **p.353** © World History Archive/Alamy Stock Photo.

Orders: please contact Bookpoint Ltd, 130 Park Drive, Milton Park, Abingdon, Oxon OX14 4SE. Telephone: +44 (0)1235 827827. Fax: +44 (0)1235 400401. Email education@bookpoint.co.uk Lines are open from 9 a.m. to 5 p.m., Monday to Saturday, with a 24-hour message answering service. You can also order through our website: www.hoddereducation.co.uk

ISBN: 978 1 4718 8332 3

© Ben Sparks, Claire Baldwin, Roger Porkess, Heather Davis, Owen Toller, Jan Dangerfield and MEI 2018

First published in 2018 by

Hodder Education,
An Hachette UK Company
Carmelite House
50 Victoria Embankment
London EC4Y 0DZ
www.hoddereducation.co.uk

Impression number   10 9 8 7 6 5 4 3 2 1
Year   2022 2021 2020 2019 2018

Cover photo © 3alda/iStock/Thinkstock/Getty Images

Typeset in Bembo Std, 11/13 pts. by Aptara, Inc.

Printed in Italy

A catalogue record for this title is available from the British Library.

Contents

Contents

Getting the most from this book

Mathematics is not only a beautiful and exciting subject in its own right, but also one that underpins many other branches of learning. It is consequently fundamental to our national well-being.

This book covers the compulsory core content of Advanced Level Further Mathematics study, following on from the Year 1/AS Further Mathematics book. The optional applied content is covered in the Mechanics and Statistics books, and the remaining option in the Discrete book.

Between 2014 and 2016 A-level Mathematics and Further Mathematics were substantially revised, for first teaching in 2017. Major changes included increased emphasis on:

- Problem solving
- Mathematical proof
- Use of ICT
- Modelling
- Working with large data sets in statistics.

This book embraces these ideas. A large number of exercise questions involve elements of problem solving. The ideas of **mathematical proof**, rigorous logical argument and mathematical modelling are also included in suitable exercise questions throughout the book.

The use of **technology**, including graphing software, spreadsheets and high specification calculators, is encouraged wherever possible, for example in the Activities used to introduce some of the topics. In particular, readers are expected to have access to a calculator which handles matrices up to order 3×3. Places where ICT can be used are highlighted by a **T** icon. Margin boxes highlight situations where the use of technology – such as graphical calculators or graphing software – can be used to further explore a particular topic.

Throughout the book the emphasis is on understanding and interpretation rather than mere routine calculations, but the various exercises do nonetheless provide plenty of scope for practising basic techniques. The exercise questions are split into three bands. Band 1 questions are designed to reinforce basic understanding; Band 2 questions are broadly typical of what might be expected in an examination, but you should refer to the AQA SAMs and practice papers; Band 3 questions explore around the topic and some of them are rather more demanding. In addition, extensive online support, including further questions, is available by subscription to MEI's Integral website, integralmaths.org.

In addition to the exercise questions, there are three sets of Practice questions, covering groups of chapters. These include identified questions requiring **problem solving** **PS**, **mathematical proof** **MP**, **use of ICT** **T** and **modelling** **M**.

This book is written on the assumption that readers are studying or have studied A-level Mathematics. It can be studied alongside the Year 2/A-level Mathematics book, or after studying A-level Mathematics. There are also places where an understanding of the topics depends on knowledge from earlier in the book or in the Year 1/AS Mathematics book, the Year 2 Mathematics book or the Year 1/AS Further Mathematics book, and this is flagged up in the Prior knowledge boxes (as well as in specific review sections – as explained in the paragraph below). This should be seen as an invitation to those who have problems with the particular topic to revisit it. At the end of each chapter there is a list of key points covered as well as a summary of the new knowledge (learning outcomes) that readers should have gained.

This book follows on from *AQA A-level Further Mathematics Year 1 (AS)* and most readers will be familiar with the material covered in it. However, there may be occasions when they want to check on topics in the earlier book: we have therefore included three short Review chapters to provide a condensed summary of the work that was covered in the earlier book, including one or more exercises. In addition there are a number of chapters that begin with a Review section and exercise, and then go on to new work based on it. Confident readers may choose to miss out the Review material, and only refer to these parts of the book when they are uncertain about particular topics. Others, however, will find it helpful to work through some or all of the Review material to consolidate their understanding of the first year work.

Two common features of the book are Activities and Discussion points. These serve rather different purposes. The Activities are designed to help readers get into the thought processes of the new work that they are about to meet; having done an Activity, what follows will seem much easier. The Discussion points invite readers to talk about particular points with their fellow students and their teacher and so enhance their understanding. Another feature is a Caution icon ❶, highlighting points where it is easy to go wrong.

Answers to all exercise questions and practice questions are provided at the back of the book, and also online at www.hoddereducation.co.uk/AQAFurtherMathsYear2

This is a 4th edition AQA textbook so much of the material is well tried and tested. However, as a consequence of the changes to A Level requirements in further mathematics, large parts of the book are either new material or have been very substantially rewritten.

Catherine Berry

Roger Porkess

Prior knowledge

This book follows on from AQA A-level Further Mathematics Year 1 (AS). It is designed so that it can be studied alongside AQA A-level Mathematics Year 2. Knowledge of the work in AQA A-level Mathematics Year 1 (AS) is assumed.

- **Chapter 1: Vectors 1** reviews and develops the work in Chapter 12 of AQA A-level Further Mathematics Year 1.

- **Review: Matrices and transformations** reviews the work covered in Chapter 1 of AQA A-level Further Mathematics Year 1.

- **Chapter 2: Matrices 1** reviews and builds on the work in Chapter 11 of AQA A-level Further Mathematics Year 1.

- **Chapter 3: Conics** reviews and builds on the work in Chapter 4 of AQA A-level Further Mathematics Year 1.

- **Chapter 4: Series and induction** reviews and develops the work introduced in Chapter 6 of AQA A-level Further Mathematics Year 1. It requires knowledge of partial fractions which is covered in chapter 7 of AQA A-level Mathematics Year 2.

- **Chapter 5: Further algebra and graphs** reviews and builds on work in Chapter 10 of AQA A-level Further Mathematics Year 1.

- **Chapter 6: Further calculus** assumes knowledge of the calculus from AQA A-level Mathematics Year 1 (Chapters 10 and 11). You also need to be able to differentiate and integrate exponential functions, the function $\frac{1}{x}$ and related functions, and trigonometric functions (covered in Chapters 9 and 10 of AQA A-level Mathematics Year 2). You need to be able to differentiate a function implicitly (Chapter 9 of AQA A-level Mathematics Year 2). You should also be familiar with the inverse trigonometric functions (covered in Chapter 6 of AQA A-level Mathematics Year 2). You also need to have covered the work on partial fractions in Chapter 7 of AQA A-level Mathematics Year 2.

- **Chapter 7: Polar coordinates** reviews and builds on the work in Chapter 9 of AQA A-level Further Mathematics Year 1. The chapter assumes knowledge of radians (covered in Chapter 2 of AQA A-level Mathematics Year 2). You will need to be familiar with the reciprocal trigonometric functions sec and cosec (introduced in Chapter 6 of AQA A-level Mathematics Year 2), and the double angle formulae (introduced in Chapter 8 of AQA A-level Mathematics Year 2). You also need to be confident in integration (covered in Chapter 11 of AQA A-level Mathematics Year 1) and know how to integrate simple trigonometric functions (Chapter 10 of AQA A-level Mathematics Year 2).

- **Chapter 8: Series and limits** reviews and builds on the work on Maclaurin series in Chapter 6 of AQA A-level Further Mathematics Year 1. This chapter uses differentiation of simple exponential, logarithmic and trigonometric functions, covered in Chapter 9 of AQA A-level Mathematics Year 2.

- **Chapter 9: Matrices 2** builds on the work covered in Chapter 2 of this book.

- **Review: Complex numbers** reviews the work in Chapters 2 and 8 of AQA A-level Further Mathematics Year 1.

- **Chapter 10: Hyperbolic functions** reviews and builds on the work in Chapter 5 of AQA A-level Further Mathematics Year 1. It uses the ideas of the domain and range of a function, and an inverse function, covered in Chapter 4 of AQA A-level Mathematics Year 2. It uses similar techniques to those covered in Chapter 6 of this book.

- **Chapter 11: Further integration** uses all the calculus techniques covered in AQA A-level Mathematics Year 1 and AQA A-level Mathematics Year 2, and in Chapters 6 and 10 of this book.

- **Review: Roots of polynomials** reviews the work covered in Chapter 3 of AQA A-level Further Mathematics Year 1.

Prior knowledge

- **Chapter 12: First order differential equations** uses all the calculus techniques covered in AQA A-level Mathematics Year 1 and AQA A-level Mathematics Year 2, and in Chapters 6 and 10 of this book.

- **Chapter 13: Numerical methods** assumes knowledge of the work on numerical integration covered in Chapter 14 of AQA A-level Mathematics Year 2, and of first order differential equations covered in Chapters 12 of this book.

- **Chapter 14: Complex numbers** builds on the work reviewed in Review: Complex numbers. You need to be familiar with trigonometric identities such as the double angle formulae (covered in Chapter 8 of AQA A-level Mathematics Year 2) and you need to know about geometric series (covered in Chapter 3 of AQA A-level Mathematics Year 2).

- **Chapter 15: Vectors 2** builds on the work on lines and planes in Chapter 1 of this book and in chapter 12 of AQA A-level Further Mathematics Year 1.

- **Chapter 16: Second order differential equations** uses all the calculus techniques covered in AQA A-level Mathematics Year 1 and AQA A-level Mathematics Year 2, and in Chapters 6 and 10 of this book. It follows on from Chapter 9 of this book. You also need to know how to write an expression of the form $a\cos\theta + b\sin\theta$ in the form $r\sin(\theta + \alpha)$. This is covered in Chapter 9 of AQA A-level Mathematics Year 2.

1 Vectors 1

I pulled out, on the spot, a pocket book, which still exists, and made an entry, on which, at the very moment, I felt it might be worth my while to expend the labour of at least ten (or it might be fifteen) years to come. But then it is fair to say that this was because I felt a problem to have been at that moment solved, an intellectual want relieved, which had haunted me for at least fifteen years before.

William R. Hamilton, writing on 16th October 1858, about his invention of quarternions on 16th October 1843

Discussion points

→ A taut zip wire can be modelled as a straight line. How can you find the equation of a line in three dimensions?

→ How could you work out the distance between the two zip wires?

Review: Working with vectors

Prior knowledge

You need to be able to use the language of vectors, including the terms magnitude, direction and position vector. You should also be able to find the distance between two points represented by position vectors and be able to add and subtract vectors and multiply a vector by a scalar.

- A vector quantity has magnitude and direction.
- A scalar quantity has magnitude only.
- Vectors are typeset in bold, **a** or **OA**, or in the form $\overrightarrow{OA}$.
  They are handwritten either in the underlined form $\underline{a}$, or as $\overrightarrow{OA}$.

- Unit vectors in the x, y and z directions are denoted by $\mathbf{i}$, $\mathbf{j}$ and $\mathbf{k}$ respectively.
- The resultant of two (or more) vectors is found by the sum of the vectors. A resultant vector is usually denoted by a double-headed arrow.
- The position vector $\overrightarrow{OP}$ of a point P is the vector joining the origin to P.
- The vector $\overrightarrow{AB} = \mathbf{b} - \mathbf{a}$, where $\mathbf{a}$ and $\mathbf{b}$ are the position vectors of A and B.
- The length (or modulus or magnitude) of the vector $\mathbf{r}$ is written as r or as $|\mathbf{r}|$.

$$\mathbf{r} = a\mathbf{i} + b\mathbf{j} + c\mathbf{k} \Rightarrow |\mathbf{r}| = \sqrt{a^2 + b^2 + c^2}$$

The scalar product

Figure 1.1 shows two vectors $\begin{pmatrix} a_1 \\ a_2 \end{pmatrix}$ and $\begin{pmatrix} b_1 \\ b_2 \end{pmatrix}$. The angle between these two vectors can be found using the **scalar product**:

$$\mathbf{a} \cdot \mathbf{b} = |\mathbf{a}||\mathbf{b}| \cos\theta$$

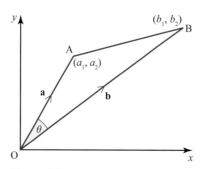

Figure 1.1

Using the column format, the scalar product can be written as:

$$\mathbf{a} \cdot \mathbf{b} = \begin{pmatrix} a_1 \\ a_2 \end{pmatrix} \cdot \begin{pmatrix} b_1 \\ b_2 \end{pmatrix} = a_1 b_1 + a_2 b_2$$

In three dimensions this is extended to:

$$\mathbf{a} \cdot \mathbf{b} = \begin{pmatrix} a_1 \\ a_2 \\ a_3 \end{pmatrix} \cdot \begin{pmatrix} b_1 \\ b_2 \\ b_3 \end{pmatrix} = a_1 b_1 + a_2 b_2 + a_3 b_3$$

Example 1.1

Find the acute angle between the vectors:

(i) $\mathbf{a} = \begin{pmatrix} 2 \\ -3 \end{pmatrix}$ and $\mathbf{b} = \begin{pmatrix} 4 \\ 5 \end{pmatrix}$

(ii) $\mathbf{c} = \begin{pmatrix} 1 \\ 2 \\ 1 \end{pmatrix}$ and $\mathbf{d} = \begin{pmatrix} 0 \\ 4 \\ -3 \end{pmatrix}$

Solution

(i)

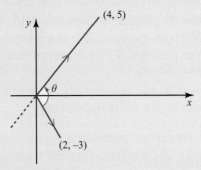

Figure 1.2

$$|\mathbf{a}| = \sqrt{2^2 + (-3)^2} = \sqrt{13} \quad \text{and} \quad |\mathbf{b}| = \sqrt{4^2 + 5^2} = \sqrt{41}$$

Then

$$\mathbf{a} \cdot \mathbf{b} = |\mathbf{a}||\mathbf{b}| \cos\theta$$

$$\Rightarrow (2 \times 4) + (-3 \times 5) = \sqrt{13}\sqrt{41} \cos\theta$$

$$\Rightarrow \frac{-7}{\sqrt{13}\sqrt{41}} = \cos\theta$$

$$\Rightarrow \theta = 107.7°$$

So the acute angle between the vectors is $180° - 107.7° = 72.3°$.

(ii) $\quad |\mathbf{c}| = \sqrt{1^2 + 2^2 + 1^2} = \sqrt{6} \quad \text{and} \quad |\mathbf{d}| = \sqrt{0^2 + 4^2 + (-3)^2} = 5$

Then

$$\mathbf{c} \cdot \mathbf{d} = |\mathbf{c}||\mathbf{d}| \cos\theta$$

$$\Rightarrow (1 \times 0) + (2 \times 4) + (1 \times -3) = \sqrt{6} \times 5 \times \cos\theta$$

$$\Rightarrow 5 = 5\sqrt{6} \cos\theta$$

$$\Rightarrow \frac{1}{\sqrt{6}} = \cos\theta$$

$$\Rightarrow \theta = 65.9°$$

Notes

1 The scalar product, unlike a vector, has size but no direction.

2 The scalar product of two vectors is **commutative.** This is because multiplication of numbers is commutative. For example:

$$\begin{pmatrix} 3 \\ -4 \end{pmatrix} \cdot \begin{pmatrix} 1 \\ 5 \end{pmatrix} = (3 \times 1) + (-4 \times 5) = (1 \times 3) + (5 \times -4) = \begin{pmatrix} 1 \\ 5 \end{pmatrix} \cdot \begin{pmatrix} 3 \\ -4 \end{pmatrix}$$

3 When finding the scalar product, using the vectors $\overrightarrow{AB}$ and $\overrightarrow{CB}$ (both directed towards the point B), or the vectors $\overrightarrow{BA}$ and $\overrightarrow{BC}$ (both directed away from the point B) gives the angle θ. This angle could be acute or obtuse.

However, if you use vectors $\overrightarrow{AB}$ (directed towards B) and $\overrightarrow{BC}$ (directed away from B), then you will obtain the angle $180° - \theta$ instead, as shown in Figure 1.3.

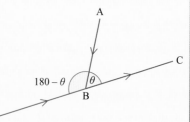

Figure 1.3

4 If two vectors are **perpendicular**, then the angle between them is $90°$.

Since $\cos 90° = 0$, it follows that if vectors **a** and **b** are perpendicular then $\mathbf{a} \cdot \mathbf{b} = 0$. Conversely, if the scalar product of two non-zero vectors is zero, they are perpendicular.

Example 1.2

A, B and C have position vectors $-4\mathbf{i} + 17\mathbf{j} - \mathbf{k}$, $5\mathbf{i} + 2\mathbf{j} - 2\mathbf{k}$ and $\mathbf{i} + 5\mathbf{j} - 9\mathbf{k}$ respectively.

(i)   Prove that triangle ABC is right angled.

(ii) Find the area of ABC correct to three significant figures.

Solution

(i) $\overrightarrow{AB} = \begin{pmatrix} 5 \\ 2 \\ -2 \end{pmatrix} - \begin{pmatrix} -4 \\ 17 \\ -1 \end{pmatrix} = \begin{pmatrix} 9 \\ -15 \\ -1 \end{pmatrix}$

$\overrightarrow{BC} = \begin{pmatrix} 1 \\ 5 \\ -9 \end{pmatrix} - \begin{pmatrix} 5 \\ 2 \\ -2 \end{pmatrix} = \begin{pmatrix} -4 \\ 3 \\ -7 \end{pmatrix}$

$\overrightarrow{AC} = \begin{pmatrix} 1 \\ 5 \\ -9 \end{pmatrix} - \begin{pmatrix} -4 \\ 17 \\ -1 \end{pmatrix} = \begin{pmatrix} 5 \\ -12 \\ -8 \end{pmatrix}$

$\overrightarrow{AB} \cdot \overrightarrow{BC} = \begin{pmatrix} 9 \\ -15 \\ -1 \end{pmatrix} \cdot \begin{pmatrix} -4 \\ 3 \\ -7 \end{pmatrix} = (9 \times -4) + (-15 \times 3) + (-1 \times -7) = -74$

> The scalar product of vectors $\overrightarrow{AB}$ and $\overrightarrow{BC}$ is not zero, so these two vectors are not perpendicular.

$\overrightarrow{BC} \cdot \overrightarrow{AC} = \begin{pmatrix} -4 \\ 3 \\ -7 \end{pmatrix} \cdot \begin{pmatrix} 5 \\ -12 \\ -8 \end{pmatrix} = (-4 \times 5) + (3 \times -12) + (-7 \times -8) = 0$

So vectors $\overrightarrow{BC}$ and $\overrightarrow{AC}$ are perpendicular and angle C is $90°$.

(ii) The area of triangle ABC is $\frac{1}{2} \times |\overrightarrow{BC}| \times |\overrightarrow{AC}|$

$= \frac{1}{2} \times \sqrt{(-4)^2 + 3^2 + (-7)^2} \times \sqrt{5^2 + (-12)^2 + (-8)^2}$

$= \frac{1}{2} \times \sqrt{74} \times \sqrt{233}$

$= 65.7$ square units (to 3 significant figures)

The vector equation of a line

In 2 or 3 dimensions you can write the equation of a line in vector form

$$\mathbf{r} = \mathbf{a} + \lambda \mathbf{d}$$

where

$\mathbf{r}$ is the position vector of a general point on the line

$\mathbf{a}$ is the position vector of a given point on the line

$\mathbf{d}$ is the direction vector of the line ·

λ is a scalar parameter; every different value of λ corresponds to a different point on the line.

The diagram in Figure 1.4 illustrates this for a particular case in two dimensions but the diagram could just as well be drawn in three dimensions showing three perpendicular axes were it not for the fact that pages of books are two dimensional!

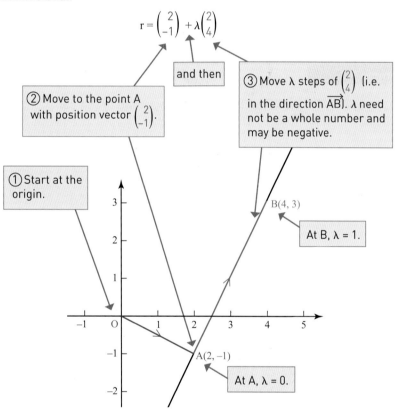

$$r = \begin{pmatrix} 2 \\ -1 \end{pmatrix} + \lambda \begin{pmatrix} 2 \\ 4 \end{pmatrix}$$

and then

② Move to the point A with position vector $\begin{pmatrix} 2 \\ -1 \end{pmatrix}$.

③ Move λ steps of $\begin{pmatrix} 2 \\ 4 \end{pmatrix}$ (i.e. in the direction $\overrightarrow{AB}$). λ need not be a whole number and may be negative.

① Start at the origin.

B(4, 3)

At B, $\lambda = 1$.

A(2, −1)

At A, $\lambda = 0$.

Figure 1.4

You should also have noticed that:

$\lambda = 0$	corresponds to the point A
$\lambda = 1$	corresponds to the point B, one 'step' along the line away from A
$0 < \lambda < 1$	corresponds to points lying between A and B
$\lambda > 1$	corresponds to points lying beyond B
$\lambda < 0$	corresponds to points beyond A, in the opposite direction to B.

This is shown by the green part of the line in Figure 1.4.

The vector equation of a line is not unique. In this case, any vector parallel, or in the opposite direction, so $\begin{pmatrix} 2 \\ 4 \end{pmatrix}$ could be used as the direction vector, for example, $\begin{pmatrix} 1 \\ 2 \end{pmatrix}, \begin{pmatrix} -3 \\ -6 \end{pmatrix}$ or $\begin{pmatrix} 20 \\ 40 \end{pmatrix}$. Similarly, you can 'step' on to the line at any point, such as B(4, 3).

So the line $\mathbf{r} = \begin{pmatrix} 2 \\ -1 \end{pmatrix} + \lambda \begin{pmatrix} 2 \\ 4 \end{pmatrix}$ could also have equation

$$\mathbf{r} = \begin{pmatrix} 2 \\ -1 \end{pmatrix} + \lambda \begin{pmatrix} 1 \\ 2 \end{pmatrix} \quad \text{or} \quad \mathbf{r} = \begin{pmatrix} 4 \\ 3 \end{pmatrix} + \lambda \begin{pmatrix} 2 \\ 4 \end{pmatrix}, \text{ for example.}$$

The Cartesian equation of a line

In either two or three dimensions you can find the Cartesian equation of a straight line from its vector form by equating the x and y (and z in three dimensions) components and eliminating λ.

Example 1.3

Find the Cartesian equation of the line $\mathbf{r} = \begin{pmatrix} 2 \\ 1 \\ -3 \end{pmatrix} + \lambda \begin{pmatrix} 2 \\ 1 \\ -1 \end{pmatrix}$.

Solution

Let $\mathbf{r} = \begin{pmatrix} x \\ y \\ z \end{pmatrix}$

$$\therefore x = 2 + 2\lambda$$
$$y = 1 + \lambda$$
$$z = -3 - \lambda$$

> You can also write the equation in the apparently more complicated form
> $$\lambda = \frac{x-2}{2} = \frac{y-1}{1} = \frac{z+3}{-1}$$ as in this form it is easier to read off the direction vector.

So

$$\lambda = \frac{x-2}{2} = y - 1 = -z - 3 \text{ is the three-dimensional Cartesian form.}$$

If one or two of the components of the direction vector are zero then the Cartesian form looks slightly different. For example, if the z component were zero then there would just be an equation relating x and y, and z would be a constant; the line would lie in a plane parallel to the x–y plane. If both y and z components were zero then both y and z would be constant and x would be arbitrary; the line would be parallel to the x-axis.

Example 1.4

Find the Cartesian equation of the line through the points $(4, -2, 7)$ and $(9, -2, 1)$.

Solution

The vector equation of the line is

$$\mathbf{r} = \begin{pmatrix} 4 \\ -2 \\ 7 \end{pmatrix} + \lambda \begin{pmatrix} 5 \\ 0 \\ -6 \end{pmatrix}$$

$\therefore x = 4 + 5\lambda, y = -2, z = 7 - 6\lambda.$

So from the x and z equations

$$\lambda = \frac{x-4}{5} = \frac{z-7}{-6}$$

and the Cartesian equation of the line is written

$$\frac{x-4}{5} = \frac{z-7}{-6}, y = -2$$

The angle between two lines

The angle between two lines is the acute angle between their two direction vectors. So to find the angle between two lines, all you have to do is to identify the direction vectors and use the dot product to find the angle between them. If this comes out obtuse then simply subtract it from $180°$.

Example 1.5

Find the angle between the two lines

$$l_1: \frac{x}{3} = y - 1 = 3 - 2z \text{ and } l_2: \mathbf{r} = \begin{pmatrix} 7 \\ 0 \\ -8 \end{pmatrix} + \lambda \begin{pmatrix} -2 \\ 2 \\ 3 \end{pmatrix}.$$

Solution

$$l_1: \frac{x}{3} = y - 1 = 3 - 2z$$

> Firstly, rewrite the equation of l_1 so that you can write it in vector form.

$$\therefore l_1: \frac{x-0}{3} = \frac{y-1}{1} = \frac{z - \frac{3}{2}}{-\frac{1}{2}}$$

> Note that you must **not** use the symbol λ for the parameter since this is already used for l_2.

$$\therefore l_1: \mathbf{r} = \begin{pmatrix} 0 \\ -1 \\ -\frac{3}{2} \end{pmatrix} + \mu \begin{pmatrix} 3 \\ 1 \\ -\frac{1}{2} \end{pmatrix}$$

> To avoid fractions you can double the direction vector. Strictly speaking, the μ here is actually half the μ on the previous line, but you can use the same letter for both. Note that you **cannot**

$$\therefore l_1: \mathbf{r} = \begin{pmatrix} 0 \\ -1 \\ -\frac{3}{2} \end{pmatrix} + \mu \begin{pmatrix} 6 \\ 2 \\ -1 \end{pmatrix}$$

> scale up the position vector $\begin{pmatrix} 0 \\ -1 \\ -\frac{3}{2} \end{pmatrix}$
>
> in the same way. With care you could simplify it by adding any multiple of the direction vector but this is not actually necessary in this case.

So the direction vectors of the two lines are $\mathbf{d}_1 = \begin{pmatrix} 6 \\ 2 \\ -1 \end{pmatrix}$ and $\mathbf{d}_2 = \begin{pmatrix} -2 \\ 2 \\ 3 \end{pmatrix}$.

$$\mathbf{d}_1 \cdot \mathbf{d}_2 = \begin{pmatrix} 6 \\ 2 \\ -1 \end{pmatrix} \cdot \begin{pmatrix} -2 \\ 2 \\ 3 \end{pmatrix} = -12 + 4 - 3 = -11$$

and

$$|\mathbf{d}_1| = \sqrt{6^2 + 2^2 + (-1)^2} = \sqrt{41} \text{ and } |\mathbf{d}_2| = \sqrt{2^2 + (-2)^2 + 3^2} = \sqrt{17}.$$

> Remember, it's best to use all the figures on your calculator to prevent rounding errors.

$$\therefore \cos\theta = \frac{-11}{\sqrt{41}\sqrt{17}} = -0.41665...$$

> Using $\mathbf{a} \cdot \mathbf{b} = |\mathbf{a}||\mathbf{b}|\cos\theta$.
> Note that this answer must work out to be between −1 and 1. If it does not you have made a calculation error.

$$\therefore \theta = \cos^{-1}(-0.41665...)$$

$$\therefore \theta = 114.62...°$$

> Angles in degrees are usually quoted to 1 decimal place.

> This is obtuse so subtract it from 180°. Note that the same effect could have been achieved by stripping the minus sign off −0.41665... but this is not good practice.

$$180 - 114.62... = 65.376...$$

So the angle between the two lines is 65.4° (1 d.p.).

> **Note**
> ----
> It is not necessary that the lines intersect – the angle between the lines is still properly defined.

Configuration of two lines in three dimensions

In two dimensions, two distinct lines either intersect at a point or are parallel. However, if you consider two distinct lines in three dimensions then there are three different possibilities:

- the lines intersect
- the lines are parallel
- the lines are skew.

> Recall that skew lines are not parallel and do not intersect. Imagine any line drawn on the floor and any line drawn on the ceiling but in a different direction from the line on the floor. These are skew.

Given the equations of two lines you should be able to determine whether they intersect, are parallel or are skew and, if they intersect, you should be able to find the point of intersection.

Example 1.6

The lines l_1 and l_2 are represented by the equations

$$l_1: \frac{x-1}{1} = \frac{y+6}{2} = \frac{z+1}{3} \qquad l_2: \frac{x-9}{2} = \frac{y-7}{3} = \frac{z-2}{-1}$$

(i) Write these lines in vector form.

(ii) Hence find whether the lines meet and, if so, the coordinates of their point of intersection.

Solution

(i) The equation of l_1 is $\mathbf{r} = \begin{pmatrix} 1 \\ -6 \\ -1 \end{pmatrix} + \lambda \begin{pmatrix} 1 \\ 2 \\ 3 \end{pmatrix}$

The equation of l_2 is $\mathbf{r} = \begin{pmatrix} 9 \\ 7 \\ 2 \end{pmatrix} + \mu \begin{pmatrix} 2 \\ 3 \\ -1 \end{pmatrix}$

(ii) If there is a point $\begin{pmatrix} X \\ Y \\ Z \end{pmatrix}$ that is common to both lines then

$$\begin{pmatrix} X \\ Y \\ Z \end{pmatrix} = \begin{pmatrix} 1 \\ -6 \\ -1 \end{pmatrix} + \lambda \begin{pmatrix} 1 \\ 2 \\ 3 \end{pmatrix} = \begin{pmatrix} 9 \\ 7 \\ 2 \end{pmatrix} + \mu \begin{pmatrix} 2 \\ 3 \\ -1 \end{pmatrix}$$

for some parameters λ and μ.

This gives the three equations

$$X = \lambda + 1 = 2\mu + 9 \quad \text{①}$$
$$Y = 2\lambda - 6 = 3\mu + 7 \quad \text{②}$$
$$Z = 3\lambda - 1 = -\mu + 2 \quad \text{③}$$

Now solve any two of the three equations simultaneously.

$$\left. \begin{matrix} \lambda - 2\mu = 8 \\ 2\lambda - 3\mu = 13 \end{matrix} \right\} \Leftrightarrow \left. \begin{matrix} 2\lambda - 4\mu = 16 \\ 2\lambda - 3\mu = 13 \end{matrix} \right\} \Leftrightarrow \mu = -3, \lambda = 2$$

> Using ① and ②

If these values for λ and μ also satisfy equation ③, then the lines meet.

Using equation ③, when $\lambda = 2$, $Z = 6 - 1 = 5$ and when $\mu = -3$, $Z = 3 + 2 = 5$.

As both values of Z are equal this proves the lines intersect.

Using either $\lambda = 2$ or $\mu = -3$ in equations ①, ② and ③ gives $X = 3$, $Y = -2$, $Z = 5$ so the lines meet at the point $(3, -2, 5)$.

Example 1.7

Prove that the lines l_1 and l_2 are skew, where

$$l_1 : \begin{pmatrix} 1 \\ -6 \\ -1 \end{pmatrix} + \lambda \begin{pmatrix} 1 \\ 2 \\ 3 \end{pmatrix}$$

$$l_2 : \begin{pmatrix} 9 \\ 8 \\ 2 \end{pmatrix} + \mu \begin{pmatrix} 2 \\ 3 \\ -1 \end{pmatrix}.$$

Solution

If there is a point (X, Y, Z) common to both lines then

$$\begin{pmatrix} X \\ Y \\ Z \end{pmatrix} = \begin{pmatrix} 1 \\ -6 \\ -1 \end{pmatrix} + \lambda \begin{pmatrix} 1 \\ 2 \\ 3 \end{pmatrix} = \begin{pmatrix} 9 \\ 8 \\ 2 \end{pmatrix} + \mu \begin{pmatrix} 2 \\ 3 \\ -1 \end{pmatrix}$$

for some parameters λ and μ.

$$X = \lambda + 1 = 2\mu + 9 \quad \text{①}$$
$$Y = 2\lambda - 6 = 3\mu + 8 \quad \text{②}$$
$$Z = 3\lambda - 1 = -\mu + 2 \quad \text{③}$$

> Solving equations ① and ② simultaneously

$$\left. \begin{array}{l} \lambda - 2\mu = 8 \\ 2\lambda - 3\mu = 14 \end{array} \right\} \Leftrightarrow \left. \begin{array}{l} 2\lambda - 4\mu = 16 \\ 2\lambda - 3\mu = 14 \end{array} \right\} \Leftrightarrow \mu = -2, \lambda = 4$$

When $\lambda = 4$, $Z = 12 - 1 = 11$
and when $\mu = -2$, $Z = 2 + 2 = 4$.

> Substitute these values into equation ③

Therefore the values $\mu = -2$, $\lambda = 4$ do not satisfy the third equation and so the lines do not meet. As the lines are distinct, the only other alternatives are that the lines are parallel or skew.

Look at the direction vectors of the lines: $\begin{pmatrix} 1 \\ 2 \\ 3 \end{pmatrix}$ and $\begin{pmatrix} 2 \\ 3 \\ -1 \end{pmatrix}$. Neither of

these is a multiple of the other so they are not parallel and hence the two lines are not parallel. So, lines l_1 and l_2 are skew.

① (i) Find the following:

(a) $\begin{pmatrix} 2 \\ 3 \end{pmatrix} \cdot \begin{pmatrix} 1 \\ 5 \end{pmatrix}$

(b) $\begin{pmatrix} 5 \\ 0 \end{pmatrix} \cdot \begin{pmatrix} -2 \\ 3 \end{pmatrix}$

(c) $\begin{pmatrix} -1 \\ -1 \end{pmatrix} \cdot \begin{pmatrix} 2 \\ 3 \end{pmatrix}$

(d) $\begin{pmatrix} 3 \\ 4 \end{pmatrix} \cdot \begin{pmatrix} -4 \\ 3 \end{pmatrix}$

(e) $\begin{pmatrix} 1 \\ 1 \\ 3 \end{pmatrix} \cdot \begin{pmatrix} 5 \\ 2 \\ 4 \end{pmatrix}$

(f) $\begin{pmatrix} 2 \\ 3 \\ -2 \end{pmatrix} \cdot \begin{pmatrix} 3 \\ 0 \\ -4 \end{pmatrix}$

(g) $\begin{pmatrix} 1 \\ -2 \\ 5 \end{pmatrix} \cdot \begin{pmatrix} 1 \\ -2 \\ 5 \end{pmatrix}$

(h) $\begin{pmatrix} 5 \\ 1 \\ -2 \end{pmatrix} \cdot \begin{pmatrix} 3 \\ -1 \\ 7 \end{pmatrix}$

(ii) Which pairs of vectors in part (i) are perpendicular?

② The vectors $\mathbf{a} = \begin{pmatrix} p \\ -2 \\ 2 \end{pmatrix}$ and $\mathbf{b} = \begin{pmatrix} -2 \\ -2 \\ q \end{pmatrix}$ are perpendicular and $|\mathbf{a}| = \sqrt{33}$.

Find all possible pairs of values of p and q.

③ Find the equations of the following lines in vector form:

(i) through $(5, 6)$ in the direction $\begin{pmatrix} 2 \\ 1 \end{pmatrix}$

(ii) through $(3, -2)$ in the direction $\begin{pmatrix} 1 \\ -4 \end{pmatrix}$

(iii) through $(2, 3)$ and $(7, 5)$

(iv) through $(-3, 3)$ and $(-2, -5)$.

④ Find the equations of the following lines in vector form:

(i) through $(1, 5, 2)$ in the direction $\begin{pmatrix} 1 \\ 2 \\ 1 \end{pmatrix}$

(ii) through $(-2, -2, 2)$ in the direction $\begin{pmatrix} -3 \\ 0 \\ 4 \end{pmatrix}$

(iii) through $(4, 5, 2)$ and $(7, 6, 6)$

(iv) through $(-3, 0, -2)$ and $(-1, -5, -5)$.

⑤ Write the equation of the following lines in Cartesian form.

(i) $\mathbf{r} = \begin{pmatrix} 4 \\ 3 \\ 2 \end{pmatrix} + \lambda \begin{pmatrix} 1 \\ 1 \\ 4 \end{pmatrix}$

(ii) $\mathbf{r} = \begin{pmatrix} -1 \\ 0 \\ 2 \end{pmatrix} + \lambda \begin{pmatrix} -2 \\ 2 \\ 7 \end{pmatrix}$

(iii) $\mathbf{r} = \begin{pmatrix} 2 \\ 3 \\ -6 \end{pmatrix} + \lambda \begin{pmatrix} 3 \\ 0 \\ -2 \end{pmatrix}$

(iv) $\mathbf{r} = \begin{pmatrix} 1 \\ 1 \\ 1 \end{pmatrix} + \lambda \begin{pmatrix} 1 \\ 0 \\ 0 \end{pmatrix}$

⑥ Write the equations of the following lines in vector form.

(i) $\dfrac{x-3}{5} = \dfrac{y-2}{4} = z - 8$

(ii) $\dfrac{x+1}{2} = \dfrac{3-y}{3} = \dfrac{z}{3}$

(iii) $\dfrac{2x-1}{4} = \dfrac{3y}{5} = z$

(iv) $2x - 3 = \dfrac{5-2y}{8}$ and $z = -2$

⑦ Write down both the vector and Cartesian equations of the line through the point $(3, -2, -5)$ which is parallel to the z-axis.

⑧ For each of the following pairs of lines decide whether they intersect or not. If they do intersect then find the point of intersection; if they do not then state whether they are parallel or skew.

(i) $l_1: \mathbf{r} = \begin{pmatrix} 3 \\ 2 \end{pmatrix} + \lambda \begin{pmatrix} 1 \\ -2 \end{pmatrix}$ and $l_2: \mathbf{r} = \begin{pmatrix} 10 \\ 0 \end{pmatrix} + \mu \begin{pmatrix} -1 \\ -1 \end{pmatrix}$

(ii) $l_1: \mathbf{r} = \begin{pmatrix} -5 \\ 7 \end{pmatrix} + \lambda \begin{pmatrix} -8 \\ 4 \end{pmatrix}$ and $l_2: \mathbf{r} = \begin{pmatrix} 2 \\ 2 \end{pmatrix} + \mu \begin{pmatrix} 2 \\ -1 \end{pmatrix}$

(iii) $l_1: \mathbf{r} = \begin{pmatrix} 1 \\ 2 \\ 3 \end{pmatrix} + \lambda \begin{pmatrix} 1 \\ 1 \\ -2 \end{pmatrix}$ and $l_2: \mathbf{r} = \begin{pmatrix} 13 \\ -2 \\ 3 \end{pmatrix} + \mu \begin{pmatrix} 5 \\ -3 \\ 2 \end{pmatrix}$

(iv) $l_1: \mathbf{r} = \begin{pmatrix} -3 \\ 5 \\ 6 \end{pmatrix} + \lambda \begin{pmatrix} 2 \\ 1 \\ 1 \end{pmatrix}$ and $l_2: \mathbf{r} = \begin{pmatrix} 1 \\ 1 \\ 1 \end{pmatrix} + \mu \begin{pmatrix} 3 \\ -4 \\ 2 \end{pmatrix}$

(v) $l_1: \mathbf{r} = \begin{pmatrix} 5 \\ 5 \\ -2 \end{pmatrix} + \lambda \begin{pmatrix} -9 \\ 12 \\ -6 \end{pmatrix}$ and $l_2: \mathbf{r} = \begin{pmatrix} -2 \\ -2 \\ -2 \end{pmatrix} + \mu \begin{pmatrix} 3 \\ -4 \\ 2 \end{pmatrix}$

⑨ Find the size of the acute angle (or the right angle) between these pairs of lines.

(i) $l_1: \mathbf{r} = \begin{pmatrix} 3 \\ 2 \end{pmatrix} + \lambda \begin{pmatrix} 1 \\ -2 \end{pmatrix}$ and $l_2: \mathbf{r} = \begin{pmatrix} 10 \\ 0 \end{pmatrix} + \mu \begin{pmatrix} -1 \\ -1 \end{pmatrix}$

(ii) $l_1: \mathbf{r} = \begin{pmatrix} -5 \\ 7 \end{pmatrix} + \lambda \begin{pmatrix} -8 \\ 4 \end{pmatrix}$ and $l_2: \mathbf{r} = \begin{pmatrix} 2 \\ 2 \end{pmatrix} + \mu \begin{pmatrix} 2 \\ -1 \end{pmatrix}$

(iii) $l_1: \mathbf{r} = \begin{pmatrix} 1 \\ 2 \\ 3 \end{pmatrix} + \lambda \begin{pmatrix} 1 \\ 1 \\ -2 \end{pmatrix}$ and $l_2: \mathbf{r} = \begin{pmatrix} 13 \\ -2 \\ 3 \end{pmatrix} + \mu \begin{pmatrix} 5 \\ -3 \\ 2 \end{pmatrix}$

(iv) $l_1: \mathbf{r} = \begin{pmatrix} -3 \\ 5 \\ 6 \end{pmatrix} + \lambda \begin{pmatrix} 2 \\ 1 \\ 1 \end{pmatrix}$ and $l_2: \mathbf{r} = \begin{pmatrix} 1 \\ 1 \\ 1 \end{pmatrix} + \mu \begin{pmatrix} 3 \\ -4 \\ 2 \end{pmatrix}$

(v) $l_1: \mathbf{r} = \begin{pmatrix} 5 \\ 5 \\ -2 \end{pmatrix} + \lambda \begin{pmatrix} -9 \\ 12 \\ -6 \end{pmatrix}$ and $l_2: \mathbf{r} = \begin{pmatrix} -2 \\ -2 \\ -2 \end{pmatrix} + \mu \begin{pmatrix} 3 \\ -4 \\ 2 \end{pmatrix}$

1 Using the scalar product to express the equation of a plane

You can write the equation of a plane in either vector or Cartesian form. The Cartesian form is used more often but to see where it comes from it is helpful to start with the vector form.

Discussion points

→ Lay a sheet of paper on a flat horizontal table and mark several straight lines on it. Now take a pencil and stand it upright on the sheet of paper (see Figure 1.5).

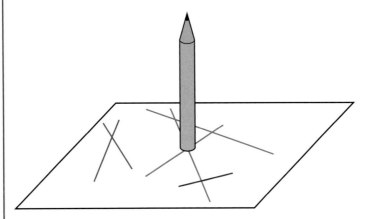

Figure 1.5

→ What angle does the pencil make with any individual line?

→ Would it make any difference if the table were tilted at an angle (apart from the fact that you could no longer balance the pencil)?

The discussion above shows you that there is a direction (that of the pencil) which is at right angles to every straight line in the plane. A line in that direction is said to be perpendicular to the plane and is referred to as a **normal** to the plane.

It is often denoted by $\mathbf{n} = \begin{pmatrix} n_1 \\ n_2 \\ n_3 \end{pmatrix}$.

In Figure 1.6 the point A is on the plane and the vector $\mathbf{n}$ is perpendicular to the plane. This information allows you to find an expression for the position vector $\mathbf{r}$ of a general point R on the plane.

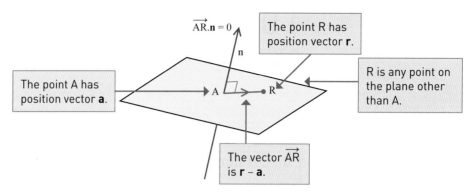

$\overrightarrow{AR}.\mathbf{n} = 0$

The point R has position vector **r**.

The point A has position vector **a**.

R is any point on the plane other than A.

The vector $\overrightarrow{AR}$ is **r** − **a**.

Figure 1.6

The vector $\overrightarrow{AR}$ is a line in the plane, and so it follows that $\overrightarrow{AR}$ is at right angles to the direction **n**.

$$\overrightarrow{AR}\cdot\mathbf{n} = 0$$

The vector $\overrightarrow{AR}$ is given by $\overrightarrow{AR} = \mathbf{r} - \mathbf{a}$ and so

$$(\mathbf{r} - \mathbf{a})\cdot\mathbf{n} = 0.$$

This is the vector equation of the plane.

Expanding the brackets lets you write this in an alternative form as

$$\mathbf{r}\cdot\mathbf{n} - \mathbf{a}\cdot\mathbf{n} = 0. \quad\longleftarrow\quad \boxed{\text{This can also be written as } \mathbf{r}\cdot\mathbf{n} = \mathbf{a}\cdot\mathbf{n}}$$

Although the vector equation of a plane is very compact, it is more common to use the Cartesian form. This is derived from the vector form as follows.

Write the normal vector **n** as $\begin{pmatrix} n_1 \\ n_2 \\ n_3 \end{pmatrix}$ and the position vector of A as

$\mathbf{a} = \begin{pmatrix} a_1 \\ a_2 \\ a_3 \end{pmatrix}$. The position vector of the general point R is $\mathbf{r} = \begin{pmatrix} x \\ y \\ z \end{pmatrix}$.

So the equation $\mathbf{r}\cdot\mathbf{n} - \mathbf{a}\cdot\mathbf{n} = 0$

can be written as $\begin{pmatrix} x \\ y \\ z \end{pmatrix} \cdot \begin{pmatrix} n_1 \\ n_2 \\ n_3 \end{pmatrix} - \begin{pmatrix} a_1 \\ a_2 \\ a_3 \end{pmatrix} \cdot \begin{pmatrix} n_1 \\ n_2 \\ n_3 \end{pmatrix} = 0.$

This is the same as $n_1 x + n_2 y + n_3 z + d = 0$ where $d = -(a_1 n_1 + a_2 n_2 + a_3 n_3)$.

The following example shows you how to use this.

Notice that *d* is a constant and is a scalar.

TECHNOLOGY

If you have access to 3D graphing software, experiment with planes in the form $ax + by + cx + d = 0$, varying the values of *a*, *b*, *c* and *d*.

Example 1.8

The point A$(2, 3, -5)$ lies on a plane. The vector $\mathbf{n} = \begin{pmatrix} -4 \\ 2 \\ 1 \end{pmatrix}$ is perpendicular to the plane.

(i)   Find the Cartesian equation of the plane.

(ii)   Investigate whether the points P$(5, 3, -2)$ and Q$(3, 5, -5)$ lie in the plane.

Solution

(i)   The Cartesian equation of the plane is
$n_1 x + n_2 y + n_3 z + d = 0.$

$$\mathbf{n} = \begin{pmatrix} -4 \\ 2 \\ 1 \end{pmatrix} \text{ so } n_1 = -4, n_2 = 2 \text{ and } n_3 = 1$$

The equation of the plane is $-4x + 2y + z + d = 0.$

It remains to find d. There are two ways of doing this.

Either:
$d = -\mathbf{a} \cdot \mathbf{n}$

where $\mathbf{a}$ is the position vector of A, $(2, 3, -5)$.

So $\mathbf{a} = \begin{pmatrix} 2 \\ 3 \\ -5 \end{pmatrix}$ and $\mathbf{n} = \begin{pmatrix} -4 \\ 2 \\ 1 \end{pmatrix}$

Or:
The point A is $(2, 3, -5)$.

Substituting for x, y and z in $-4x + 2y + z + d = 0$ gives,
$-4 \times 2 + 2 \times 3 - 5 + d = 0$
so, $d = 8 - 6 + 5 = 7.$

$$d = -\begin{pmatrix} 2 \\ 3 \\ -5 \end{pmatrix} \cdot \begin{pmatrix} -4 \\ 2 \\ 1 \end{pmatrix}$$

$$= -[(2 \times -4) + (3 \times 2) + (-5 \times 1)]$$

$$= -[-8 + 6 - 5] = 7$$

So the equation of the plane is $-4x + 2y + z + 7 = 0.$

(ii)   P is $(5, 3, -2)$.

Substituting in the left-hand side of the equation of the plane gives $(-4 \times 5) + (2 \times 3) - 2 + 7 = -9.$
Since this is not equal to 0, P does not lie on the plane.
Q is $(3, 5, -5)$.
Substituting in the left-hand side of the equation of the plane gives $(-4 \times 3) + (2 \times 5) - 5 + 7 = 0$
Since this is equal to 0, Q lies on the plane.

Look carefully at the equation of the plane in Example 1.8. You can see at once that the vector $\begin{pmatrix} -4 \\ 2 \\ 1 \end{pmatrix}$, formed from the coefficients of x, y and z, is perpendicular to the plane.

In general the vector $\begin{pmatrix} n_1 \\ n_2 \\ n_3 \end{pmatrix}$ is perpendicular to all planes of the form

$n_1 x + n_2 y + n_3 z + d = 0$, whatever the value of d (see Figure 1.7).

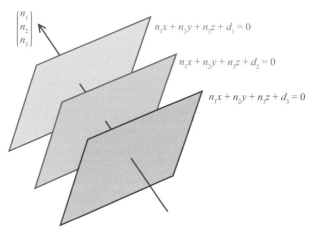

Figure 1.7

Consequently, all planes of that form are parallel; the coefficients of x, y and z determine the direction of the plane, the value of d its location.

Example 1.9

Find the Cartesian equation of the plane which is parallel to the plane $3x - y + 2z + 5 = 0$ and contains the point $(1, 0, -2)$.

Solution

The normal to the plane $3x - y + 2z + 5 = 0$ is $\begin{pmatrix} 3 \\ -1 \\ 2 \end{pmatrix}$.

Any plane parallel to this plane has the same normal vector, so the required plane has an equation of the form $3x - y + 2z + d = 0$.

The plane contains the point $(1, 0, -2)$, so $(3 \times 1) - 0 + (2 \times -2) + d = 0$

$\Rightarrow d = 1$

The equation of the plane is $3x - y + 2z + 1 = 0$.

Discussion point

→ Given the coordinates of three points A, B and C in a plane, how could you find the equation of the plane?

Notation

So far the Cartesian equation of a plane has been written as
$n_1x + n_2y + n_3z + d = 0$.

Another common way of writing it is $ax + by + cz + d = 0$.

In this case the vector $\begin{pmatrix} a \\ b \\ c \end{pmatrix}$ is normal to the plane.

ACTIVITY 1.1

A plane $ax + by + cz + d = 0$ contains the points $(1, 1, 1)$, $(1, -1, 0)$ and $(-1, 0, 2)$. Use this information to write down three simultaneous equations and solve these. Hence, by taking $a = 1$, find the equation of the plane.

Exercise 1.1

① A plane has equation $5x - 3y + 2x + 1 = 0$.

   (i)   Write down the normal vector to this plane.

   (ii) Show that the point $(1, 4, 3)$ lies on the plane.

② Find, in vector form, the equation of the planes which contain the point with position vector **a** and are perpendicular to the vector **n**.

   (i) $\mathbf{a} = 3\mathbf{i} + 5\mathbf{j} - 2\mathbf{k}$ $\mathbf{n} = \mathbf{i} + \mathbf{j} + \mathbf{k}$

   (ii) $\mathbf{a} = -3\mathbf{i} + 2\mathbf{j} + \mathbf{k}$ $\mathbf{n} = \mathbf{i} + \mathbf{j} + \mathbf{k}$

   (iii) $\mathbf{a} = 3\mathbf{i} + 5\mathbf{j} - 2\mathbf{k}$ $\mathbf{n} = -\mathbf{i} - \mathbf{j} - \mathbf{k}$

   (iv) $\mathbf{a} = 2\mathbf{i} + 7\mathbf{j} - \mathbf{k}$ $\mathbf{n} = 2\mathbf{i} + 2\mathbf{j} + 2\mathbf{k}$

③ Find the Cartesian equation of the planes in question 2.

   Comment on your answers.

④ The plane π_1 has equation $-x + 3y - 2z - 13 = 0$.

   Find the Cartesian and vector equations of the plane π_2 that is parallel to π_1 and passes through the point $(3, 0, -4)$.

⑤ Find the Cartesian equation of the plane which contains the point $(0, 1, -4)$ and is parallel to the plane $\left(\mathbf{r} - (4\mathbf{i} + 2\mathbf{j} - \mathbf{k})\right) \cdot (4\mathbf{i} - 5\mathbf{j} + 6\mathbf{k}) = 0$.

⑥ The points A, B and C have coordinates $(0, -1, 2)$, $(2, 1, 0)$ and $(5, 1, 1)$.

   (i)   Write down the vectors $\overrightarrow{AB}$ and $\overrightarrow{AC}$.

   (ii) Show that $\overrightarrow{AB} \cdot \begin{pmatrix} 1 \\ -4 \\ -3 \end{pmatrix} = \overrightarrow{AC} \cdot \begin{pmatrix} 1 \\ -4 \\ -3 \end{pmatrix} = 0$.

   (iii) Find the equation of the plane containing the points A, B and C.

⑦ (i)   Show that the points $A(1, 1, 1)$, $B(3, 0, 0)$ and $C(2, 0, 2)$ all lie in the plane $2x + 3y + z = 6$.

   (ii) Show that $\overrightarrow{AB} \cdot \begin{pmatrix} 2 \\ 3 \\ 1 \end{pmatrix} = \overrightarrow{AC} \cdot \begin{pmatrix} 2 \\ 3 \\ 1 \end{pmatrix} = 0$.

(iii) The point D has coordinates $(7, 6, 2)$ and lies on a line perpendicular to the plane through one of the points A, B or C.

Through which of these points does the line pass?

⑧ A plane π has Cartesian equation $2x - 3y + 2z + 10 = 0$.

(i) Assuming that the equation of the plane in normal form is $\mathbf{r \cdot n} - \mathbf{a \cdot n} = 0$, write down the normal vector $\mathbf{n}$ and the value of $d = -\mathbf{a \cdot n}$.

(ii) Find a possible position vector $\mathbf{a}$ to represent a point A in the plane.

(iii) Use your answers to parts (i) and (ii) to write down a vector equation for the plane π in the form $(\mathbf{r} - \mathbf{a}) \cdot \mathbf{n} = 0$.

⑨ The points A, B and C have coordinates $(2, -3, 2)$, $(3, 1, 3)$ and $(5, 5, 4)$ respectively.

(i) Find, as column vectors, $\overrightarrow{AB}$ and $\overrightarrow{AC}$.

(ii) Show that the points A, B and C are not collinear.

(iii) By trying a vector $\mathbf{n}$ of the form $\mathbf{n} = \begin{pmatrix} 1 \\ p \\ q \end{pmatrix}$ (and then $\mathbf{n} = \begin{pmatrix} 0 \\ 1 \\ t \end{pmatrix}$ if the first form fails), find a vector which is perpendicular to both $\overrightarrow{AB}$ and $\overrightarrow{AC}$.

(iv) Find, in scalar product form, the equation of the plane through A, B and C.

(v) Hence show that the Cartesian equation of the plane is $y - 4z = -11$.

(vi) Describe the orientation of the plane in relation to the x-y, x-z and y-z planes.

2 Intersecting planes

Using the scalar product to find the angle between two planes

The angle between two planes can be found by using the scalar product. As Figures 1.8 and 1.9 show, the angle between planes π_1 and π_2 is the same as the angle between their normal, $\mathbf{n}_1$ and $\mathbf{n}_2$.

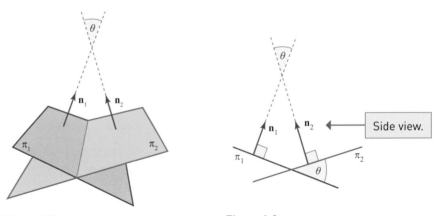

Figure 1.8   Figure 1.9

It might be more accurate to say that in fact there are usually two different angles between two planes and that the acute angle is the one which is required.

So to find the angle between two planes, identify the normal vectors of the planes and use the scalar product to find the angle between these two vectors in the usual way. It is possible for the angle between the two vectors to be obtuse (imagine that one of the normals in Figure 1.9 pointed the other way). However, the acute angle is still the angle required so if this turns out to be the case then simply subtract your answer from 180°, as shown in Example 1.10.

Discussion points

➔ Is there always a properly defined angle between any two planes?

➔ Is it always the case that there is one acute angle and one obtuse angle between any two planes?

Note that if $\mathbf{n}_1 \cdot \mathbf{n}_2 = 0$ then the planes are perpendicular to each other.

Example 1.10

Find, to 1 decimal place, the acute angle between the planes

$$\pi_1 : 2x + 3y + 5z = 8 \text{ and } \pi_2 : 5x + y - 4z = 12.$$

Solution

The planes have normals $\mathbf{n}_1 = \begin{pmatrix} 2 \\ 3 \\ 5 \end{pmatrix}$ and $\mathbf{n}_2 = \begin{pmatrix} 5 \\ 1 \\ -4 \end{pmatrix}$ so the angle

between the planes is given by $\mathbf{n}_1 \cdot \mathbf{n}_2 = |\mathbf{n}_1||\mathbf{n}_2| \cos\theta$.

$$\Rightarrow \begin{pmatrix} 2 \\ 3 \\ 5 \end{pmatrix} \cdot \begin{pmatrix} 5 \\ 1 \\ -4 \end{pmatrix} = \sqrt{4+9+25}\sqrt{25+1+16}\cos\theta$$

$$\Rightarrow 10 + 3 - 20 = \sqrt{38}\sqrt{42}\cos\theta$$

$$\Rightarrow \theta = \cos^{-1}\left(\frac{-7}{\sqrt{38}\sqrt{42}}\right) = 100.1°$$

So the acute angle between the planes is 79.9°.

Using the scalar product to find the angle between a line and a plane

If they are not perpendicular, the acute angle between a line and a plane is the acute angle θ between the line and its *orthogonal projection* onto the plane, shown by the dotted line AB in Figure 1.10.

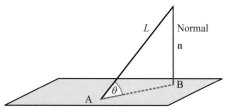

Figure 1.10

You can find the angle θ by first finding the angle α between the direction vector **d** of the straight line L and a normal vector **n** to the plane, as shown in Figure 1.11.

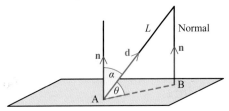

Figure 1.11

The angle θ can then be found by calculating $90 - \alpha$.

This method is illustrated in the following example.

Example 1.11

Find the angle between the line $\mathbf{r} = \begin{pmatrix} 2 \\ 0 \\ -3 \end{pmatrix} + \lambda \begin{pmatrix} 1 \\ 3 \\ 2 \end{pmatrix}$ and the plane $3x - y + z = 4$.

Solution

For this line, the direction vector $\mathbf{d} = \begin{pmatrix} 1 \\ 3 \\ 2 \end{pmatrix}$ and a normal to the plane is

$\mathbf{n} = \begin{pmatrix} 3 \\ -1 \\ 1 \end{pmatrix}$.

The angle α between the normal vector and the direction vector satisfies $\mathbf{d.n.} = |\mathbf{d}||\mathbf{n}|\cos \alpha$.

$\Rightarrow \begin{pmatrix} 1 \\ 3 \\ 2 \end{pmatrix}.\begin{pmatrix} 3 \\ -1 \\ 1 \end{pmatrix} = \sqrt{14}\sqrt{11}\cos\theta$

$\Rightarrow \cos\alpha = \dfrac{2}{\sqrt{14}\sqrt{11}}$

$\Rightarrow \alpha = 80.7°$

So the angle between the line and plane $\theta = 90 - 80.71 = 9.3°$.

Note that, as with the angle between two planes, it is possible for the angle between the direction vector of the line and the normal to the plane to be obtuse; imagine, for example, that the vector **n** in Figure 1.10 pointed downwards rather than upwards. If this is the case, again take the supplementary angle (i.e. subtract it from $180°$) before subtracting it from $90°$.

Discussion points

➜ Suppose that $\mathbf{n.d} = 0$. What are the possible cases?

➜ How could you tell whether a given line is perpendicular to a plane?

The intersection of two planes

Two planes might be parallel or even identical.

> **Discussion point**
>
> → For what values of d are the following planes (i) identical, (ii) parallel?
>
> $3x - 6y + 9z = 12$
>
> $-2x + 4y - 6z = d$

Two planes that are neither identical nor parallel meet in a line, as shown in Figure 1.12.

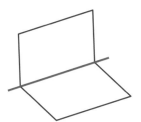

Figure 1.12

How can you find the equation of the line in which they meet?

There are several possible methods, but the one used here is to find the equation of the line in terms of a parameter of the line, by eliminating one variable. The method is shown in Example 1.12.

Example 1.12

Find the equation of the line in which the following planes meet:

$x - y + z = 2$

$2x - 5y + z = -5$

Solication

Subtract one equation from the other: ◄—— First eliminate one variable from the two simultaneous equations.

$x - 4y = -7$

Now let $y = \lambda$. ◄—— Let one variable in the resulting equation equal λ. Choose the variable that minimises the need for fractions.

$\therefore \quad x = 4y - 7$

$\quad = 4\lambda - 7.$

From the first equation,

$x - y + z = 2$

$\therefore \quad z = 2 - x + y$

$\quad = 2 - (4\lambda - 7) + \lambda$

$\quad = 30 - 3\lambda$

Use the equations to write the other variables in terms of λ.

$$\therefore \quad \begin{pmatrix} x \\ y \\ z \end{pmatrix} = \begin{pmatrix} 4\lambda - 7 \\ \lambda \\ 30 - 3\lambda \end{pmatrix}$$

Write the results as a vector.

or, writing this in the form $\mathbf{r} = \mathbf{a} + \lambda\mathbf{d}$:

$$\mathbf{r} = \begin{pmatrix} -7 \\ 0 \\ 30 \end{pmatrix} + \lambda \begin{pmatrix} 4 \\ 1 \\ -3 \end{pmatrix}$$

This can now easily be put into Cartesian form if required.

Discussion points

→ Does it matter which variable you choose as λ?

→ Will this process ever fail to work?

Exercise 1.2

① Find, to 1 decimal place, the smaller angle between the planes:

(i) $\mathbf{r} \cdot \begin{pmatrix} 2 \\ 2 \\ -3 \end{pmatrix} = 4$ and $\mathbf{r} \cdot \begin{pmatrix} 3 \\ -3 \\ -1 \end{pmatrix} = 2$

(ii) $\mathbf{r} \cdot \begin{pmatrix} 1 \\ 2 \\ 3 \end{pmatrix} = 4$ and $\mathbf{r} \cdot \begin{pmatrix} 3 \\ -3 \\ -1 \end{pmatrix} = 2$

(iii) $x + y - 4z = 4$ and $5x - 2y + 3z = 13$

② (i) Two planes have equations

$x + y + z = 5$

$x + 2y + 3z = 11$

(a) Eliminate x from these two equations.

(b) Let $z = \lambda$. Express y in terms of λ.

(c) Write x in terms of λ.

(d) Hence write down the vector equation of the line of intersection of the two planes.

(ii) Repeat part (i) for the two planes

$x + 2y + 2z = 6$

$3x + y - z = 6$

③ Find a Cartesian equation of the line in which the following planes meet.

$x + 4y - 5z = 2$

$2x - 3y + z = 4$

④ The planes $x - 3y - 2z = 5$ and $k^2x + ky + 2z = 3$ are perpendicular. Find the possible values of k.

⑤ Two sloping roof structures can be modelled as planes given by the equations

$x + 2y + 2z = 5$

$ax + y + z = 2$

where a is a positive constant. The roof structures must meet at an angle of exactly $60°$.

Find the exact value of a.

⑥ Find the equation of the plane π which is perpendicular to the planes

$3x - y - z + 4 = 0$

$x + 2y + z + 3 = 0$

and which passes through the point P$(4, 3, 5)$.

⑦ Find a vector equation of the line where the following planes meet:

(i) $x + y - z = 3$

$x + y + z = 5$

(ii) $2x + y + z = 4$

$3x + y + z = 5$

(iii) $3x + z = 4$

$3x - z = 2$

⑧ (i) Find the vector equation of the line l in which the following planes meet:

$x + 3y - 2z = 6$

$2x - y + z = 1$

(ii) Show that the point $(1, 3, 2)$ lies on l and also lies in the plane with equation

$3x + 2y + z = 11$

(iii) Hence write down the solution to the three simultaneous equations

$x + 3y - 2z = 6$

$2x - y + z = 1$

$3x + 2y + z = 11$

⑨ Three planes have equations

$2x + 5y + z = 2$

$3x + 2y - 2z = 1$

$x - 3y - 3z = a$

(i) Show that if $a = -4$ the three planes all meet in the same straight line, and find its Cartesian equation.

(ii) Show that if $a = 0$ the three planes meet in three parallel lines, and find their Cartesian equations.

⑩ Three planes have equations

$\pi_1: ax + 2y + z = 3$

$\pi_2: x + ay + z = 4$

$\pi_3: x + y + az = 5$

Given that the angle between planes π_1 and π_2 is equal to the angle between the planes π_2 and π_3, show that a must satisfy the quartic equation:

$5a^4 + 2a^3 - 2a^2 - 8a - 3 = 0$.

 Four planes are give by the equations

$\pi_1: 2x - 3y + 5z + 4 = 0$

$\pi_2: 2x + 3y + z + 4 = 0$

$\pi_3: 4x - 6y + 10z + 4 = 0$

$\pi_4: 2x - 3y + z + 4 = 0$

Determine whether each *pair* of planes is parallel, perpendicular or neither.

LEARNING OUTCOMES

Now you have finished this chapter, you should be able to:

➤ find the scalar product of two vectors in two and three dimensions

➤ know that two vectors are perpendicular if and only if their scalar product is zero

➤ form the equation of a line in two or three dimensions in vector or Cartesian form

➤ find the angle between two lines

➤ know the different ways in which two lines can intersect or not in three-dimensional space

➤ find out whether two lines in three dimensions are parallel, skew or intersect, and find the point of intersection if there is one

➤ find the equation of a plane in vector or Cartesian form

➤ identify a vector normal to a plane, given the equation of the plane

➤ find the angle between two planes

➤ find the angle between a line and a plane

➤ find the equation of the line in which two non-parallel planes meet.

KEY POINTS

1 The scalar product $\mathbf{a} \cdot \mathbf{b}$ of two vectors is defined as

$\mathbf{a} \cdot \mathbf{b} = |\text{length of } \mathbf{a}| \times |\text{length of } \mathbf{b}| \times \cos(\text{angle between } \mathbf{a} \text{ and } \mathbf{b})$

2 In component form, the scalar product is calculated in two dimensions as

$$\begin{pmatrix} a_1 \\ a_2 \end{pmatrix} \cdot \begin{pmatrix} b_1 \\ b_2 \end{pmatrix} = a_1 b_1 + a_2 b_2$$

and in three dimensions as

$$\begin{pmatrix} a_1 \\ a_2 \\ a_3 \end{pmatrix} \cdot \begin{pmatrix} b_1 \\ b_2 \\ b_3 \end{pmatrix} = a_1 b_1 + a_2 b_2 + a_3 b_3$$

3 The equation of a line through a point with position vector **a** and direction vector **d** is given in vector form as $\mathbf{r} = \mathbf{a} + \lambda\mathbf{d}$
and in Cartesian form as

$$\lambda = \frac{x - a_1}{d_1} = \frac{y - a_2}{d_2} = \frac{z - a_3}{d_3}$$ (unless one or two of d_1, d_2 and d_3 are zero)

4 The acute angle between lines with vector equations $\mathbf{r} = \mathbf{a}_1 + \lambda\mathbf{d}_1$ and $\mathbf{r} = \mathbf{a}_2 + \mu\mathbf{d}_2$ is given by

$$\cos^{-1}\left(\frac{|\mathbf{d}_1 \cdot \mathbf{d}_2|}{|\mathbf{d}_1||\mathbf{d}_2|}\right)$$

5 In three dimensions, two lines can meet, or they can be identical, parallel or skew.

6 The equation of a plane which is normal to a vector **n** and which contains the point with position vector **p** has vector equation $(\mathbf{r} - \mathbf{p}) \cdot \mathbf{n} = \mathbf{0}$

and Cartesian equation $n_1 x + n_2 y + n_3 z = d$ where
$d = n_1 p_1 + n_2 p_2 + n_3 p_3$

7 A vector normal to the plane with Cartesian equation $ax + by + cz = d$ is

$$\begin{pmatrix} a \\ b \\ c \end{pmatrix}.$$

8 The angle between two planes is given by the angle between their normal vectors.

9 The angle between a plane with normal to vector **n** and a line with direction vector **d** is

$$\sin^{-1}\left(\frac{|\mathbf{n} \cdot \mathbf{d}|}{|\mathbf{n}||\mathbf{d}|}\right)$$

10 To find the equation of the line in which two planes meet:
- eliminate one variable from the equations of the two planes
- write one of the remaining variables as λ
- write the other two variables in terms of λ.

FUTURE USES

- You will learn more about working with lines and planes in Vectors 2.

Review: Matrices and transformations

1 Matrices

What is a matrix?

A matrix is an array of numbers (the plural is matrices), usually written inside curved brackets, for example:

$$\mathbf{M} = \begin{pmatrix} 2 & 3 \\ 0 & -1 \\ 4 & 5 \end{pmatrix}$$

It is usual to represent matrices by capital letters, often in bold print.

A matrix consists of rows and columns, and the entries in the various cells are known as **elements**. $\mathbf{M}$ has 6 elements which are arranged in 3 rows and 2 columns, so $\mathbf{M}$ is called a 3×2 matrix. The matrix

$$\mathbf{N} = \begin{pmatrix} 3 & 1 & -4 \\ -2 & 5 & 0 \end{pmatrix}$$

also has 6 elements and is described as a 2×3 matrix as it has 2 rows and 3 columns. The number of rows and columns is called the **order** of the matrix – a matrix with r rows and c columns has order $r \times c$.

Some matrices are described by special names which relate to the number of rows and columns or the nature of the elements.

■ Matrices such as $\begin{pmatrix} 4 & 2 \\ 1 & 0 \end{pmatrix}$ and $\begin{pmatrix} 3 & 5 & 1 \\ 2 & 0 & -4 \\ 1 & 7 & 3 \end{pmatrix}$ which have the same

 number of rows as columns are called **square matrices**.

■ The matrix $\begin{pmatrix} 1 & 0 \\ 0 & 1 \end{pmatrix}$ is called the 2×2 **identity matrix** or **unit matrix**,

 and similarly $\begin{pmatrix} 1 & 0 & 0 \\ 0 & 1 & 0 \\ 0 & 0 & 1 \end{pmatrix}$ is called the 3×3 identity matrix. Identity

 matrices must be square and are usually denoted by $\mathbf{I}$.

■ The matrix $\mathbf{O} = \begin{pmatrix} 0 & 0 \\ 0 & 0 \end{pmatrix}$ is called the 2×2 **zero matrix**. Zero matrices

 can be of any order.

Two matrices are said to be **equal** if and only if they have the same order and each element in one matrix is equal to the corresponding element in the other matrix. So, for example, the matrices **A** and **C** below are equal, but **B** and **D** are not equal to any of the other matrices.

$$\mathbf{A} = \begin{pmatrix} 2 & 3 \\ -1 & 0 \\ 4 & 4 \end{pmatrix} \quad \mathbf{B} = \begin{pmatrix} 2 & 3 \\ -1 & 0 \end{pmatrix} \quad \mathbf{C} = \begin{pmatrix} 2 & 3 \\ -1 & 0 \\ 4 & 4 \end{pmatrix} \quad \mathbf{D} = \begin{pmatrix} 2 & 3 & 4 \\ -1 & 0 & 4 \end{pmatrix}$$

Working with matrices

Matrices can be added or subtracted if they are of the same order.

> Add the elements in corresponding positions.

$$\begin{pmatrix} 1 & 2 & -2 \\ 11 & 2 & -5 \end{pmatrix} + \begin{pmatrix} -5 & 6 & 4 \\ -5 & 0 & -5 \end{pmatrix} = \begin{pmatrix} -4 & 8 & 2 \\ 6 & 2 & -10 \end{pmatrix}$$

$$\begin{pmatrix} -7 & -2 \\ 3 & 12 \end{pmatrix} - \begin{pmatrix} 4 & -7 \\ -1 & 2 \end{pmatrix} = \begin{pmatrix} -11 & 5 \\ 4 & 10 \end{pmatrix}$$

> Subtract the elements in corresponding positions.

But $\begin{pmatrix} 1 & -2 & 7 \\ 0 & 4 & 5 \end{pmatrix} + \begin{pmatrix} -5 & -3 \\ -2 & 1 \end{pmatrix}$ cannot be evaluated because the matrices are

not of the same order. These matrices are **non-conformable for addition**.

You can also multiply a matrix by a number, or **scalar**:

> Multiply each of the elements by 3.

$$3 \begin{pmatrix} 4 & -2 \\ -5 & 1 \end{pmatrix} = \begin{pmatrix} 12 & -6 \\ -15 & 3 \end{pmatrix}$$

Matrix addition is **associative** and **commutative** as it is always true that:

> Commutative as the addition can take place in any order to obtain the same answer.

$$\mathbf{A} + (\mathbf{B} + \mathbf{C}) = (\mathbf{A} + \mathbf{B}) + \mathbf{C}$$
$$\mathbf{A} + \mathbf{B} = \mathbf{B} + \mathbf{A}$$

> Associative as the matrices can be grouped in different ways to obtain the same answer

Matrix subtraction is neither associative nor commutative. It is generally the case that

$$\mathbf{A} - (\mathbf{B} - \mathbf{C}) \neq (\mathbf{A} - \mathbf{B}) - \mathbf{C}$$
$$\mathbf{A} - \mathbf{B} \neq \mathbf{B} - \mathbf{A}$$

Matrix multiplication

Two matrices can be multiplied together if their orders satisfy the following rule:

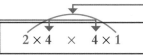

> The two 'outside' numbers give you the order of the product matrix, in this case 2×1.

$$2 \times 4 \quad \times \quad 4 \times 1$$

Figure R.1

TECHNOLOGY

You should check that you can use your calculator to input matrices of different orders, add and subtract conformable matrices and multiply a matrix by a scalar.

> The two 'middle' numbers, in this case 4, must be the same for it to be possible to multiply two matrices. If two matrices can be multiplied, they are **conformable for multiplication**.

So, for example, the product $\begin{pmatrix} 1 & 0 & -1 \\ 3 & 2 & 5 \end{pmatrix} \begin{pmatrix} 4 \\ -1 \\ 5 \end{pmatrix}$ can be calculated because

the orders of the matrices are 2×3 and 3×1; the resulting product would be a

2×1 matrix. However the product $\begin{pmatrix} 4 \\ -1 \\ 5 \end{pmatrix} \begin{pmatrix} 1 & 0 & -1 \\ 3 & 2 & 5 \end{pmatrix}$ is not possible as

the orders 3×1 and 2×3 do not have the same 'middle' numbers. This also illustrates that, generally, matrix multiplication is **not commutative**.

The example which follows shows how you multiply two matrices which are conformable.

Example R.1

Find $\begin{pmatrix} 6 & -2 \\ 4 & 3 \end{pmatrix} \begin{pmatrix} 8 \\ -1 \end{pmatrix}$.

Solution

The matrices have orders 2×2 and 2×1, so the matrices are conformable and the product will have order 2×1.

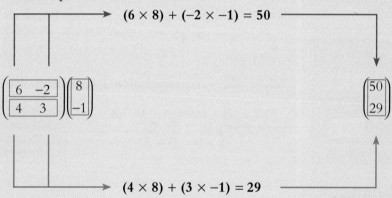

Figure R.2

TECHNOLOGY

In addition to using this method, you should check that you are able to multiply conformable matrices using your calculator.

Note that when a square matrix **A** is multiplied by an identity matrix **I** of the same size, the result is unchanged:

AI = **IA** = **A**

Matrix multiplication is **associative**. This means that for three conformable matrices **A**, **B** and **C**:

(**AB**)**C** = **A**(**BC**)

This result is important for later work on matrices.

① (i) Write down the orders of the matrices:

$$\mathbf{A} = \begin{pmatrix} 2 & 3 \\ 0 & 7 \end{pmatrix} \qquad \mathbf{B} = \begin{pmatrix} 5 & -2 & 4 \end{pmatrix}$$

$$\mathbf{C} = \begin{pmatrix} 1 \\ -4 \end{pmatrix} \qquad \mathbf{D} = \begin{pmatrix} 2 & -5 & 1 \\ 0 & 2 & 1 \end{pmatrix}$$

$$\mathbf{E} = \begin{pmatrix} 3 & 0 \\ -1 & -7 \\ 3 & 2 \end{pmatrix} \qquad \mathbf{F} = \begin{pmatrix} 5 & -2 & 1 \\ 3 & -2 & 1 \\ 0 & 5 & 0 \end{pmatrix}$$

$$\mathbf{G} = (-4) \qquad \mathbf{H} = \begin{pmatrix} 1 & -2 & 3 & -4 & 5 \end{pmatrix}$$

(ii) Without a calculator find, where possible, the matrix products:

(a) **AC**   (b) **EC**   (c) **BF**   (d) **FB**

(e) **GH**   (f) **CD**   (g) **DF**

(iii) Given also the matrices:

$$\mathbf{I} = \begin{pmatrix} -6 & 0 \\ 1 & 3 \end{pmatrix} \qquad \mathbf{J} = \begin{pmatrix} 2 & 5 & -3 \\ 0 & 1 & 1 \\ 2 & 0 & -1 \end{pmatrix} \qquad \mathbf{K} = \begin{pmatrix} 2 \\ -3 \end{pmatrix}$$

find, where possible:

(a) $\mathbf{A} + \mathbf{I}$ (b) $\mathbf{A} - \mathbf{I}$ (c) $2\mathbf{C} + 3\mathbf{K}$ (d) $\mathbf{I} + \mathbf{D}$

(e) $\mathbf{F} - 2\mathbf{J}$ (f) $\mathbf{K} - 3\mathbf{G}$ (g) $\mathbf{J} + \dfrac{1}{2}\mathbf{E}$

② Given the matrices $\mathbf{M} = \begin{pmatrix} 2 & 1 \\ -3 & 0 \end{pmatrix}$ and $\mathbf{N} = \begin{pmatrix} 1 & -1 \\ 4 & 2 \end{pmatrix}$

(i) find **MN** and **NM**

(ii) What property of matrices does your result from (i) illustrate?

③ Given the matrices $\mathbf{P} = \begin{pmatrix} 2 & 1 \\ 0 & -1 \end{pmatrix}, \mathbf{Q} = \begin{pmatrix} 0 & 1 & -2 \\ 3 & -2 & 1 \end{pmatrix}$ and

$$\mathbf{R} = \begin{pmatrix} 1 & 3 & 0 \\ -2 & 1 & 1 \\ 0 & 2 & -1 \end{pmatrix}$$

(i) find **PQ** and **(PQ)R**

(ii) find **QR** and **P(QR)**.

(iii) What property of matrices do your answers to (i) and (ii) illustrate?

④ Find the possible values of a and b such that:

$$\begin{pmatrix} 5a^2 & 4 \\ 0 & 18 \end{pmatrix} - \begin{pmatrix} 7(2a+1) & 7 \\ 5 & b^2 \end{pmatrix} = \begin{pmatrix} -4 & -3 \\ -5 & 2 \end{pmatrix}$$

⑤ For the matrix $\mathbf{A} = \begin{pmatrix} 5 & 1 \\ 0 & 1 \end{pmatrix}$ find, without using a calculator:

(i) $\mathbf{A}^2$

(ii) $\mathbf{A}^3$

(iii) $\mathbf{A}^4$

Proof by induction was covered in AQA A-level Further Mathematics Year 1 and will be revised in Chapter 4.

(iv) Suggest a general form for the matrix $\mathbf{A}^n$ in terms of n.

(v) Find $\mathbf{A}^6$ on your calculator and confirm that it gives the same answer as using (iv).

(vi) Use proof by induction to prove the result you found in (iv).

⑥ For the matrices $\mathbf{A} = \begin{pmatrix} 5 & x & 0 \\ -1 & -1 & 3 \end{pmatrix}$ and $\mathbf{B} = \begin{pmatrix} 2 & -3 \\ 3 & x \\ x & 5 \end{pmatrix}$:

(i) Find the product $\mathbf{AB}$ in terms of x.

> A **symmetric** matrix is one in which the entries are symmetrical about the leading diagonal,
>
> for example $\begin{pmatrix} 2 & 5 \\ 5 & 0 \end{pmatrix}$ and $\begin{pmatrix} 3 & 4 & -6 \\ 4 & 2 & 5 \\ -6 & 5 & 1 \end{pmatrix}$.

(ii) Given that the matrix $\mathbf{AB}$ is symmetric, find the possible values of x.

(iii) Write down the possible matrices $\mathbf{AB}$.

⑦ If $\mathbf{A} = \begin{pmatrix} a & 0 & -b \\ 2 & a & 3 \end{pmatrix}$, $\mathbf{B} = \begin{pmatrix} a & 6 \\ 1 & 2a \\ b & 4 \end{pmatrix}$ and $\mathbf{AB} = \begin{pmatrix} 45 & -50 \\ -15 & 122 \end{pmatrix}$, find the possible values of a and b.

2 Using matrices to represent transformations

A transformation maps an object according to a rule and can be represented by a matrix.

Figure R.3 and Figure R.4 show two possible transformations of a triangle OAB (shown in red). The vertices of the image (shown in blue) are denoted by the same letters with a dash e.g. A′, B′.

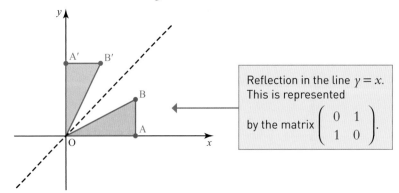

Reflection in the line $y = x$. This is represented by the matrix $\begin{pmatrix} 0 & 1 \\ 1 & 0 \end{pmatrix}$.

Figure R.3

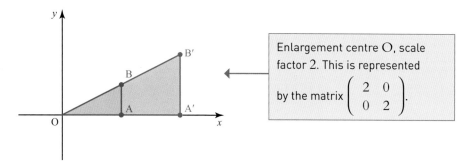

Enlargement centre O, scale factor 2. This is represented by the matrix $\begin{pmatrix} 2 & 0 \\ 0 & 2 \end{pmatrix}$.

Figure R.4

The transformation represented by a matrix can be found by looking at the effect of multiplying the unit vectors $\mathbf{i} = \begin{pmatrix} 1 \\ 0 \end{pmatrix}$ and $\mathbf{j} = \begin{pmatrix} 0 \\ 1 \end{pmatrix}$ by the matrix.

Figure R.3 shows reflection in the line $y = x$ which is represented by the matrix $\begin{pmatrix} 0 & 1 \\ 1 & 0 \end{pmatrix}$. The images of the unit vectors $\mathbf{i}$ and $\mathbf{j}$ are:

$$\begin{pmatrix} 0 & 1 \\ 1 & 0 \end{pmatrix}\begin{pmatrix} 1 \\ 0 \end{pmatrix} = \begin{pmatrix} 0 \\ 1 \end{pmatrix}$$

> The image of the unit vector $\mathbf{i}$ is $\mathbf{j}$.

$$\begin{pmatrix} 0 & 1 \\ 1 & 0 \end{pmatrix}\begin{pmatrix} 0 \\ 1 \end{pmatrix} = \begin{pmatrix} 1 \\ 0 \end{pmatrix}$$

> The image of the unit vector $\mathbf{j}$ is $\mathbf{i}$.

The interchange of the unit vectors $\mathbf{i}$ and $\mathbf{j}$ can also be determined by thinking about the transformation geometrically. Notice that $\begin{pmatrix} 0 \\ 1 \end{pmatrix}$ and $\begin{pmatrix} 1 \\ 0 \end{pmatrix}$ form the columns of the transformation matrix.

Figure R.4 shows an enlargement centre O scale factor 2 which is represented by the matrix $\begin{pmatrix} 2 & 0 \\ 0 & 2 \end{pmatrix}$. The images of the unit vectors $\mathbf{i}$ and $\mathbf{j}$ are:

$$\begin{pmatrix} 2 & 0 \\ 0 & 2 \end{pmatrix}\begin{pmatrix} 1 \\ 0 \end{pmatrix} = \begin{pmatrix} 2 \\ 0 \end{pmatrix}$$

> The image of the unit vector $\mathbf{i}$ is $\begin{pmatrix} 2 \\ 0 \end{pmatrix}$.

> The image of the unit vector $\mathbf{j}$ is $\begin{pmatrix} 0 \\ 2 \end{pmatrix}$.

$$\begin{pmatrix} 2 & 0 \\ 0 & 2 \end{pmatrix}\begin{pmatrix} 0 \\ 1 \end{pmatrix} = \begin{pmatrix} 0 \\ 2 \end{pmatrix}$$

Again $\begin{pmatrix} 2 \\ 0 \end{pmatrix}$ and $\begin{pmatrix} 0 \\ 2 \end{pmatrix}$ form the columns of the transformation matrix.

This connection between the images of the unit vectors $\mathbf{i}$ and $\mathbf{j}$ and the matrix representing the transformation provides a quick method for finding the matrix representing a transformation.

> You may find it easier to see what the transformation is when you use a shape, like the unit square, rather than points or lines.

It is common to use the unit square with coordinates O(0, 0), I(1, 0), P(1, 1) and J(0, 1). You can think about the images of the points I and J, and from this you can write down the images of the unit vectors $\mathbf{i}$ and $\mathbf{j}$.

This is done in the next example.

Example R.2

By drawing a diagram to show the image of the unit square, find the matrices which represent each of the following transformations:

(i)   a rotation of 90° clockwise about the origin

(ii)   a stretch scale factor 3 parallel to the y-axis.

Solution

(i)   Figure R.5 shows the effect of a rotation of 90° clockwise about the origin on the unit square.

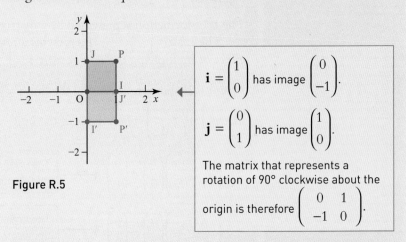

Figure R.5

$\mathbf{i} = \begin{pmatrix} 1 \\ 0 \end{pmatrix}$ has image $\begin{pmatrix} 0 \\ -1 \end{pmatrix}$.

$\mathbf{j} = \begin{pmatrix} 0 \\ 1 \end{pmatrix}$ has image $\begin{pmatrix} 1 \\ 0 \end{pmatrix}$.

The matrix that represents a rotation of 90° clockwise about the origin is therefore $\begin{pmatrix} 0 & 1 \\ -1 & 0 \end{pmatrix}$.

(ii)   Figure R.6 shows the effect of a stretch scale factor 3 parallel to the y-axis.

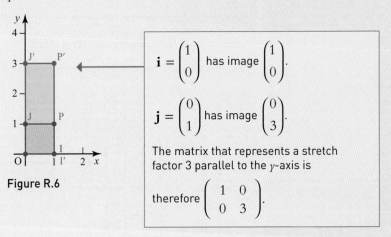

Figure R.6

$\mathbf{i} = \begin{pmatrix} 1 \\ 0 \end{pmatrix}$ has image $\begin{pmatrix} 1 \\ 0 \end{pmatrix}$.

$\mathbf{j} = \begin{pmatrix} 0 \\ 1 \end{pmatrix}$ has image $\begin{pmatrix} 0 \\ 3 \end{pmatrix}$.

The matrix that represents a stretch factor 3 parallel to the y-axis is therefore $\begin{pmatrix} 1 & 0 \\ 0 & 3 \end{pmatrix}$.

In summary:

- The matrix $\begin{pmatrix} k & 0 \\ 0 & k \end{pmatrix}$ represents an enlargement of scale factor k, centre the origin.

- The matrix $\begin{pmatrix} m & 0 \\ 0 & 1 \end{pmatrix}$ represents a stretch of scale factor m parallel to the x-axis.

- The matrix $\begin{pmatrix} 1 & 0 \\ 0 & n \end{pmatrix}$ represents a stretch of scale factor n parallel to the y-axis.

- The matrix $\begin{pmatrix} 0 & 1 \\ -1 & 0 \end{pmatrix}$ represents a rotation of 90° clockwise about the origin. This is a special case of the matrix $\begin{pmatrix} \cos\theta & -\sin\theta \\ \sin\theta & \cos\theta \end{pmatrix}$ which represents a rotation through angle θ degrees about the origin.

- The matrices $\begin{pmatrix} 1 & 0 \\ 0 & -1 \end{pmatrix}$ and $\begin{pmatrix} -1 & 0 \\ 0 & 1 \end{pmatrix}$ represent reflections in the x-axis and y-axis respectively.

- The matrix $\begin{pmatrix} 0 & 1 \\ 1 & 0 \end{pmatrix}$ represents a reflection in the line $y = x$.

- The matrix $\begin{pmatrix} 0 & -1 \\ -1 & 0 \end{pmatrix}$ represents a reflection in the line $y = -x$.

Figure R.7 shows the unit square and its image under the transformation represented by the matrix $\begin{pmatrix} 1 & 5 \\ 0 & 1 \end{pmatrix}$ on the unit square. The matrix $\begin{pmatrix} 1 & 5 \\ 0 & 1 \end{pmatrix}$ transforms the unit vector $\mathbf{i} = \begin{pmatrix} 1 \\ 0 \end{pmatrix}$ to the vector $\begin{pmatrix} 1 \\ 0 \end{pmatrix}$ and transforms the unit vector $\mathbf{j} = \begin{pmatrix} 0 \\ 1 \end{pmatrix}$ to the vector $\begin{pmatrix} 5 \\ 1 \end{pmatrix}$.

The point with position vector $\begin{pmatrix} 1 \\ 1 \end{pmatrix}$ is transformed to the point with position vector $\begin{pmatrix} 6 \\ 1 \end{pmatrix}$. $\longleftarrow$ As $\begin{pmatrix} 1 & 5 \\ 0 & 1 \end{pmatrix}\begin{pmatrix} 1 \\ 1 \end{pmatrix} = \begin{pmatrix} 6 \\ 1 \end{pmatrix}$.

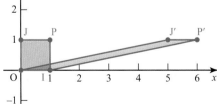

Figure R.7

This transformation is called a **shear**. The points on the x-axis stay the same and the points J and P move parallel to the x-axis to the right.

This shear can be described fully by saying that the x-axis is fixed, and giving the image of one point that is not on the x-axis, e.g. $(0, 1)$ is mapped to $(5, 1)$.

Generally, a shear with the x-axis fixed has the form $\begin{pmatrix} 1 & k \\ 0 & 1 \end{pmatrix}$

and a shear with the y-axis fixed has the form $\begin{pmatrix} 1 & 0 \\ k & 1 \end{pmatrix}$.

For each point, calculating the quantity

$$\frac{\text{distance between the point and its image}}{\text{distance of original point from }x\text{-axis}}$$

produces the same numerical value, which is the same as the number in the top right of the matrix. This is called the **shear factor** for the shear.

 There are different conventions about the sign of a shear factor and, for this reason, shear factors are not used to define a shear in this book. It is possible to show the effect of matrix transformations using some geometrical computer software packages. You might find that some packages use different approaches towards shears and define them in different ways.

Transformations in three dimensions

A plane is an infinite two-dimensional flat surface with no thickness. Figure R.8 below illustrates some common planes in three dimensions – the XY plane, the XZ plane and the YZ plane. The plane XY can also be referred to as $z = 0$, since the z coordinate would be zero for all points on the XY plane. Similarly, the XZ plane is referred to as $y = 0$ and the YZ plane as $x = 0$.

Figure R.8

As in two dimensions, the matrix can be found algebraically or by considering the effect of the transformation on the three unit vectors

$$\mathbf{i} = \begin{pmatrix} 1 \\ 0 \\ 0 \end{pmatrix}, \mathbf{j} = \begin{pmatrix} 0 \\ 1 \\ 0 \end{pmatrix} \text{ and } \mathbf{k} = \begin{pmatrix} 0 \\ 0 \\ 1 \end{pmatrix}.$$

So, for example, Figure R.9 shows the effect of a reflection in the plane $y = 0$.

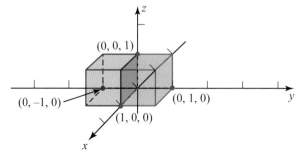

Figure R.9

$$\mathbf{i} = \begin{pmatrix} 1 \\ 0 \\ 0 \end{pmatrix} \text{ maps to } \begin{pmatrix} 1 \\ 0 \\ 0 \end{pmatrix}, \ \mathbf{j} = \begin{pmatrix} 0 \\ 1 \\ 0 \end{pmatrix} \text{ maps to } \begin{pmatrix} 0 \\ -1 \\ 0 \end{pmatrix} \text{ and } \mathbf{k} = \begin{pmatrix} 0 \\ 0 \\ 1 \end{pmatrix} \text{ maps to } \begin{pmatrix} 0 \\ 0 \\ 1 \end{pmatrix}.$$

The images of **i**, **j** and **k** form the columns of the 3 × 3 transformation matrix.

$$\begin{pmatrix} 1 & 0 & 0 \\ 0 & -1 & 0 \\ 0 & 0 & 1 \end{pmatrix}$$

When rotating an object about an axis, the rotation is taken to be anticlockwise about the axis of rotation when looking along the axis from the positive end towards the origin. Figure R.10 shows a rotation of 90° anticlockwise about the x-axis.

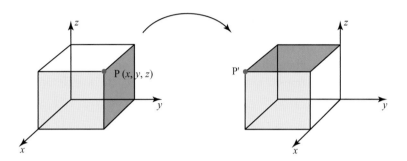

Figure R.10

Looking at the effect of the transformation on the unit vectors **i**, **j** and **k** shows

that $\mathbf{i} = \begin{pmatrix} 1 \\ 0 \\ 0 \end{pmatrix}$ maps to $\begin{pmatrix} 1 \\ 0 \\ 0 \end{pmatrix}$, $\mathbf{j} = \begin{pmatrix} 0 \\ 1 \\ 0 \end{pmatrix}$ maps to $\begin{pmatrix} 0 \\ 0 \\ 1 \end{pmatrix}$ and $\mathbf{k} = \begin{pmatrix} 0 \\ 0 \\ 1 \end{pmatrix}$ maps to $\begin{pmatrix} 0 \\ -1 \\ 0 \end{pmatrix}$.

The images of **i**, **j** and **k** form the columns of the 3 × 3 transformation

matrix $\begin{pmatrix} 1 & 0 & 0 \\ 0 & 0 & -1 \\ 0 & 1 & 0 \end{pmatrix}$.

Successive transformations

Figure R.11 shows the effect of two successive transformations on a triangle. The transformation A represents a reflection in the x-axis. A maps the point P to the point A(P).

The transformation B represents a rotation of 90° anticlockwise about O. When you apply B to the image formed by A, the point A(P) is mapped to the point B(A(P)). This is abbreviated to BA(P).

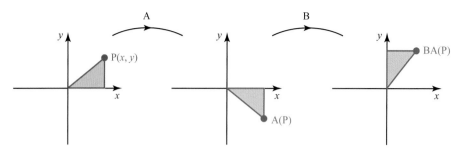

Figure R.11

Notice that a transformation written as BA means 'carry out A, then carry out B'.

This process is sometimes called **composition of transformations**.

In general, the matrix for a composite transformation is found by multiplying the matrices of the individual transformations in reverse order. So, for two transformations the matrix representing the first transformation is on the right and the matrix for the second transformation is on the left. For n transformations $T_1, T_2, \ldots, T_{n-1}, T_n$ the matrix product would be $\mathbf{A}_n \mathbf{A}_{n-1} \ldots \mathbf{A}_2 \mathbf{A}_1$.

Review exercise R.2

① A triangle has vertices at the origin, A(3, 0) and B(3, 1).

For each of the transformations below, draw a diagram to show the triangle OAB and its image OA′B′ and find the matrix which represents the transformation.

   (i)   Reflection in the line $y = x$

   (ii) Rotation 180° anticlockwise about O

   (iii) Enlargement scale factor 4, centre O

   (iv) Shear with the x-axis fixed, which maps the point (0, 1) to (2, 1)

② Describe the geometrical transformations represented by these matrices:

   (i) $\begin{pmatrix} 0 & -1 \\ -1 & 0 \end{pmatrix}$
   (ii) $\begin{pmatrix} 4 & 0 \\ 0 & 1 \end{pmatrix}$
   (iii) $\begin{pmatrix} 4 & 0 \\ 0 & 4 \end{pmatrix}$

   (iv) $\begin{pmatrix} 1 & 0 \\ 0 & -1 \end{pmatrix}$
   (v) $\begin{pmatrix} 0 & 1 \\ -1 & 0 \end{pmatrix}$

③ Figure R.12 shows a square with vertices at the points A(1, 1), B(1, −1), C(−1, −1) and D(−1, 1).

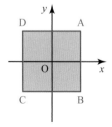

Figure R.12

(i) Draw a diagram to show the image of this square of the transformation matrix $\mathbf{M} = \begin{pmatrix} 1 & 0 \\ 5 & 1 \end{pmatrix}$.

(ii) Describe fully the transformation represented by the matrix $\mathbf{M}$. State the fixed line and the image of the point A.

④ Find the matrix that represents each of the following transformations in three dimensions:

(i) Rotation of $270°$ anticlockwise about the x-axis

(ii) Reflection in the plane $z = 0$

(iii) Rotation of $180°$ about the y-axis

⑤ Describe the transformations represented by these matrices:

(i) $\begin{pmatrix} 2 & 0 & 0 \\ 0 & 2 & 0 \\ 0 & 0 & 2 \end{pmatrix}$ (ii) $\begin{pmatrix} 1 & 0 & 0 \\ 0 & 1 & 0 \\ 0 & 0 & -1 \end{pmatrix}$

(iii) $\begin{pmatrix} 1 & 0 & 0 \\ 0 & 0 & 1 \\ 0 & -1 & 0 \end{pmatrix}$ (iv) $\begin{pmatrix} 4 & 0 & 0 \\ 0 & \frac{1}{2} & 0 \\ 0 & 0 & 3 \end{pmatrix}$

⑥ The 2×2 matrix $\mathbf{P}$ represents a rotation of $90°$ anticlockwise about the origin.

The 2×2 matrix $\mathbf{Q}$ represents a reflection in the line $y = -x$.

(i) Write down the matrices $\mathbf{P}$ and $\mathbf{Q}$.

(ii) Find the matrix $\mathbf{PQ}$ and describe the transformation it represents.

(iii) Find the matrix $\mathbf{QP}$ and describe the transformation it represents.

(iv) Show algebraically that there is only one point $r = \begin{pmatrix} x \\ y \end{pmatrix}$ which has the same image under the transformations represented by $\mathbf{PQ}$ and $\mathbf{QP}$ and state this point.

⑦ (i) Write down the matrix $\mathbf{A}$ which represents an anticlockwise rotation of $120°$ about the origin.

(ii) Write down the matrices $\mathbf{B}$ and $\mathbf{C}$ which represent anticlockwise rotations of $30°$ and $90°$ respectively about the origin. Find the matrix $\mathbf{BC}$ and verify that $\mathbf{A} = \mathbf{BC}$.

(iii) Calculate the matrix $\mathbf{B}^3$ and comment on your answer.

⑧ In three dimensions, the four matrices $\mathbf{J}, \mathbf{K}, \mathbf{L}$ and $\mathbf{M}$ represent transformations as follows:

$\mathbf{J}$ represents a reflection in the plane $y = 0$.

$\mathbf{K}$ represents an anticlockwise rotation of $90°$ about the z-axis.

$\mathbf{L}$ represents a reflection in the plane $x = 0$.

$\mathbf{M}$ represents a rotation of $180°$ about the x-axis.

(i) Write down the matrices $\mathbf{J}, \mathbf{K}, \mathbf{L}$ and $\mathbf{M}$.

(ii) Write down matrix products which would represent the single transformations obtained by each of the following combinations of transformations. Find the matrix in each case:

(a) a reflection in the plane $y = 0$ followed by a reflection in the plane $x = 0$

(b) a reflection in the plane $y = 0$ followed by an anticlockwise rotation of $90°$ about the z-axis

(c) an anticlockwise rotation of $90°$ about the z-axis followed by a second anticlockwise rotation of $90°$ about the z-axis

(d) a rotation of $180°$ about the x-axis followed by a reflection in the plane $x = 0$ followed by a reflection in the plane $y = 0$.

⑨ (i) Write down the matrix **P** which represents a stretch of scale factor 3 parallel to the y-axis.

(ii) The matrix $\mathbf{Q} = \begin{pmatrix} 2 & 0 \\ 0 & -1 \end{pmatrix}$. Write down the two single transformations which are represented by the matrix **Q**.

(iii) Find the matrix **PQ**. Write a list of the three transformations which are represented by the matrix **PQ**. In how many different orders could the three transformations occur?

(iv) Find the matrix **R** for which the matrix product **RPQ** would transform an object to its original position.

⑩ (i) Write down the matrix **A** representing a rotation about the origin through angle θ, and the matrix **B** representing a rotation about the origin through angle ϕ.

(ii) Find the matrix **BA**, representing a rotation about the origin through angle θ, followed by a rotation about the origin through angle ϕ.

(iii) Write down the matrix **C** representing a rotation about the origin through angle $\theta + \phi$.

(iv) By equating **C** to **BA**, write down expressions for $\sin(\theta + \phi)$ and $\cos(\theta + \phi)$.

(v) Explain why **BA** = **AB** in this case.

⑪ The matrix **R** represents a reflection in the line $y = mx$.

Show that $\mathbf{R}^2 = \begin{pmatrix} 1 & 0 \\ 0 & 1 \end{pmatrix}$ and explain geometrically why this is the case.

⑫ The matrix **A** is $\begin{pmatrix} 0 & -1 \\ 1 & 0 \end{pmatrix}$.

(i) Explain in terms of transformations why $\mathbf{A}^4 = \mathbf{I}$.

(ii) Using geometrical considerations, find the matrix **B** such that **BA** = **I**.

(iii) Write down the matrix **C** which represents a rotation of $60°$ anticlockwise about the origin.

(iv) Write down the smallest positive integers m and n such that $\mathbf{A}^m = \mathbf{C}^n$, explaining your answer in terms of transformations.

(v) Find **AC** and explain in terms of transformations why **AC** = **CA**.

3 Invariance

Points which map to themselves under a transformation are called **invariant points**. The origin is always an invariant point under a transformation that can be represented by a matrix, as the following statement is always true:

$$\begin{pmatrix} a & b \\ c & d \end{pmatrix}\begin{pmatrix} 0 \\ 0 \end{pmatrix} = \begin{pmatrix} 0 \\ 0 \end{pmatrix}$$

More generally, a point (x, y) is invariant if it satisfies the matrix equation:

$$\begin{pmatrix} a & b \\ c & d \end{pmatrix}\begin{pmatrix} x \\ y \end{pmatrix} = \begin{pmatrix} x \\ y \end{pmatrix}$$

For example, the point $(-2, 2)$ is invariant under the transformation represented by the matrix $\begin{pmatrix} 6 & 5 \\ 2 & 3 \end{pmatrix}$:

$$\begin{pmatrix} 6 & 5 \\ 2 & 3 \end{pmatrix}\begin{pmatrix} -2 \\ 2 \end{pmatrix} = \begin{pmatrix} -2 \\ 2 \end{pmatrix}$$

Example R.3

M is the matrix $\begin{pmatrix} 10 & -3 \\ 3 & 0 \end{pmatrix}$.

(i)   Show that $(-2, -6)$ is an invariant point under the transformation represented by **M**.

(ii) What can you say about the invariant points under this transformation?

Solution

(i) $\begin{pmatrix} 10 & -3 \\ 3 & 0 \end{pmatrix}\begin{pmatrix} -2 \\ -6 \end{pmatrix} = \begin{pmatrix} -2 \\ -6 \end{pmatrix}$

so $(-2, -6)$ is an invariant point under the transformation represented by **M**.

(ii) Suppose the point $\begin{pmatrix} x \\ y \end{pmatrix}$ maps to itself. Then:

$$\begin{pmatrix} 10 & -3 \\ 3 & 0 \end{pmatrix}\begin{pmatrix} x \\ y \end{pmatrix} = \begin{pmatrix} x \\ y \end{pmatrix}$$

$$\begin{pmatrix} 10x - 3y \\ 3x \end{pmatrix} = \begin{pmatrix} x \\ y \end{pmatrix}$$

> These points all have the form $\lambda, 3\lambda$. The point $(-2, -6)$ is just one of the points on this line.

$$\Leftrightarrow 10x - 3y = x \text{ and } 3x = y.$$

> Both equations simplify to $y = 3x$.

So the invariant points of the transformation are all the points on the line $y = 3x$.

The simultaneous equations in Example R.3 were equivalent and so all the invariant points were on a straight line. Generally, any matrix equation set up to find the invariant points will lead to two equations of the form $ax + by = 0$, which can also be expressed in the form $y = -\dfrac{ax}{b}$. These equations may be equivalent, in which case this is a line of invariant points. If the two equations are not equivalent, the origin is the only point which satisfies both equations, and so this is the only invariant point.

Invariant lines

A line AB is known as an **invariant line** under a transformation if the image of every point on AB is also on AB. It is important to note that it is not necessary for each of the points to map to itself; it can map to itself or to some other point on the line AB.

Sometimes it is easy to spot which lines are invariant. For example, in Figure R.13 the position of the points A − F and their images A′ − F′ show that the transformation is a reflection in the line *l*. So every point on *l* maps onto itself and *l* is a line of invariant points.

Look at the lines perpendicular to the mirror line in Figure R.13 , for example the line ABB′A′. Any point on one of these lines maps onto another point on the same line. Such a line is invariant but it is not a line of invariant points.

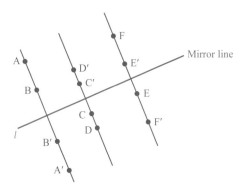

Figure R.13

| **Example R.4** | Find the invariant lines of the transformation given by the matrix |

$$\mathbf{M} = \begin{pmatrix} 4 & -1 \\ -2 & 1 \end{pmatrix}.$$

Solution

Suppose the invariant line has the form $y = mx + c$.

> Let the original point be (x, y) and the image point be (x', y').

$$\begin{pmatrix} x' \\ y' \end{pmatrix} = \begin{pmatrix} 4 & -1 \\ -2 & 1 \end{pmatrix} \begin{pmatrix} x \\ y \end{pmatrix} \Leftrightarrow x' = 4x - y \text{ and } y' = -2x + y.$$

$$\Leftrightarrow \begin{cases} x' = 4x - (mx + c) = (4 - m)x - c \quad \longleftarrow \boxed{\text{Using } y = mx + c} \\ \\ y' = -2x + (mx + c) = (-2 + m)x + c \end{cases}$$

As the line is invariant, (x', y') also lies on the line, so $y' = mx' + c$.

Therefore:

$$(-2 + m)x + c = m\big[(4 - m)x - c\big] + c$$

$$\Leftrightarrow 0 = \big(m^2 - 3m - 2\big)x + mc$$

For the right hand side to equal zero, both $m^2 - 3m - 2 = 0$ and $mc = 0$.

$$(m - 1)(m - 2) = 0 \Leftrightarrow m = 1 \text{ or } m = 2.$$

and

$$mc = 0 \Leftrightarrow m = 0 \text{ or } c = 0. \longleftarrow$$

$m = 0$ is not a viable solution as $m^2 - 3m - 2 \neq 0$.

So, there are two possible solutions for the invariant line:

$$m = 1, c = 0 \Rightarrow y = x$$
$$\text{or } m = 2, c = 0 \Rightarrow y = 2x$$

Review exercise R.3

① Find the invariant points under the transformations represented by the following matrices:

(i) $\begin{pmatrix} 3 & 3 \\ 1 & 1 \end{pmatrix}$ (ii) $\begin{pmatrix} 2 & 3 \\ 1 & 2 \end{pmatrix}$ (iii) $\begin{pmatrix} 3 & 2 \\ 1 & 2 \end{pmatrix}$ (iv) $\begin{pmatrix} 7 & -4 \\ 3 & -1 \end{pmatrix}$

② What lines, if any, are invariant under the following transformations?

(i) Enlargement, centre the origin

(ii) Rotation through 180° about the origin

(iii) Rotation through 90° about the origin

(iv) Reflection in the line $y = x$

(v) Reflection in the line $y = -x$

(vi) Shear, x-axis fixed

③ For the matrix $\mathbf{M} = \begin{pmatrix} 2 & 7 \\ 7 & 2 \end{pmatrix}$:

(i) show that the origin is the only invariant point

(ii) find the invariant lines of the transformation represented by $\mathbf{M}$.

④ For the matrix $\mathbf{M} = \begin{pmatrix} 1 & 0 \\ 3 & -1 \end{pmatrix}$:

(i) find the line of invariant points of the transformation given by $\mathbf{M}$

(ii) find the invariant lines of the transformation

(iii) draw a diagram to show the effect of the transformation on the unit square.

⑤ A reflection in a line l is represented by the matrix $\mathbf{A} = \begin{pmatrix} -0.6 & 0.8 \\ 0.8 & 0.6 \end{pmatrix}$.

(i) Find the image of the point $(3, 6)$ and hence write down the equation of the mirror line l.

(ii) The matrix $\mathbf{B} = \begin{pmatrix} 0 & -1 \\ 1 & 0 \end{pmatrix}$ represents a rotation. By considering the point $(3, 2)$, find the centre and angle of rotation.

(iii) Find the matrix $\mathbf{BA}$.

(iv) Show that under the transformation $\mathbf{BA}$ the point $(1, -3)$ is invariant. Hence state the equation of the line of invariant points under the transformation $\mathbf{BA}$.

⑥ The matrix $\begin{pmatrix} \dfrac{1-m^2}{1+m^2} & \dfrac{2m}{1+m^2} \\[3mm] \dfrac{2m}{1+m^2} & \dfrac{m^2-1}{1+m^2} \end{pmatrix}$ represents a reflection in the line $y = mx$.

Prove that the line $y = mx$ is a line of invariant points.

KEY POINTS

1 A matrix is a rectangular array of numbers or letters.
2 The shape of a matrix is described by its order. A matrix with r rows and c columns has order $r \times c$.
3 A matrix with the same number of rows and columns is called a **square matrix**.
4 The matrix $\mathbf{O} = \begin{pmatrix} 0 & 0 \\ 0 & 0 \end{pmatrix}$ is known as the 2×2 **zero matrix**. Zero matrices can be of any order.
5 A matrix of the form $\mathbf{I} = \begin{pmatrix} 1 & 0 \\ 0 & 1 \end{pmatrix}$ is known as an **identity matrix**. All identity matrices are square, with 1s on the leading diagonal and zeros elsewhere.
6 Matrices can be added or subtracted if they have the same order.
7 Two matrices $\mathbf{A}$ and $\mathbf{B}$ can be multiplied to give matrix $\mathbf{AB}$ if their orders are of the form $p \times q$ and $q \times r$ respectively. The resulting matrix will have the order $p \times r$.
8 Matrix multiplication

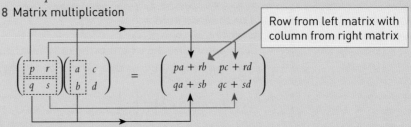

Row from left matrix with column from right matrix

Figure R.14

9 Matrix addition and multiplication are **associative**.
$$\mathbf{A} + (\mathbf{B} + \mathbf{C}) = (\mathbf{A} + \mathbf{B}) + \mathbf{C}$$
$$\mathbf{A}(\mathbf{BC}) = (\mathbf{AB})\mathbf{C}$$

10 Matrix addition is **commutative** but matrix multiplication is generally not commutative.
$$\mathbf{A} + \mathbf{B} = \mathbf{B} + \mathbf{A}$$
$$\mathbf{AB} \neq \mathbf{BA}$$

11 The matrix $\mathbf{M} = \begin{pmatrix} a & b \\ c & d \end{pmatrix}$ represents the transformation which maps the point with position vector $\begin{pmatrix} x \\ y \end{pmatrix}$ to the point with position vector $\begin{pmatrix} ax + by \\ cx + dy \end{pmatrix}$.

12 Under the transformation represented by **M**, the image of $\mathbf{i} = \begin{pmatrix} 1 \\ 0 \end{pmatrix}$ is the first

column of **M** and the image of $\mathbf{j} = \begin{pmatrix} 0 \\ 1 \end{pmatrix}$ is the second column of **M**.

Similarly, in three dimensions the images of the unit vectors

$\mathbf{i} = \begin{pmatrix} 1 \\ 0 \\ 0 \end{pmatrix}$, $\mathbf{j} = \begin{pmatrix} 0 \\ 1 \\ 0 \end{pmatrix}$ and $\mathbf{k} = \begin{pmatrix} 0 \\ 0 \\ 1 \end{pmatrix}$ are the first, second and third columns of the

transformation matrix.

13 Summary of transformations in two dimensions

$\begin{pmatrix} 1 & 0 \\ 0 & -1 \end{pmatrix}$ Reflection in the x-axis

$\begin{pmatrix} -1 & 0 \\ 0 & 1 \end{pmatrix}$ Reflection in the y-axis

$\begin{pmatrix} 0 & 1 \\ 1 & 0 \end{pmatrix}$ Reflection in the line $y = x$

$\begin{pmatrix} 0 & -1 \\ -1 & 0 \end{pmatrix}$ Reflection in the line $y = -x$

$\begin{pmatrix} \cos\theta & -\sin\theta \\ \sin\theta & \cos\theta \end{pmatrix}$ Rotation anticlockwise about the origin through angle θ

$\begin{pmatrix} k & 0 \\ 0 & k \end{pmatrix}$ Enlargement centre the origin, scale factor k

$\begin{pmatrix} k & 0 \\ 0 & 1 \end{pmatrix}$ Stretch parallel to the x-axis, scale factor k

$\begin{pmatrix} 1 & 0 \\ 0 & k \end{pmatrix}$ Stretch parallel to the y-axis, scale factor k

$\begin{pmatrix} 1 & k \\ 0 & 1 \end{pmatrix}$ Shear, x-axis invariant, with $(0, 1)$ mapped to $(k, 1)$

$\begin{pmatrix} 1 & 0 \\ k & 1 \end{pmatrix}$ Shear, y-axis invariant, with $(1, 0)$ mapped to $(1, k)$

14 Examples of transformations in three dimensions

$\begin{pmatrix} -1 & 0 & 0 \\ 0 & 1 & 0 \\ 0 & 0 & 1 \end{pmatrix}$ Reflection in plane $x = 0$

$\begin{pmatrix} 1 & 0 & 0 \\ 0 & -1 & 0 \\ 0 & 0 & 1 \end{pmatrix}$ Reflection in plane $y = 0$

$$\begin{pmatrix} 1 & 0 & 0 \\ 0 & 1 & 0 \\ 0 & 0 & -1 \end{pmatrix}$$ Reflection in plane $z = 0$

$$\begin{pmatrix} 1 & 0 & 0 \\ 0 & 0 & -1 \\ 0 & 1 & 0 \end{pmatrix}$$ Rotation of 90° about the x-axis

$$\begin{pmatrix} 0 & 0 & 1 \\ 0 & 1 & 0 \\ -1 & 0 & 0 \end{pmatrix}$$ Rotation of 90° about the y-axis

$$\begin{pmatrix} 0 & -1 & 0 \\ 1 & 0 & 0 \\ 0 & 0 & 1 \end{pmatrix}$$ Rotation of 90° about the z-axis

When rotating an object about an axis, the rotation is taken to be anticlockwise about the axis of rotation when looking along the axis from the positive end towards the origin.

15 The composite of the transformation represented by **M** followed by that represented by **N** is represented by the matrix product **NM**.

16 If (x, y) is an **invariant point** under a transformation represented by the matrix **M** then $\mathbf{M} \begin{pmatrix} x \\ y \end{pmatrix} = \begin{pmatrix} x \\ y \end{pmatrix}$.

17 A line AB is known as an **invariant line** under a transformation if the image of every point on AB is also on AB.

2 Matrices 1

Search engines' algorithms determine which order to display results by solving huge systems of simultaneous equations. In practice this happens using matrices. Matrices are also used extensively in computer games – particularly to apply rotations and translations to cameras and objects whose positions and orientations will be described by coordinates and vectors in three dimensional worlds.

Review: Determinant and inverse of a 2 × 2 matrix

Figure 2.1 shows the parallelogram produced when the unit square is transformed by the matrix $\begin{pmatrix} a & b \\ c & d \end{pmatrix}$.

> You can find the area of the red parallelogram by subtracting the areas of the yellow rectangles and the green and blue triangles from the area of the large rectangle.

Figure 2.1

The area of the parallelogram is $(ad - bc)$ units².

Since the area of the unit square is one unit, the quantity $(ad - bc)$ is the area scale factor associated with the transformation matrix $\begin{pmatrix} a & b \\ c & d \end{pmatrix}$. It is called the **determinant** of the matrix and is denoted by det **M** or |**M**|. Another notation that is sometimes used as an alternative is Δ.

Example 2.1

For each of the matrices $\mathbf{A} = \begin{pmatrix} 3 & 4 \\ 1 & 2 \end{pmatrix}$ and $\mathbf{B} = \begin{pmatrix} 3 & 4 \\ 2 & 1 \end{pmatrix}$:

(i)   Draw a diagram to show the image of the unit square OIPJ under the transformation represented by the matrix.

(ii)   Find the determinant of the matrix.

(iii)   Use your answer to (ii) to find the area of the transformed shape.

Solution

(a)   (i)

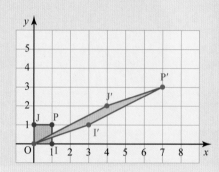

Figure 2.2

(ii) $\det \mathbf{A} = (3 \times 2) - (1 \times 4) = 2$

(iii)   Area of quadrilateral OI′P′J′ is $1 \times 2 = 2$.

(b)   (i)

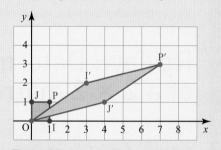

Figure 2.3

(ii) $\det \mathbf{B} = (3 \times 1) - (4 \times 2) = -5$

(iii)   Area of quadrilateral OI′P′J′ is $1 \times 5 = 5$.

> Notice that the determinant is negative. Since area cannot be negative, the area of the transformed shape is 5 square units.

In Example 2.1 the sign of the determinant has significance. In part (i), if you move anticlockwise around the original unit square you come to vertices O, I, P, J in that order. Moving anticlockwise around the image gives O, I′, P′, J′, i.e. the order is unchanged.

However, in part (ii) moving anticlockwise about the image reverses the order of the vertices, i.e. O, J', P', I'. This reversal in the order of the vertices produces the negative determinant.

The same principle applies in three dimensions.

Important results about determinants

1. For square matrices $\mathbf{P}$ and $\mathbf{Q}$, $\det \mathbf{PQ} = \det \mathbf{P} \times \det \mathbf{Q}$.

> A special case is the zero matrix which maps all points to the origin.

2. If a matrix $\mathbf{M}$ has zero determinant, the area scale factor of the transformation is zero, so all points are mapped to a shape with zero area. In fact the matrix maps all points in the plane to a straight line.

Finding the inverse of a 2 × 2 matrix

The inverse of a 2 × 2 matrix $\mathbf{M} = \begin{pmatrix} a & b \\ c & d \end{pmatrix}$ is $\mathbf{M}^{-1} = \dfrac{1}{ad - bc} \begin{pmatrix} d & -b \\ -c & a \end{pmatrix}$.

If the determinant is zero then the inverse matrix does not exist and the matrix is said to be **singular**. If $\det \mathbf{M} \neq 0$ the matrix is said to be **non-singular**.

If a matrix is singular, then it maps an infinite number of points in the plane to the same point on the straight line. It is therefore not possible to find the inverse of the transformation, because it would need to map a point on that straight line to just one other point, not to an infinite number of them.

Example 2.2

$$\mathbf{A} = \begin{pmatrix} 5 & -3 \\ 9 & 1 \end{pmatrix}$$

(i) Find $\mathbf{A}^{-1}$.

(ii) The point P is mapped to the point Q(−22, −14) under the transformation represented by $\mathbf{A}$. Find the coordinates of P.

Solution

(i) $\det \mathbf{A} = (5 \times 1) - (-3 \times 9) = 32$

$$\mathbf{A}^{-1} = \frac{1}{32} \begin{pmatrix} 1 & 3 \\ -9 & 5 \end{pmatrix}$$

> A maps P to Q, so $\mathbf{A}^{-1}$ maps Q to P.

(ii) $\mathbf{A}^{-1} \begin{pmatrix} -22 \\ -14 \end{pmatrix} = \frac{1}{32} \begin{pmatrix} 1 & 3 \\ -9 & 5 \end{pmatrix} \begin{pmatrix} -22 \\ -14 \end{pmatrix} = \frac{1}{32} \begin{pmatrix} -64 \\ 128 \end{pmatrix} = \begin{pmatrix} -2 \\ 4 \end{pmatrix}$

So P has coordinates (−2, 4).

An important result about a matrix and its inverse is that $\mathbf{MM}^{-1} = \mathbf{M}^{-1}\mathbf{M} = \mathbf{I}$. This is true for all square matrices, not just 2 × 2 matrices.

The inverse of a product of matrices

Suppose you want to find the inverse of the product **MN**, where **M** and **N** are non-singular matrices. This means that you need to find a matrix **X** such that **X(MN) = I**.

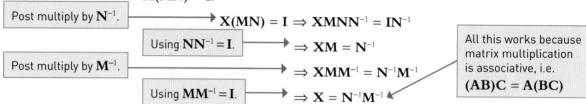

Post multiply by $\mathbf{N}^{-1}$.	$\longrightarrow$	$\mathbf{X(MN) = I} \Rightarrow \mathbf{XMNN^{-1} = IN^{-1}}$
Using $\mathbf{NN^{-1} = I}$.	$\longrightarrow$	$\Rightarrow \mathbf{XM = N^{-1}}$
Post multiply by $\mathbf{M}^{-1}$.	$\longrightarrow$	$\Rightarrow \mathbf{XMM^{-1} = N^{-1}M^{-1}}$
Using $\mathbf{MM^{-1} = I}$.	$\longrightarrow$	$\Rightarrow \mathbf{X = N^{-1}M^{-1}}$

All this works because matrix multiplication is associative, i.e.
$$\mathbf{(AB)C = A(BC)}$$

So $\mathbf{(MN)^{-1} = N^{-1}M^{-1}}$ for matrices **M** and **N** of the same order. This means that when working backwards, you must reverse the second transformation before reversing the first transformation.

Using matrices to solve simultaneous equations

There are a number of methods to solve a pair of linear simultaneous equations of the form

$$3x + 4y = 7$$
$$2x - y = 12$$

such as elimination, substitution or graphical methods.

Another method involves the use of inverse matrices. This method has the advantage that it can more easily be extended to solving a set of n equations in n variables.

Example 2.3

Use a matrix method to solve the simultaneous equations

$$3x + 4y = 7$$
$$2x - y = 12$$

Solution

$$\begin{pmatrix} 3 & 4 \\ 2 & -1 \end{pmatrix}\begin{pmatrix} x \\ y \end{pmatrix} = \begin{pmatrix} 7 \\ 12 \end{pmatrix}$$

Write the equations in matrix form.

The inverse of the matrix $\begin{pmatrix} 3 & 4 \\ 2 & -1 \end{pmatrix}$ is $-\dfrac{1}{11}\begin{pmatrix} -1 & -4 \\ -2 & 3 \end{pmatrix} = \dfrac{1}{11}\begin{pmatrix} 1 & 4 \\ 2 & -3 \end{pmatrix}$.

Pre-multiply both sides of the matrix equation by the inverse matrix.

$$\dfrac{1}{11}\begin{pmatrix} 1 & 4 \\ 2 & -3 \end{pmatrix}\begin{pmatrix} 3 & 4 \\ 2 & -1 \end{pmatrix}\begin{pmatrix} x \\ y \end{pmatrix} = \dfrac{1}{11}\begin{pmatrix} 1 & 4 \\ 2 & -3 \end{pmatrix}\begin{pmatrix} 7 \\ 12 \end{pmatrix}$$

$$\begin{pmatrix} x \\ y \end{pmatrix} = \dfrac{1}{11}\begin{pmatrix} 55 \\ -22 \end{pmatrix} = \begin{pmatrix} 5 \\ -2 \end{pmatrix}$$

As $\mathbf{M^{-1}M = I}$ the left hand side simplifies to $\begin{pmatrix} x \\ y \end{pmatrix}$.

The solution is $x = 5$, $y = -2$.

Geometrical interpretation in two dimensions

Two equations in two unknowns can be represented in a plane by two straight lines. The number of points of intersection of the lines determines the number of solutions to the equations.

There are three different possibilities.

Case 1

Example 2.3 shows that two simultaneous equations can have a unique solution. Graphically, this is represented by a single point of intersection, as shown in Figure 2.4.

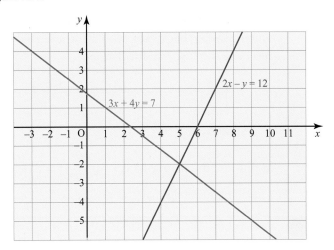

Figure 2.4

This is the case where $\det \mathbf{M} \neq 0$ and so the inverse matrix $\mathbf{M}^{-1}$ exists, allowing the equations to be solved.

Case 2

If two lines are parallel they do not have a point of intersection. For example, the lines

$$x + 2y = 10$$
$$x + 2y = 4$$

are parallel (see Figure 2.5).

> The equations can be written in matrix form as $\begin{pmatrix} 1 & 2 \\ 1 & 2 \end{pmatrix}\begin{pmatrix} x \\ y \end{pmatrix} = \begin{pmatrix} 10 \\ 4 \end{pmatrix}$.

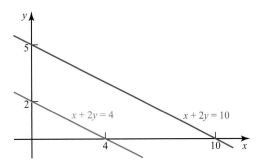

Figure 2.5

The matrix $\mathbf{M} = \begin{pmatrix} 1 & 2 \\ 1 & 2 \end{pmatrix}$ has determinant zero and hence the inverse matrix does not exist.

Case 3

The equations can be written in matrix form as

$$\begin{pmatrix} 1 & 2 \\ 3 & 6 \end{pmatrix}\begin{pmatrix} x \\ y \end{pmatrix} = \begin{pmatrix} 10 \\ 30 \end{pmatrix}.$$

More than one solution is possible in cases where the lines are coincident, i.e. lie on top of each other. For example, the two lines

$$x + 2y = 10$$
$$3x + 6y = 30$$

are coincident (see Figure 2.6). You can see this because the equations are multiples of each other.

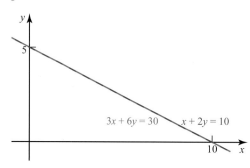

Figure 2.6

In this case the matrix **M** is $\begin{pmatrix} 1 & 2 \\ 3 & 6 \end{pmatrix}$ and det **M** = 0.

There are infinitely many solutions to these equations.

Review exercise

① The matrix $\begin{pmatrix} 2x - 7 & 2 \\ 9 - 8x & x + 1 \end{pmatrix}$ has determinant −4.

   Find the possible values of x.

② (i) Write down the matrices **A** and **B** which represent:

   A a reflection in the y-axis

   B a reflection in the line $y = -x$

   (ii) Show that the matrices **A** and **B** each have determinant of −1.

   (iii) Draw diagrams for each of the transformations **A** and **B** to demonstrate that the images of the vertices labelled anticlockwise on the unit square OIPJ are reversed to a clockwise labelling.

③ Figure 2.7 shows the unit square transformed by a shear.

   (i) Write down the matrix which represents this transformation.

   (ii) Show that under this transformation the area of the image is always equal to the area of the object.

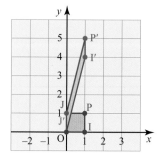

Figure 2.7

④ For the matrix $\begin{pmatrix} 3 & -5 \\ 1 & 5 \end{pmatrix}$:

   (i) find the image of the point $(2, -1)$

   (ii) find the inverse matrix

   (iii) find the point which maps to the image $(3, 1)$.

⑤ The matrix $\begin{pmatrix} 3-k & k \\ 2 & 2-k \end{pmatrix}$ is singular.

Find the possible values of k.

⑥ $\mathbf{M} = \begin{pmatrix} -2 & 3 \\ -1 & 6 \end{pmatrix}$ and $\mathbf{N} = \begin{pmatrix} 4 & 7 \\ 9 & -1 \end{pmatrix}$.

(i) Find the determinants of $\mathbf{M}$ and $\mathbf{N}$.

(ii) Find the matrix $\mathbf{MN}$ and show that $\det(\mathbf{MN}) = \det \mathbf{M} \times \det \mathbf{N}$.

⑦ Given that $\mathbf{M} = \begin{pmatrix} 1 & -3 \\ 2 & 1 \end{pmatrix}$ and $\mathbf{MN} = \begin{pmatrix} 6 & -1 & 3 & 6 \\ 5 & -2 & 6 & 12 \end{pmatrix}$, find the matrix $\mathbf{N}$.

⑧ The plane is transformed by the matrix $\mathbf{M} = \begin{pmatrix} 3 & -9 \\ -2 & 6 \end{pmatrix}$.

(i) Draw a diagram to show the image of the unit square under the transformation represented by $\mathbf{M}$.

(ii) Describe the effect of the transformation and explain this with reference to the determinant of $\mathbf{M}$.

⑨ Use matrices to solve the following pairs of simultaneous equations:

(i) $5x - 3y = 13$
     $2x - y = 5$

(ii) $4x - y = -16$
     $x - 3y = -15$

⑩ Find the two values of k for which the equations

$2x + ky = 3$
$kx + 8y = 6$

do not have a unique solution.

How many solutions are there in each case?

⑪ $\mathbf{M} = \begin{pmatrix} a & b \\ c & d \end{pmatrix}$ is a singular matrix.

(i) Show that $\mathbf{M}^2 = (a + d)\mathbf{M}$.

(ii) Find a formula which expresses $\mathbf{M}^n$ in terms of $\mathbf{M}$, where n is a positive integer.

(iii) Prove the formula you found in part (ii) by induction.

⑫ The plane is transformed using the matrix $\begin{pmatrix} a & b \\ c & d \end{pmatrix}$ where $ad - bc = 0$.

Prove that the general point $P(x, y)$ maps to P' on the line $cx - ay = 0$.

⑬ Triangle T has vertices at $(1, 0)$, $(0, 2)$ and $(-3, 0)$.

It is transformed to triangle T$'$ by the matrix $\mathbf{M} = \begin{pmatrix} 4 & 1 \\ 1 & 1 \end{pmatrix}$.

(i) Find the coordinates of the vertices of T$'$.
    Show the triangles T and T$'$ on a single diagram.

(ii) Find the ratio of the area of T$'$ to the area of T.
     Comment on your answer in relation to the matrix $\mathbf{M}$.

(iii) Find $\mathbf{M}^{-1}$ and verify that this matrix maps the vertices of T′ to the vertices of T.

⑭ (i) The matrix $\mathbf{S} = \begin{pmatrix} -1 & 2 \\ -3 & 4 \end{pmatrix}$ represents a transformation.

(a) Show that the point (1, 1) is invariant under this transformation.

(b) Calculate $\mathbf{S}^{-1}$.

(c) Verify that (1, 1) is also invariant under the transformation represented by $\mathbf{S}^{-1}$.

(ii) Part (i) can be generalised as follows:

If (x, y) is an invariant point under a transformation represented by the non-singular matrix $\mathbf{T}$, it is also invariant under the transformation represented by $\mathbf{T}^{-1}$.

Starting with $\mathbf{T}\begin{pmatrix} x \\ y \end{pmatrix} = \begin{pmatrix} x \\ y \end{pmatrix}$ prove this result algebraically.

⑮ The simultaneous equations

$$2x - y = 1$$
$$3x + ky = b$$

are represented by the matrix $\mathbf{M}\begin{pmatrix} x \\ y \end{pmatrix} = \begin{pmatrix} 1 \\ b \end{pmatrix}$.

(i) Write down the matrix $\mathbf{M}$.

(ii) State the value of k for which $\mathbf{M}^{-1}$ does not exist and find $\mathbf{M}^{-1}$ in terms of k when $\mathbf{M}^{-1}$ exists.

Use $\mathbf{M}^{-1}$ to solve the simultaneous equations when $k = 5, b = 21$.

(iii) What can you say about the solutions of the equations when $k = -\dfrac{3}{2}$?

(iv) The two equations can be interpreted as representing two lines in the $x–y$ plane.

Describe the relationship between the two lines when:

(a) $k = 5, b = 21$

(b) $k = -\dfrac{3}{2}, b = 1$

(c) $k = -\dfrac{3}{2}, b = \dfrac{3}{2}$

⑯ Matrices $\mathbf{M}$ and $\mathbf{N}$ are given by $\mathbf{M} = \begin{pmatrix} 3 & 2 \\ 0 & 1 \end{pmatrix}$ and $\mathbf{N} = \begin{pmatrix} 1 & -3 \\ 1 & 4 \end{pmatrix}$.

(i) Find $\mathbf{M}^{-1}$ and $\mathbf{N}^{-1}$.

(ii) Find $\mathbf{MN}$ and $(\mathbf{MN})^{-1}$. Verify that $(\mathbf{MN})^{-1} = \mathbf{N}^{-1}\mathbf{M}^{-1}$.

(iii) The result $(\mathbf{PQ})^{-1} = \mathbf{Q}^{-1}\mathbf{P}^{-1}$ is true for any two 2 × 2 non-singular matrices $\mathbf{P}$ and $\mathbf{Q}$.

The first two lines of a proof of this general result are given below.

Beginning with these two lines, complete the general proof.

$$(\mathbf{PQ})^{-1}\,\mathbf{PQ} = \mathbf{I}$$

$$(\mathbf{PQ})^{-1}\,\mathbf{PQQ}^{-1} = \mathbf{IQ}^{-1}$$

1 Determinant and inverse of a 3 × 3 matrix

The determinant of a 3 × 3 matrix is sometimes denoted $|\mathbf{a}\ \mathbf{b}\ \mathbf{c}|$.

In this section you will find the determinant and inverse of 3 × 3 matrices using the calculator facility and also using a non-calculator method.

Finding the inverse of a 3 × 3 matrix using a calculator

ACTIVITY 2.1

Using a calculator, find the determinant and inverse of the matrix

$$\mathbf{A} = \begin{pmatrix} 3 & -2 & 1 \\ 0 & 1 & 2 \\ 4 & 0 & 1 \end{pmatrix}.$$

Still using a calculator, find out which of the following matrices are non-singular and find the inverse in each of these cases.

$$\mathbf{B} = \begin{pmatrix} 5 & 5 & 5 \\ 2 & 2 & 2 \\ 2 & 4 & -3 \end{pmatrix} \qquad \mathbf{C} = \begin{pmatrix} 1 & 3 & 2 \\ -1 & 0 & 1 \\ 2 & 1 & 4 \end{pmatrix} \qquad \mathbf{D} = \begin{pmatrix} 0 & 3 & -2 \\ 1 & -1 & 2 \\ 3 & 0 & 3 \end{pmatrix}$$

Finding the determinant of a 3 × 3 matrix without using a calculator

It is also possible to find the determinant and inverse of a 3 × 3 matrix without using a calculator. This is useful in cases where some of the elements of the matrix are algebraic rather than numerical.

If $\mathbf{M}$ is the 3 × 3 matrix $\begin{pmatrix} a_1 & b_1 & c_1 \\ a_2 & b_2 & c_2 \\ a_3 & b_3 & c_3 \end{pmatrix}$ then the determinant of $\mathbf{M}$ is defined by

$$\det \mathbf{M} = a_1 \begin{vmatrix} b_2 & c_2 \\ b_3 & c_3 \end{vmatrix} - a_2 \begin{vmatrix} b_1 & c_1 \\ b_3 & c_3 \end{vmatrix} + a_3 \begin{vmatrix} b_1 & c_1 \\ b_2 & c_2 \end{vmatrix},$$

which is sometimes referred to as the **expansion of the determinant by the first column**.

Notice that you do not really need to calculate $\begin{vmatrix} -2 & 1 \\ 0 & 1 \end{vmatrix}$ as it is going to be multiplied by zero. Keeping an eye open for helpful zeros can reduce the number of calculations needed.

For example, to find the determinant of the matrix $\mathbf{A} = \begin{pmatrix} 3 & -2 & 1 \\ 0 & 1 & 2 \\ 4 & 0 & 1 \end{pmatrix}$ from Activity 2.1:

$$\det \mathbf{A} = 3 \begin{vmatrix} 1 & 2 \\ 0 & 1 \end{vmatrix} - 0 \begin{vmatrix} -2 & 1 \\ 0 & 1 \end{vmatrix} + 4 \begin{vmatrix} -2 & 1 \\ 1 & 2 \end{vmatrix}$$
$$= 3(1 - 0) - 0(-2 - 0) + 4(-4 - 1)$$
$$= 3 - 20$$
$$= -17$$

This is the same answer as you will have obtained earlier using your calculator.

The 2×2 determinant $\begin{vmatrix} b_2 & c_2 \\ b_3 & c_3 \end{vmatrix}$ is called the **minor** of the

element a_1. It is obtained by deleting the row and column containing a_1:

$$\begin{vmatrix} a_1 & b_1 & c_1 \\ a_2 & b_2 & c_2 \\ a_3 & b_3 & c_3 \end{vmatrix}$$

Other minors are defined in the same way, for example the minor of a_2 is

$$\begin{vmatrix} a_1 & b_1 & c_1 \\ a_2 & b_2 & c_2 \\ a_3 & b_3 & c_3 \end{vmatrix} = \begin{vmatrix} b_1 & c_1 \\ b_3 & c_3 \end{vmatrix}$$

> ### Note
>
> As an alternative to using the first column, you could use the **expansion of the determinant by the second column**:
>
> $$\det \mathbf{M} = -b_1 \begin{vmatrix} a_2 & c_2 \\ a_3 & c_3 \end{vmatrix} + b_2 \begin{vmatrix} a_1 & c_1 \\ a_3 & c_3 \end{vmatrix} - b_3 \begin{vmatrix} a_1 & c_1 \\ a_2 & c_2 \end{vmatrix},$$
>
> or the **expansion of the determinant by the third column**:
>
> $$\det \mathbf{M} = c_1 \begin{vmatrix} a_2 & b_2 \\ a_3 & b_3 \end{vmatrix} - c_2 \begin{vmatrix} a_1 & b_1 \\ a_3 & b_3 \end{vmatrix} + c_3 \begin{vmatrix} a_1 & b_1 \\ a_2 & b_2 \end{vmatrix}.$$
>
> It is fairly easy to show that all three expressions above for $\det \mathbf{M}$ simplify to:
> $$a_1 b_2 c_3 + a_2 b_3 c_1 + a_3 b_1 c_2 - a_3 b_2 c_1 - a_1 b_3 c_2 - a_2 b_1 c_3$$

You may have noticed that in the expansions of the determinant, the signs on the minors alternate as shown:

$$\begin{vmatrix} + & - & + \\ - & + & - \\ + & - & + \end{vmatrix}$$

A minor, together with its correct sign, is known as a **cofactor** and is denoted by the corresponding capital letter; for example, the cofactor of a_3 is A_3. This means that the expansion by the first column, say, can be written as

$$a_1 A_1 + a_2 A_2 + a_3 A_3.$$

Example 2.4

Find the determinant of the matrix $\mathbf{M} = \begin{pmatrix} 3 & 0 & -4 \\ 7 & 2 & -1 \\ -2 & 1 & 3 \end{pmatrix}$.

Solution

> To find the determinant you can also expand by rows. So, for example, expanding by the top row would give:
>
> $$3\begin{vmatrix} 2 & -1 \\ 1 & 3 \end{vmatrix} - 0\begin{vmatrix} 7 & -1 \\ -2 & 3 \end{vmatrix} + (-4)\begin{vmatrix} 7 & 2 \\ -2 & 1 \end{vmatrix}$$
>
> which also gives the answer −23.

Expanding by the first column using the expression:

$$\det \mathbf{M} = a_1 \begin{vmatrix} b_2 & c_2 \\ b_3 & c_3 \end{vmatrix} - a_2 \begin{vmatrix} b_1 & c_1 \\ b_3 & c_3 \end{vmatrix} + a_3 \begin{vmatrix} b_1 & c_1 \\ b_2 & c_2 \end{vmatrix}$$

gives:

$$\det \mathbf{M} = 3\begin{vmatrix} 2 & -1 \\ 1 & 3 \end{vmatrix} - 7\begin{vmatrix} 0 & -4 \\ 1 & 3 \end{vmatrix} + (-2)\begin{vmatrix} 0 & -4 \\ 2 & -1 \end{vmatrix}$$

$$= 3(6 - (-1)) - 7(0 - (-4)) - 2(0 - (-8))$$

$$= 21 - 28 - 16$$

$$= -23$$

> Notice that expanding by the top row would be quicker here as it has a zero element.

Earlier you saw that the determinant of a 2 × 2 matrix represents the area scale factor of the transformation represented by the matrix. In the case of a 3 × 3 matrix the determinant represents the volume scale factor. For example, the

matrix $\begin{pmatrix} 2 & 0 & 0 \\ 0 & 2 & 0 \\ 0 & 0 & 2 \end{pmatrix}$ has determinant 8; this matrix represents an enlargement of

scale factor 2, centre the origin, so the volume scale factor of the transformation is $2 \times 2 \times 2 = 8$.

As was the case with 2 × 2 matrices, the sign of the determinant of a 3 × 3 matrix also has significance. A negative determinant means that the transformation reverses the orientation of the original object.

Finding the inverse of a 3 × 3 matrix without using a calculator

> Recall that a minor, together with its correct sign, is known as a cofactor and is denoted by the corresponding capital letter; for example the cofactor of a_3 is A_3.

The matrix $\begin{pmatrix} A_1 & A_2 & A_3 \\ B_1 & B_2 & B_3 \\ C_1 & C_2 & C_3 \end{pmatrix}$ is known as the **adjugate** or **adjoint** of **M**,

denoted adj **M**.

The adjugate of **M** is formed by

- replacing each element of **M** by its cofactor;
- then transposing the matrix (i.e. changing rows into columns and columns into rows).

The unique inverse of a 3 × 3 matrix can be calculated as follows:

$$\mathbf{M}^{-1} = \frac{1}{\det \mathbf{M}} \operatorname{adj} \mathbf{M} = \frac{1}{\det \mathbf{M}} \begin{pmatrix} A_1 & A_2 & A_3 \\ B_1 & B_2 & B_3 \\ C_1 & C_2 & C_3 \end{pmatrix}, \ \det \mathbf{M} \neq 0$$

The steps involved in the method are shown in the following example.

| **Example 2.5** | Find the inverse of the matrix $\mathbf{M}$ without using a calculator, where |

$$\mathbf{M} = \begin{pmatrix} 2 & 3 & 4 \\ 2 & -5 & 2 \\ -3 & 6 & -3 \end{pmatrix}.$$

Solution

Step 1: Find the determinant Δ and check $\Delta \neq 0$

Expanding by the first column

$$\Delta = 2\begin{vmatrix} -5 & 2 \\ 6 & -3 \end{vmatrix} - 2\begin{vmatrix} 3 & 4 \\ 6 & -3 \end{vmatrix} + (-3)\begin{vmatrix} 3 & 4 \\ -5 & 2 \end{vmatrix}$$

$$= (2 \times 3) - (2 \times -33) - (3 \times 26) = -6$$

Therefore the inverse matrix exists.

Step 2: Evaluate the cofactors

> You can evaluate the determinant Δ using these cofactors to check your earlier arithmetic is correct:
>
> 2nd column:
> $\Delta = 3B_1 - 5B_2 + 6B_3$
> $= (3 \times 0) - (5 \times 6)$
> $\quad + (6 \times 4) = -6$
>
> 3rd column:
> $\Delta = 4C_1 + 2C_2 - 3C_3$
> $= (4 \times -3) + (2 \times -21)$
> $\quad - (3 \times -16) = -6$

$$A_1 = \begin{vmatrix} -5 & 2 \\ 6 & -3 \end{vmatrix} = 3 \qquad B_1 = -\begin{vmatrix} 2 & 2 \\ -3 & -3 \end{vmatrix} = 0 \qquad C_1 = \begin{vmatrix} 2 & -5 \\ -3 & 6 \end{vmatrix} = -3$$

$$A_2 = -\begin{vmatrix} 3 & 4 \\ 6 & -3 \end{vmatrix} = 33 \qquad B_2 = \begin{vmatrix} 2 & 4 \\ -3 & -3 \end{vmatrix} = 6 \qquad C_2 = -\begin{vmatrix} 2 & 3 \\ -3 & 6 \end{vmatrix} = -21$$

$$A_3 = \begin{vmatrix} 3 & 4 \\ -5 & 2 \end{vmatrix} = 26 \qquad B_3 = -\begin{vmatrix} 2 & 4 \\ 2 & 2 \end{vmatrix} = 4 \qquad C_3 = \begin{vmatrix} 2 & 3 \\ 2 & -5 \end{vmatrix} = -16$$

Step 3: Form the matrix of cofactors and transpose it, then multiply by $\frac{1}{\Delta}$

$$\mathbf{M}^{-1} = \frac{1}{-6}\begin{pmatrix} 3 & 0 & -3 \\ 33 & 6 & -21 \\ 26 & 4 & -16 \end{pmatrix}^{\mathrm{T}}$$

> The capital T indicates the matrix is to be transposed.

> Matrix of cofactors.

> Multiply by $\frac{1}{\Delta}$.

$$= \frac{1}{-6}\begin{pmatrix} 3 & 33 & 26 \\ 0 & 6 & 4 \\ -3 & -21 & -16 \end{pmatrix}$$

$$= \frac{1}{6}\begin{pmatrix} -3 & -33 & -26 \\ 0 & -6 & -4 \\ 3 & 21 & 16 \end{pmatrix}$$

The final matrix could then be simplified and written as

$$\mathbf{M}^{-1} = \begin{pmatrix} -\dfrac{1}{2} & -\dfrac{11}{2} & -\dfrac{13}{3} \\ 0 & -1 & -\dfrac{2}{3} \\ \dfrac{1}{2} & \dfrac{7}{2} & \dfrac{8}{3} \end{pmatrix}$$

$$\text{Check: } \mathbf{MM}^{-1} = \begin{pmatrix} 2 & 3 & 4 \\ 2 & -5 & 2 \\ -3 & 6 & -3 \end{pmatrix}\frac{1}{6}\begin{pmatrix} -3 & -33 & -26 \\ 0 & -6 & -4 \\ 3 & 21 & 16 \end{pmatrix}$$

$$= \frac{1}{6}\begin{pmatrix} 6 & 0 & 0 \\ 0 & 6 & 0 \\ 0 & 0 & 6 \end{pmatrix} = \begin{pmatrix} 1 & 0 & 0 \\ 0 & 1 & 0 \\ 0 & 0 & 1 \end{pmatrix}$$

This adjugate method for finding the inverse of a 3×3 matrix is reasonably straightforward but it is important to check your arithmetic as you go along, as it is very easy to make mistakes. You can use your calculator to check that you have calculated the inverse correctly.

As shown in the example above, you might also multiply the inverse by the original matrix and check that you obtain the 3×3 identity matrix.

Exercise 2.1

① Evaluate these determinants without using a calculator. Check your answers using your calculator.

(i) (a) $\begin{vmatrix} 1 & 1 & 3 \\ -1 & 0 & 2 \\ 3 & 1 & 4 \end{vmatrix}$

(b) $\begin{vmatrix} 1 & -1 & 3 \\ 1 & 0 & 1 \\ 3 & 2 & 4 \end{vmatrix}$

(ii) (a) $\begin{vmatrix} 1 & -5 & -4 \\ 2 & 3 & 3 \\ -2 & 1 & 0 \end{vmatrix}$

(b) $\begin{vmatrix} 1 & 2 & -2 \\ -5 & 3 & 1 \\ -4 & 3 & 0 \end{vmatrix}$

(iii) (a) $\begin{vmatrix} 2 & 1 & 2 \\ 3 & 5 & 3 \\ 1 & -1 & 1 \end{vmatrix}$

(b) $\begin{vmatrix} 1 & 5 & 0 \\ 1 & 5 & 0 \\ 2 & 1 & -2 \end{vmatrix}$

What do you notice about the determinants?

② Find the inverses of the following matrices, if they exist, without using a calculator.

(i) $\begin{pmatrix} 1 & 2 & 4 \\ 2 & 4 & 5 \\ 0 & 1 & 2 \end{pmatrix}$

(ii) $\begin{pmatrix} 3 & 2 & 6 \\ 5 & 3 & 11 \\ 7 & 4 & 16 \end{pmatrix}$

(iii) $\begin{pmatrix} 5 & 5 & -5 \\ -9 & 3 & -5 \\ -4 & -6 & 8 \end{pmatrix}$

(iv) $\begin{pmatrix} 6 & 5 & 6 \\ -5 & 2 & -4 \\ -4 & -6 & -5 \end{pmatrix}$

③ Find the inverse of the matrix $\begin{pmatrix} 4 & -5 & 3 \\ 3 & 3 & -4 \\ 5 & 4 & -6 \end{pmatrix}$ and hence solve the simultaneous equations:

$4x - 5y + 3z = 3$

$3x + 3y - 4z = 48$

$5x + 4y - 6z = 74$

④ Find the inverse of the matrix $\mathbf{M} = \begin{pmatrix} 1 & 3 & -2 \\ k & 0 & 4 \\ 2 & -1 & 4 \end{pmatrix}$ where $k \neq 0$.

For what value of k is the matrix $\mathbf{M}$ singular?

⑤ (i) Investigate the relationship between the matrices

$$\mathbf{A} = \begin{pmatrix} 0 & 3 & 1 \\ 2 & 4 & 2 \\ -1 & 3 & 5 \end{pmatrix} \quad \mathbf{B} = \begin{pmatrix} 1 & 0 & 3 \\ 2 & 2 & 4 \\ 5 & -1 & 3 \end{pmatrix} \quad \mathbf{C} = \begin{pmatrix} 3 & 1 & 0 \\ 4 & 2 & 2 \\ 3 & 5 & -1 \end{pmatrix}$$

(ii) Find $\det \mathbf{A}$, $\det \mathbf{B}$ and $\det \mathbf{C}$ and comment on your answer.

⑥ Show that $x = 1$ is one root of the equation $\begin{vmatrix} 2 & 2 & x \\ 1 & x & 1 \\ x & 1 & 4 \end{vmatrix} = 0$ and find the other roots.

⑦ Find the values of x for which the matrix $\begin{pmatrix} 3 & -1 & 1 \\ 2 & x & 4 \\ x & 1 & 3 \end{pmatrix}$ is singular.

⑧ Given that the matrix $\mathbf{M} = \begin{pmatrix} k & 2 & 1 \\ 0 & -k & 2 \\ 2k & 1 & 3 \end{pmatrix}$ has determinant greater than 5, find the range of possible values for k.

⑨ (i) $\mathbf{P}$ and $\mathbf{Q}$ are non-singular matrices. Prove that $(\mathbf{PQ})^{-1} = \mathbf{Q}^{-1}\mathbf{P}^{-1}$.

(ii) Find the inverses of the matrices $\mathbf{P} = \begin{pmatrix} 0 & 3 & -1 \\ -2 & 2 & 2 \\ -3 & 0 & 1 \end{pmatrix}$ and $\mathbf{Q} = \begin{pmatrix} 2 & 1 & 2 \\ 1 & 0 & 1 \\ 4 & -3 & 2 \end{pmatrix}$.

Using the result from part (i), find $(\mathbf{PQ})^{-1}$.

⑩ (i) Prove that $\begin{vmatrix} ka_1 & b_1 & c_1 \\ ka_2 & b_2 & c_2 \\ ka_3 & b_3 & c_3 \end{vmatrix} = k \begin{vmatrix} a_1 & b_1 & c_1 \\ a_2 & b_2 & c_2 \\ a_3 & b_3 & c_3 \end{vmatrix}$, where k is a constant.

(ii) Explain in terms of volumes why multiplying all the elements in the first column by a constant k multiplies the value of the determinant by k.

(iii) What would happen if you multiplied a different column by k?

⑪ Given that $\begin{vmatrix} 1 & 2 & 3 \\ 6 & 4 & 5 \\ 7 & 5 & 1 \end{vmatrix} = 43$, write down the values of the determinants:

(i) $\begin{vmatrix} 10 & 2 & 3 \\ 60 & 4 & 5 \\ 70 & 5 & 1 \end{vmatrix}$

(ii) $\begin{vmatrix} 4 & 10 & -21 \\ 24 & 20 & -35 \\ 28 & 25 & -7 \end{vmatrix}$

(iii) $\begin{vmatrix} x & 4 & 3y \\ 6x & 8 & 5y \\ 7x & 10 & y \end{vmatrix}$

(iv) $\begin{vmatrix} x^4 & \dfrac{1}{x} & 12y \\ 6x^4 & \dfrac{2}{x} & 20y \\ 7x^4 & \dfrac{5}{2x} & 4y \end{vmatrix}$

2 Solving simultaneous equations in three unknowns and its geometrical interpretation

In most cases, two planes intersect in a straight line. The exception occurs when they are parallel. It is also possible that they are one and the same plane. However in what follows, it is assumed that all the planes being considered are distinct.

In this section you will look at the different possibilities for how *three* planes can be arranged in three-dimensional space.

You saw in Chapter 1 that the Cartesian equation of a plane is of the form $ax + by + cz + d = 0$. Solving a system of three equations in three unknowns x, y and z is equivalent to finding where the three planes represented by those equations intersect.

There are five ways in which three distinct planes π_1, π_2 and π_2 can intersect in 3D space.

If two of the planes are parallel, there are two possibilities for the third:

- it can be parallel to the other two (see Figure 2.8); or
- it can cut the other two (see Figure 2.9).

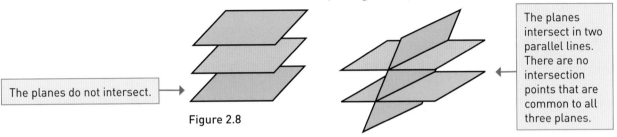

The planes do not intersect.

Figure 2.8

The planes intersect in two parallel lines. There are no intersection points that are common to all three planes.

Figure 2.9

If none of the planes are parallel, there are three possibilities:

- The planes intersect in a single point (see Figure 2.10)

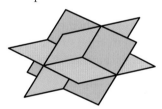

Figure 2.10

- The planes form a **sheaf** (see Figure 2.11)

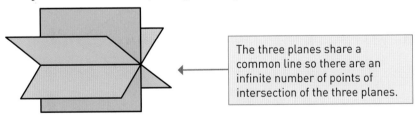

The three planes share a common line so there are an infinite number of points of intersection of the three planes.

Figure 2.11

- The planes form a **triangular prism** (see Figure 2.12)

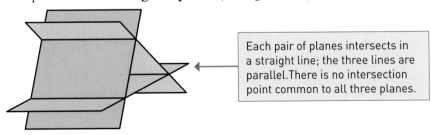

Each pair of planes intersects in a straight line; the three lines are parallel. There is no intersection point common to all three planes.

Figure 2.12

The diagrams above show that three planes either intersect in a unique point, intersect in an infinite number of points or do not have a common intersection point.

Finding the unique point of intersection of three planes

3×3 matrices can be used to find the point of intersection of three planes that intersect in a unique point or to determine that the planes have a different arrangement.

Suppose the three planes have equations

$$a_1x + b_1y + c_1z = d_1$$
$$a_2x + b_2y + c_2z = d_2$$
$$a_3x + b_3y + c_3z = d_3$$

They can be written as $\mathbf{M} \begin{pmatrix} x \\ y \\ z \end{pmatrix} = \begin{pmatrix} d_1 \\ d_2 \\ d_3 \end{pmatrix}$, where $\mathbf{M} = \begin{pmatrix} a_1 & b_1 & c_1 \\ a_2 & b_2 & c_2 \\ a_3 & b_3 & c_3 \end{pmatrix}$.

Figure 2.13 summarises the decisions that need to be made and Example 2.6 shows how these decision are carried out. Note that the trivial situation where the matrix $\mathbf{M}$ consists entirely of zeroes has been ignored, and also cases where two or three of the equations are the same (i.e. two or three of the planes are coincident).

If the determinant of $\mathbf{M}$ is zero then there is no unique solution to the equations i.e. no unique point of intersection of the planes represented by the matrix. In this case one of the other arrangements of the planes would be relevant.

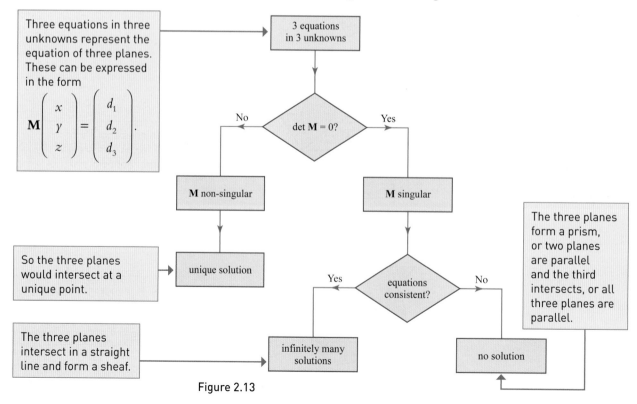

Figure 2.13

To summarise:

If **M** is non-singular, the planes intersect in a unique point.

If **M** is singular, the planes must be arranged in one of the other four possible arrangements:

- three parallel planes
- two parallel planes that are cut by the third to form two parallel lines
- a sheaf of planes that intersect in a common line
- a prism of planes in which each pair of planes meet in a straight line but there are no common points of intersection between the three planes.

Example 2.6

Three planes have equations

$\pi_1: x + 3y - 2z = 7$ ①

$\pi_2: 2x - 2y + z = 3$ ②

$\pi_3: 3x + y - z = k$ ③

(i) Explain how you know that none of the planes are parallel to any of the other planes.

(ii) Investigate the intersection of the planes when:

   (a) $k = 10$ (b) $k = 12$

Solution

> You cannot multiply any of the equations by a number to obtain one of the other equations.

(i) If planes are parallel they can be written as a scalar multiple of each other. That is not true in this case, so none of the planes are parallel to any of the others.

> Writing the equation of the planes in the matrix format
>
> $$\mathbf{M}\begin{pmatrix} x \\ y \\ z \end{pmatrix} = \begin{pmatrix} d_1 \\ d_2 \\ d_3 \end{pmatrix}.$$

(ii) $$\begin{pmatrix} 1 & 3 & -2 \\ 2 & -2 & 1 \\ 3 & 1 & -1 \end{pmatrix}\begin{pmatrix} x \\ y \\ z \end{pmatrix} = \begin{pmatrix} 7 \\ 3 \\ k \end{pmatrix}$$

Using a calculator, $\det \mathbf{M} = 0$ so the matrix is singular and hence the planes do not intersect at a unique point.

> To investigate the intersection of the planes in a case where the matrix is similar, try to solve the equations algebraically.

Eliminating the variable z produces two equations in x and y:

① $- 2 \times$ ③: $-5x + y = 7 - 2k$ ←

② $+$ ③: $5x - y = 3 + k$ ←

> For the equations to be consistent, the value of $5x - y$ must be the same in each equation.

These equations are consistent if $2k - 7 = 3 + k$

$$\Leftrightarrow k = 10$$

and in this case both equations reduce to $5x - y = 13$.

(a) The equations are consistent in the case $k = 10$ and so there are infinitely many solutions. Since none of the planes are coincident, they intersect in a straight line and form a sheaf.

(b) When $k = 12$ the two equations in x and y are inconsistent:

$$-5x + y = -17$$

> $5x - y$ cannot be equal to both 17 and 15.

$$5x - y = 15$$

Therefore there are no solutions. Since the planes are not parallel they must form a prism of planes.

Exercise 2.2

① (i) Without using a calculator, find the determinant and inverse of the

matrix $\mathbf{M} = \begin{pmatrix} 2 & -3 & 1 \\ 3 & 0 & 4 \\ 1 & -1 & 3 \end{pmatrix}$.

(ii) Using your answer to part (i), show that the planes

$$2x - 3y + z = 10$$
$$3x + 4z = 25$$
$$x - y + 3z = 20$$

intersect in the unique point $(0, -1.25, 6.25)$.

② Without carrying out any calculations, describe the arrangements of the following sets of planes:

(i) $\pi_1: 2x + y - z = 4$

$\pi_2: x + y - z = 2$

$\pi_3: 2x + y - z = 6$

(ii) $\pi_1: 2x + y - z = 4$

$\pi_2: 6x + 3y - 3z = 12$

$\pi_3: 2x + y - z = 6$

(iii) $\pi_1: 2x + y - z = 4$

$\pi_2: 10x + 5y - 5z = 15$

$\pi_3: 2x + y - z = 6$

③ (i) Express the equations of the three planes

$$\pi_1: 5x + 3y - 2z = 6$$
$$\pi_2: 6x + 2y + 3z = 11$$
$$\pi_3: 7x + y + 8z = 12$$

in the form $\mathbf{M} \begin{pmatrix} x \\ y \\ z \end{pmatrix} = \begin{pmatrix} d_1 \\ d_2 \\ d_3 \end{pmatrix}$ and show that $\det \mathbf{M} = 0$.

(ii) Eliminate the variable y from the three equations and hence show that the three planes form a prism of planes.

④ Determine the arrangement of the following planes in three dimensions. You should find the determinant and inverse of matrices without using a calculator.

(i) $\pi_1: x + 2y + 4z = 7$

$\pi_2: 3x + 2y + 5z = 21$

$\pi_3: 4x + y + 2z = 14$

(ii) $\pi_1: x + y + z = 4$

$\pi_2: 2x + 3y - 4z = 3$

$\pi_3: 5x + 8y - 13z = 8$

(iii) $\pi_1: 2x - y = 1$

$\pi_2: 3x + 2z = 13$

$\pi_3: 3y + 4z = 23$

(iv) $\pi_1: 3x + 2y + z = 2$

$\pi_2: 5x + 3y - 4z = 1$

$\pi_3: x + y + 4z = 5$

(v) $\pi_1: 2x + y - z = 5$

$\pi_2: 8x + 4y - 4z = 20$

$\pi_3: -2x - y + z = -5$

⑤ Solve, where possible, the equation

$$\begin{pmatrix} 1 & 3 & -2 \\ -3 & 1 & m \\ -3 & 11 & -4 \end{pmatrix} \begin{pmatrix} x \\ y \\ z \end{pmatrix} = \begin{pmatrix} -2 \\ 6 \\ k \end{pmatrix}$$

in each of these cases:

(i) $m = 2$, $k = 3$ (ii) $m = 1$, $k = 3$ (iii) $m = 1$, $k = 6$

In each case interpret the solution geometrically with reference to three planes in three dimensions.

⑥ (i)   Obtain an expression for the inverse of the matrix $\begin{pmatrix} k & -7 & 4 \\ 2 & -2 & 3 \\ 1 & -3 & -2 \end{pmatrix}$ in terms of k.

State the value of k for which the planes

$$kx - 7y + 4z = 3$$
$$2x - 2y + 3z = 7$$
$$x - 3y - 2z = -1$$

would not intersect in a unique point.

(ii)   Describe the intersection of the planes
$$4x - 7y + 4z = p$$
$$2x - 2y + 3z = 1$$
$$x - 3y - 2z = 2$$

giving your answer in terms of p.

(iii)   Find the value of p for which the planes
$$5x - 7y + 4z = p$$
$$2x - 2y + 3z = 1$$
$$x - 3y - 2z = 2$$

have at least one point of intersection and describe the arrangement of the planes geometrically in three dimensions.

3 Factorisation of determinants using row and column operations

You should note that this section is concerned solely with 3×3 matrices.

You sometimes need to find the determinant of a matrix whose entries may be unknowns. This could be done straightforwardly using the methods encountered earlier in Chapter 2, in effect using the formula

> Here the determinant has been found by 'expanding by the first column'. This was covered on page 53.

$$\begin{vmatrix} a_1 & b_1 & c_1 \\ a_2 & b_2 & c_2 \\ a_3 & b_3 & c_3 \end{vmatrix} = a_1\left(b_2 c_3 - b_3 c_2\right) - b_1\left(a_2 c_3 - a_3 c_2\right) + c_1\left(a_2 b_3 - a_3 b_2\right)$$

The problem with this is that you will end up with six terms in an unfactorised expression which might well be cubic or higher in one or more of the unknowns. It is usually easier to work with a factorised form, typically so that you can easily find the values of the unknown for which the matrix is singular.

In this section you will learn some techniques which will help you to find the determinant of a matrix in factorised form.

Find an expression for the determinant of matrix $\mathbf{M}$ where $\mathbf{M} = \begin{vmatrix} a & b & c \\ 0 & d & e \\ 0 & 0 & f \end{vmatrix}$.

Solution

Expanding by the first column,

> Note that the answer is simply the product of the entries in the leading diagonal of $\mathbf{M}$.

$$|\mathbf{M}| = a \begin{vmatrix} d & e \\ 0 & f \end{vmatrix} - 0 \begin{vmatrix} b & c \\ 0 & f \end{vmatrix} + 0 \begin{vmatrix} b & c \\ d & e \end{vmatrix} = a(df - 0 \times e) = adf$$

In Example 2.7, $\mathbf{M}$ is an example of an **upper triangular matrix**, a matrix all of whose non-zero entries are on or above the leading diagonal. Similarly, a

matrix such as $\begin{pmatrix} 7 & 0 & 0 \\ -2 & -1 & 0 \\ 8 & 7 & 2.3 \end{pmatrix}$ is called a **lower triangular matrix**.

Triangular matrices are very important in mathematics for many reasons. In this chapter it is important simply to note that the determinant of any triangular matrix can be easily found by multiplying together the entries in the leading diagonal. Notice that this gives the determinant in ready-factorised form. This provides you with one possible strategy although, as you will see, there are others.

> These rules are often referred to as 'row and column operations'.

Formal methods have been developed for triangularising a matrix. However, to solve the problems that you will encounter in this section, all that you will require is a few simple rules on manipulating matrices and the consequent effect that this has on the determinant.

The rules which follow are illustrated by looking at a particular numerical

matrix, $\mathbf{A}$, where $\mathbf{A} = \begin{pmatrix} 1 & -4 & 3 \\ 2 & 2 & 6 \\ -2 & 8 & -3 \end{pmatrix}$.

The determinant of $\mathbf{A}$ is given by

$$|\mathbf{A}| = \begin{vmatrix} 1 & -4 & 3 \\ 2 & 2 & 6 \\ -2 & 8 & -3 \end{vmatrix}$$

$$= 1 \begin{vmatrix} 2 & 6 \\ 8 & -3 \end{vmatrix} - 4 \begin{vmatrix} 2 & 6 \\ -2 & -3 \end{vmatrix} + 3 \begin{vmatrix} 2 & 2 \\ -2 & 8 \end{vmatrix}$$

$$= -6 - 48 + 4(-6 + 12) + 3(16 + 4) = -54 + 24 + 60 = 30$$

Rule 1

If you swap any two rows (or any two columns) of a matrix then you change the sign of the determinant.

For example, swapping columns 1 and 2 of **A** and recalculating:

$$\begin{vmatrix} -4 & 1 & 3 \\ 2 & 2 & 6 \\ 8 & -2 & -3 \end{vmatrix} = -4\begin{vmatrix} 2 & 6 \\ -2 & -3 \end{vmatrix} - 1\begin{vmatrix} 2 & 6 \\ 8 & -3 \end{vmatrix} + 3\begin{vmatrix} 2 & 2 \\ 8 & -2 \end{vmatrix}$$

$$= -4(-6 + 12) - (-6 - 48) + 3(-16 - 4)$$
$$= -24 + 54 - 60$$
$$= -30$$

Rule 2

If you multiply every entry in a row (or a column) by the same number then you multiply the determinant by that number.

For example, multiplying the third row of **A** by 5 and recalculating:

$$\begin{vmatrix} 1 & -4 & 3 \\ 2 & 2 & 6 \\ -10 & 40 & -15 \end{vmatrix} = 1\begin{vmatrix} 2 & 6 \\ 40 & -15 \end{vmatrix} - (-4)\begin{vmatrix} 2 & 6 \\ -10 & -15 \end{vmatrix} + 3\begin{vmatrix} 2 & 2 \\ -10 & 40 \end{vmatrix}$$

$$= -30 - 240 + 4(-30 + 60) + 3(80 + 20)$$
$$= -270 + 120 + 300$$
$$= 150$$
$$= 5 \times 30$$

Again, if you divide every entry of a row (or column) by the same non-zero number then you divide the determinant by that number. This rule is very useful as it can be used to pull out common factors in a row or column.

Rule 3

This of course is simply applying Rule 2 three times, once to each row.

If you multiply every entry of a matrix by the same number then you multiply the determinant by the cube of the number.

For example, multiplying every entry of **A** by 3 and recalculating:

$$\begin{vmatrix} 3 & -12 & 9 \\ 6 & 6 & 18 \\ -6 & 24 & -9 \end{vmatrix} = 3\begin{vmatrix} 6 & 18 \\ 24 & -9 \end{vmatrix} - (-12)\begin{vmatrix} 6 & 18 \\ -6 & -9 \end{vmatrix} + 9\begin{vmatrix} 6 & 6 \\ -6 & 24 \end{vmatrix}$$

$$= 3(-54 - 432) + 12(-54 + 108) + 9(144 + 36)$$
$$= -1458 + 648 + 1620$$
$$= 810$$
$$= 27 \times 30$$
$$= 3^3 \times 30$$

Of course, if you divide every entry by the same number then you would divide the determinant by the cube of the number.

Rule 4

If you add the entries of one row to the corresponding entries of another row then the determinant is unaffected (and similarly for columns).

For example, adding the contents of the first column of **A** to the third column of **A** and recalculating:

$$\begin{vmatrix} 1 & -4 & 4 \\ 2 & 2 & 8 \\ -2 & 8 & -5 \end{vmatrix} = 1\begin{vmatrix} 2 & 8 \\ 8 & -5 \end{vmatrix} - (-4)\begin{vmatrix} 2 & 8 \\ -2 & -5 \end{vmatrix} + 4\begin{vmatrix} 2 & 2 \\ -2 & 8 \end{vmatrix}$$

$$= -10 - 64 + 4(-10 + 16) + 4(16 + 4)$$

$$= -74 + 24 + 80$$

$$= 30$$

Rule 5

If any row is the same as, or a multiple of, any other row then the determinant is zero (and similarly for columns).

So, for example, all of the following matrices are singular.

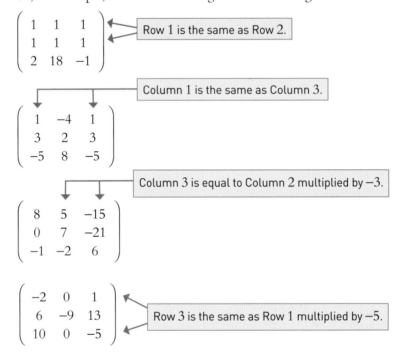

$$\begin{pmatrix} 1 & 1 & 1 \\ 1 & 1 & 1 \\ 2 & 18 & -1 \end{pmatrix}$$ Row 1 is the same as Row 2.

Column 1 is the same as Column 3.

$$\begin{pmatrix} 1 & -4 & 1 \\ 3 & 2 & 3 \\ -5 & 8 & -5 \end{pmatrix}$$

Column 3 is equal to Column 2 multiplied by −3.

$$\begin{pmatrix} 8 & 5 & -15 \\ 0 & 7 & -21 \\ -1 & -2 & 6 \end{pmatrix}$$

$$\begin{pmatrix} -2 & 0 & 1 \\ 6 & -9 & 13 \\ 10 & 0 & -5 \end{pmatrix}$$ Row 3 is the same as Row 1 multiplied by −5.

Rule 6

If any row is any linear combination of the other two rows then the determinant is zero (and similarly for columns).

So, for example, the following matrix is singular:

$$\begin{pmatrix} 1 & 3 & -2 \\ 6 & -4 & 2 \\ 7 & -1 & 0 \end{pmatrix}$$ Row 1 added to Row 2 = Row 3.

To check: $$\begin{vmatrix} 1 & 3 & -2 \\ 6 & -4 & 2 \\ 7 & -1 & 0 \end{vmatrix} = 1\begin{vmatrix} -4 & 2 \\ -1 & 0 \end{vmatrix} - 3\begin{vmatrix} 6 & 2 \\ 7 & 0 \end{vmatrix} + -2\begin{vmatrix} 6 & -4 \\ 7 & -1 \end{vmatrix}$$

$$= 2 - 3(-14) - 2(-6 + 28)$$

$$= 2 + 42 - 44$$

$$= 0$$

Note that if one row is a linear combination of the two other rows then any column must also be a linear combination of the two other columns. In this example, Column 2 multiplied by 7 subtracted from Column 3 multiplied by −11 will give Column 1; clearly, this relationship is a lot harder to see.

Example 2.8

$$\mathbf{A} = \begin{pmatrix} 1 & -4 & 3 \\ 2 & 2 & 6 \\ -2 & 8 & -3 \end{pmatrix}$$. Using manipulation of rows and columns, find $|\mathbf{A}|$.

Solution

The trick is to triangularise matrix **A** by using the rules given above. For example, you may notice that multiplying **1** and −4 (the first two numbers of Row 1) with 2 (the first two numbers of Row 2) gives 2 and −8 which is exactly the negative of the first two numbers of the Row 3. So the first operation can be 'Add (2 × Row 1) to Row 3'. This operation leaves the determinant unchanged.

$$|\mathbf{A}| = \begin{vmatrix} 1 & -4 & 3 \\ 2 & 2 & 6 \\ -2 & 8 & -3 \end{vmatrix} = \begin{vmatrix} 1 & -4 & 3 \\ 2 & 2 & 6 \\ 0 & 0 & 3 \end{vmatrix}$$

Now all that has to be done is to remove the 2 which is the first number in Row 2. This can be easily done by the operation 'Subtract (2 × Row 1) from Row 2'.

$$|\mathbf{A}| = \begin{vmatrix} 1 & -4 & 3 \\ 2 & 2 & 6 \\ 0 & 0 & 3 \end{vmatrix} = \begin{vmatrix} 1 & -4 & 3 \\ 0 & 10 & 0 \\ 0 & 0 & 3 \end{vmatrix}$$

$$|\mathbf{A}| = \begin{vmatrix} 1 & -4 & 3 \\ 0 & 10 & 0 \\ 0 & 0 & 3 \end{vmatrix} = 1 \times 10 \times 3 = 30$$

Now that the matrix is upper triangular, it is easy to find the determinant; it is simply the product of the entries in the leading diagonal.

Note that in no sense is it claimed that **A** and the matrix $\begin{pmatrix} 1 & -4 & 3 \\ 0 & 10 & 0 \\ 0 & 0 & 3 \end{pmatrix}$ are

the *same matrix*, merely that the *determinant* of **A** happens to be the same as the determinant of that triangular matrix. Note also that there are many different operations that could be carried out to achieve the same end.

Example 2.9

Using row and column operations find $\Delta = \begin{vmatrix} 2 & 3 & 5 \\ 3 & 7 & 6 \\ 6 & 2 & -3 \end{vmatrix}$.

Solution

You could see this as bringing out a factor of $\frac{1}{5}$ from Column 1 and $\frac{1}{2}$ from Column 3.

Multiply Column 1 by 5 and Column 3 by 2 to make the first entry the same in both:

$$\Delta = \frac{1}{5} \times \frac{1}{2} \begin{vmatrix} 10 & 3 & 10 \\ 15 & 7 & 12 \\ 30 & 2 & -6 \end{vmatrix}$$

There are no useful 1s. Finding the lowest common multiple allows fractions within the matrix to be avoided.

Subtract Column 1 from Column 3:

$$\Delta = \frac{1}{5} \times \frac{1}{2} \begin{vmatrix} 10 & 3 & 0 \\ 15 & 7 & -3 \\ 30 & 2 & -36 \end{vmatrix}$$

Multiply Column 1 by $\frac{1}{5}$ and then by 3 and multiply Column 2 by 2

$$\Delta = \frac{1}{3} \times \frac{1}{2} \times \frac{1}{2} \begin{vmatrix} 6 & 6 & 0 \\ 9 & 14 & -3 \\ 18 & 4 & -36 \end{vmatrix}$$

Subtract Column 1 from Column 2:

$$\Delta = \frac{1}{12} \begin{vmatrix} 6 & 0 & 0 \\ 9 & 5 & -3 \\ 18 & -14 & -36 \end{vmatrix}$$

Multiply Column 2 by 3 and Column 3 by 5 to make the second entry the same (except for the sign) in both:

$$\Delta = \frac{1}{12} \times \frac{1}{3} \times \frac{1}{5} \begin{vmatrix} 6 & 0 & 0 \\ 9 & 15 & -15 \\ 18 & -42 & -180 \end{vmatrix}$$

> Note that it is often easier, even with numbers, to leave expressions in factorised form because you can then easily cancel.

Add Column 2 to Column 3:

$$\Delta = \frac{1}{12} \times \frac{1}{3} \times \frac{1}{5} \begin{vmatrix} 6 & 0 & 0 \\ 9 & 15 & 0 \\ 18 & -42 & -222 \end{vmatrix} = \frac{1}{12} \times \frac{1}{3} \times \frac{1}{5} \times 6 \times 15 \times -222 = -111$$

Factorising the denominator when the matrix contains unknowns

Using the above rules, and some general guiding principles, you should now be able to find the factorised determinant of matrices that contain unknowns. In Example 2.10 you will see how you can use row and column operations to do this. Note that in the example if you expand the determinant directly then you would get a cubic expression in x which might prove difficult to factorise. Even though the question indicates that there will be real linear factors these might be difficult to find (for example, the factorised form could be $(3x - 157)(5x - 219)(2x + 335)$ which you could not find using the factor theorem).

Example 2.10

Find, in factorised form, $|\mathbf{M}|$ where $\mathbf{M} = \begin{pmatrix} x - 2 & 4 & 1 \\ x - 4 & x + 2 & 3 \\ x - 2 & x + 6 & x \end{pmatrix}$. Hence find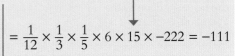

the values of x for which $\mathbf{M}$ is a singular matrix.

> You may be tempted by the fact that rows 1 and 3 both have $x - 2$ as their first entry. But in fact it is usually easiest to work with numbers, and 1s are always useful.

Solution 1

Subtract $(3 \times \text{Row } 1)$ from Row 2:

$$|\mathbf{M}| = \begin{vmatrix} x - 2 & 4 & 1 \\ -2x + 2 & x - 10 & 0 \\ x - 2 & x + 6 & x \end{vmatrix}$$

> Use the 1 at the top of column 3 to clear the rest of the column.

Subtract $(x \times \text{Row } 1)$ from Row 3:

$$|\mathbf{M}| = \begin{vmatrix} x - 2 & 4 & 1 \\ -2x + 2 & x - 10 & 0 \\ (1 - x)(x - 2) & 6 - 3x & 0 \end{vmatrix}$$

> Try to keep expressions in factorised form where possible.

> Now one of the entries in the lower left 2×2 array needs to be made 0 while keeping the two existing 0s. It should be clear that Row 1 and Column 3 cannot help to achieve this. One strategy is to make two of the entries equal.

Multiply Row 2 by $(6 - 3x)$:

> Best to simplify $(6 - 3x)$ and $(2 - 2x)$.

$$(6 - 3x)|\mathbf{M}| = \begin{vmatrix} x - 2 & 4 & 1 \\ (6 - 3x)(2 - 2x) & (6 - 3x)(x - 10) & 0 \\ (1 - x)(x - 2) & 6 - 3x & 0 \end{vmatrix}$$

Multiply Row 3 by $x - 10$:

$$3(x - 10)(2 - x)|\mathbf{M}| = \begin{vmatrix} x - 2 & 4 & 1 \\ 6(2 - x)(1 - x) & 3(2 - x)(x - 10) & 0 \\ (1 - x)(x - 2)(x - 10) & 3(2 - x)(x - 10) & 0 \end{vmatrix}$$

Subtract Row 2 from Row 3:

$$3(x - 10)(2 - x)|\mathbf{M}| = \begin{vmatrix} x - 2 & 4 & 1 \\ 6(2 - x)(1 - x) & 3(2 - x)(x - 10) & 0 \\ (1 - x)(x - 2)(x - 4) & 0 & 0 \end{vmatrix}$$

Note that $(1 - x)(x - 2)(x - 10) - 6(2 - x)(1 - x)$
$= (1 - x)\left[(x - 2)(x - 10) + 6(x - 2)\right] = (1 - x)(x - 2)(x - 4)$

Exchange Columns 1 and 3:

> Note that this introduces a minus sign.

$$-3(x - 10)(2 - x)|\mathbf{M}| = \begin{vmatrix} 1 & 4 & x - 2 \\ 0 & 3(2 - x)(x - 10) & 6(2 - x)(1 - x) \\ 0 & 0 & (1 - x)(x - 2)(x - 4) \end{vmatrix}$$

> This may at first sight look complicated but the steps are actually straightforward and if you keep a few basic principles in mind it is not difficult to do.

The matrix is now a triangular matrix so this leads to:
$$-3(x - 10)(2 - x)|\mathbf{M}| = 3(2 - x)(x - 10)(1 - x)(x - 2)(x - 4)$$

$$\therefore |\mathbf{M}| = (2 - x)(1 - x)(x - 4) = (x - 1)(x - 2)(x - 4)$$

and the values of x which make $\mathbf{M}$ is singular are therefore $1, 2$ or 4.

Note that this is not actually the most efficient way of achieving the answer, as you will see below.

Another strategy is to try to make one row or one column have a common factor. If this can be factorised out then either this will make triangularisation easier to see, or in practice it may not be required since the determinant of what remains may be quadratic.

Solution 2

Starting again from the original matrix, $\mathbf{M} = \begin{pmatrix} x-2 & 4 & 1 \\ x-4 & x+2 & 3 \\ x-2 & x+6 & x \end{pmatrix}$

Add Column 3 to Column 1:

$$|\mathbf{M}| = \begin{vmatrix} x-1 & 4 & 1 \\ x-1 & x+2 & 3 \\ 2x-2 & x+6 & x \end{vmatrix} = \begin{vmatrix} x-1 & 4 & 1 \\ x-1 & x+2 & 3 \\ 2(x-1) & x+6 & x \end{vmatrix}$$

> It is not necessarily easy to see this but one strategy you may develop is simply to try adding each column to each other (and similarly for rows) to see if this will produce common factors.

Factorise:

$$|\mathbf{M}| = (x-1)\begin{vmatrix} 1 & 4 & 1 \\ 1 & x+2 & 3 \\ 2 & x+6 & x \end{vmatrix}$$

There are now two possible approaches.

Either you can now simply find the determinant, which is quadratic, and factorise, preserving the all-important factor of $(x-1)$:

> Once you get down to a case where the determinant is quadratic in an unknown then, provided that you have not been specifically asked to triangularise the matrix or to use row and column operations, it is often easier simply to find it directly.

$$|\mathbf{M}| = (x-1)\begin{vmatrix} 1 & 4 & 1 \\ 1 & x+2 & 3 \\ 2 & x+6 & x \end{vmatrix}$$

$$= (x-1)\left[(x+2)x - 3(x+6) - 4(x-6) + (x+6) - 2(x+2)\right]$$

$$= (x-1)\left(x^2 + 2x - 3x - 18 - 4x + 24 + x + 6 + 2x - 4\right)$$

$$= (x-1)\left(x^2 - 6x + 8\right)$$

$$= (x-1)(x-2)(x-4)$$

Or you can continue to carry out row and column operations, which are generally much simpler with a factor removed:

Subtract (Row 1 + Row 2) from Row 3:

$$|\mathbf{M}| = (x-1)\begin{vmatrix} 1 & 4 & 1 \\ 1 & x+2 & 3 \\ 0 & 0 & x-4 \end{vmatrix}$$

Subtract Row 1 from Row 2:

$$|\mathbf{M}| = (x-1)\begin{vmatrix} 1 & 4 & 1 \\ 0 & x-2 & 2 \\ 0 & 0 & x-4 \end{vmatrix}$$

> **Discussion point**
> → How can you quickly check that the answer is correct?

Hence, $|\mathbf{M}| = (x-1)(x-2)(x-4)$.

A summary of general principles:

1. Use the rules given above to manipulate the matrix, keeping track of the effect on the determinant. It is best to make a clear note of each manipulation.

2. If a row or column has a common factor, especially if it is unknown, then factorise.

3. The general aim should be *either* to give a row or column a common factor which is an unknown, or to produce a triangular matrix.

4. Try to keep entries in factorised form where possible. Some standard results are given in the box below.

5. '1' can be very useful in eliminating values while keeping algebraic manipulation as simple as possible. Typically, you can use a 1 in a row (or column) to 'clear' the rest of that row (or column). Then use one column containing a zero to create another zero in another column containing a zero. This is easier to do than it sounds!

6. Try to avoid fractions in the matrix. They make manipulation harder.

7. Look for entries that are the same or simple multiples of each other occupying the same rows in different columns (or the same columns in different rows).

8. If you are stuck then you can try adding or subtracting each row from each other (or each column from each other) to see if this makes the matrix look simpler.

> Remember, you can always change the order of rows and columns using the rules.

Some standard results:

$$
\begin{aligned}
a^2 - b^2 &= (a - b)(a + b) \\
a^3 - b^3 &= (a - b)(a^2 + ab + b^2) \\
a^3 + b^3 &= (a + b)(a^2 - ab + b^2) \\
a^4 - b^4 &= (a - b)(a + b)(a^2 + b^2)
\end{aligned}
$$

Exercise 2.3

① Without using a calculator, find $\begin{vmatrix} 1 & -2 & 3 \\ 50 & -90 & 140 \\ -30 & 60 & 100 \end{vmatrix}$.

> Hint: You can use the first row to triangularise the matrix.

② Given that $\begin{vmatrix} 2 & 3 & -5 \\ 1 & -5 & 4 \\ -3 & 7 & -5 \end{vmatrix} = 13$, find the following without using a calculator.

(i) $\begin{vmatrix} -5 & 3 & 2 \\ 4 & -5 & 1 \\ -5 & 7 & -3 \end{vmatrix}$ (ii) $\begin{vmatrix} 1 & -5 & 4 \\ 2 & 3 & -5 \\ -3 & 7 & -5 \end{vmatrix}$ (iii) $\begin{vmatrix} 2 & 3 & -5 \\ 2 & -10 & 8 \\ -3 & 7 & -5 \end{vmatrix}$

(iv) $\begin{vmatrix} 2 & 3 & 15 \\ 1 & -5 & -12 \\ -3 & 7 & 15 \end{vmatrix}$ (v) $\begin{vmatrix} -5 & 1 & 4 \\ 3 & 2 & -5 \\ 7 & -3 & -5 \end{vmatrix}$ (vi) $\begin{vmatrix} 2 & 3 & 15 \\ 1 & -5 & -12 \\ -6 & 14 & 30 \end{vmatrix}$

(vii) $\begin{vmatrix} 2 & 3 & -5 \\ 3 & -2 & -1 \\ -3 & 7 & -5 \end{vmatrix}$ (viii) $\begin{vmatrix} 2 & 3 & -5 \\ 0 & 5 & -6 \\ -3 & 7 & -5 \end{vmatrix}$

③ Explain how you can tell by inspection that the following are true.

(i) $\begin{vmatrix} 3 & 2 & 3 \\ 5 & 1 & 5 \\ 2 & 3 & 2 \end{vmatrix} = 0$

(ii) $\begin{vmatrix} 4 & -3 & 2 \\ 2 & 0 & -3 \\ 7 & 4 & 5 \end{vmatrix} = -\begin{vmatrix} 4 & 2 & -3 \\ 2 & -3 & 0 \\ 7 & 5 & 4 \end{vmatrix}$

(iii) $\begin{vmatrix} 2 & 5 & -7 \\ 3 & 2 & 0 \\ -1 & 4 & 3 \end{vmatrix} = \begin{vmatrix} 5 & -7 & 2 \\ 2 & 0 & 3 \\ 4 & 3 & -1 \end{vmatrix}$

(iv) $(x-3)$ is a factor of $\begin{vmatrix} x & 3x & 3 \\ x+2 & 8 & 2x-1 \\ 11 & 5-x & x^2+2 \end{vmatrix}$.

④ Given that $\begin{vmatrix} 1 & 2 & 3 \\ 6 & 4 & 5 \\ 7 & 5 & 1 \end{vmatrix} = 43$, use the row and column operation rules to evaluate the following without expanding the determinant.

(i) $\begin{vmatrix} 10 & 2 & 3 \\ 60 & 4 & 5 \\ 70 & 5 & 1 \end{vmatrix}$ (ii) $\begin{vmatrix} 4 & 10 & -21 \\ 24 & 20 & -35 \\ 28 & 25 & -7 \end{vmatrix}$ (iii) $\begin{vmatrix} x^4 & \dfrac{1}{x} & 12y \\ 6x^4 & \dfrac{2}{x} & 20y \\ 7x^4 & \dfrac{5}{2x} & 4y \end{vmatrix}$

⑤ Using row and column operations and showing detailed reasoning, evaluate the following.

(i) $\begin{vmatrix} 21 & 2 & 3 \\ 46 & 4 & 5 \\ 57 & 5 & 1 \end{vmatrix}$ (ii) $\begin{vmatrix} 19 & 14 & 20 \\ 23 & 27 & 25 \\ 15 & 26 & 17 \end{vmatrix}$ (iii) $\begin{vmatrix} 25 & 17 & 51 \\ 38 & 33 & 78 \\ 25 & 32 & 52 \end{vmatrix}$

⑥ In a Fibonacci sequence, the third and subsequent terms satisfy $u_{n+2} = u_{n+1} + u_n$. Show that if $u_1, u_2, u_3,...$ is a Fibonacci sequence

then $\begin{vmatrix} u_1 & u_2 & u_3 \\ u_4 & u_5 & u_6 \\ u_7 & u_8 & u_9 \end{vmatrix} = 0$.

⑦ Express $\begin{vmatrix} 2x-3 & 3 & 2 \\ 2x-2 & 3x+11 & 3 \\ 6x-1 & 6x-1 & 6x-1 \end{vmatrix}$ in factorised form.

⑧ Show that $\begin{vmatrix} 0 & 0 & a \\ 0 & b & d \\ c & e & f \end{vmatrix} = -abc$

(i) by direct calculation

(ii) using row and column operations.

Note that the given matrix is NOT triangular since the zeroes are above the wrong diagonal.

⑨ (i) Explain why $\begin{vmatrix} x+1 & 4 & 4 \\ 5 & x+2 & 7 \\ 2 & 2 & x-3 \end{vmatrix} = 0$ is a cubic equation.

(ii) Show that $x = 3$ is one root, and find the other two roots.

⑩ Prove that $\begin{vmatrix} a & b & c \\ c & a & b \\ b & c & a \end{vmatrix} = (a+b+c)\begin{vmatrix} 1 & b & c \\ 1 & a & b \\ 1 & c & a \end{vmatrix}$.

Hence deduce that

$$a^3 + b^3 + c^3 - 3abc = (a+b+c)(a^2+b^2+c^2-bc-ca-ab).$$

⑪ Let $\mathbf{M} = \begin{pmatrix} a & b & c \\ a^2 & b^2 & c^2 \\ a^3 & b^3 & c^3 \end{pmatrix}$. By using row and column operations prove

that $|\mathbf{M}| = abc(b-a)(c-a)(c-b)$.

Hence find all the conditions on a, b and c for $\mathbf{M}$ to be a singular matrix.

⑫ Factorise the following determinants.

(i) $\begin{vmatrix} 1 & a & bc \\ 1 & b & ca \\ 1 & c & ab \end{vmatrix}$

(ii) $\begin{vmatrix} 1 & 1 & 1 \\ x & y & z \\ x^2 & y^2 & z^2 \end{vmatrix}$

(iii) $\begin{vmatrix} 1 & 1 & 1 \\ x^2 & y^2 & z^2 \\ yz & zx & xy \end{vmatrix}$

(iv) $\begin{vmatrix} x & y & z \\ x^2 & y^2 & z^2 \\ yz & zx & xy \end{vmatrix}$

⑬ Show that x, $(x-1)$ and $(x+1)$ must be factors of $\Delta = \begin{vmatrix} 1 & 1 & 1 \\ 1 & x & x^2 \\ 1 & x^2 & x^4 \end{vmatrix}$.

Hence or otherwise find Δ in fully factorised form.

⑭ Consider the matrix $\mathbf{M} = \begin{pmatrix} 1 & 1 & 1 \\ \sin^2 x & \cos^2 x & 1 \\ \sin^4 x & \cos^4 x & 1 \end{pmatrix}$.

(i) Prove that $|\mathbf{M}| = \dfrac{1}{8}\sin 4x \sin 2x$.

(ii) Show that if $0 \leqslant x \leqslant \dfrac{\pi}{2}$ then $\mathbf{M}$ is singular if and only if $x \in \left\{0, \dfrac{\pi}{4}, \dfrac{\pi}{2}\right\}$.

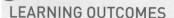

LEARNING OUTCOMES

Now you have finished this chapter, you should be able to:

➤ find the determinant of a 2×2 and a 3×3 matrix and explain their geometrical significance

➤ find the inverse of a non-singular 2×2 or 3×3 matrix

➤ use matrices to solve simultaneous linear equations

➤ use matrices to determine how three planes intersect in three dimensions

➤ apply row and column operations to 3×3 matrices and understand the effect that this has on the determinant

➤ factorise the determinant of 3×3 matrices.

KEY POINTS

1 If $\mathbf{M} = \begin{pmatrix} a & b \\ c & d \end{pmatrix}$ then the determinant of $\mathbf{M}$, written det $\mathbf{M}$ or $|\mathbf{M}|$ or Δ is

given by det $\mathbf{M} = ad - bc$.

2 The determinant of a 2×2 matrix represents the area scale factor of the transformation.

3 If $\mathbf{M} = \begin{pmatrix} a & b \\ c & d \end{pmatrix}$ then $\mathbf{M}^{-1} = \dfrac{1}{ad - bc} \begin{pmatrix} d & -b \\ -c & a \end{pmatrix}$.

4 The determinant of a 3×3 matrix $\mathbf{M} = \begin{pmatrix} a_1 & b_1 & c_1 \\ a_2 & b_2 & c_2 \\ a_3 & b_3 & c_3 \end{pmatrix}$ is given by

$$\det \mathbf{M} = a_1 \begin{vmatrix} b_2 & c_2 \\ b_3 & c_3 \end{vmatrix} - a_2 \begin{vmatrix} b_1 & c_1 \\ b_3 & c_3 \end{vmatrix} + a_3 \begin{vmatrix} b_1 & c_1 \\ b_2 & c_2 \end{vmatrix}.$$

5 The determinant of a 3×3 matrix represents the volume scale factor of the transformation. A negative determinant means that the orientation has been reversed.

6 For a 3×3 matrix $\begin{pmatrix} a_1 & b_1 & c_1 \\ a_2 & b_2 & c_2 \\ a_3 & b_3 & c_3 \end{pmatrix}$ the **minor** of an element is formed by

crossing out the row and column containing that element and finding the determinant of the resulting 2×2 matrix.

7 A minor, together with its correct sign, given by the matrix $\begin{vmatrix} + & - & + \\ - & + & - \\ + & - & + \end{vmatrix}$

is known as a **cofactor** and is denoted by the corresponding capital letter; for example the cofactor of a_3 is A_3.

8 The inverse of a 3×3 matrix $\mathbf{M} = \begin{pmatrix} a_1 & b_1 & c_1 \\ a_2 & b_2 & c_2 \\ a_3 & b_3 & c_3 \end{pmatrix}$ can be found using a calculator or using the formula

$$\mathbf{M}^{-1} = \frac{1}{\det \mathbf{M}} \operatorname{adj} \mathbf{M} = \frac{1}{\det \mathbf{M}} \begin{pmatrix} A_1 & A_2 & A_3 \\ B_1 & B_2 & B_3 \\ C_1 & C_2 & C_3 \end{pmatrix}, \Delta \neq 0$$

The matrix $\begin{pmatrix} A_1 & A_2 & A_3 \\ B_1 & B_2 & B_3 \\ C_1 & C_2 & C_3 \end{pmatrix}$ is the **adjoint** or **adjugate** matrix, denoted

adj $\mathbf{M}$, formed by replacing each element of $\mathbf{M}$ by its cofactor and then transposing (i.e. changing rows into columns and columns into rows).

9 $(\mathbf{MN})^{-1} = \mathbf{N}^{-1}\mathbf{M}^{-1}$

10 A matrix is **singular** if the determinant is zero. If the determinant is non-zero the matrix is said to be **non-singular**.

11 If the determinant of a matrix is zero, all points are mapped to either a straight line (in two dimensions) or to a plane (three dimensions).

12 If $\mathbf{A}$ is a non-singular matrix, $\mathbf{A}\mathbf{A}^{-1} = \mathbf{A}^{-1}\mathbf{A} = \mathbf{I}$.

13 When solving 2 simultaneous equations in 2 unknowns, the equations can be

written as a matrix equation $\mathbf{M}\begin{pmatrix} x \\ y \end{pmatrix} = \begin{pmatrix} a \\ b \end{pmatrix}$.

When solving 3 simultaneous equations in 3 unknowns, the equations can be

written as a matrix equation $\mathbf{M}\begin{pmatrix} x \\ y \\ z \end{pmatrix} = \begin{pmatrix} a \\ b \\ c \end{pmatrix}$.

In both cases, if $\det \mathbf{M} \neq 0$ there is a unique solution to the equations which can be found by pre-multiplying both sides of the equation by the inverse matrix $\mathbf{M}^{-1}$.

If $\det \mathbf{M} = 0$ there is no unique solution to the equations. In this case there is either no solution or an infinite number of solutions.

14 Three distinct planes in three dimensions will be arranged in one of five ways:

- They meet in a unique point of intersection
- All three planes are parallel and therefore do not meet
- Two of the planes are parallel, and these are cut by the third plane to form two parallel lines
- The planes form a sheaf of planes that intersect in a common line
- The planes form a prism of planes in which each pair of planes meet in a straight line but there are no common points of intersection between the three planes

15 Three distinct planes

$$a_1x + b_1y + c_1z = d_1$$
$$a_2x + b_2y + c_2z = d_2$$
$$a_3x + b_3y + c_3z = d_3$$

can be expressed in the form

$$\mathbf{M}\begin{pmatrix} x \\ y \\ z \end{pmatrix} = \begin{pmatrix} d_1 \\ d_2 \\ d_3 \end{pmatrix}$$

where $\mathbf{M} = \begin{pmatrix} a_1 & b_1 & c_1 \\ a_2 & b_2 & c_2 \\ a_3 & b_3 & c_3 \end{pmatrix}$.

If $\mathbf{M}$ is non-singular, the unique point of intersection is given by $\mathbf{M}^{-1}\begin{pmatrix} d_1 \\ d_2 \\ d_3 \end{pmatrix}$.

Otherwise, the planes meet in one of the other four possible arrangements. In the case of a sheaf of planes, the equations have an infinite number of possible solutions, and in the other three cases the equations have no solutions.

16 The determinant of a triangular matrix is the product of the entries in its leading diagonal.

17 You can use the following rules (or row and column operations) for manipulating a matrix in order to help find or factorise its determinant.

Rule 1: If you swap any two rows (or columns) then you change the sign of the determinant.

Rule 2: If you multiply every entry of a matrix by the same number then you multiply the determinant by the cube of the number.

Rule 3: If you multiply every entry in a row (or a column) by the same number then you multiply the determinant by that number.

Rule 4: If you add the entries of one row to the corresponding entries of another row then the determinant is unaffected (and similarly for columns).

Rule 5: If any row is the same as, or a multiple of, any other row then the determinant is 0 (and similarly for columns).

Rule 6: If any row is any linear combination of the other two rows then the determinant is 0 (and similarly for columns).

3 Conics

'... he seemed to approach the grave as a hyperbolic curve approaches a line, less directly as he got nearer, till it was doubtful if he would ever reach it at all.'
Far from the Madding Crowd
Thomas Hardy

Discussion point
→ Find out why a satellite dish's cross-section is shaped as a parabola.

Review: Conic sections

You will recall from the Year 1 Further Mathematics book that you met a family of curves called conics or conic sections. These are the curves that result when you slice through a double cone with a plane section as illustrated in Figures 3.1, 3.2 and 3.3.

Figure 3.1

Taking a horizontal slice gives you a circle.

Figure 3.2

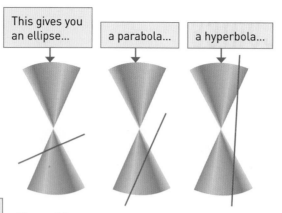

This gives you an ellipse...

a parabola...

a hyperbola...

Figure 3.3

As you will see later, these are not totally general, since the curves can be translated or otherwise transformed; in these forms the curves are 'centred' on the origin (although strictly a parabola does not have a centre).

The names and general equations of these curves are:

Circle: $\qquad\qquad x^2 + y^2 = a^2$

Ellipse: $\qquad\qquad \dfrac{x^2}{a^2} + \dfrac{y^2}{b^2} = 1$

Parabola: $\qquad\qquad y^2 = 4ax$

Hyperbola: $\qquad\qquad \dfrac{x^2}{a^2} - \dfrac{y^2}{b^2} = 1$ or $\dfrac{y^2}{b^2} - \dfrac{x^2}{a^2} = 1$

Rectangular hyperbola asymptotic to the x- and y-axes: $\quad xy = c^2$

Sketching conic sections

Circle

The equation $x^2 + y^2 = a^2$ represents a circle centre O, radius a as illustrated in Figure 3.4.

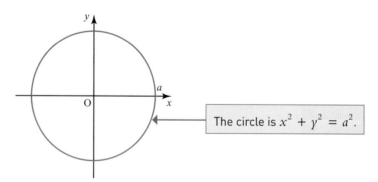

The circle is $x^2 + y^2 = a^2$.

Figure 3.4

Ellipse

Historical note

It is interesting to note that, contrary to popular belief, planets do not move in a circular orbit around the sun. As Johannes Kepler (1571–1630) showed, the paths that they actually follow are elliptical.

As you will see later on, an ellipse is a stretched circle. In the equation of the ellipse $\frac{x^2}{a^2} + \frac{y^2}{b^2} = 1$, the lengths a and b are called the semi-axes. If $a > b$ then a is the semi-major axis and b is the semi-minor axis and the ellipse is a circle which has been stretched in the x-direction, as shown in Figure 3.5.

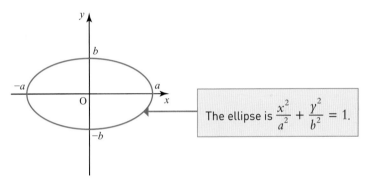

The ellipse is $\frac{x^2}{a^2} + \frac{y^2}{b^2} = 1$.

Figure 3.5

You should be able to see from the equation that the x-intercepts are $(-a, 0)$ and $(a, 0)$ and the y-intercepts are $(0, -b)$ and $(0, b)$.

Discussion point

➜ What does the ellipse look like if (i) $b > a$, (ii) $b = a$?

ACTIVITY

Firmly fix two ends of a piece of string to a piece of paper, for example using tin tacks. Pull the string taut using a pencil and, keeping both sections of the string taut and moving the pencil along, draw a curve until the portions of the string are along the same straight line. Put the pencil on the other side of the string and repeat. What curve have you traced out? What happens if you move the tin tacks closer together?

Look up the terms *focus* and *eccentricity*.

Parabola

You will already be familiar with a parabola (for example $y = x^2$) although when presented as a conic section the standard form of its equation is $y^2 = 4ax$, which means that the x-axis, rather than the y-axis, is its line of symmetry. The graph is shown in Figure 3.6.

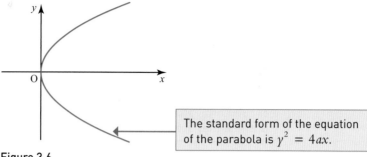

Figure 3.6

The origin is the only place where the graph intersects either axis.

Hyperbola

The standard form of the equation of a hyperbola is $\dfrac{x^2}{a^2} - \dfrac{y^2}{b^2} = 1$ although you should be aware that the equation $\dfrac{y^2}{b^2} - \dfrac{x^2}{a^2} = 1$ also represents a hyperbola. The graph is shown in Figure 3.7.

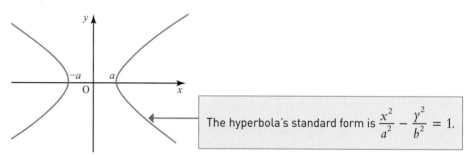

Figure 3.7

Again, you should be able to see from the equation that the x-intercepts are $(a, 0)$ and $(-a, 0)$.

> The hyperbola is the only conic section that has asymptotes.

The hyperbola has asymptotes: straight lines that the curve gets closer to as x gets large in value. These are $y = \pm\dfrac{b}{a}x$.

You should be careful when you are sketching the curve to show that the tails are indicated as straightening, in contrast to a parabola.

A **rectangular hyperbola** has asymptotes that are at right angles to each other.

In sketching the graph of a hyperbola, it often helps if you draw the asymptotes in first. This is illustrated in Figure 3.8.

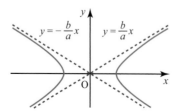

Figure 3.8

Discussion points

➜ Why does the graph of $\dfrac{x^2}{a^2} - \dfrac{y^2}{b^2} = 1$ not have any y-intercepts?

➜ What does the graph of $\dfrac{y^2}{b^2} - \dfrac{x^2}{a^2} = 1$ look like?

A special case of the hyperbola is the rectangular hyperbola in which the asymptotes are the x- and y-axes. It is usual to draw this curve so that it is in the first and third quadrants as illustrated in Figure 3.9.

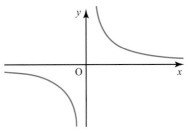

Figure 3.9

In this form, the equation of the rectangular hyperbola is $xy = c^2$. It has no axes intercepts.

Discussion point

➜ What happens if you try to find the x- or y-intercepts of the graph with equation $xy = c^2$?

Discussion points

➜ Do the above curves represent every single possibility for a plane intersecting a double cone?

➜ What is the most systematic way of considering every possible case?

➜ Are there any 'degenerate' solutions?

This is the graph of an **explicitly defined** function; y is given explicitly in terms of x, so for a given x value you can work out the corresponding y value directly.

Transformations of the graph of $y = f(x)$

In the A-level Mathematics Year 1 book you met rules for how to translate, stretch or reflect a graph defined by $y = f(x)$. These can be applied to the standard equations of conics to give other conics.

The rules are summarised below:

Transformation	Function
Translation $\begin{pmatrix} a \\ b \end{pmatrix}$	$y = f(x - a) + b$
Reflection in x-axis	$y = -f(x)$
Reflection in y-axis	$y = f(-x)$
One-way stretch, parallel to y-axis, scale factor a	$y = af(x)$
One-way stretch, parallel to x-axis, scale factor a	$y = f\left(\dfrac{x}{a}\right)$

Table 3.1

Review exercise

① Write down the equation of a circle, centre O and radius 10.

② An ellipse cuts the x-axis at $(\pm 5, 0)$ and the y-axis at $(0, \pm\frac{1}{3})$. Write down the equation of the ellipse.

③ Sketch the equations of the following curves. Show the coordinates of the points of intersection with the coordinate axes.

(i) $\dfrac{x^2}{9} + \dfrac{y^2}{4} = 1$ 　　　 (ii) $\dfrac{x^2}{9} - \dfrac{y^2}{4} = 1$

(iii) $\dfrac{y^2}{9} - \dfrac{x^2}{4} = 1$ 　　　 (iv) $xy = -2$

④ The rectangular hyperbola $xy = c^2$ passes through the point $(5, 10)$. Find, in its simplest form, the exact value of c.

⑤ A hyperbola has asymptotes $y = \pm\frac{1}{2}x$. It cuts the y-axis at the points $(0, \pm 5)$. Find the equation of the hyperbola.

⑥ (i) The rectangular hyperbola $xy = 13$ is reflected in the x-axis. What is the equation of the reflected curve?

(ii) The rectangular hyperbola $xy = 13$ is reflected in the y-axis. What is the equation of the reflected curve?

(iii) Comment on your answers to parts (i) and (ii).

⑦ The ellipse $\dfrac{x^2}{3} + \dfrac{y^2}{4} = 1$ is translated by the vector $\begin{pmatrix} 3 \\ 0 \end{pmatrix}$.

(i) Write down the equation of the transformed curve.

(ii) Find the coordinates of any axes intercepts.

⑧ The hyperbola C with equation $\dfrac{x^2}{36} - \dfrac{y^2}{16} = 1$ is stretched with scale factor 9 in the x-direction to give the curve C'.

(i) Write down the equation of C'.

(ii) Find the x-intercepts of C and C'.

(iii) Show that the major axis of C' is 9 times bigger than the major axis of C.

⑨ (i) On the same set of axes sketch the ellipse E defined by the equation
$\dfrac{x^2}{25} + \dfrac{y^2}{4} = 1$ and the hyperbola H defined by the equation $xy = 4$.

(ii) Find, in exact form, the coordinates of all the points of intersection of E and H.

(iii) The points of intersection are joined, in anticlockwise order starting from the one with the largest x-coordinate, to form a closed, convex polygon. What is the type of polygon formed?

1 Conics and further transformations

When you learned the rules for transforming the graph of $y = f(x)$ you may have wondered why it appears that x and y 'behave' differently in these rules; it seems that if something applies to x then the opposite applies to y. This seems counter-intuitive; you would expect them to behave in the same way as each other.

In this chapter you will learn an alternative way of looking at transformations which will demonstrate that, in fact, there is no difference in the way that x and y behave in such transformations. This new way of looking at transformations is much more efficient when applied to graphs of implicitly defined functions, and it will enable you to apply some more complicated transformations, including composite transformations, far more easily.

The basis of the new way of looking at this topic is to think about what each variable, either x or y, is being replaced with.

Transformations of the graph defined by $g(x, y) = 0$

In Chapter 1 of the AS Further Mathematics book you learnt how to find the images of points under transformations that could be described by matrices. If the matrix for the transformation was $\mathbf{A}$, then the coordinates of the image (x', y') are related to the coordinates of the object (x, y) by the equation $\begin{pmatrix} x' \\ y' \end{pmatrix} = \mathbf{A} \begin{pmatrix} x \\ y \end{pmatrix}$. To obtain a new equation relating x' and y', you need to replace x and y in the original equation by expressions involving x' and y', so you need to use the **inverse** matrix: $\begin{pmatrix} x \\ y \end{pmatrix} = \mathbf{A}^{-1} \begin{pmatrix} x' \\ y' \end{pmatrix}$. The replacement therefore involves the **inverse transformation**. This is true even when the transformation cannot be described by a matrix – for example, in the case of a translation.

Example 3.1

Find the image of the graph of $y = x^2$ under the translation $\begin{pmatrix} 3 \\ 2 \end{pmatrix}$.

You know how to answer this question using your previous knowledge, but the use of inverse transformations explains features that may have been puzzling, and the method can be readily extended to other transformations.

Solution

The translation can be written as $\begin{pmatrix} x' \\ y' \end{pmatrix} = \begin{pmatrix} x \\ y \end{pmatrix} + \begin{pmatrix} 3 \\ 2 \end{pmatrix}$.

The inverse transformation can therefore be written as

$\begin{pmatrix} x \\ y \end{pmatrix} = \begin{pmatrix} x' \\ y' \end{pmatrix} - \begin{pmatrix} 3 \\ 2 \end{pmatrix}$, or $x = x' - 3$, $y = y' - 2$.

Replacing x by $x' - 3$ and y by $y' - 2$ gives

$$y' - 2 = (x' - 3)^2 \quad \text{or} \quad y' = (x' - 3)^2 + 2$$

It is usual to write the final answer in terms of x and y rather than x' and y', so the equation of the transformed curve is

$$y = (x - 3)^2 + 2$$

which you will recognise from your previous knowledge as the right answer.

Table 3.2 shows you what happens to the graph of the function $g(x, y) = 0$ when x or y are replaced by simple variants. You will have met the transformations in the left-hand column in GCSE or in A-level Mathematics.

Transformations involving x		Transformations involving y	
Translate $\begin{pmatrix} a \\ 0 \end{pmatrix}$	$g(x - a, y) = 0$	Translate $\begin{pmatrix} 0 \\ a \end{pmatrix}$	$g(x, y - a) = 0$
Reflect in the y-axis	$g(-x, y) = 0$	Reflect in the x-axis	$g(x, -y) = 0$
Stretch parallel to the x-axis, scale factor a	$g(\frac{x}{a}, y) = 0$	Stretch parallel to the y-axis, scale factor a	$g(x, \frac{y}{a}) = 0$

Table 3.2

The first thing you should notice about these rules is that they are totally symmetrical in x and y. The second thing you should notice is that in fact these are entirely equivalent to the rules for an explicitly defined graph. For example, think about the graph defined by $y = f(x)$ (which could be rewritten as, for example, $y - f(x) = 0$ if the form $g(x, y) = 0$ is required). To carry out a translation by a units in the positive y-direction, the new rule says 'replace y by $y - a$' and so the equation of the transformed graph is $y - a = f(x)$. This is simply $y = f(x) + a$ which is exactly the same answer that first set of rules gave.

It is not necessary for the implicitly defined function to be of the form $g(x, y) = 0$. All that is required is that x and y are related by an equation.

Discussion point

➜ Show that all of the 'new' rules are the equivalent to the 'old' rules. From a symmetry viewpoint, which set of rules feels more natural?

Example 3.2

The ellipse with equation $\dfrac{x^2}{4} + \dfrac{y^2}{9} = 1$ is translated by vector $\begin{pmatrix} 2 \\ -3 \end{pmatrix}$. Find the equation of the new ellipse.

Note that the question does not specify a particular form, so you can leave the equation like this, which is, in any case, the most natural form. In other questions, subsequent algebraic manipulation may be necessary.

Solution

This is a translation of 2 units in the positive x direction and -3 units in the positive y-direction so replace x by $x - 2$ and y by $y - (-3) = y + 3$ (using brackets where necessary) wherever they occur.

So the equation of the transformed graph is $\dfrac{(x-2)^2}{4} + \dfrac{(y+3)^2}{9} = 1$.

Some further rules can now be added straightforwardly. They can all be obtained by using the inverse matrix for the appropriate transformation. For example, the

This is given in the AQA Formula Book.

matrix for a rotation of θ anticlockwise about O is $\begin{pmatrix} \cos\theta & -\sin\theta \\ \sin\theta & \cos\theta \end{pmatrix}$.

So, using the inverse matrix, $\begin{pmatrix} x \\ y \end{pmatrix} = \begin{pmatrix} \cos\theta & \sin\theta \\ -\sin\theta & \cos\theta \end{pmatrix}\begin{pmatrix} x' \\ y' \end{pmatrix}$. This means

that x is replaced by $x \cos\theta + y \sin\theta$ and y is replaced by $-x \sin\theta + y \cos\theta$, which is the result in the last row of the table.

Transformation	Function
Reflection in the line $y = x$	$g(y, x)$
Reflection in the line $y = -x$	$g(-y, -x)$
Enlargement by scale factor a, centre O	$g(x/a, y/a)$
Rotation by 90° anticlockwise about O	$g(y, -x)$
Rotation by 180° about O	$g(-x, -y)$
Rotation by 270° anticlockwise about O	$g(y, -x)$
Rotation θ anticlockwise about O	$g(x \cos\theta + y \sin\theta, y \cos\theta - x \sin\theta)$

Table 3.3

Composite transformations

You can also carry out several of these transformations in sequence. This is called **composition of transformations** and the resulting transformation is called a **composite transformation**. You should recall, however, that when transformations are composed, the order in which they are applied is important, so great care should be taken (that is, composition of transformations is not commutative).

Example 3.3

The rectangular hyperbola whose equation is $xy = 16$ is translated by vector $\begin{pmatrix} -1 \\ 4 \end{pmatrix}$ and then rotated by 90° clockwise about the origin. Find the equation of the resulting curve.

Again, there is no instruction as to the form of the final answer so it can be left as it is although many other forms would also do, for example, $(1 - y)(x - 4) = 16$ or $(y - 1)(x - 4) = -16$.

Solution

First, perform the translation so x is replaced by $x + 1$ and y by $y - 4$ in the original equation. This gives the equation of the translated hyperbola: $(x + 1)(y - 4) = 16$.

For the rotation, note that a 90° clockwise rotation is equivalent to a 270° anticlockwise rotation. So to form the new equation replace x by $-y$ and y by x to give $(-y + 1)(x - 4) = 16$.

Discussion point

→ How would the answer to this example have differed if the transformations had been performed in the reverse order?

Algebraic manipulation, such as completing the square, can sometimes be required in order to put the equation into a recognisable form.

Example 3.4

Show that the curve with equation $x^2 + 32y = 27 + 4y^2 - 2x$ is a translation of the hyperbola $\dfrac{y^2}{9} - \dfrac{x^2}{36} = 1$, and find the translation vector.

Bring all the x terms and y terms together, and complete the square for x and y in the usual way.

Solution

$$x^2 + 32y = 27 + 4y^2 - 2x$$
$$x^2 + 2x = 4y^2 - 32y + 27$$
$$(x + 1)^2 - 1 = 4(y^2 - 8y) + 27$$
$$(x + 1)^2 = 4((y - 4)^2 - 16) + 28$$
$$(x + 1)^2 = 4(y - 4)^2 - 36$$

Rearrange it so that it is written in the standard form.

$$\frac{(y - 4)^2}{9} - \frac{(x + 1)^2}{36} = 1$$

You should now recognise this as the equation of the hyperbola $\dfrac{y^2}{9} - \dfrac{x^2}{36} = 1$ but with x replaced by $x + 1$ and y replaced by $y - 4$.

So this represents a translation of the hyperbola $\dfrac{y^2}{9} - \dfrac{x^2}{36} = 1$ by the vector $\begin{pmatrix} -1 \\ 4 \end{pmatrix}$.

Exercise 3.1

① Copy and complete the following table:

	Equation of original curve	Equation of transformed curve	Description of transformation
(i)	$x^2 + y^2 = 25$		Enlargement centre O scale factor 3
(ii)	$\dfrac{x^2}{3} - \dfrac{y^2}{5} = 1$		Reflection in the line $y = -x$
(iii)		$\dfrac{x^2}{36} + \dfrac{y^2}{324} = 1$	Stretch of scale factor 3 in the y-direction
(iv)	$xy = 4$	$(x - 3)(y - 2) = 4$	
(v)	$xy = 9$		Reflection in the x-axis
(vi)	$xy = 9$		Reflection in the y-axis
(vii)	$\dfrac{x^2}{a^2} + \dfrac{y^2}{b^2} = 1$	$4x^2 + y^2 + 16 = 2y + 16x$	Translation by vector $\begin{pmatrix} p \\ q \end{pmatrix}$ where a, b, p and q are to be found
(viii)		$\dfrac{y^2}{10} - \dfrac{x^2}{5} = 1$	Rotation by 90° anticlockwise about O

② (i) Prove that the hyperbola with equation $\dfrac{x^2}{a^2} - \dfrac{y^2}{b^2} = 1$ is symmetrical in both the x- and y-axes.

  (ii) Prove that the hyperbola with equation $xy = c^2$ is not symmetrical in either the x- or y-axis.

  (iii) Prove that the lines $y = x$ and $y = -x$ are both lines of symmetry of the hyperbola with equation $xy = c^2$.

③ Prove that the ellipse with equation $\dfrac{x^2}{a^2} + \dfrac{y^2}{b^2} = 1$ is a stretch of the circle with equation $x^2 + y^2 = a^2$. State the scale factor and direction of the stretch.

④ A curve C has equation $\dfrac{x^2}{3} + \dfrac{y^2}{4} = 1$. T represents a translation by vector $\begin{pmatrix} 4 \\ -2 \end{pmatrix}$. S represents a stretch of scale factor 3 parallel to the y-axis. Find the equation of the image of C after the following transformations:

  (i) T

  (ii) S^2

  (iii) TS

  (iv) ST

⑤ Consider the hyperbola H defined by the equation $\dfrac{x^2}{25} - \dfrac{y^2}{169} = 1$. The curve H′ is the image of the curve H after it is rotated by 90° anticlockwise about the origin.

  (i) Find the equation of the curve H′.

  (ii) Find the coordinates of the points of intersection of H and H′.

    (iii)   Give the name of the shape formed when the points of intersection are joined by straight lines in anticlockwise order and find its area.

⑥ The parabola $y = 3x(x - 1)$ is transformed to the parabola $y^2 = 4ax$ by first carrying out a rotation and then a translation.

    (i)   Specify the required rotation.

    (ii)   Find the translation vector.

    (iii) Determine the value of a.

⑦ The parabola $y = 3 - 2x^2 - 8x$ is transformed to the parabola $y^2 = 4ax$ by first carrying out a translation and then a reflection.

    (i)   Find the translation vector.

    (ii)   State the equation of the mirror line for the reflection.

    (iii) Determine the value of a.

⑧ (i)   A hyperbola has equation $\dfrac{x^2}{a^2} - \dfrac{y^2}{b^2} = 1$ where a and b are positive constants. By considering the equations of the asymptotes, show that it is a rectangular hyperbola if and only if $a = b$. Hence show that its equation can be written $x^2 - y^2 = a^2$.

    (ii)   By using your knowledge of the shape of each graph, write down the angle and direction of the rotation about the origin which is required to transform the rectangular hyperbola $x^2 - y^2 = a^2$ to the rectangular hyperbola $xy = c^2$ where c is a positive constant.

    (iii) Use the rule for a rotation to confirm that the rectangular hyperbola $x^2 - y^2 = a^2$ can be mapped onto the rectangular hyperbola $xy = c^2$ using a rotation. Express c in terms of a.

    (iv) Hence find an expression in terms of c for the shortest distance between the two branches of the rectangular hyperbola $xy = c^2$.

⑨ The equation of a curve C can be written as $g(x^2, y^2) = 0$.

Use the formulae for transformations of the curve to explain how you know that:

    (i)   C is symmetrical about both the x-axis and the y-axis

    (ii)   the image of C under a rotation of 90° anticlockwise about O is identical to the image of C under a reflection in the line $y = x$.

LEARNING OUTCOMES

Now you have finished this chapter, you should be able to:

➤ know what the conic curves are, their shapes and their standard forms

➤ be able to sketch the graphs of conic curves with equations given in standard forms

➤ find the intercepts for conic curves from their equations, and the asymptotes for a hyperbola

➤ find the equation of a conic which has been transformed by a translation, a stretch parallel to an axis, a reflection in either axis, a reflection in the line $y = x$ or $y = -x$, an enlargement, or a rotation about the origin

➤ find the equation of a conic after composition of such transformations.

KEY POINTS

1 If you cut a double cone with one plane cut, the cross-section can be:

   a circle (standard form $x^2 + y^2 = a^2$)

   an ellipse (standard form $\dfrac{x^2}{a^2} + \dfrac{y^2}{b^2} = 1$)

   a parabola (standard form $y^2 = 4ax$)

   a hyperbola (standard form $\dfrac{x^2}{a^2} - \dfrac{y^2}{b^2} = 1$)

   (A pair of intersecting straight lines or a single point are also possible.)

2 The asymptotes for the hyperbola $\dfrac{x^2}{a^2} - \dfrac{y^2}{b^2} = 1$ are $y = \pm\dfrac{b}{a}x$.

3 If you start with a curve defined by the implicitly defined function $g(x, y) = 0$ then you should be aware of the following transformations:

Transformation	Function
Translation by a units in the positive x-direction	$g(x - a, y) = 0$
Translation by a units in the positive y-direction	$g(x, y - a) = 0$
Reflection in the y-axis	$g(-x, y) = 0$
Reflection in the x-axis	$g(x, -y) = 0$
Stretch parallel to the x-axis, scale factor a	$g(\frac{x}{a}, y) = 0$
Stretch parallel to the y-axis, scale factor a	$g(x, \frac{y}{a}) = 0$
Reflection in the line $y = x$	$g(y, x) = 0$
Reflection in the line $y = -x$	$g(-y, -x) = 0$
Enlargement by scale factor a, centre O	$g(\frac{x}{a}, \frac{y}{a}) = 0$
Rotation by $90°$ anticlockwise about O	$g(y, -x) = 0$
Rotation by $180°$ about O	$g(-x, -y) = 0$
Rotation by $270°$ anticlockwise about O	$g(-y, x) = 0$
Rotation by angle θ anticlockwise about O	$g(x \cos\theta + y \sin\theta, y \cos\theta - x \sin\theta) = 0$

Table 3.4

4 You **compose** transformations by carrying out one followed by another. The process can be repeated. In general if you change the order in which you perform the composition then the resulting curve will be different (that is, composition of transformations is not commutative).

5 Completing the square can often help to find the transformation that has been applied.

4

Series and induction

The essence of mathematics is not to make simple things complicated, but to make complicated things simple.

S. Gudder

Discussion points

The image of Pascal's triangle shown here has the odd numbers coloured. This results in a pattern similar to the Sierpinksy triangle, which is an example of a fractal.

→ Investigate the patterns produced by colouring multiples of 3, 4, etc.

→ Investigate the sum of the numbers in the first *n* rows of Pascal's triangle.

→ How could you prove your result?

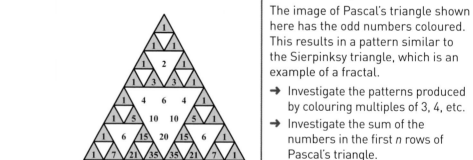

Figure 4.1

Review: Summing series

Terminology and notation

A **sequence** is an ordered set of objects with an underlying rule.

- The terms of a sequence are often written as $a_1, a_2, a_3, \ldots$ or $u_1, u_2, u_3, \ldots$.
- The general term of a sequence may be written as a_r or u_r. (Sometimes the letters k or i are used instead of r.)
- The last term is usually written as a_n or u_n.

A **series** is the sum of the terms of a numerical sequence.

- The sum of the first n terms of a sequence is often denoted by S_n.
- $S_n = a_1 + a_2 + \ldots + a_n = \displaystyle\sum_{r=1}^{n} a_r$.

Types of sequences

- A sequence is **increasing** if each term is greater than the previous term.
- A sequence is **decreasing** if each term is smaller than the previous term.
- In an **oscillating** sequence, the terms lie above and below a middle number.
- In an **arithmetic sequence**, the difference between each term and the next is constant. It is called the **common difference** and denoted by d.
- In a **geometric sequence**, the ratio of each term to the next is constant. It is called the **common ratio** and denoted by r.
- The terms of a **convergent** sequence get closer and closer to a limiting value. A geometric sequence is convergent if $-1 < r < 1$.

Summing series using standard formulae

Many series can be summed by using the standard formulae:

$$\sum_{r=1}^{n} 1 = n$$

> Remember that $\displaystyle\sum_{r=1}^{n} 1 = 1 + 1 + 1 + \ldots + 1$, so this means that 1 is added together n times, giving a total of n.

$$\sum_{r=1}^{n} r = \tfrac{1}{2}n(n + 1)$$

$$\sum_{r=1}^{n} r^2 = \tfrac{1}{6}n(n + 1)(2n + 1)$$

$$\sum_{r=1}^{n} r^3 = \tfrac{1}{4}n^2(n + 1)^2$$

> This result is precisely the square of the result $\sum r$.

ACTIVITY 4.1

Prove that $1 + 2 + 3 + \ldots + n = \tfrac{1}{2}n(n + 1)$ by using the formula for the sum of an arithmetic series.

You will prove the formula for $\displaystyle\sum_{r=1}^{n} r^2$ later in this chapter.

Example 4.1

Find the sum of the series $(2 \times 3) + (3 \times 4) + (4 \times 5) + \ldots + (n+1)(n+2)$.

Solution

The series can be written in the form $\displaystyle\sum_{r=1}^{n}(r+1)(r+2)$.

> To simplify this expression, look for common factors. It's usually helpful to take out any fractions as factors.

$$\sum_{r=1}^{n}(r+1)(r+2) = \sum_{r=1}^{n}r^2 + 3\sum_{r=1}^{n}r + 2\sum_{r=1}^{n}1$$

$$= \tfrac{1}{6}n(n+1)(2n+1) + 3 \times \tfrac{1}{2}n(n+1) + 2n$$

$$= \tfrac{1}{6}n[(n+1)(2n+1) + 9(n+1) + 12]$$

$$= \tfrac{1}{6}n[2n^2 + 3n + 1 + 9n + 9 + 12]$$

$$= \tfrac{1}{6}n[2n^2 + 12n + 22]$$

$$= \tfrac{1}{3}n(n^2 + 6n + 11)$$

Summing series using the method of differences

Sometimes the general term of a sequence can be written so that most of the terms cancel out.

Example 4.2

(i) Show that $(r+1)^2 - (r-1)^2 = 4r$.

(ii) Hence find $\displaystyle\sum_{r=1}^{n} 4r$.

(iii) Deduce that $\displaystyle\sum_{r=1}^{n} r = \frac{1}{2}n(n+1)$.

Solution

(i) $(r+1)^2 - (r-1)^2 = r^2 + 2r - 1 - (r^2 - 2r + 1)$

$\qquad\qquad\qquad\qquad\quad = 4r$

(ii) $\displaystyle\sum_{r=1}^{n} 4r = \sum_{r=1}^{n}\left[(r+1)^2 - (r-1)^2\right]$ ← Using the result of (i).

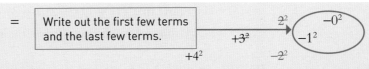

$= $ | Write out the first few terms and the last few terms. |

Most of the terms cancel out in pairs, leaving only two at the start and two at the end.

$= (n + 1)^2 + n^2 - 1^2$

$= n^2 + 2n + 1 + n^2 - 1$

$= 2n^2 + 2n$

$= 2n(n + 1)$

(iii) From part (ii), $\displaystyle\sum_{r=1}^{n} 4r = 2n(n + 1)$

This is the standard formula for the sum of the integers, given on page 91.

so $\displaystyle\sum_{r=1}^{n} r = \frac{1}{4} \times 2n(n + 1) = \frac{1}{2}n(n + 1)$

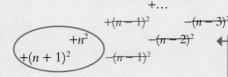

① Determine which of the following sequences
  (a) converge   (b) diverge   (c) decrease   (d) oscillate.
  (Some may do more than one of these.)

   (i) $a_{n+1} = 1.2a_n(1 - a_n), a_1 = 0.5$

   (ii) $a_{n+1} = 2.2a_n(1 - a_n), a_1 = 0.5$

   (iii) $a_{n+1} = 3.2a_n(1 - a_n), a_1 = 0.5$

   (iv) $a_{n+1} = 4.2a_n(1 - a_n), a_1 = 0.5$

② Find $\displaystyle\sum_{r=1}^{n}(4r - 1)$.

③ Find $\displaystyle\sum_{r=1}^{n}(3r^2 + r)$.

④ Find $\displaystyle\sum_{r=1}^{n}\left(\frac{1}{r^2} - \frac{1}{(r + 1)^2}\right)$.

⑤ (i) Simplify $r^2(r + 1)^2 - r^2(r - 1)^2$.

   (ii) Hence prove that $\displaystyle\sum_{r=1}^{n} r^3 = \frac{1}{4}n^2(n + 1)^2$.

⑥ (i) Write $\dfrac{2}{r(r + 2)}$ in the form $\dfrac{A}{r} + \dfrac{B}{r + 2}$.

   (ii) Hence find $\displaystyle\sum_{r=1}^{n}\frac{2}{r(r + 2)}$.

⑦ (i) Find $\displaystyle\sum_{r=1}^{n} r(r + 1)(r + 2)$.

   (ii) Hence find $1 \times 2 \times 3 + 2 \times 3 \times 4 + 3 \times 4 \times 5 + \ldots + 100 \times 101 \times 102$.

⑧ (i) Show that $\frac{1}{3}(r+1)(r+2)(r+3) - \frac{1}{3}r(r+1)(r+2) = (r+1)(r+2)$.

(ii) Using the result from (i) and the method of differences,

find $\displaystyle\sum_{r=1}^{n}(r+1)(r+2)$.

(iii) Use standard results to find $\displaystyle\sum_{r=1}^{n}(r+1)(r+2)$ and show that this is the same as the result from (ii).

⑨ A sum is given by $S = 1 - 2\left(\frac{1}{3}\right) + 3\left(\frac{1}{3}\right)^{2} - 4\left(\frac{1}{3}\right)^{3} + \dots$

(i) Write down an expression for $\frac{1}{3}S$.

(ii) Add S and $\frac{1}{3}S$. Describe the resulting series and find its sum to infinity. Hence find the value of S.

(iii) Show that S is the binomial expansion of $\left(1 + \frac{1}{3}\right)^{-2}$. Use this result to confirm the value of S you found in part (ii).

Prior knowledge

You need to know how to use partial fractions. This is covered in Chapter 7 of the AQA A-level Mathematics Year 2 book.

1 Applying the method of differences with partial fractions

Sometimes the general term of the series is an algebraic fraction with a factorisable denominator. If this is the case then the use of partial fractions often allows cancellation within the sum to allow it to be reduced to a simple form. This is illustrated in the following examples.

Example 4.3

Find $\displaystyle\sum_{r=1}^{n}\frac{1}{r(r+1)}$.

Solution

$\dfrac{1}{r(r+1)} = \dfrac{A}{r} + \dfrac{B}{r+1}$ ← First, write the expression using partial fractions.

$1 = A(r+1) + Br$

Let $r = 0 \implies A = 1$
Let $r = -1 \implies B = -1$

$\displaystyle\sum_{r=1}^{n}\frac{1}{r(r+1)} = \sum_{r=1}^{n}\left(\frac{1}{r} - \frac{1}{r+1}\right)$

$= \dfrac{1}{1} \quad -\dfrac{1}{2}$

$+\dfrac{1}{2} \quad -\dfrac{1}{3}$ ← Write out the first few terms and the last few terms. Most of the terms cancel, leaving just one term at the start and one at the end.

$+\dfrac{1}{3} \quad -\dfrac{1}{4}$

$+\dots$

$+\dfrac{1}{n-2} \quad -\dfrac{1}{n-1}$

$+\dfrac{1}{n-1} \quad -\dfrac{1}{n}$

$+\dfrac{1}{n} \quad \boxed{-\dfrac{1}{n+1}}$

$$= 1 - \frac{1}{n+1}$$

$$= \frac{n+1-1}{n+1}$$

$$= \frac{n}{n+1}$$

Sometimes the partial fractions involve three terms, as in the next example. This means that it is particularly important to lay out your work carefully, so that you can see clearly which terms cancel.

Example 4.4

(i) Find $\displaystyle\sum_{r=1}^{n} \frac{2}{r(r+1)(r+2)}$.

(ii) Hence state the value of $\displaystyle\sum_{r=1}^{\infty} \frac{2}{r(r+1)(r+2)}$.

Solution

(i) $\displaystyle\frac{2}{r(r+1)(r+2)} = \frac{A}{r} + \frac{B}{r+1} + \frac{C}{r+2}$

$2 = A(r+1)(r+2) + Br(r+2) + Cr(r+1)$

Write the expression using partial fractions.

Let $r = 0 \Rightarrow 2 = 2A \Rightarrow A = 1$

Let $r = -1 \Rightarrow 2 = -B \Rightarrow B = -2$

Let $r = -2 \Rightarrow 2 = 2C \Rightarrow C = 1$

$$\sum_{r=1}^{n} \frac{2}{r(r+1)(r+2)} = \sum_{r=1}^{n}\left(\frac{1}{r} - \frac{2}{r+1} + \frac{1}{r+2}\right)$$

$$= \frac{1}{1} - \frac{2}{2} + \frac{1}{3}$$

Most of the terms cancel out in groups of 3.

$$+ \frac{1}{2} - \frac{2}{3} + \frac{1}{4}$$

$$+ \frac{1}{3} - \frac{2}{4} + \frac{1}{5}$$

$$+ \dots - \dots + \dots$$

$$+ \frac{1}{n-2} - \frac{2}{n-1} + \frac{1}{n}$$

$$+ \frac{1}{n-1} - \frac{2}{n} + \frac{1}{n+1}$$

$$+ \frac{1}{n} - \frac{2}{n+1} + \frac{1}{n+2}$$

There are three terms left at the beginning and three at the end. Notice the symmetrical pattern.

$$= 1 - 1 + \frac{1}{2} + \frac{1}{n+1} - \frac{2}{n+1} + \frac{1}{n+2}$$

$$= \frac{1}{2} - \frac{1}{n+1} + \frac{1}{n+2}$$

(ii) As $n \to \infty$, $\dfrac{1}{n+1} \to 0$ and $\dfrac{1}{n+2} \to 0$

so $\displaystyle\sum_{r=1}^{\infty} \frac{2}{r(r+1)(r+2)} = \frac{1}{2}$

Exercise 4.1

① (i) Write $\dfrac{1}{(2r-1)(2r+1)}$ in partial fractions.

Hence

(i) find a formula for $\displaystyle\sum_{r=1}^{n}\dfrac{1}{(2r-1)(2r+1)}$

(ii) evaluate $\displaystyle\sum_{r=1}^{\infty}\dfrac{1}{(2r-1)(2r+1)}$.

② (i) Show that $\displaystyle\sum_{r=1}^{n}\left(\dfrac{1}{r^2}-\dfrac{1}{(r+1)^2}\right)=1-\dfrac{1}{(n+1)^2}$.

(ii) Show that $\dfrac{1}{r^2}-\dfrac{1}{(r+1)^2}=\dfrac{2r+1}{r^2(r+1)^2}$.

(iii) Hence find $\displaystyle\sum_{r=1}^{n}\dfrac{2r+1}{r^2(r+1)^2}$.

(iv) Explain why $\displaystyle\sum_{r=1}^{\infty}\dfrac{2r+1}{r^2(r+1)^2}=1$.

③ Find $\displaystyle\sum_{r=1}^{n}\dfrac{1}{r(r+1)(r+2)}$ and $\displaystyle\sum_{r=1}^{\infty}\dfrac{1}{r(r+1)(r+2)}$.

④ (i) Write $\dfrac{2}{r(r+2)}$ in the form $\dfrac{A}{r}+\dfrac{B}{r+2}$.

(ii) Hence find $\displaystyle\sum_{r=1}^{n}\dfrac{2}{r(r+2)}$ and $\displaystyle\sum_{r=1}^{\infty}\dfrac{2}{r(r+2)}$.

⑤ (i) Write $\dfrac{1}{(r-1)(r+2)}$ in partial fractions.

Hence

(i) find a formula for $\displaystyle\sum_{r=2}^{n}\dfrac{1}{(r-1)(r+2)}$

(ii) evaluate $\displaystyle\sum_{r=2}^{\infty}\dfrac{1}{(r-1)(r+2)}$.

⑥ By writing $\dfrac{r}{(r+1)(r+2)(r+3)}$ as $\dfrac{1}{(r+2)(r+3)}-\dfrac{1}{(r+1)(r+2)(r+3)}$,

evaluate $\displaystyle\sum_{r=1}^{n}\dfrac{r}{(r+1)(r+2)(r+3)}$.

⑦ (i) Show that $\dfrac{2r+1}{r(r+1)(r+2)}\equiv\dfrac{\frac{1}{2}}{r}-\dfrac{\frac{1}{2}}{r+1}+\dfrac{\frac{3}{2}}{r+1}-\dfrac{\frac{3}{2}}{r+2}$.

(ii) Hence find an expression for the sum to n terms of the series

$$\dfrac{3}{1\times2\times3}+\dfrac{5}{2\times3\times4}+\dfrac{7}{3\times4\times5}+\ldots$$

Review: Proof by induction

When you are solving a mathematical problem, you may sometimes make a **conjecture**. You might, for example, find a formula which seems to work in the cases you have investigated. You would then want to prove your conjecture.

Mathematical induction is a very powerful method that can be used to prove a conjecture or a given result, such as for the sum of a series.

The principle of proof by induction is to show that:

> **if** the result is true for the case $n = k$
>
> **then** it must be true for the case $n = k + 1$.

If you also show that it is true for an initial case, say $n = 1$, you can then deduce that it must be true for $n = 2$, and therefore it must be true for $n = 3$, and so on. You can then state that it is true for all positive integer values of n.

Steps in mathematical induction

To prove something by mathematical induction you need to state a conjecture to start with. Then there are five elements needed to try to prove the conjecture is true.

- Proving that it is true for a starting value (e.g. $n = 1$).
- Finding the target expression: using the result for $n = k$ to find the equivalent result for $n = k + 1$. ← To find the target expression you replace k with $k + 1$ in the result for $n = k$.

- Proving that: **if** it is true for $n = k$, **then** it is true for $n = k + 1$.

  This can be done before or after finding the target expression, but you may find it easier to find the target expression first so that you know what you are working towards.

- Arguing that since it is true for $n = 1$, it is also true for $n = 1 + 1 = 2$, and so for $n = 2 + 1 = 2$ and for all subsequent values of n.
- Concluding the argument by writing down the result and stating that it has been proved. ← This ensures the argument is properly rounded off. You will often use the word 'therefore'.

Example 4.5

Finding the sum of a series

Prove by induction that, for all positive integers n

$$1 + 3 + 5 + \ldots + (2n - 1) = n^2$$

Solution

When $n = 1$,   L.H.S. $= 1$ R.H.S. $= 1$

So it is true for $n = 1$.

Assume the result is true for $n = k$, so

$$1 + 3 + 5 + \ldots + 2k - 1 = k^2$$

You want to prove that the result is true for $n = k + 1$ (if the assumption is true).

> **Target expression**
> $$1 + 3 + 5 + \ldots + (2k - 1)$$
> $$+ [2(k + 1) - 1] = (k + 1)^2$$

Look at the L.H.S. of the result you want to prove:

$$1 + 3 + 5 + \ldots + 2k - 1 + [2(k + 1) - 1]$$

Use the assumed result for $n = k$, to replace the first k terms

The first k terms → $= k^2 + [2(k + 1) - 1]$ ← The $(k + 1)$th term

$$= k^2 + 2k + 2 - 1 \quad \leftarrow \text{Expand and simplify.}$$

$$= k^2 + 2k + 1$$

$$= (k + 1)^2 \quad \leftarrow \text{This is the same as the target expression.}$$

as required.

If the result is true for $n = k$, **then** it is true for $n = k + 1$.

Since it is true for $n = 1$, it is true for all positive integer values of n. Therefore the result that $1 + 3 + 5 + \ldots + (2n - 1) = n^2$ is true.

The method of proof by induction is often used in the context of the sum of a series, as in the example above. However, it has a number of other applications as well.

Induction can be used in divisibility proofs, as shown in the next example. In proofs like these, there is no 'target expression'; instead your target is to express the result in a form which shows the divisibility property that you are proving.

Example 4.6

Divisibility

Prove that $u_n = 4^n + 6n - 1$ is divisible by 9 for all $n \geqslant 1$.

Solution

When $n = 1$, $u_1 = 4^1 + 6 - 1 = 9$ which is divisible by 9. So it is true for $n = 1$.

Assume the result is true for $n = k$, so

$$u_k = 4^k + 6k - 1 \text{ is divisible by } 9$$

You want to prove that u_{k+1} is divisible by 9 (if the assumption is true).

$$u_{k+1} = 4^{k+1} + 6(k+1) - 1$$
$$= 4 \times 4^k + 6k + 5$$
$$= 4(u_k - 6k + 1) + 6k + 5$$
$$= 4u_k - 24k + 4 + 6k + 5$$
$$= 4u_k - 18k + 9$$
$$= 4u_k - 9(2k + 1)$$

> You want to express u_{k+1} in terms of u_k.

> Substituting $4^k = u_k - 6k + 1$.

> You have assumed that u_k is divisible by 9, and $9(2k+1)$ is divisible by 9, so u_{k+1} is divisible by 9.

If u_k is divisible by 9, then u_{k+1} is divisible by 9.

Since it is true for $n = 1$, it is true for all positive integer values of n. Therefore the result that $u_n = 4^n + 6n - 1$ is divisible by 9 is true.

Example 4.7

Matrix powers

Given that $\mathbf{A} = \begin{pmatrix} -3 & 8 \\ -2 & 5 \end{pmatrix}$, prove by induction that $\mathbf{A}^n = \begin{pmatrix} 1 - 4n & 8n \\ -2n & 1 + 4n \end{pmatrix}$.

Solution

When $n = 1$, $\mathbf{A}^1 = \begin{pmatrix} 1 - 4 & 8 \\ -2 & 1 + 4 \end{pmatrix} = \begin{pmatrix} -3 & 8 \\ -2 & 5 \end{pmatrix} = \mathbf{A}$

so the result is true for $n = 1$.

Assume the result is true for $n = k$, so

$$\mathbf{A}^k = \begin{pmatrix} 1 - 4k & 8k \\ -2k & 1 + 4k \end{pmatrix}$$

You want to prove that the result is true for $n = k + 1$ (if the assumption is true).

> Target expression
>
> $$\mathbf{A}^{k+1} = \begin{pmatrix} 1 - 4(k+1) & 8(k+1) \\ -2(k+1) & 1 + 4(k+1) \end{pmatrix}$$
> $$= \begin{pmatrix} -3 - 4k & 8k + 8 \\ -2k - 2 & 4k + 5 \end{pmatrix}$$

$$\mathbf{A}^{k+1} = \mathbf{A}^k \mathbf{A}$$

$$= \begin{pmatrix} 1 - 4k & 8k \\ -2k & 1 + 4k \end{pmatrix} \begin{pmatrix} -3 & 8 \\ -2 & 5 \end{pmatrix}$$

> Multiply the assumed result for $\mathbf{A}^k$ by the matrix $\mathbf{A}$.

$$= \begin{pmatrix} -3(1 - 4k) - 16k & 8(1 - 4k) + 40k \\ 6k - 2(1 + 4k) & -16k + 5(1 + 4k) \end{pmatrix}$$

$$= \begin{pmatrix} -3 - 4k & 8k + 8 \\ -2k - 2 & 4k + 5 \end{pmatrix}$$

> This is the same as the target expression.

If the result is true for $n = k$, **then** it is true for $n = k + 1$.

Since it is true for $n = 1$, it is true for all positive integer values of n.

Therefore the result that $\mathbf{A}^n = \begin{pmatrix} 1 - 4n & 8n \\ -2n & 1 + 4n \end{pmatrix}$ is true.

Example 4.8

nth term of a sequence

A sequence is defined by $u_1 = 1$ and $u_{n+1} = 3u_n - 4$.

Prove by induction that $u_n = 2 - 3^{n-1}$.

Solution

When $n = 1, u_n = 2 - 3^0 = 2 - 1 = 1$

so the result is true for $n = 1$.

Assume the result is true for $n = k$, so

$$u_k = 2 - 3^{k-1}$$

> **Target expression**
> $$u_{k+1} = 2 - 3^{(k+1)-1}$$
> $$= 2 - 3^k$$

You want to prove that the result is true for $n = k + 1$
(if the assumption is true).

$$u_{k+1} = 3u_k - 4 \longleftarrow$$

$$= 3(2 - 3^{k-1}) - 4 \longleftarrow$$

$$= 6 - 3 \times 3^{k-1} - 4$$

$$= 2 - 3^k \longleftarrow$$

> Use the given relationship between u_{n+1} and u_n.

> Substitute the assumed result for u_k.

> This is the same as the target expression.

If the result is true for $n = k$, **then** it is true for $n = k + 1$.

Since it is true for $n = 1$, it is true for all positive integer values of n.
Therefore the result that $u_n = 2 - 3^{n-1}$ is true.

Review exercise

① Given that $\mathbf{A} = \begin{pmatrix} -2 & 9 \\ -1 & 4 \end{pmatrix}$, you are going to prove by induction that

$$\mathbf{A}^n = \begin{pmatrix} 1 - 3n & 9n \\ -n & 1 + 3n \end{pmatrix}.$$

(i) Show that the result is true for $n = 1$.

(ii) If the result is true, write down the target expression for $\mathbf{A}^{k+1}$.

(iii) Assuming that the result is true for $n = k$, so $\mathbf{A}^k = \begin{pmatrix} 1 - 3k & 9k \\ -k & 1 + 3k \end{pmatrix}$,

use matrix multiplication to find an expression for $\mathbf{A}^{k+1}$.

(iv) Show that your answers to (ii) and (iii) are the same, and write a conclusion for your proof.

② You are going to prove by induction that $\sum_{r=1}^{n}(3r - 1) = \frac{1}{2}n(3n + 1)$.

(i) Show that the result is true for $n = 1$.

(ii) If the result is true, write down a target expression for $\sum_{r=1}^{k+1}(3r - 1)$.

(iii) Assuming that the result is true for $n = k$, so $\sum_{r=1}^{k}(3r - 1) = \frac{1}{2}k(3k + 1)$,

find an expression for $\sum_{r=1}^{k+1}(3r - 1)$ by adding the $(k + 1)$th term to the sum of the first k terms.

(iv) Show that your answers to (ii) and (iii) are the same, and write a conclusion for your proof.

③ A sequence is defined by $u_1 = 3$ and $u_{n+1} = 2u_n + 1$.
You are going to prove by induction that $u_n = 2^{n+1} - 1$.

(i) Show that the result is true for $n = 1$.

(ii) If the result is true, write down a target expression for u_{k+1}.

(iii) Assuming that the result is true for $n = k$, so $u_k = 2^{k-1} - 1$, find an expression for u_{k+1} by applying the rule $u_{k+1} = 2u_k + 1$.

(iv) Show that your answers to (ii) and (iii) are the same, and write a conclusion for your proof.

④ Prove by induction that $1^2 + 2^2 + 3^2 + ... + n^2 = \frac{1}{6}n(n + 1)(2n + 1)$.

⑤ Given that $\mathbf{A} = \begin{pmatrix} 6 & 5 \\ -5 & -4 \end{pmatrix}$, prove by induction that

$\mathbf{A}^n = \begin{pmatrix} 1 + 5n & 5n \\ -5n & 1 - 5n \end{pmatrix}$.

⑥ Prove by induction that $u_n = n^3 + 2n$ is a multiple of 3 for any positive integer n.

⑦ A sequence is defined by $u_1 = 3$ and $u_{n+1} = u_n + 2^n$.
Prove by induction that $u_n = 2^n + 1$.

⑧ Prove by induction that $u_n = 8^n - 3^n$ is divisible by 5 for any positive integer n.

⑨ Prove by induction that
$1 \times 3 + 2 \times 5 + 3 \times 7 + ... + n(2n + 1) = \frac{1}{6}n(n + 1)(4n + 5)$

⑩ Given that $\mathbf{P} = \begin{pmatrix} 1 & 0 \\ -1 & 2 \end{pmatrix}$, prove by induction that $\mathbf{P}^n = \begin{pmatrix} 1 & 0 \\ 1 - 2^n & 2^n \end{pmatrix}$.

⑪ Prove by induction that $u_n = 2^{4n+1} + 3$ is a multiple of 5 for any positive integer n.

⑫ A sequence is defined by $u_1 = 2$ and $u_{n+1} = 2u_n + 5$.
Prove by induction that $u_n = 7 \times 2^{n-1} - 5$.

⑬ Prove by induction that $u_n = 11^{n+2} + 12^{2n+1}$ is divisible by 133 for $n \geqslant 0$.

⑭ Given that $\mathbf{M} = \begin{pmatrix} 3 & 2 & -1 \\ 0 & 3 & 0 \\ 0 & 6 & 0 \end{pmatrix}$

(i) use a calculator to find $\mathbf{M}^2$, $\mathbf{M}^3$ and $\mathbf{M}^4$

(ii) make a conjecture about the matrix $\mathbf{M}^n$

(iii) prove your conjecture by induction.

LEARNING OUTCOMES

Now you have finished this chapter, you should be able to:

➤ sum a simple series using standard formulae for $\sum r$, $\sum r^2$ and $\sum r^3$

➤ sum a simple series using the method of differences

➤ sum a simple series using partial fractions

➤ construct and present a proof using mathematical induction for given results for a formula for the nth term of a simple sequence, the sum of a simple series, the nth power of a matrix, or a divisibility result.

KEY POINTS

1 Some series can be expressed as combinations of these standard results:

$$\sum_{r=1}^{n} r = \tfrac{1}{2}n(n + 1) \qquad \sum_{r=1}^{n} r^2 = \tfrac{1}{6}n(n + 1)(2n + 1) \qquad \sum_{r=1}^{n} r^3 = \tfrac{1}{4}n^2(n + 1)^2$$

2 Some series can be summed by using the method of differences. If the terms of the series can be written as the difference of terms of another series, then many terms may cancel out. This is called a telescoping sum.

3 Some series that involve fractions can be summed by writing the general term as partial fractions, and then using the method of differences.

4 To prove by induction that a statement involving an integer n is true for all $n \geqslant n_0$, you need to:

■ Prove that the result is true for an initial value of n, typically $n = 1$ for $n = n_0$.

■ Find the target expression: use the result for $n = k$ to find the equivalent result for $n = k + 1$.

■ Prove that: **if** it is true for $n = k$, **then** it is true for $n = k + 1$.

■ Argue that since it is true for $n = 1$, it is also true for $n = 1 + 1 = 2$, and so for $n = 2 + 1 = 3$ and for all subsequent values of n.

■ Conclude the argument with a precise statement about what has been proved.

FUTURE USES

■ You will use proof by induction to prove de Moivre's theorem in Chapter 14 Complex numbers.

T **PS** ① A 2×2 matrix is given by $\mathbf{M} = \begin{pmatrix} a & b \\ c & d \end{pmatrix}$.

Figure 1 was drawn using graphing software. It shows an arrowhead A and its images under $\mathbf{M}, \mathbf{M}^2, \mathbf{M}^3, \mathbf{M}^4$ and $\mathbf{M}^5$.

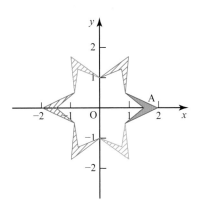

Figure 1

Find two sets of values for a, b, c and d. [4 marks]

M ② You are given the three related sequences $a_1, a_2, a_3, \ldots, b_1, b_2, b_3, \ldots$ and $c_1, c_2, c_3, \ldots$ where

$a_1 = 1$ $b_1 = 3$ $c_1 = \dfrac{a_1}{b_1}$

$a_2 = 1 + 3$ $b_2 = 5 + 7$ $c_2 = \dfrac{a_2}{b_2}$

$a_3 = 1 + 3 + 5$ $b_3 = 7 + 9 + 11$ $c_3 = \dfrac{a_3}{b_3}$

$a_4 = 1 + 3 + 5 + 7$ $b_4 = 9 + 11 + 13 + 15$ $c_4 = \dfrac{a_4}{b_4}$

and this pattern continues.

Prove that c_n is independent of n and state its value. [6 marks]

③ It is given that $y = xe^{2x}$. Prove by induction that, for all positive

integers n, $\dfrac{d^n y}{dx^n} = (2^n x + n \times 2^{n-1})e^{2x}$. [7 marks]

④ The plane π_1 passes through the points $A(3, 0, 1), B(5, -2, 1)$ and $C(4, 1, 5)$.

(i) Show that $\overrightarrow{AB} \bullet \begin{pmatrix} 2 \\ 2 \\ -1 \end{pmatrix} = 0$ and that $\overrightarrow{AC} \bullet \begin{pmatrix} 2 \\ 2 \\ -1 \end{pmatrix} = 0$. [2 marks]

(ii) Write down the Cartesian equation of π_1. [2 marks]

The plane π_2 has equation $3x - 4y + 12z = 42$.

(iii) Find the acute angle between π_1 and π_2. [3 marks]

⑤ (i) Express $\dfrac{1}{(2r+3)(2r+5)}$ in partial fractions. [3 marks]

(ii) Using the method of differences, find $\displaystyle\sum_{r=1}^{n}\dfrac{1}{(2r+3)(2r+5)}$,

expressing your answer as a single fraction. [4 marks]

(iii) Evaluate $\displaystyle\sum_{r=1}^{\infty}\dfrac{1}{(2r+3)(2r+5)}$. [1 mark]

⑥ Curve C_1 has equation $\dfrac{x^2}{16}-\dfrac{y^2}{9}=1$.

Curve C_2 has equation $9x^2-y^2-54x-10y+20=0$.

(i) Find a sequence of transformations that map C_1 onto C_2. [5 marks]

(ii) Find the equations of the asymptotes of C_2. [3 marks]

⑦ (i) Find the inverse of the matrix $\begin{pmatrix} k & 1 & 1 \\ 1 & k & 0 \\ 1 & 0 & k \end{pmatrix}$.

State any values of k for which the matrix has no inverse. [6 marks]

(ii) Describe how the following three planes intersect when $k=0$.

$kx+y+z=5$

$x+ky\ \ \ \ =0$

$x\ \ \ \ +kz=3$ [2 marks]

Ⓜ (iii) Prove that, for all values of k for which the following three planes intersect in one point, the coordinates of the point of intersection are independent of k.

$kx+y+z=k$

$x+ky\ \ \ =1$

$x\ \ \ +kz=1$ [3 marks]

(iv) Three planes have equations

$\sqrt{2}x+y+z=p$

$x+\sqrt{2}y\ \ \ =3$

$x\ \ \ +\sqrt{2}z=5$

Describe how the three planes meet in each of these cases.

(a) $p=4$ [2 marks]

(b) $p=4\sqrt{2}$ [2 marks]

⑧ Factorise fully $\begin{vmatrix} x+2 & x-1 & (x-1)^2 \\ y+2 & y-1 & (y-1)^2 \\ z+2 & z-1 & (z-1)^2 \end{vmatrix}$. [5 marks]

Further algebra and graphs

Algebra is nothing more than geometry, in words; geometry is nothing more than algebra, in pictures.

Sophie Germain

Discussion point

➔ How could you model the sides of the Eiffel Tower using an equation?

Review: Graphs of rational functions and inequalities

In the Further Mathematics Year 1 book you were introduced to the idea of a **rational function**, which is a function of the form $\dfrac{\mathrm{N}(x)}{\mathrm{D}(x)}$ where N and D are both polynomials and D is not the zero polynomial.

You can sketch the graph of a function of the form $y = \dfrac{ax + b}{cx + d}$ using two different methods; firstly, by using general graph-sketching principles and, secondly, by recognising it as a sequence of transformations of the graph of $y = \dfrac{1}{x}$.

For example, the following steps demonstrate how you can sketch the graph
$y = \dfrac{x + 3}{x - 2}$.

Sketching a graph using general graph-sketching principles

Step 1: Find where the graph cuts the axes

> Note that in general if $d = 0$ the graph will not have a y-intercept but instead the y-axis will be a vertical asymptote.

When $x = 0$, $y = -\dfrac{3}{2}$, which is the y-intercept.

$y = 0$ when $\dfrac{x + 3}{x - 2} = 0$. But a rational function will only be zero if the numerator is zero, since the denominator, being a polynomial, must be finite.

> Again, for the general function, if $a = 0$ then there is no x-intercept and the x-axis is a horizontal asymptote.

So the x-intercept is -3.

Step 2: Find the vertical asymptotes and examine the behaviour of the graph on either side of them

There will be a vertical asymptote where the denominator is zero. So the line $x = 2$ is a vertical asymptote. If x is close to 2 then $x + 3$ will be close to 5 and so positive. If x is just less than 2 (e.g. 1.99) then $x - 2$ will be negative and so the function will tend towards negative infinity to the left of the asymptote. Similarly, if x is just greater than 2 then $x - 2$ will be positive and so the function will tend towards positive infinity to the right of the asymptote.

Step 3: Examine the behaviour as x tends to infinity

When x is large in magnitude you can see that the function value will be approximately 1 since both $x + 3$ and also $x - 2$ will be almost the same as x. The line $y = 1$ is therefore a horizontal asymptote. In fact, it can sometimes help to analyse further; if x is a large positive number (e.g. 100) then the function value will be just greater than 1 since $x + 3$ will be greater than $x - 2$ and so to the right the function will tend to 1 from above. Similar considerations show that to the left the function will tend to 1 from below.

Step 4: Completing the sketch

It is often helpful to consider other issues such as turning points. In this case, if you differentiate you find that $\dfrac{dy}{dx} = \dfrac{-5}{(x - 2)^2}$ which can never equal zero and so the graph has no turning points.

You can therefore deduce the shape of the graph and it is shown on Figure 5.1.

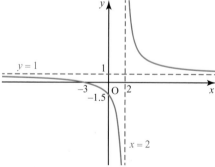

Figure 5.1

The same principles can be applied to more complicated graphs such as
$y = \dfrac{ax^2 + bx + c}{dx^2 + ex + f}$.

Sketching a graph using transformations

You can rewrite $y = \dfrac{x+3}{x-2}$ as $y = \dfrac{x-2+5}{x-2} = 1 + \dfrac{5}{x-2}$.

You can therefore think of the graph of $y = \dfrac{x+3}{x-2}$ as the result of a sequence

of transformations of the graph of $y = \dfrac{1}{x}$ as follows:

■ Stretch $y = \dfrac{1}{x}$ in the y-direction with scale factor of 5. The equation of the

image is $y = \dfrac{5}{x}$.

■ Translate $y = \dfrac{5}{x}$ by 2 units in the positive x-direction. The equation of the

image is $y = \dfrac{5}{x-2}$.

■ Translate $y = \dfrac{5}{x-2}$ by 1 unit in the positive y-direction. The equation of

the image is $y = 1 + \dfrac{5}{x-2} = \dfrac{x+3}{x-2}$. You should look at Figure 5.1 and

convince yourself that it is plausible as the image of these transformations

on the graph of $y = \dfrac{1}{x}$.

Discussion points

➜ What would it mean if the numerator and denominator of a rational function
were zero for the same x-value in the following cases?

(i) $y = \dfrac{ax+b}{cx+d}$ (ii) $y = \dfrac{N(x)}{D(x)}$

➜ Are there any cases where the graph of $y = \dfrac{ax+b}{cx+d}$ does not have a vertical
asymptote?

Range of values taken by a function

You can use quadratic theory to find the range of values taken by functions of
the form $y = \dfrac{ax^2 + bx + c}{dx^2 + ex + f}$, along with the exact locations of turning points as
you will see in the next example.

Example 5.1

(i) Use quadratic theory to find the coordinates of any turning points on
the graph of $y = \dfrac{6x+30}{x^2-9}$.

(ii) Hence sketch the graph $y = \dfrac{6x+30}{x^2-9}$.

(iii) State the range of the function $f(x) = \dfrac{6x+30}{x^2-9}$.

Solution

(i) Suppose $\dfrac{6x + 30}{x^2 - 9} = c$.

$6x + 30 = cx^2 - 9c$

$cx^2 - 6x - (9c + 30) = 0$

$x = \dfrac{6 \pm \sqrt{36 + 4c(9c + 30)}}{2c}$

$x = \dfrac{6 \pm \sqrt{36c^2 + 120c + 36}}{2c}$

$x = \dfrac{3 \pm \sqrt{3(3c^2 + 10c + 3)}}{c}$

$x = \dfrac{3 \pm \sqrt{3(3c + 1)(c + 3)}}{c}$

> Think about the intersection points of the curve and any horizontal line (i.e. the line $y = c$ for some constant c). A horizontal line will be a tangent to the curve at any turning points, so if you try to find intersection points by solving the equations simultaneously, at turning points you will find repeated roots. You can therefore find the values of c for which this happens; these will be y-coordinates of the turning points.

So, for this curve in general, horizontal lines $y = c$ will intersect the graph twice. However, at turning points there will just be one intersection point. So $c = -3$ and $c = -\dfrac{1}{3}$ are the y-coordinates of the turning points since when c is either of these values there is only one solution for x.

If $c = -3$ then $x = \dfrac{3 \pm 0}{-3} = -1$ and if $c = -\dfrac{1}{3}$ then $x = \dfrac{3 \pm 0}{-\dfrac{1}{3}} = -9$

so the turning points are then $(-1, -3)$ and $(-9, -\dfrac{1}{3})$.

> To sketch the graph you should use the answer to part (i) and apply the standard graph-sketching methodology.

(ii) **Step 1: Find where the graph cuts the axes**

When $x = 0$, $y = -\dfrac{10}{3}$ which is the y-intercept.

$y = 0$ when $\dfrac{6x + 30}{x^2 - 9} = 0$ so the x-intercept is -5.

Step 2: Find the vertical asymptotes and examine the behaviour of the graph on either side of them

There will be a vertical asymptote when $x^2 - 9 = (x - 3)(x + 3) = 0$ so the lines $x = 3$ and $x = -3$ are vertical asymptotes.

Note that $6x + 30 = 6(x + 5)$. If x is close to 3 then $x + 5$ will be close to 8 and so positive, while $x + 3$ will be close to 6 and so also positive. If x is just less than 3 (e.g. 2.99) then $x - 3$ will be negative and so the function will tend towards negative infinity to the left of the asymptote at $x = 3$. Similarly, if x is just greater than 3 then $x - 3$ will be positive and so the function will tend towards positive infinity to the right of this asymptote.

If x is close to -3 then $x + 5$ will be close to 2 and so positive, while $x - 3$ will be close to -6 and so negative. If x is just less than -3 (e.g. -3.01) then $x + 3$ will be negative and so the function will tend towards positive infinity to the left of the asymptote at $x = -3$ because of the signs of the other two factors. Similarly, if x is just greater than 3 then $x - 3$ will be positive and so the function will tend towards negative infinity to the right of this asymptote.

Step 3: Examine the behaviour as x tends to infinity

When x is large in magnitude you should be able to see that the function value will be approximately zero since the top will be almost the same as x while the bottom will be almost the same as x^2; the function will behave approximately like $\dfrac{1}{x}$. The x-axis is therefore a horizontal asymptote.

Step 4: Completing the sketch

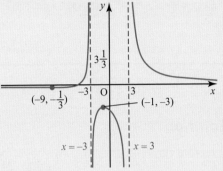

> Note that it is impossible for y values to lie between $-\dfrac{1}{3}$ and -3. This agrees with the comment at the end of part (i) of the solution. You can see this 'forbidden zone' on the graph in Figure 5.2.

Figure 5.2

> You should be familiar with set notation as a method for expressing intervals.

(iii)   From the graph, the range of $f(x)$ is

$$\left\{ y : y \in \mathbb{R}, y \geqslant -\frac{1}{3} \right\} \cup \{ y : y \in \mathbb{R}, y \leqslant -3 \}$$

Inequalities

In the Further Mathematics AS book you used some basic rules for manipulating inequalities algebraically, along with a variety of methods for solving them.

The rules are shown in Table 5.1.

Rule	Example (Based on 'is greater than')
1 You may add the same number to each side of an inequality.	$x > y \Leftrightarrow x + a > y + a$
2 You may multiply (or divide) both sides of an inequality by the same positive number.	If p is positive: $x > y \Leftrightarrow px > py$
3 If both sides of an inequality are multiplied (or divided) by the same negaive number, the inequality is reversed.	If n is negative: $x > y \Leftrightarrow nx < ny$
4 You may add (but not subtract) corresponding sides of inequalities of the same type.	$a > b$ and $x > y \Rightarrow a + x > b + y$
5 Inequalities of the same type are **transitive**.	$x > y$ and $y > z \Rightarrow x > z$

Table 5.1

The various methods are illustrated in the worked examples that follow. Which method you use will depend on the particular question and, perhaps, what work might have led up to it. For example, if the question requires the drawing of a sketch before the inequality is to be solved then this suggests that a partly graphical method is simplest, while if instead the function looks difficult to sketch then this might suggest an analytical method.

Example 5.2

Solve the inequality $\dfrac{6x + 30}{x^2 - 9} \geqslant 0$

Solution

Look carefully at the graph to decide whether the inequalities need to be strict or not.

The graph of $y = \dfrac{6x + 30}{x^2 - 9}$ is shown in Figure 5.2 on page 109. From the graph you can see that the solution (i.e. the set of x-values where $y \geqslant 0$) is $\{x : x \in \mathbb{R}, -5 \leqslant x < -3\} \cup \{x : x \in \mathbb{R}, x > 3\}$.

Example 5.3

Solve the inequality $x^3 - 2x^2 - 15x < 0$.

Solution

To find the critical values, solve $x^3 - 2x^2 - 15x = 0$.

$x(x^2 - 2x - 15) = 0$

$x(x - 5)(x + 3) = 0$

So the critical values are $-3, 0$ and 5.

Note the use of the word 'or' here (analogous to the symbol $\cup$ in set notation). A common mistake is either to omit the word, or worse, to use 'and'; there is no number which is both less than -3 and also between 0 and 5. You *must* use 'or' or $\cup$.

	$x < -3$	$-3 < x < 0$	$0 < x < 5$	$x > 5$
$(x - 5)$	$-$	$-$	$-$	$+$
x	$-$	$-$	$+$	$+$
$(x + 3)$	$-$	$+$	$+$	$+$
$x(x - 5)(x + 3)$	$-$	$+$	$-$	$+$

Table 5.2

So the solution is $x < -3$ or $0 < x < 5$.

Example 5.4

Solve the inequality $x^3 - 4x^2 - 3x + 14 \geqslant \dfrac{8}{x}$.

Solution 1

For the points of intersection of the graphs (i.e. the critical values)

$x^3 - 4x^2 - 3x + 14 = \dfrac{8}{x}$.

To solve a quartic equation you use the factor theorem.

$x^4 - 4x^3 - 3x^2 + 14x - 8 = 0$

$1^4 - 4 \times 1^3 - 3 \times 1^2 + 14 \times 1 - 8 = 1 - 4 - 3 + 14 - 8 = 0$

so $(x - 1)$ is a factor.

You need to find a factor for a cubic so it's back to the factor theorem!

$x^4 - 4x^3 - 3x^2 + 14x - 8 = (x - 1)(x^3 - 3x^2 - 6x + 8)$

For the cubic, $1^3 - 3 \times 1^2 - 6 \times 1 + 8 = 1 - 3 - 6 + 8 = 0$

so $(x - 1)$ is a factor again.

Note that the multiplicity of the root 1 is 2 (i.e. it is a repeated root) and so the graphs just touch each other at that point.

$x^3 - 3x^2 - 6x + 8 = (x - 1)(x^2 - 2x - 8)$

$x^4 - 4x^3 - 3x^2 + 14x - 8 = (x - 1)^2(x^2 - 2x - 8) = (x - 1)^2(x - 4)(x + 2)$

and so the critical values are $-2, 1$ and 4.

Solution 2

Sketching the graphs gives Figure 5.3.

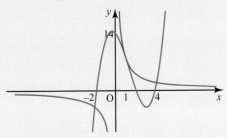

> This is the set of x-values where the red graph (the cubic) is above (or the same level as) the blue graph (the hyperbola). Note that this does NOT include $x = 0$ because the blue graph is not defined there. It is NOT the case that the red graph is above the blue graph along this vertical line.

Figure 5.3

> Be careful not to miss the solution $x = 1$. If the inequality had been strict (i.e. > rather than ⩾) then this would not have been in the solution set.

So the solution set is $-2 \leqslant x < 0$ or $x = 1$ or $x \geqslant 4$.

Example 5.5

Solve the inequality $\dfrac{6x + 8}{x - 2} \leqslant x + 5.$

> This is always positive, provided that $x \neq 2$, and so the inequality is preserved. $x = 2$ cannot be in the solution set anyway since the left-hand side is undefined when $x = 2$.

Solution

Multiply both sides by $(x - 2)^2$.

> It's easiest NOT to expand but to use the known factor.

$(6x + 8)(x - 2) \leqslant (x + 5)(x - 2)^2$

> Bring everything over to one side (the cubic side is easiest)...

$0 \leqslant (x + 5)(x - 2)^2 - (6x + 8)(x - 2)$

> ...and factorise.

$0 \leqslant (x - 2)\big((x + 5)(x - 2) - (6x + 8)\big)$

$0 \leqslant (x - 2)\big(x^2 + 3x - 10 - 6x - 8\big)$

$(x - 2)\big(x^2 - 3x - 18\big) \geqslant 0$

$(x - 2)(x + 3)(x - 6) \geqslant 0$

> Note that 2 is excluded from the solution set. It is always a good idea to look back at the original form of the inequality to ensure that you have not included any 'forbidden' points in your solution set.

So the critical values, in increasing order, are $-3, 2$ and 6. Therefore, because the left-hand side is a positive cubic, the solution set is $-3 \leqslant x < 2$ or $x \geqslant 6$.

Review exercise

① For each of the following expressions, write down

   (i)   the values of x for which $y = 0$

   (ii) the equations of any vertical asymptotes.

    (a) $y = \dfrac{x - 3}{x + 4}$ (b) $y = \dfrac{x^2 - 2x - 15}{x^2 - 16}$ (c) $y = \dfrac{x^2 - 4x}{x^2 + 4}$

② State the value of y to which the following expressions tend as x tends to infinity.

   (i) $y = \dfrac{x}{x^2 - 3}$ (ii) $y = \dfrac{x^2 + 4}{x^2 - 2}$ (iii) $y = \dfrac{2 + x - 3x^2}{4 - 2x + x^2}$

③ (i) Write the following functions $f(x)$ in the form $a + \dfrac{b}{cx + d}$:

(a) $\dfrac{x + 5}{x - 2}$ (b) $\dfrac{3x - 1}{x + 4}$ (c) $\dfrac{x + 1}{2x + 1}$ (d) $\dfrac{1 - 2x}{1 + 2x}$

(ii) Hence describe a series of transformations that will map the graph of $y = \dfrac{1}{x}$ onto the graph of $y = f(x)$ for each $f(x)$ in part (i).

(iii) Write down the equations of the asymptotes of the graphs of $y = f(x)$ for each $f(x)$ in part (i).

④ (i) Sketch the graph of $y = \dfrac{x + 2}{x - 1}$.

(ii) Hence solve the inequality $\dfrac{x + 2}{x - 1} \geqslant 0$.

⑤ (i) On the same axes, sketch the graphs of $y = x^2$ and $y = \dfrac{8}{x}$.

(ii) Hence solve the inequality $x^2 \geqslant \dfrac{8}{x}$.

⑥ The equation $\dfrac{2x^2 + 9x + 11}{x + 3} = c$ has no real solutions for x. State the range of possible values of c.

⑦ Find the range of possible values of y when

(i) $y = \dfrac{2x^2 - x - 1}{x^2 + x + 1}$ (ii) $y = \dfrac{2x^2 + 5}{x^2 - 6x - 4}$

⑧ In this question $f(x) = \dfrac{x^2}{x^2 - x - 2}$.

(i) State the equation of the horizontal asymptote of the graph of $y = f(x)$.

(ii) Determine whether the graph approaches the asymptote from above or from below as x tends to

(a) $+\infty$ (b) $-\infty$

⑨ Sketch the graphs of the following. Show the equations of any asymptotes.

(i) $y = \dfrac{2x + 5}{x - 2}$ (ii) $y = \dfrac{4 - x}{3 + x}$

⑩ Sketch the graphs of the following. Show the equations of any asymptotes.

(i) $y = \dfrac{(x - 2)(x + 3)}{(x + 2)(x - 3)}$ (ii) $y = \dfrac{(x - 3)(x + 4)}{(x - 1)(x + 2)}$

⑪ Solve the inequality $\dfrac{2x + 3}{3x - 1} < 4x - 3$.

⑫ (i) Solve the equation $x^2 - x = x - x^2 + 2$.

(ii) Solve the inequality $x^2 - x < x - x^2 + 2$.

⑬ (i) Show that $1 \leqslant \dfrac{9x^2 + 8x + 3}{x^2 + 1} \leqslant 11$.

(ii) Sketch the graph of $y = \dfrac{9x^2 + 8x + 3}{x^2 + 1}$, giving the coordinates of the turning points.

⑭ Solve the following inequalities.

(i) $\dfrac{x + 3}{2x - 1} \geqslant 2$ (ii) $\dfrac{2x - 1}{x + 3} \leqslant \dfrac{1}{2}$

⑮ (i) Describe the graph of $y = \dfrac{6x - 4}{2 - 3x}$.

(ii) What can be said about the function $f(x) = \dfrac{x^2 - 3x + 2}{x^2 + 2x - 8}$ at the following values?

(a) $x = 2$ (b) $x \neq 2$

(iii) Explain why the graph of $y = \dfrac{x^n - a^n}{x - a}$, $n \in \mathbb{N}$, does not have a vertical asymptote at $x = a$.

⑯ Sketch the graph of $y = \dfrac{x^4}{x^2 - 4}$, indicating its asymptotic behaviour.

1 Graphs with oblique asymptotes

So far all of the graphs of rational functions that you have seen in this book have had asymptotes that are either vertical or horizontal. However, some rational functions have asymptotes that are neither horizontal nor vertical. Asymptotes like this are called **oblique** (or slant) asymptotes. For example, the graph of $y = \dfrac{x^2 + 1}{x + 1}$ is shown in Figure 5.4 below. The line $y = x - 1$ is also shown and, as you can see, as x gets large in magnitude the curve approaches the straight line; the line $y = x - 1$ is an oblique asymptote of the graph of $y = \dfrac{x^2 + 1}{x + 1}$.

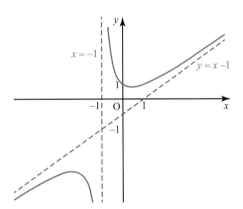

Figure 5.4

Discussion points

➜ Does the graph of $y = \dfrac{x^2 + 1}{x + 1}$ have a horizontal asymptote? Other than $x = 1$, does the graph have a vertical asymptote?

➜ If x is very large in magnitude, which terms of the expressions $x^2 + 1$ and $x + 1$ dominate?

➜ Find a simple approximate expression for $\dfrac{x^2 + 1}{x + 1}$ when x is very large in magnitude. How can you tell that it does have an asymptote?

At the beginning of this chapter you reviewed sketching graphs of the form $y = \dfrac{ax^2 + bx + c}{dx^2 + ex + f}$ and you learned how to sketch them. Now you need to recognise when such graphs have an oblique asymptote. If so, you need to include it in your sketch.

In general the graph of $y = \dfrac{ax^2 + bx + c}{dx^2 + ex + f}$ will have an oblique asymptote if

> Assuming that the denominator is not a factor of the numerator.

$d = 0$ and $a \neq 0$ (that is, if it is a quadratic function over a linear function and so of the form $y = \dfrac{ax^2 + bx + c}{ex + f}$ with $a \neq 0$).

The equation of the asymptote can be found by division; this can be done formally, using symbolic long division, or by algebraic manipulation which is usually simplest. The technique is illustrated in Example 5.6

Example 5.6

Sketch the graph $y = \dfrac{x^2 + 2x - 8}{x - 1}$, giving the equations of any asymptotes. You may assume that the graph has no turning points.

Solution

Step 1: Find where the graph cuts the axes

$x = 0$ when $y = 8$, which is the y-intercept.

$y = 0$ when $\dfrac{x^2 + 2x - 8}{x - 1} = 0 \Rightarrow (x + 4)(x - 2) = 0$ so the x-intercepts are -4 and 2.

Step 2: Find the vertical asymptotes and examine the behaviour of the graph on either side of them

There will be a vertical asymptote when $x - 1 = 0$ so the line $x = 1$ is a vertical asymptote.

If x is close to 1 then $x^2 + 2x - 8$ will be close to -5 and so negative. If x is just less than 1 (e.g. 0.99) then $x - 1$ will be negative and so the function will tend towards positive infinity to the left of the asymptote. Similarly, if x is just greater than 1 then $x - 1$ will be positive and so the function will tend towards negative infinity to the right of the asymptote.

Step 3: Examine the behaviour as x tends to infinity

> Find the equation of the oblique asymptote by rewriting the function.

$\dfrac{x^2 + 2x - 8}{x - 1}$ is a quadratic function over a linear so there is an oblique asymptote.

> You would get the same result using long division but this method produces the answer more quickly when the numbers are simple.

$$\frac{x^2 + 2x - 8}{x - 1} = \frac{x^2 - x + 3x - 8}{x - 1} = \frac{x(x - 1) + 3x - 3 - 5}{x - 1}$$
$$= \frac{x(x - 1) + 3(x - 1) - 5}{x - 1} = x + 3 - \frac{5}{x - 1}$$

When x is large in magnitude you should be able to see that the function value will be approximately the same as the value of $x + 3$ since the algebraic fraction will be almost zero. The equation of the oblique asymptote is therefore $y = x + 3$.

Step 4: Completing the sketch

There are no turning points and so there are no forbidden y values. The equations of the asymptotes are $x = 1$ and $y = x + 3$ and the sketch of the graph is shown in Figure 5.5.

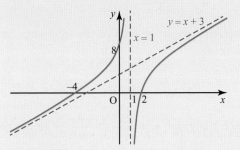

Figure 5.5

---

① (i) Write $\dfrac{x^2 + 3x + 5}{x - 2}$ in the form $Q(x) + \dfrac{a}{x + b}$, where $Q(x)$ is a polynomial and a and b are constants.

(ii) Write down the equations of the asymptotes of the corresponding graph.

② (i) Write $\dfrac{-x^2 + 4x + 1}{x + 3}$ in the form $Q(x) + \dfrac{a}{x + b}$, where $Q(x)$ is a polynomial and a and b are constants.

(ii) Write down the equations of the asymptotes of the corresponding graph.

③ (i) Write $\dfrac{x^3 + 3x + 5}{x^2 - 3x + 2}$ in the form $Q(x) + \dfrac{ax + b}{D(x)}$, where $Q(x)$ and $D(x)$ are polynomials and a and b are constants.

(ii) Write down the equations of the asymptotes of the corresponding graph.

④ (i) Write $\dfrac{-x^3 + 4x^2 + 1}{x^2 + 3x - 10}$ in the form $Q(x) + \dfrac{ax + b}{D(x)}$, where $Q(x)$ and $D(x)$ are polynomials and a and b are constants.

(ii) Write down the equations of the asymptotes of the corresponding graph.

⑤ Which of the following functions have graphs with an oblique asymptote?

(i) $y = \dfrac{3x^2 + 7x - 5}{x^2 + 3}$

(ii) $y = \dfrac{6 - 2x}{4x + 5}$

(iii) $y = \dfrac{x^2}{1 - x}$

(iv) $y = \dfrac{x^2 - 4x + 3}{1 - x}$

⑥ Sketch the graphs of each of the following. State the equations of any asymptotes.

(i) $y = \dfrac{x^2 + x + 1}{x + 2}$ 　　　　(ii) $y = \dfrac{2x^2 + 9x - 5}{x + 3}$

(iii) $y = \dfrac{1 + 3x - 2x^2}{x - 1}$

⑦ Sketch the graphs of each of the following. State the equations of any asymptotes.

(i) $y = \dfrac{x^3}{x^2 + 3x - 4}$ 　　　　(ii) $y = \dfrac{2x^3 + x + 1}{(1 + x)(2 - x)}$

⑧ Sketch the graph $y = \dfrac{2x^2 + 9x - 5}{3 - x}$ giving the equations of any asymptotes and the coordinates of any turning points.

⑨ (i) For which value(s) of a does the graph $y = \dfrac{ax^2 + 6x + 8}{x + 2}$ *not* have an oblique asymptote?

　　(ii) For which value(s) of b does the graph $y = \dfrac{x^2 + bx + 6}{x - 3}$ *not* have an oblique asymptote?

⑩ (i) Find the equation of the oblique asymptote of the graph of
$y = \dfrac{x^3 - 6x^2 + 6x + 10}{x^2 - 7x + 12}$.

　　(ii) In how many places does the graph of $y = \dfrac{x^3 - 6x^2 + 6x + 10}{x^2 - 7x + 12}$ intersect the line of its oblique asymptote?

　　(iii) What does this tell you about the graph of $y = \dfrac{x^3 - 6x^2 + 6x + 10}{x^2 - 7x + 12}$?

2 Modulus and reciprocal graphs

By now you should be familiar with the idea of sketching graphs. You will remember that the aim is *not* to plot points and join them up, but to deduce the key geometrical features of the graph from the algebraic formula, and to display these key features on your sketch, which is not meant to be accurately to scale. The key features you are looking for are:

- Where does the graph cut the *x*-axis and the *y*-axis?
- Is there a restricted domain? If so, what happens at the end-points of the domain?
- Are there any vertical asymptotes?
- Where is the graph positive? Where is it negative?
- How does the graph behave as *x* becomes very large and positive, or very large and negative?
- If there are any non-vertical asymptotes, does the graph approach them from above or from below?

It is sometimes helpful to find the turning points of the graph, but this is often not necessary as you can usually work out where they must be from the information you have found from the key features listed above. In any case it is best to try to avoid doing complicated algebra when sketching curves.

You should also know the graphs of the following functions and know how to sketch them when they are transformed by translations, stretches or reflections:

- $y = x^n$ for $n = -1, -2, \frac{1}{2}$, and positive integers
- $y = \sin x$, $y = \cos x$, $y = \tan x$
- $y = a^x$ (and in particular $y = e^x$)
- $y = \log_a x$ (and in particular $y = \ln x$)

You should also be able to sketch the graphs of polynomial functions, especially when factorised, for instance $y = x(x - 1)(x - 2)^2$.

In this section you will see how to extend your existing knowledge of graph sketching to graphs of the form $y = |f(x)|$ and $y = \dfrac{1}{f(x)}$. In both cases it will be assumed that you are able to sketch the graph of $y = f(x)$, so if you are not confident about graph sketching it is important that you review this important skill before continuing.

Sketching modulus graphs

Prior knowledge

You should be familiar with the modulus function, covered in AQA A-level Mathematics.

The modulus function is defined as follows:

$$|x| = \begin{cases} x \text{ if } x \geq 0 \\ -x \text{ if } x < 0 \end{cases}$$

So, for example,

$$\left|7.3\right| = 7.3, \ \left|-2\tfrac{1}{4}\right| = 2\tfrac{1}{4} \text{ and } |0| = 0.$$

The graph of $y = |x|$ is shown in Figure 5.6.

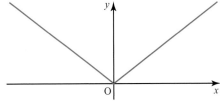

Figure 5.6

ACTIVITY 5.1

Use a graphing calculator or graphing software to help you to do this activity.

1 When $f(x) = x^3 - 3x^2 + 2x$:
   (i) Plot the graph of $y = f(x)$.
   (ii) On the same axes, plot the graph of $y = |f(x)|$.
   (iii) Describe any relationships between the two graphs. In particular, where do the graphs intersect? Where do they change signs? What happens when $f(x) = 0$ and where $f(x) \to \infty$? What happens as one of them increases? What happens to the turning points?
   (iv) In the regions where the graphs differ describe the relationship between them.

Repeat parts (i) to (iv) for the following functions:

2 $f(x) = 1 + x^2$
3 $f(x) = 1 - x^2$
4 $f(x) = 2^x$

Activity 5.1 shows the relationship between the graph of $y = f(x)$ and the graph of $y = |f(x)|$. You can use this relationship to sketch the graph of $y = |f(x)|$, as follows:

1 Sketch the graph of $y = f(x)$.

2 Leave alone any portion of the graph on or above the x-axis, as this is part of the graph of $y = |f(x)|$.

→3 Reflect in the x-axis any portion of the graph which is below the x-axis.

The graph in your final sketch should all lie on or above the x-axis. It is very important to realise that in most cases wherever the graph touches the x-axis there is a **cusp**: a sudden change of gradient. You should resist any temptation to smooth this.

> Once you have drawn the reflected portion either erase the portion below the x-axis or make it clear (for example, by labelling it $y = f(x)$ and the reflected portion $y = |f(x)|$) that this is part of your working, not part of your answer.

Example 5.7

> Note that no domain has been specified in the question. If this is the case in a question you are given then it is conventional to assume the largest possible domain (this is called the natural domain of a function), in this case the set of real numbers.

Sketch the graph of $y = |x^2 - 5x + 4|$.

Solution

$y = x^2 - 5x + 4 = (x - 1)(x - 4)$

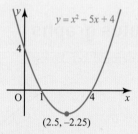

Figure 5.7

Now simply reflect (or 'bounce') any portion of the graph which is below the x-axis.

> Note the cusps (remember, a cusp is a point where there is a sudden change of gradient so it is sharp rather than smooth) at $x = 1$ and $x = 4$ on the x-axis. The graph should be roughly equally steep either side of any cusp, but sloping up on one side and down on the other.

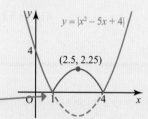

Figure 5.8

Discussion points

→ What condition would the graph of $y = |f(x)|$ have to satisfy for it to touch the x-axis but not have a cusp there? Find some examples of such graphs.

→ What is the value of the gradient at a cusp? What happens if you try to differentiate the function at a cusp? What happens to the gradient function as you move from one side of a cusp to another?

Sketching reciprocal graphs

> **ACTIVITY 5.2**
>
> Use a graphing calculator or graphing software to help you to do this activity.
> 1 When $f(x) = x^3 - 3x^2 + 2x$:
>   (i)   Plot the graph of $y = f(x)$.
>   (ii) On the same axes, plot the graph of $y = \dfrac{1}{f(x)}$.
>   (iii) Describe any relationships between the two graphs. In particular, where do the graphs intersect? Where do they change signs? What happens when $f(x) = 0$ and where $f(x) \to \infty$? What happens as one of them increases? What happens to the turning points?
>   (iv) What happens to the gradient of the graph of $y = \dfrac{1}{f(x)}$ at points where $f(x) = 0$?
>
> Repeat parts (i) to (iv) for the following functions:
> 2 $f(x) = 1 + x^2$
> 3 $f(x) = 1 - x^2$
> 4 $f(x) = 2^x$

$\dfrac{1}{x}$ is called the **reciprocal** of x.

To help draw graphs of this form (reciprocal graphs) it is worth firstly thinking about some general properties of the reciprocal function:

1 The reciprocal of a positive number is positive. The reciprocal of a negative number is negative.

2 The reciprocal of 0 is undefined.

3 The reciprocal of 1 is 1. The reciprocal of -1 is -1.

4 The reciprocal of a big number (bigger than 1 or smaller than -1) is small (between -1 and 1). The reciprocal of a small number is big.

5 As x approaches positive infinity, the reciprocal of x approaches 0 from above. As x approaches negative infinity, the reciprocal of x approaches 0 from below.

6 If x reaches a (non-zero) maximum then $\dfrac{1}{x}$ reaches a minimum. If x reaches a (non-zero) minimum then $\dfrac{1}{x}$ reaches a maximum.

You can use these properties to sketch the graph of $y = \dfrac{1}{f(x)}$ as follows:

First sketch the graph of $y = f(x)$.

Then indicate the features listed in Table 5.3 faintly in pencil:

Feature of $y = f(x)$	Feature of $y = \dfrac{1}{f(x)}$
Meets x-axis at $x = a$	Vertical asymptote at $x = a$
Vertical asymptote at $x = a$	Meets x-axis at $x = a$
Tends to $+\infty/-\infty$ for large x	Tends to 0 from above/below for large x
Tends to 0 from above/below for large x	Tends to $+\infty/-\infty$ for large x
Horizontal asymptote $y = k$	Horizontal asymptote $y = \dfrac{1}{k}$
y positive/negative	y positive/negative
$y = +1; y = -1$	$y = +1; y = -1$
$y > 1; 0 < y < 1$	$0 < y < 1; y > 1$ (and similarly for negative y)
Maximum/minimum at $x = a$ (except for turning points on the x-axis)	Minimum/maximum at $x = a$

Table 5.3

With the information about these features, you should now be able to draw the graph of $y = \dfrac{1}{f(x)}$.

Example 5.8

If $f(x) = 2x^2 - 6x + 4$, sketch the graph of $y = \dfrac{1}{f(x)}$.

Solution

1. Draw the graph of $y = f(x)$	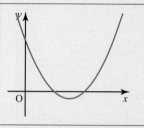
2. Where the graph of $y = f(x)$ intersects the x-axis, draw vertical asymptotes.	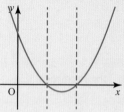
3. As x becomes very large (positive or negative) $f(x)$ tends to $+\infty$, so the graph of $\dfrac{1}{f(x)}$ is asymptotic to the x-axis from above.	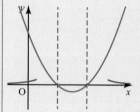
4. The graph of $y = f(x)$ intersects the line $y = 1$ at two points, and does not intersect the line $y = -1$.	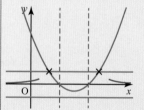
5. The graph of $y = f(x)$ has a minimum for a value of y which is negative but greater than -1. So the graph of $y = \dfrac{1}{f(x)}$ has a maximum for a value of y which is negative and less than -1.	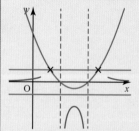
6. Where the graph of $y = f(x)$ is close to 0 but positive, the graph of $y = \dfrac{1}{f(x)}$ tends to $+\infty$. Where the graph of $y = f(x)$ is close to 0 but negative, the graph of $y = \dfrac{1}{f(x)}$ tends to $-\infty$.	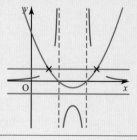

It is now easy to fill in the gaps. The graph of $y = \dfrac{1}{f(x)}$ is shown in Figure 5.9.

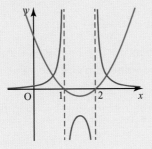

Figure 5.9

Exercise 5.2

① Sketch the graph of $y = |3 - x|$.

② Sketch the graph of $y = |x^2 - 3x - 10|$.

③ On the same axes sketch the graphs of $y = x - 4$ and $y = \dfrac{1}{x - 4}$.

④ Sketch the graph of $y = |12 - 5x - 2x^2|$.

⑤ Sketch the graph of $y = |e^x - 1|$.

⑥ Sketch the graph of $y = |(x - 1)(x - 2)(x - 3)|$.

⑦ Sketch the graph of $y = |\sin x|$ in $0 \leqslant x \leqslant 2\pi$.

⑧ Sketch the graph of $y = \dfrac{1}{e^x}$ and comment on the shape of your graph.

⑨ The graph of $y = f(x)$ is shown in Figure 5.10. The vertical asymptote and the lines $y = 1$ and $y = -1$ are also shown.

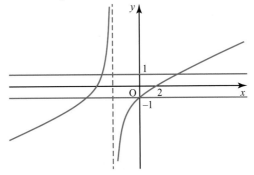

Figure 5.10

On separate copies of the diagram, and showing all important features, sketch the graphs of

(i) $y = |f(x)|$ (ii) $y = \dfrac{1}{f(x)}$.

⑩ Sketch the graph of $y = \operatorname{cosec} x = \dfrac{1}{\sin x}$ for $-\pi \leqslant x \leqslant 3\pi$.

⑪ Sketch the graph of $y = \cot x = \dfrac{1}{\tan x}$ for $-\pi \leqslant x \leqslant \pi$.

⑫ On the same axes, sketch the graphs of $y = \dfrac{x + 1}{x - 2}$ and $y = \dfrac{x - 2}{x + 1}$. Label each graph clearly.

⑬ Sketch the graphs of

(i) $y = \dfrac{1}{x^2 + 3x - 4}$ (ii) $y = \dfrac{1}{x^2 + 4x + 5}$.

⑭ Sketch the graph of $y = \dfrac{x^2}{x^2 + 1}$.

⑮ The graph of $y = a\,|\ln(x + 3)| + b$ passes through the points $(-2, 5)$ and $(e^{-1} - 3, 7)$. Find the value of a and the value of b.

⑯ Sketch on the same axes the graphs of $y = x^2 - 5x + 6$ and $y = x^2 - 5|x| + 6$. In general, what is the relationship between the graphs of $y = f(x)$ and $y = f(|x|)$?

3 Equations and inequalities

You can use the principles you have met in this chapter to solve equations and inequalities involving the modulus function. You will almost always find that a sketch is useful.

 Using algebraic methods to solve equations involving the modulus function can result in false roots. This is shown in Example 5.9.

Example 5.9

Solve the equation $|2x - 7| = 2 - x$.

Solution

Squaring both sides gives $(2x - 7)^2 = (2 - x)^2$.

$4x^2 - 28x + 49 = 4 - 4x + x^2$

$3x^2 - 24x + 45 = 0$

$x^2 - 8x + 15 = 0$

$(x - 3)(x - 5) = 0$

So $x = 3$ or $x = 5$.

> You can see that while the statement $1 = -1$ is false, squaring both sides turned it into a true statement, giving rise to the false root.

Checking $x = 3$: LHS $= |2 \times 1 - 7| = |-1| = 1$, while RHS $= 2 - 3 = -1$ so $x = 3$ is not, in fact, a root of the equation.

Checking $x = 5$: LHS $= |2 \times 5 - 7| = |3| = 3$, while RHS $= 2 - 5 = -3$ so $x = 5$ is also not a root.

So the equation has no roots.

The graphs of $y = |2x - 7|$ and $y = 2 - x$ are shown in Figure 5.11. These reveal why the equation Example 5.9 has no real roots.

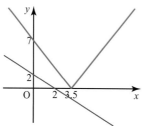

Figure 5.11

When equations are more complicated more care must be taken.

Example 5.10

Solve the equation $|3x + 2| = 2 + |x|$.

Solution 1 (algebraic)

Because both sides are definitely positive you can square both sides without introducing rogue roots.

$(3x + 2)^2 = \left(2 + |x|\right)^2$

Expand...

$\Rightarrow 9x^2 + 12x + 4 = 4 + 4|x| + x^2$

...and simplify.

$\Rightarrow 8x^2 + 12x = 4|x|$

$\Rightarrow |x| = 2x^2 + 3x$

You should try not to square again because, firstly, the right-hand side is not necessarily positive and, secondly, this would make the equation a quartic.

So either $x = 2x^2 + 3x$ if $x \geqslant 0$ or $-x = 2x^2 + 3x$ if $x < 0$.

$x \geqslant 0 \Rightarrow x = 2x^2 + 3x$

$\Rightarrow 2x^2 + 2x = 0$

$\Rightarrow x(x + 1) = 0$

Note that $= -1$ is NOT a valid solution since to derive it, you have assumed that $x \geqslant 0$.

$\Rightarrow x = 0$

$x < 0 \Rightarrow -x = 2x^2 + 3x$

$\Rightarrow 2x^2 + 4x = 0$

$\Rightarrow x(x + 2) = 0$

$\Rightarrow x = -2$

So the solution set is $x = 0$ or $x = -2$.

Solution 2 (geometric)

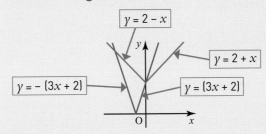

| $y = 2 - x$ |

| $y = 2 + x$ |

| $y = -(3x + 2)$ |

| $y = (3x + 2)$ |

Figure 5.12

The graphs of $y = |3x + 2|$ and $y = 2 + |x|$ are drawn in Figure 5.12.

Note that the two halves of each graph are labelled with their equations written without modulus signs.

It is clear that the two graphs intersect where $-(3x + 2) = 2 - x$ and where $3x + 2 = 2 - x$.

If $-(3x + 2) = 2 - x$ then $x = -2$.

If $3x + 2 = 2 - x$ then $x = 0$.

Therefore, $|3x + 2| = 2 + |x|$ at $x = -2$ and at $x = 0$.

It is certainly possible to solve modulus equations algebraically provided sufficient care is taken. However, Example 5.10 should convince you that it is generally easier to sketch the graphs to help you find the solution.

In general, if you take the modulus of a general function of x, $f(x)$, then:

$$\left| f(x) \right| = \begin{cases} f(x) & \text{if } x \geqslant 0 \\ -f(x) & \text{if } x < 0 \end{cases}$$

exactly as before. It is, however, sometimes harder to see that a function is negative.

So, for example, $\left| \cos \dfrac{3\pi}{4} \right| = \left| -\dfrac{1}{\sqrt{2}} \right| = \dfrac{1}{\sqrt{2}}$ and $\left| \ln \dfrac{1}{e} \right| = \left| -1 \right| = 1$.

Example 5.11

Solve the equation $\left| \sin x \right| = \dfrac{1}{2}$ for $0 \leqslant x \leqslant 2\pi$.

Solution

Figure 5.13 shows a sketch of the graphs of $y = \left| \sin x \right|$ and the line $y = \dfrac{1}{2}$.

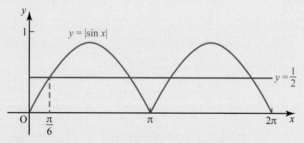

Figure 5.13

$$\left| \sin x \right| = \frac{1}{2} \Rightarrow \pm \sin x = \frac{1}{2}$$

If $\sin x = \dfrac{1}{2}$ then $x = \dfrac{\pi}{6}$ or $x = \dfrac{5\pi}{6}$.

If $-\sin x = \dfrac{1}{2}$ then $x = \dfrac{7\pi}{6}$ or $x = \dfrac{11\pi}{6}$.

Inequalities

One important point to remember is that you can only square both sides of an inequality if you know that both sides are non-negative (i.e. positive or zero). A typical path to the solution of an inequality with modulus signs is to sketch the graph, find the critical values and then use the graph to provide the solution.

Example 5.12

(i) On the same set of coordinate axes sketch the graphs $y = x + 1$ and $y = |2x + 1| - 2$.

(ii) Hence solve the inequality $|2x + 1| - 2 < x + 1$.

Solution

(i)

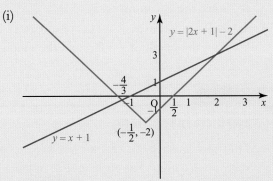

To sketch the graph of the modulus function you can consider it as a series of transformations of the graph $y = |x|$.

Figure 5.14

(ii) To find the critical values, solve $|2x + 1| - 2 = x + 1$.

$$\therefore |2x + 1| = x + 3$$

$$\therefore \pm(2x + 1) = x + 3$$

$$\therefore 2x + 1 = x + 3 \text{ or } -2x - 1 = x + 3$$

$$\therefore x = 2 \text{ or } x = -\frac{4}{3}$$

From Figure 5.14 you can tell that these are both valid solutions.

So the solution of the inequality is $-\frac{4}{3} < x < 2$.

In Figure 5.14 it is where the straight line graph is above the modulus graph.

Exercise 5.3

① Solve the equation $|\ln x| = 2$.

② Solve the equation $|\tan x| = 1$ for $-\pi \leqslant x \leqslant \pi$.

③ Solve the equation $|e^{2x+1}| = 5$, giving your answer correct to 3 significant figures.

④ Solve the equation $|\sin x| = \cos x$ for $-\pi \leqslant x \leqslant \pi$.

⑤ Solve the equation $|4x + 5| = |x| + 5$.

⑥ Solve the equation $|2x + 1| = |2x - 2| + 1$.

⑦ Solve the inequality $|2x + 1| < 9$.

⑧ Solve the inequality $|2x + 1| \leqslant |x| + 4$.

⑨ (i) Sketch the graph of $y = |\cos x|$ for $0 \leqslant x \leqslant 2\pi$.

(ii) Hence solve the inequality $|\cos x| < \frac{1}{2}$ for $0 \leqslant x \leqslant 2\pi$.

⑩ (i) Sketch the graph of $y = |\log_2(x^2)|$.

(ii) Solve the equation $|\log_2(x^2)| = 2$.

(iii) Hence solve the inequality $|\log_2(x^2)| > 2$.

⑪ Solve the equation $\left|\sin\left(2x + \dfrac{\pi}{4}\right)\right| = \dfrac{\sqrt{3}}{2}$ for $0 \leqslant x \leqslant \pi$.

⑫ Solve the equation $|x^2 - 5| = 4$.

⑬ The graph of $y = a\,|\,(x^2 - 5x - 6)\,|$ passes through the point $(2, 6)$.

   (i) Find the value of a.

   (ii) Solve the equation $y = 5$.

⑭ Solve the equation $|\,(x^2 - 4\,|x|\,)\,| = 3$. What in general is the relationship between the graph of $y = \mathrm{f}(x)$ and the graph of $y = \mathrm{f}(\,|x|\,)$?

LEARNING OUTCOMES

Now you have finished this chapter, you should be able to:

➤ know when a graph has an oblique asymptote

➤ find the equation of an oblique asymptote

➤ use oblique asymptotes to sketch graphs

➤ sketch the graph of $y = |\,\mathrm{f}(x)\,|$ when you are given $y = \mathrm{f}(x)$ as a formula or as a graph

➤ sketch the graph of $y = \dfrac{1}{\mathrm{f}(x)}$ when you are given $y = \mathrm{f}(x)$ as a formula or as a graph

➤ solve equations and inequalities involving the modulus function

➤ solve inequalities involving factorised expressions and algebraic fractions.

KEY POINTS

1 To find oblique asymptotes of graphs with equations such as $y = \dfrac{\mathrm{f}(x)}{\mathrm{g}(x)}$ where $\mathrm{f}(x)$ is of higher degree than $\mathrm{g}(x)$:

   ■ write the equation in the form $y = Q(x) + \dfrac{R(x)}{\mathrm{g}(x)}$ where $Q(x)$ is the quotient

   ■ the oblique asymptote then has equation $y = Q(x)$.

2 The graph of $y = |\,\mathrm{f}(x)\,|$ is obtained from the graph of $y = \mathrm{f}(x)$ as follows:

   ■ Where the graph of $y = \mathrm{f}(x)$ is above the x-axis, it remains unchanged.

   ■ Where the graph of $y = \mathrm{f}(x)$ is below the x-axis, it is reflected in the x-axis.

   ■ If the graph of $y = \mathrm{f}(x)$ cuts the axis at $x = a$, and $x = a$ is not a turning point, the graph of $y = |\,\mathrm{f}(x)\,|$ has a cusp.

3 The graph of $y = \dfrac{1}{f(x)}$ is obtained from the graph of $y = f(x)$ using the following features:

Feature of $y = f(x)$	Feature of $y = \dfrac{1}{f(x)}$
Meets x-axis at $x = a$	Vertical asymptote at $x = a$
Vertical asymptote at $x = a$	Meets x-axis at $x = a$
Tends to $+\infty/-\infty$ for large x	Tends to 0 from above/below, for large x
Tends to 0 from above/below, for large x	Tends to $+\infty/-\infty$ for large x
Horizontal asymptote $y = k$	Horizontal asymptote $y = \dfrac{1}{k}$
y positive/negative	y positive/negative
$y = +1; y = -1$	$y = +1; y = -1$
$y > 1; 0 < y < 1$	$0 < y < 1; y > 1$ (and similarly for negative y)
Maximum/minimum at $x = a$ (except for turning points on the x-axis)	Minimum/maximum at $x = a$

4 To solve equations and inequalities involving the modulus function:
- it is usually best to sketch a graph
- find the critical values
- if the equation involves $|f(x)|$, use an equation with $f(x)$ and a separate equation with $-f(x)$.

5 To solve inequalities using algebraic fractions:
- clear the fractions by multiplying by the *squares* of denominators, to ensure that the inequality does not change round.

6 Further calculus

Discussion point

➔ How could you estimate the number of birds in this picture?

Prior knowledge

You should be confident in all the integration methods you have covered previously.

1 Improper integrals

All the definite integrals you have calculated so far have been **proper** integrals. In this section you will meet some examples of **improper integrals**.

Discussion point

➔ You have drawn a curve and want to use integration to find the area between the curve and the x-axis. What features of a curve would warn you of possible difficulties?

Example 6.1

(i) Sketch the graph of $y = \dfrac{1}{x^2}$ for $x > 0$ and shade the area represented by the integral $\displaystyle\int_1^\infty \dfrac{1}{x^2}\,dx$. What features of this curve warn you of possible difficulties in evaluating this integral?

(ii) Evaluate $\displaystyle\int_1^\infty \dfrac{1}{x^2}\,dx$.

Solution

(i)

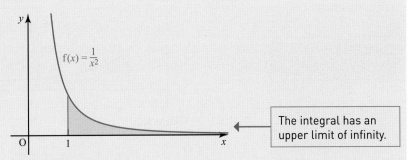

The integral has an upper limit of infinity.

Figure 6.1

(ii) The integral can be rewritten as the limit of an integral with finite limits – replacing the upper limit of infinity by the letter a, and then letting it tend towards infinity.

$$\int_1^\infty \frac{1}{x^2}\,dx = \lim_{a\to\infty}\int_1^a \frac{1}{x^2}\,dx$$

$$= \lim_{a\to\infty}\left[-\frac{1}{x}\right]_1^a$$

$$= \lim_{a\to\infty}\left(-\frac{1}{a} + 1\right)$$

As $a \to \infty$, $\dfrac{1}{a} \to 0$

As a tends to infinity, i.e. a gets very large, $\dfrac{1}{a}$ gets very small, so $\dfrac{1}{a}$ tends to zero.

So

$$\lim_{a\to\infty}\left(-\frac{1}{a} + 1\right) = 1$$

The area under the graph of $\dfrac{1}{x^2}$, from 1 to infinity, is 1 square unit.

The integral in Example 6.1 is said to be **convergent** and it converges to a value of 1. Not all integrals of this type are convergent, as the following example shows.

Example 6.2

(i) Sketch the graph of $y = \dfrac{1}{x}$ for $x > 0$, and shade the area represented by the integral $\displaystyle\int_1^\infty \frac{1}{x}\,\mathrm{d}x$.

(ii) What happens if you try to evaluate $\displaystyle\int_1^\infty \frac{1}{x}\,\mathrm{d}x$?

Solution

(i)

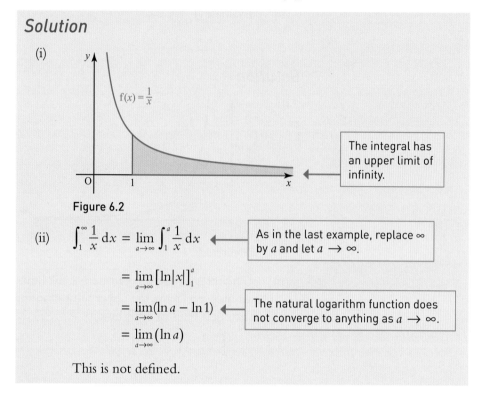

$f(x) = \dfrac{1}{x}$

The integral has an upper limit of infinity.

Figure 6.2

(ii) $\displaystyle\int_1^\infty \frac{1}{x}\,\mathrm{d}x = \lim_{a\to\infty} \int_1^a \frac{1}{x}\,\mathrm{d}x$ ← As in the last example, replace ∞ by a and let $a \to \infty$.

$\displaystyle = \lim_{a\to\infty}\Big[\ln|x|\Big]_1^a$

$\displaystyle = \lim_{a\to\infty}(\ln a - \ln 1)$ ← The natural logarithm function does not converge to anything as $a \to \infty$.

$\displaystyle = \lim_{a\to\infty}(\ln a)$

This is not defined.

Integrals like the one in Example 6.2, where there is no numerical answer, are **divergent**.

Example 6.3

(i) Sketch the graph of $y = \dfrac{1}{\sqrt{x}}$ and shade the area represented by the integral $\displaystyle\int_0^1 \frac{1}{\sqrt{x}}\,\mathrm{d}x$. What feature of the curve warns you of possible difficulties in evaluating this integral?

(ii) Evaluate $\displaystyle\int_0^1 \frac{1}{\sqrt{x}}\,\mathrm{d}x$.

Solution

(i)

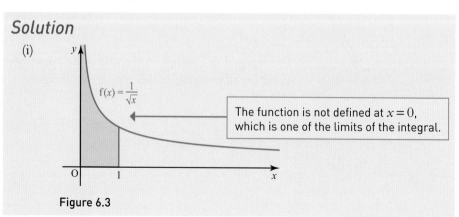

$f(x) = \dfrac{1}{\sqrt{x}}$

The function is not defined at $x = 0$, which is one of the limits of the integral.

Figure 6.3

(ii) $\displaystyle\int_0^1 \frac{1}{\sqrt{x}}\,dx = \lim_{a\to 0}\int_a^1 \frac{1}{\sqrt{x}}\,dx$ ← Notice that the variable a tends to 0 this time, not ∞.

$\displaystyle = \lim_{a\to 0}\left[2x^{\frac{1}{2}}\right]_a^1$

$\displaystyle = \lim_{a\to 0}\left(2 - 2a^{\frac{1}{2}}\right)$

As $a \to 0, a^{\frac{1}{2}} \to 0$

So $\displaystyle\lim_{a\to 0}\left(2 - 2a^{\frac{1}{2}}\right) = 2$

The area under the graph of $\dfrac{1}{\sqrt{x}}$, between 0 and 1, converges to 2 square units.

In this case you get the correct answer if you just integrate in the usual way, but it is safer to use the process shown in the example, as often you cannot be sure whether the value of the integral will converge or not.

ACTIVITY 6.1

Karen is trying to work out $\displaystyle\int_1^3 \frac{1}{(x-2)^2}\,dx$. She writes

$$\int_1^3 \frac{1}{(x-2)^2}\,dx = \left[-\frac{1}{x-2}\right]_1^3 = -1 - 1 = -2$$

How do you know that Karen's answer must be wrong?
What is the problem with Karen's working?

Example 6.4

(i) Sketch the graph of $y = \dfrac{1}{(x-2)^2}$ and shade the area represented by the integral $\displaystyle\int_1^3 \frac{1}{(x-2)^2}\,dx$. What feature of the curve warns you of possible difficulties in evaluating this integral?

(ii) What happens when you try to evaluate $\displaystyle\int_1^3 \frac{1}{(x-2)^2}\,dx$?

Solution

(i)

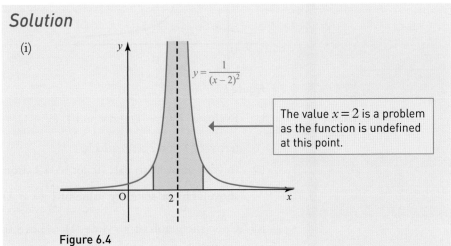

$y = \dfrac{1}{(x-2)^2}$

The value $x = 2$ is a problem as the function is undefined at this point.

Figure 6.4

(ii) $\displaystyle\int_1^3 \frac{1}{(x-2)^2}\,dx = \int_1^2 \frac{1}{(x-2)^2}\,dx + \int_2^3 \frac{1}{(x-2)^2}\,dx$ ← Split the integral at the point where it is undefined.

Now remove the problem limits $(x = 2)$ in the same way as the previous examples.

$\displaystyle = \lim_{a \to 2} \int_1^a \frac{1}{(x-2)^2}\,dx + \lim_{a \to 2} \int_a^3 \frac{1}{(x-2)^2}\,dx$

$\displaystyle = \lim_{a \to 2}\left[-\frac{1}{x-2}\right]_1^a + \lim_{a \to 2}\left[-\frac{1}{x-2}\right]_a^3$

$\displaystyle = \lim_{a \to 2}\left(-\frac{1}{a-2} - 1\right) + \lim_{a \to 2}\left(-1 + \frac{1}{a-2}\right)$

As $a \to 2$, $\dfrac{1}{a-2}$ is undefined, so the integral diverges.

The four examples above all involve **improper integrals**.

An improper integral is defined to be a definite integral in which:

■   at least one of the limits is infinite

■   or the function you wish to integrate approaches infinity at some point in the interval required.

Examples 6.1 and 6.2 both have an infinite limit, Example 6.3 has a function which approaches infinity at $x \to 0$, but 0 is one of the limits, and Example 6.4 includes the value $x = 2$ in the range required, but the function is not defined at that point.

Exercise 6.1

① (i)   Sketch the graph of $y = x^{-3}$.

(ii)   Evaluate $\displaystyle\int_2^a x^{-3}\,dx$, leaving your answer in terms of a.

(iii)   In your answer to part (ii), let $a \to \infty$, and hence state the value of the integral $\displaystyle\int_2^\infty x^{-3}\,dx$.

② The graph below shows the shape of the curve $y = (x-1)^{-\frac{2}{3}}$.

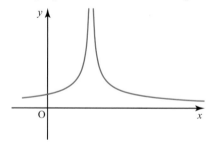

Figure 6.5

(i)   Evaluate $\displaystyle\int_0^b (x-1)^{-\frac{2}{3}}\,dx$ and $\displaystyle\int_c^3 (x-1)^{-\frac{2}{3}}\,dx$, leaving your answers in terms of b and c respectively.

(ii)   In your answers to part (i), let $b \to 1$ (from below), and let $c \to 1$ (from above). Hence state the value of $\displaystyle\int_0^3 (x-1)^{-\frac{2}{3}}\,dx$.

(iii)   Copy the graph of $y = (x-1)^{-\frac{2}{3}}$ above and indicate the area you have evaluated.

③ (i) Sketch the graph of $y = e^{-x}$.

(ii) Evaluate $\int_0^d e^{-x}\, dx$, leaving your answer in terms of d.

(iii) In your answer to part (ii), let $d \to \infty$, and hence state the value of $\int_0^\infty e^{-x}\, dx$.

Explain why each of the following integrals is improper, show whether they are convergent or divergent, and calculate their value if convergent. In each case show the area represented by the integral on a diagram.

④ $\int_1^\infty x^{-3}\, dx$ ⑤ $\int_0^\infty x^{-3}\, dx$ ⑥ $\int_0^3 \dfrac{1}{x^2}\, dx$ ⑦ $\int_2^\infty \dfrac{1}{x^2}\, dx$

⑧ $\int_{-\infty}^0 e^x\, dx$ ⑨ $\int_0^\infty e^x\, dx$ ⑩ $\int_0^{10} x\left(4 - x^2\right)^{-\frac{2}{3}}\, dx$ ⑪ $\int_0^\infty \left(e^{-2x} - e^{-x}\right) dx$

⑫ Evaluate $\int_0^\infty x e^{-x}\, dx$.

⑬ Evaluate $\int_0^\infty \dfrac{1}{x + 2} - \dfrac{1}{x + 1}\, dx$.

⑭ Evaluate $\int_{-1}^1 \left| x^{\frac{1}{3}} \right|\, dx$.

2 Calculus with inverse trigonometric functions

In this section you will see how the derivatives of the inverse trigonometric functions are very useful in integrating many functions even though they appear to be completely unrelated.

Prior knowledge

- You need to be able to differentiate and integrate trigonometric functions such as $\sin x$, $\cos x$ and $\tan x$.

- You should be familiar with the inverse trigonometric functions $\arcsin x$, $\arccos x$ and $\arctan x$ and their domains, ranges and graphs.

- You need to be able to differentiate functions defined implicitly.

> **Note**
>
> You will often see $\arcsin x$ written as $\sin^{-1} x$. They represent exactly the same function (the inverse sine function) but the second notation has the potential to be somewhat confusing when compared to, for example, $\sin^2 x$, which actually means $\left(\sin x\right)^2$. It is vital that you recognise that $\sin^{-1} x$ does NOT mean $\left(\sin x\right)^{-1}$ – which is actually $\dfrac{1}{\sin x}$ or $\operatorname{cosec} x$.

Differentiating inverse trigonometric functions

To differentiate the inverse trigonometric functions, you need to use implicit differentiation.

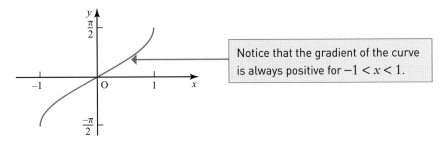

Notice that the gradient of the curve is always positive for $-1 < x < 1$.

Figure 6.6

$$y = \arcsin x$$

$$\sin y = x$$

$$\cos y \frac{dy}{dx} = 1$$

Differentiate implicitly with respect to x.

$$\frac{dy}{dx} = \frac{1}{\cos y}$$

$$= \frac{1}{\pm\sqrt{1 - \sin^2 y}}$$

$$= \frac{1}{\pm\sqrt{1 - x^2}}$$

But $y = \arcsin x$ has a range of $-\frac{\pi}{2} \leq y \leq \frac{\pi}{2}$, which implies that $\cos x \geq 0$, and then you can ignore the $\pm$ symbol since it must be positive in this case.

The conclusion is that:

$$\frac{d}{dx}(\arcsin x) = \frac{1}{\sqrt{1 - x^2}}$$

There are several things to notice with this result:

- it is positive, and only defined for $-1 < x < 1$
- it has a minimum at $x = 0$
- it tends to ∞ as $x \to \pm 1$

and these points are consistent with the graph of $y = \arcsin x$ in Figure 6.6.

ACTIVITY 6.2

Use a similar method to show that:

- $\frac{d}{dx}(\arccos x) = -\frac{1}{\sqrt{1 - x^2}}$
- $\frac{d}{dx}(\arctan x) = \frac{1}{1 + x^2}$

Integration using inverse trigonometric substitutions

You can use these results in integration. In practice it is the arcsin x and the arctan x results that are used since the arccos x one is just the negative of the arcsin x one.

$$\frac{d}{dx}(\arcsin x) = \frac{1}{\sqrt{1 - x^2}}$$

$$\frac{d}{dx}(\arctan x) = \frac{1}{1 + x^2}$$

From these results it becomes clear by integrating that:

$$\int \frac{1}{\sqrt{1 - x^2}}\, dx = \arcsin x + c$$

$$\int \frac{1}{1 + x^2}\, dx = \arctan x + c$$

ACTIVITY 6.3

Use the chain rule and the derivatives for arcsin x and arctan x given above, to show that:

- $\dfrac{d}{dx}\left(\arcsin\dfrac{x}{a}\right) = \dfrac{1}{\sqrt{a^2 - x^2}}$
- $\dfrac{d}{dx}\left(\arctan\dfrac{x}{a}\right) = \dfrac{a}{a^2 + x^2}$

The results in Activity 6.3 lead to the following results:

$$\int \frac{1}{\sqrt{a^2 - x^2}}\, dx = \arcsin\frac{x}{a} + c$$

$$\int \frac{1}{a^2 + x^2}\, dx = \frac{1}{a}\arctan\frac{x}{a} + c$$

These results can be quoted for use in integration.

Example 6.5

Calculate the value of the indefinite integral $\displaystyle\int \frac{1}{\sqrt{9 - x^2}}\, dx$.

Solution

$$\int \frac{1}{\sqrt{9 - x^2}}\, dx = \arcsin\frac{x}{3} + c$$

You can use the standard result with $a = 3$.

Notice that in the standard results, the coefficient of x^2 is 1. If you need to integrate a function of this form in which the coefficient of x^2 is not 1, then you need to first rewrite it in the standard form. This is shown in the Example 6.6.

Example 6.6

Find $\displaystyle\int \frac{1}{\sqrt{16 - 3x^2}}\, dx$.

Discussion point

→ In the example on the right, why can't you simply use the standard integral, using $a = 4$ and replacing x with $x\sqrt{3}$?

First factorise out the 3, which becomes $\sqrt{3}$ when it leaves the square root.

Solution

$$\int \frac{1}{\sqrt{16 - 3x^2}}\, dx = \frac{1}{\sqrt{3}}\int \frac{1}{\sqrt{\frac{16}{3} - x^2}}\, dx$$

This is now in the standard form, with $a = \dfrac{4}{\sqrt{3}}$.

$$= \frac{1}{\sqrt{3}}\arcsin\left(\frac{x\sqrt{3}}{4}\right) + c$$

The arctan x result can also be quoted for use in integration.

Example 6.7

Evaluate the definite integral:

$$\int_0^2 \frac{1}{4 + x^2}\, dx$$

Solution

$$\int_0^2 \frac{1}{4 + x^2}\, dx = \left[\frac{1}{2}\arctan\left(\frac{x}{2}\right)\right]_0^2 \qquad \boxed{\text{Using the standard result with } a = 2.}$$

$$= \frac{1}{2}\left(\arctan 1 - \arctan 0\right)$$

$$= \frac{\pi}{8}$$

Exercise 6.2

① State the domain and range of the arcsin, arccos and arctan functions.

② Use the standard results to evaluate the following indefinite integrals:

(i) $\displaystyle\int \frac{1}{\sqrt{25 - x^2}}\, dx$

(ii) $\displaystyle\int \frac{1}{16 + t^2}\, dt$

③ Use the standard results to evaluate the following definite integrals:

(i) $\displaystyle\int_0^1 \frac{1}{\sqrt{3 - s^2}}\, ds$

(ii) $\displaystyle\int_{-1}^1 \frac{1}{2^2 + x^2}\, dx$

④ Differentiate the following functions with respect to x:

(i) $\arcsin(3x)$

(ii) $\arccos\left(\frac{1}{2}x\right)$

(iii) $\arctan(5x)$

(iv) $\arcsin\left(3x^2\right)$

(v) $\arctan\left(e^x\right)$

(vi) $3\arctan\left(1 - x^2\right)$

⑤ Find the following indefinite integrals:

(i) $\displaystyle\int \frac{1}{4 + x^2}\, dx$

(ii) $\displaystyle\int \frac{1}{1 + 4x^2}\, dx$

(iii) $\displaystyle\int \frac{1}{\sqrt{4 - x^2}}\, dx$

(iv) $\displaystyle\int \frac{1}{\sqrt{1 - 4x^2}}\, dx$

⑥ Find the following indefinite integrals:

(i) $\displaystyle\int \frac{5}{x^2 + 36}\, dx$

(ii) $\displaystyle\int \frac{4}{25 + 4x^2}\, dx$

(iii) $\displaystyle\int \frac{1}{\sqrt{9 - 4x^2}}\, dx$

(iv) $\displaystyle\int \frac{7}{\sqrt{5 - 3x^2}}\, dx$

⑦ Evaluate the following definite integrals, leaving your answers in terms of π:

(i) $\displaystyle\int_0^3 \frac{1}{9 + x^2}\, dx$

(ii) $\displaystyle\int_0^{\sqrt{2}} \frac{1}{\sqrt{4 - x^2}}\, dx$

(iii) $\displaystyle\int_{-\frac{1}{\sqrt{3}}}^{\frac{1}{3}} \frac{1}{1 + 9x^2}\, dx$

(iv) $\displaystyle\int_0^{\frac{1}{4}} \frac{1}{\sqrt{1 - 4x^2}}\, dx$

⑧ The diagram below shows the curves $y = \dfrac{2}{1 + x^2}$ and $y = \dfrac{1}{\sqrt{4 - 3x^2}}$.

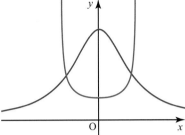

Figure 6.7

Show that the area between the curves is given by $\pi\left(1 - \dfrac{2}{3\sqrt{3}}\right)$.

⑨ Use implicit differentiation to prove:

(i) $\dfrac{\mathrm{d}}{\mathrm{d}x}\left(\arcsin\dfrac{x}{a}\right) = \dfrac{1}{\sqrt{a^2 - x^2}}$

(ii) $\dfrac{\mathrm{d}}{\mathrm{d}x}\left(\arctan\dfrac{x}{a}\right) = \dfrac{a}{a^2 + x^2}$

⑩ Differentiate the function $\mathrm{f}(x) = x\arcsin\left(x^2\right)$.

⑪ The graph below shows the curve $y = \dfrac{1}{1 + x^2}$.

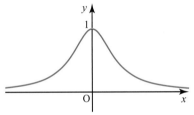

Figure 6.8

Find the total area under the curve.

⑫ Evaluate the following definite integral: $\displaystyle\int_{\sqrt{\frac{5}{6}}}^{\sqrt{\frac{5}{2}}} \dfrac{1}{5 + 2x^2}\,\mathrm{d}x$

3 Partial fractions

Partial fractions can often be used in integration.

Prior knowledge

You need to know how to find partial fractions of the following types:

$\dfrac{qx + r}{(ax + b)(cx + d)}$ can be written in the form $\dfrac{A}{ax + b} + \dfrac{B}{cx + d}$

$\dfrac{px^2 + qx + r}{(ax + b)(cx + d)(ex + f)}$ can be written in the form

$\dfrac{A}{ax + b} + \dfrac{B}{cx + d} + \dfrac{C}{ex + f}$

$\dfrac{px^2 + qx + r}{(ax + b)(cx + d)^2}$ can be written in the form

$\dfrac{A}{ax + b} + \dfrac{B}{cx + d} + \dfrac{C}{(cx + d)^2}$

Example 6.8

Find $\int \dfrac{x - 17}{(x + 1)(x - 5)}\,\mathrm{d}x$.

Solution

$$\frac{x - 17}{(x + 1)(x - 5)} = \frac{A}{x + 1} + \frac{B}{x - 5}$$

$$x - 17 = A(x - 5) + B(x + 1)$$

Let $x = 5 \implies -12 = 6B \implies B = -2$

Let $x = -1 \implies -18 = -6A \implies A = 3$

$$\int \frac{x - 17}{(x + 1)(x - 5)}\,\mathrm{d}x = \int \frac{3}{x + 1} - \frac{2}{x - 5}\,\mathrm{d}x$$

$$= 3\ln|x + 1| - 2\ln|x - 5| + c$$

Example 6.9

Find $\int \dfrac{25(4x + 1)}{(3x - 1)(2x + 1)^2}\,\mathrm{d}x$.

Solution

$$\frac{25(4x + 1)}{(3x - 1)(2x + 1)^2} = \frac{A}{3x - 1} + \frac{B}{2x + 1} + \frac{C}{(2x + 1)^2}$$

$$25(4x + 1) = A(2x + 1)^2 + B(3x - 1)(2x + 1) + C(3x - 1)$$

Let $x = \frac{1}{3} \implies 25 \times \frac{7}{3} = \frac{25}{9}A \implies A = 21$

Let $x = -\frac{1}{2} \implies -25 = -\frac{5}{2}C \implies C = 10$

Let $x = 0 \implies 25 = A - B - C \implies B = -14$

$$\int \frac{25(4x + 1)}{(3x - 1)(2x + 1)^2}\,\mathrm{d}x = \int \frac{21}{3x - 1} - \frac{14}{2x + 1} + \frac{10}{(2x + 1)^2}\,\mathrm{d}x$$

$$= 7\ln|3x - 1| - 7\ln|2x + 1| - \frac{5}{2x + 1} + c$$

The next example shows you how to extend your knowledge of partial fractions to include a quadratic expression in the denominator that cannot be factorised, and to integrate them.

An expression of the form $\dfrac{px^2 + qx + r}{(ax + b)(cx^2 + d)}$ can be written in the form $\dfrac{A}{ax + b} + \dfrac{Bx + C}{cx^2 + d}$.

Example 6.10

Find $\int \dfrac{x-2}{(x+1)\left(x^2+2\right)}\,dx$.

Solution

$$\dfrac{x-2}{(x+1)\left(x^2+2\right)} = \dfrac{A}{x+1} + \dfrac{Bx+C}{x^2+2}$$

$$x - 2 = A\left(x^2 + 2\right) + (Bx + C)(x + 1)$$

Let $x = -1 \implies -3 = 3A \implies A = -1$

Let $x = 0 \implies -2 = 2A + C \implies C = 0$

Equating coefficients of x^2: $0 = A + B \implies B = 1$

$$\int \dfrac{x-2}{(x+1)\left(x^2+2\right)}\,dx = \int \dfrac{-1}{x+1} + \dfrac{x}{x^2+2}\,dx$$

$$= -\ln|x+1| + \tfrac{1}{2}\ln\left|x^2+2\right| + c$$

$$= \ln\left|\dfrac{\sqrt{x^2+2}}{x+1}\right| + c$$

> You could substitute any value of x to find B, but equating coefficients is often easier.

> In the second term, the numerator, x, is half of the derivative of the denominator, so you can do this by inspection, or by using the substitution $u = x^2 + 2$.

In the example above, C turned out to be zero, which meant that each term could be integrated using methods you have met previously. In the next example, C is not zero.

Example 6.11

Find $\int \dfrac{9x-8}{(x+2)(x^2+9)}\,dx$.

Solution

> Write the function in partial fractions.

$$\dfrac{9x-8}{(x+2)(x^2+9)} = \dfrac{A}{x+2} + \dfrac{Bx+C}{x^2+9}$$

$$9x - 8 = A(x^2 + 9) + (Bx + C)(x + 2)$$

Let $x = -2 \implies -26 = 13A \implies A = -2$

Let $x = 0 \implies -8 = 9A + 2C \implies C = 5$

Equating coefficients of x^2: $0 = A + B \implies B = 2$

$$\int \dfrac{9x-8}{(x^2+9)(x+2)}\,dx = \int \left(\dfrac{2x+5}{x^2+9} - \dfrac{2}{x-2}\right)dx$$

$$= \int \left(\dfrac{2x}{x^2+9} + \dfrac{5}{x^2+9} - \dfrac{2}{x-2}\right)dx$$

$$= \ln(x^2+9) + \tfrac{5}{3}\arctan\tfrac{x}{3} - 2\ln|x-2| + c$$

$$= \ln\left(\dfrac{x^2+9}{(x-2)^2}\right) + \tfrac{5}{3}\arctan\tfrac{x}{3} + c$$

> The second term can be integrated using the standard arctan x result, with $a = 3$.

> The first term can be integrated by inspection, since the numerator is the derivative of the denominator.

① (i) Show that $\dfrac{3(5x+1)}{(x+1)(5x-1)} \equiv \dfrac{2}{x+1} + \dfrac{5}{5x-1}$.

(ii) Use this result to find $\displaystyle\int \dfrac{3(5x+1)}{(x+1)(5x-1)}\,dx$.

② (i) Show that $\dfrac{3x+4}{(2x+3)(x+1)^2} \equiv -\dfrac{2}{2x+3} + \dfrac{1}{x+1} + \dfrac{1}{(x+1)^2}$.

(ii) Use this result to find $\displaystyle\int \dfrac{3x+4}{(2x+3)(x+1)^2}\,dx$.

③ (i) Show that $\dfrac{2x^2 - 3x + 5}{(x-3)(x^2+5)} \equiv \dfrac{1}{x-3} + \dfrac{x}{x^2+5}$.

(ii) Use this result to find $\displaystyle\int \dfrac{2x^2 - 3x + 5}{(x-3)(x^2+5)}\,dx$.

④ (i) Write $\dfrac{17 - 5x}{(x+7)(x^2+3)}$ in the form $\dfrac{A}{x+7} + \dfrac{Bx+C}{x^2+3}$.

(ii) Find $\displaystyle\int \dfrac{17 - 5x}{(x+7)(x^2+3)}\,dx$.

⑤ Find the following integrals:

(i) $\displaystyle\int \dfrac{4}{(x^2+1)(x+1)}\,dx$ (ii) $\displaystyle\int \dfrac{x+11}{(x^2+9)(x-2)}\,dx$

(iii) $\displaystyle\int \dfrac{5x^2 + 3x + 3}{(4x^2+1)(x+2)}\,dx$

⑥ Evaluate the following definite integrals:

(i) $\displaystyle\int_0^1 \dfrac{x+3}{(x+1)(x^2+1)}\,dx$ (ii) $\displaystyle\int_2^5 \dfrac{2x^2+3}{(x-1)(x^2+4)}\,dx$

(iii) $\displaystyle\int_0^2 \dfrac{x-6}{(x+1)(3x^2+4)}\,dx$

⑦ The graph shows part of the curve $y = \dfrac{4-3x}{(x^2+4)(x+3)}$.

Find the exact area of the shaded region.

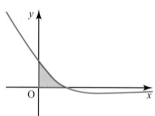

Figure 6.9

⑧ Express f(x) in a partial fraction form, where

$$f(x) = \dfrac{6 + 2x + 2x^2 - 2x^3}{(x+1)(x-1)^2(x^2+1)}.$$

Use this to show that: $\displaystyle\int f(x)\,dx = \ln\dfrac{(x^2+1)|x+1|}{|x-1|^3} - \dfrac{2}{x-1} + c$

⑨ Evaluate $\int_0^\infty \dfrac{1}{(x+1)^2(x+2)}\,dx$.

⑩ Figure 6.10 shows the start of an infinite sequence of rectangles of width 1 and height $\left(\dfrac{1}{x^2-1}\right)$ where the values of x are restricted to the integers from 2 upwards.

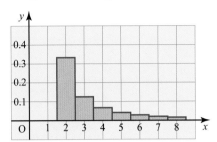

Figure 6.10

(i) Find the total area of the rectangles.

(ii) Show that the midpoints of the tops of the rectangles lie on the continuous curve $y = \dfrac{1}{x^2-1}$.

(iii) Find the exact area of the region bounded by the x-axis, the line $x = 4.5$ and the part of the curve $y = \dfrac{1}{x^2-1}$ for which $x \geqslant 1.5$.

(iv) Find the percentage error in using the area of the rectangles as an approximation for the area of the region described in part (iii).

4 Further integration

The previous section showed how differentiating the inverse trigonometric functions allows you to carry out new types of integration. These techniques can be extended to apply to less obvious integrals with a few algebraic tricks, as you will see in this section.

Example 6.12

Find $\int \dfrac{4}{x^2-2x+3}\,dx$. ◄

> The quadratic in the denominator does not factorise – but it can be manipulated slightly to end up looking like one of the standard forms previously covered.

Solution

> Complete the square on the denominator.

$$\int \frac{4}{x^2-2x+3}\,dx = 4\int \frac{1}{(x-1)^2+2}\,dx$$

> This now looks like the standard result for arctan x.

$$= 4\int \frac{1}{u^2+2}\,du \text{ where } u = x-1 \text{ and } dx = du$$

$$= 4 \times \frac{1}{\sqrt{2}}\arctan\left(\frac{u}{\sqrt{2}}\right) + c$$

$$= 2\sqrt{2}\arctan\left(\frac{x-1}{\sqrt{2}}\right) + c$$

Example 6.13

Find $\displaystyle\int \frac{5}{\sqrt{2 + 4x - 4x^2}}\, dx$.

Note

The clue as to how to proceed is that the denominator is the square root of a quadratic with negative coefficient of x^2, just as in the standard result:

$$\int \frac{1}{\sqrt{a^2 - x^2}}\, dx$$

$$= \arcsin\left(\frac{x}{a}\right) + c.$$

Solution

$$\int \frac{5}{\sqrt{2 + 4x - 4x^2}}\, dx = \frac{5}{2}\int \frac{1}{\sqrt{\frac{1}{2} - (x^2 - x)}}\, dx$$

Take out a factor of 5 from the numerator and 4 from 'inside' the square root in the denominator.

Complete the square, and adjust the constant (with care over the negative signs!).

$$= \frac{5}{2}\int \frac{1}{\sqrt{\frac{3}{4} - \left(x - \frac{1}{2}\right)^2}}\, dx$$

$$= \frac{5}{2}\int \frac{1}{\sqrt{\frac{3}{4} - u^2}}\, du \quad \text{where } u = x - \tfrac{1}{2} \text{ and } du = dx$$

$$= \frac{5}{2}\arcsin\left(\frac{2u}{\sqrt{3}}\right) + c$$

These two lines might be omitted with practice.

$$= \frac{5}{2}\arcsin\left(\frac{2x - 1}{\sqrt{3}}\right) + c$$

ACTIVITY 6.4

Try to use the method in the example above to find the following integrals:

(i) $\displaystyle\int \frac{1}{x^2 + 2x - 2}\, dx$

(ii) $\displaystyle\int \frac{1}{\sqrt{3 - 2x + x^2}}\, dx$

Explain why the method does not work.

How can you predict which integrals of this form can be done using this method, and which cannot?

Trigonometric substitutions

You have seen that functions of the form $\dfrac{1}{\sqrt{a^2 - x^2}}$ can be integrated to give an arcsin function. This gives a clue to integrating other functions that involve $\sqrt{a^2 - x^2}$. This expression might remind you that $\cos u = \sqrt{1 - \sin^2 u}$, and so a substitution of the form $x = a\sin u$ may be useful.

Example 6.14

Find $\displaystyle\int \frac{1}{(1 - x^2)^{\frac{3}{2}}}\, dx$.

Solution

Let $x = \sin u \ \Rightarrow \ \dfrac{dx}{du} = \cos u$

$$\int \frac{1}{(1 - x^2)^{\frac{3}{2}}}\, dx = \int \frac{1}{(1 - \sin^2 u)^{\frac{3}{2}}} \times \cos u\, du$$

Replace x with $\sin u$, and dx with $\cos u\, du$.

$$= \int \frac{1}{(\cos^2 u)^{\frac{3}{2}}} \times \cos u\, du$$

$\cos^2 x = 1 - \sin^2 x$

$$= \int \frac{1}{\cos^3 u} \times \cos u \, du$$

$$= \int \frac{1}{\cos^2 u} \, du$$

$$= \int \sec^2 u \, du \quad \longleftarrow \boxed{\sec^2 u \text{ is the derivative of } \tan u.}$$

$$= \tan u + c$$

$$= \frac{\sin u}{\cos u} + c$$

$$= \frac{x}{\sqrt{1 - x^2}} + c \quad \longleftarrow \boxed{\begin{array}{l}\text{Change back to the original variable}\\ \text{of } x \text{, using } \cos u = \sqrt{1 - \sin^2 u}.\end{array}}$$

Example 6.15

Evaluate $\int_0^2 \sqrt{16 - x^2} \, dx$.

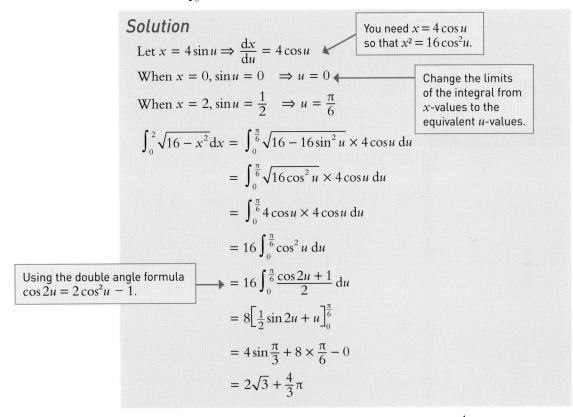

Solution

$$\boxed{\begin{array}{l}\text{You need } x = 4\cos u\\ \text{so that } x^2 = 16\cos^2 u.\end{array}}$$

Let $x = 4\sin u \Rightarrow \dfrac{dx}{du} = 4\cos u$

When $x = 0$, $\sin u = 0 \Rightarrow u = 0$

When $x = 2$, $\sin u = \dfrac{1}{2} \Rightarrow u = \dfrac{\pi}{6}$

$\boxed{\begin{array}{l}\text{Change the limits}\\ \text{of the integral from}\\ x\text{-values to the}\\ \text{equivalent } u\text{-values.}\end{array}}$

$$\int_0^2 \sqrt{16 - x^2} \, dx = \int_0^{\frac{\pi}{6}} \sqrt{16 - 16\sin^2 u} \times 4\cos u \, du$$

$$= \int_0^{\frac{\pi}{6}} \sqrt{16\cos^2 u} \times 4\cos u \, du$$

$$= \int_0^{\frac{\pi}{6}} 4\cos u \times 4\cos u \, du$$

$$= 16 \int_0^{\frac{\pi}{6}} \cos^2 u \, du$$

$\boxed{\begin{array}{l}\text{Using the double angle formula}\\ \cos 2u = 2\cos^2 u - 1.\end{array}} \longrightarrow$

$$= 16 \int_0^{\frac{\pi}{6}} \frac{\cos 2u + 1}{2} \, du$$

$$= 8\left[\frac{1}{2}\sin 2u + u\right]_0^{\frac{\pi}{6}}$$

$$= 4\sin\frac{\pi}{3} + 8 \times \frac{\pi}{6} - 0$$

$$= 2\sqrt{3} + \frac{4}{3}\pi$$

In a similar way, you have seen that functions of the form $\dfrac{1}{a^2 + x^2}$ can be integrated to give an arctan function. This gives a clue to finding other integrals that involve $a^2 + x^2$. This might remind you that $\sec^2 \theta = 1 + \tan^2 \theta$, and so a substitution of the form $x = a\tan u$ may be useful.

Example 6.16

Find $\int_0^1 \dfrac{1}{(1 + x^2)^{\frac{3}{2}}} \, dx$.

Solution

Let $x = \tan u \Rightarrow \dfrac{dx}{du} = \sec^2 u$

When $x = 0$, $\tan u = 0 \Rightarrow u = 0$

$\boxed{\begin{array}{l}\text{Change the limits of the integral from}\\ x\text{-values to the equivalent } u\text{-values.}\end{array}}$

When $x = 1, \tan u = 1 \Rightarrow u = \dfrac{\pi}{4}$

$$\int_0^1 \frac{1}{(1 + x^2)^{\frac{3}{2}}}\, dx = \int_0^{\frac{\pi}{4}} \frac{1}{(1 + \tan^2 u)^{\frac{3}{2}}} \times \sec^2 u\, du$$

$$= \int_0^{\frac{\pi}{4}} \frac{1}{(\sec^2 u)^{\frac{3}{2}}} \times \sec^2 u\, du \qquad \longleftarrow \boxed{\text{Using } \sec^2 u = 1 + \tan^2 u.}$$

$$= \int_0^{\frac{\pi}{4}} \frac{1}{\sec^3 u} \times \sec^2 u\, du$$

$$= \int_0^{\frac{\pi}{4}} \frac{1}{\sec u}\, du$$

$$= \int_0^{\frac{\pi}{4}} \cos u\, du$$

$$= \left[\sin u\right]_0^{\frac{\pi}{4}}$$

$$= \sin\frac{\pi}{4} - \sin 0$$

$$= \frac{1}{\sqrt{2}}$$

Exercise 6.4

① Using the substitution $x = 2\sin u$, find $\displaystyle\int_0^{\frac{1}{2}} \frac{1}{(4 - x^2)^{\frac{3}{2}}}\, dx$.

② Using the substitution $x = 3\sin u$, find $\displaystyle\int_0^3 \sqrt{9 - x^2}\, dx$.

③ Using the substitution $x = 2\tan u$, find $\displaystyle\int_0^2 \frac{1}{(4 + x^2)^{\frac{3}{2}}}\, dx$.

④ Find $\displaystyle\int \frac{3}{9x^2 + 6x + 5}\, dx$.

⑤ Find $\displaystyle\int \frac{1}{\sqrt{3 + 2x - x^2}}\, dx$.

⑥ (i) By writing $\arcsin x$ as $1 \times \arcsin x$, use integration by parts to find $\displaystyle\int \arcsin x\, dx$.

  (ii) Use a similar method to find the following integrals:

   (a) $\displaystyle\int \arccos x\, dx$ (b) $\displaystyle\int \arctan x\, dx$ (c) $\displaystyle\int \text{arccot}\, x\, dx$

⑦ Use a suitable substitution to evaluate:

  (i) $\displaystyle\int_0^2 \frac{1}{(16 - x^2)^{\frac{3}{2}}}\, dx$ (ii) $\displaystyle\int_{-\frac{1}{2}}^{\frac{1}{2}} \frac{1}{(1 + 4x^2)^{\frac{3}{2}}}\, dx$ (iii) $\displaystyle\int_0^{\frac{2}{5}} \sqrt{4 - 25x^2}\, dx$

⑧ (i) Find $\displaystyle\int_0^b \sqrt{a^2 - x^2}\, dx$

   where $a > b > 0$.

  (ii) Draw a sketch to show the significance of the area you calculated in part (i), and explain both terms of your answer to part (i) geometrically.

⑨ (i) Use the substitution $x = a\sin u$ to prove the result
$$\int \frac{1}{\sqrt{a^2 - x^2}}\, dx = \arcsin\frac{x}{a} + c$$

(ii) Use the substitution $x = a\tan u$ to prove the result
$$\int \frac{1}{a^2 + x^2}\, dx = \frac{1}{a}\arctan\frac{x}{a} + c \longleftarrow$$

You have already proved these results in Question 9 in Exercise 6.2, by starting from differentiation of $\arcsin\dfrac{x}{a}$ and $\arctan\dfrac{x}{a}$.

⑩ Find the following integrals:

(i) $\displaystyle\int \frac{x+1}{x^2+1}\, dx$

(ii) $\displaystyle\int \frac{x+1}{x^2+2x+3}\, dx$

(iii) $\displaystyle\int \frac{1}{x^2+2x+3}\, dx$

⑪ Find the following integrals:

(i) $\displaystyle\int \frac{2-x}{\sqrt{4-x^2}}\, dx$ 　　　　 (ii) $\displaystyle\int \frac{1}{\sqrt{4x-x^2}}\, dx$

(iii) $\displaystyle\int \frac{2-x}{\sqrt{4x-x^2}}\, dx$

⑫ Find $\dfrac{d}{dx}\left(\operatorname{arcsec} x\right)$ and $\displaystyle\int \frac{1}{x\sqrt{x^2-a^2}}\, dx$.

⑬ By considering the equation $y = \sqrt{r^2 - x^2}$ and integrating between 0 and r, prove that the area of a circle, radius r, is πr^2.

⑭ Given that
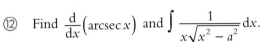
$$\int \sec^3 x\, dx = \frac{1}{2}\ln\left(\frac{\cos\left(\frac{x}{2}\right) + \sin\left(\frac{x}{2}\right)}{\cos\left(\frac{x}{2}\right) - \sin\left(\frac{x}{2}\right)}\right) + \frac{1}{2}\sec x \tan x$$

evaluate the following definite integral by performing an arctan x substitution, giving your answer to 4 d.p.
$$\int_0^1 \sqrt{1 + x^2}\, dx$$

● LEARNING OUTCOMES

Now you have finished this chapter, you should be able to:

➤ identify why a given integral is improper

➤ evaluate improper integrals where either the integrand is undefined at a value in the interval of integration or the interval of integration extends to infinity

➤ use the method of partial fractions in integration, including where the denominator has a quadratic factor of form $ax^2 + c$ and one linear term

➤ differentiate inverse trigonometric functions

➤ recognise integrals of functions of the form $\left(a^2 - x^2\right)^{-\frac{1}{2}}$ and $\left(a^2 + x^2\right)^{-1}$

and be able to integrate related functions by using trigonometric substitutions.

KEY POINTS

An improper integral is an integral in which either:

■ at least one of the limits is infinity; or

■ the function to be integrated approaches infinity at some point in the interval of integration.

1 Improper integrals involving a limit of infinity may be investigated by replacing the problem limit by the constant a and then finding, if possible, the limit of the value of the integral as $a \rightarrow \infty$. The value of the integral may be finite, in which case the integral is convergent, or it may be divergent, in which case it cannot be evaluated.

2 Improper integrals in which the functions to be integrated approach infinity at some point in the interval of integration, may be investigated by splitting the integral into two at the problem point (if the problem point is not one of the end points), and replacing the problem value with the constant a, and then finding, if possible, the limit of the value of the integral as $a \rightarrow \infty$.

3 $\displaystyle\int \frac{1}{\sqrt{a^2 - x^2}}\, dx = \arcsin \frac{x}{a} + c$

4 $\displaystyle\int \frac{1}{a^2 + x^2}\, dx = \frac{1}{a} \arctan \frac{x}{a} + c$

5 Integrals of the form $\displaystyle\int \frac{px^2 + qx + r}{(a + bx^2)(cx + d)}\, dx$ can be found by first splitting the function into partial fractions of the form $\dfrac{Ax + B}{a + bx^2} + \dfrac{C}{cx + d}$.

6 Integrals that involve functions of the form $\sqrt{a^2 - x^2}$ may often be integrated using the substitution $x = a \sin u$.

7 Integrals that involve functions of the form $a^2 + x^2$ may often be integrated using the substitution $x = a \tan u$.

FUTURE USES

■ You will meet some other integrals similar to those covered in this chapter in the Hyperbolic functions chapter, and you will use similar techniques there.

■ You will use many of the integration techniques covered in this chapter in the chapters Further integration, First order differential equations and Second order differential equations.

7 Polar coordinates

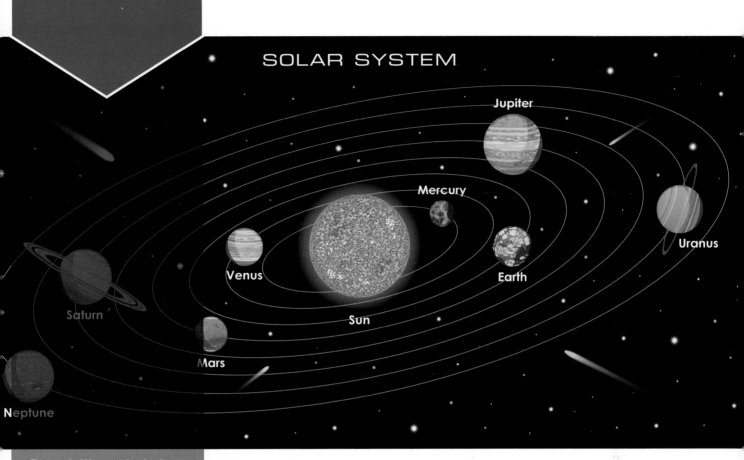

SOLAR SYSTEM

Round, like a circle in a spiral

Like a wheel within a wheel

Never ending or beginning

On an ever-spinning reel

'The Windmills of Your Mind'

Alan and Marilyn Bergman

Discussion point

➔ How can polar coordinates be used to track planets as they orbit the sun?

Review: Polar coordinates

Polar coordinates define a point in the plane using its distance from a fixed point O, called the **pole**, and the angle the line joining the point to the pole makes with the **initial line**.

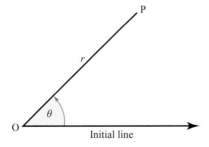

Figure 7.1

Using the notation in Figure 7.1, the coordinates of the point, P, are written (r, θ) where θ is measured in radians in an anti-clockwise direction from the initial line. The initial line is a **half line** drawn from O, horizontally, to the right.

Figure 7.2 shows the point $\left(3, \frac{\pi}{6}\right)$. It is 3 units from the pole, O, and the angle of the line joining it to the pole is $\frac{\pi}{6}$.

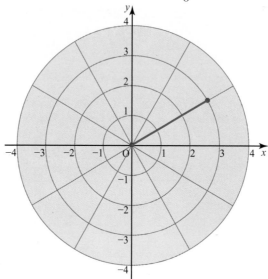

Figure 7.2

> **Note**
>
> The point $\left(-3, \frac{7\pi}{6}\right)$
>
> also coincides with $\left(3, \frac{\pi}{6}\right)$. The negative value for r is interpreted as going in the reverse direction from the pole.

Notice that the points $\left(3, \frac{13\pi}{6}\right)$ and $\left(3, \frac{25\pi}{6}\right)$ would coincide with the point in Figure 7.2 as well as infinitely many other points. It is usual to use the **principal polar coordinates**, where $r > 0$ and $-\pi < \theta \leqslant \pi$ to specify the polar coordinates of a point.

Converting between polar and Cartesian coordinates

This is a straightforward application of trigonometry in a right-angled triangle.

$$x = r\cos\theta \qquad y = r\sin\theta \qquad r = \sqrt{x^2 + y^2} \qquad \tan\theta = \frac{y}{x}$$

Be careful to check the quadrant you are in so that the value of θ is the correct one. Draw a sketch to make sure.

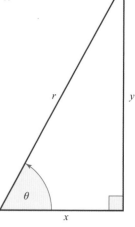

Figure 7.3

Example 7.1

(i) Find the Cartesian coordinates of $\left(5, \dfrac{11\pi}{6}\right)$.

(ii) Find the polar coordinates of $\left(-1, \sqrt{3}\right)$.

Solution

First draw a diagram to represent the coordinates of the point:

(i) $\left(5, \dfrac{11\pi}{6}\right)$

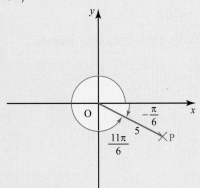

Figure 7.4

$$x = 5\cos\frac{\pi}{6} = \frac{5\sqrt{3}}{2}$$

$$y = -5\sin\frac{\pi}{6} = -\frac{5}{2}$$

So $\left(5, \dfrac{11\pi}{6}\right)$ has Cartesian coordinates $\left(\dfrac{5\sqrt{3}}{2}, -\dfrac{5}{2}\right)$.

(ii) $\left(-1, \sqrt{3}\right)$

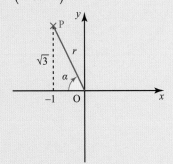

Figure 7.5

$$r = \sqrt{1^2 + \sqrt{3}^2} = 2$$

$$\tan\alpha = \frac{\sqrt{3}}{1} \Rightarrow \alpha = \frac{\pi}{3} \quad \text{so} \quad \theta = \frac{2\pi}{3}$$

So $\left(-1, \sqrt{3}\right)$ has polar coordinates $\left(2, \dfrac{2\pi}{3}\right)$.

Sketching curves with polar equations

Polar equations look different from Cartesian equations for curves.

They are expressed in the form $r = f(\theta)$. Example 7.2 shows three ways to produce the curve for the equation $r = f(\theta)$.

Example 7.2

Draw the curve $r = 6\sin\theta$.

Solution

Method 1 – Plotting points

Start with a table of values – this table uses values of θ that increase in intervals of $\frac{\pi}{12}$.

θ	0	$\frac{\pi}{12}$	$\frac{\pi}{6}$	$\frac{\pi}{4}$	$\frac{\pi}{3}$	$\frac{5\pi}{12}$	$\frac{\pi}{2}$	$\frac{7\pi}{12}$	$\frac{2\pi}{3}$	$\frac{3\pi}{4}$	$\frac{5\pi}{6}$	$\frac{11\pi}{12}$	π
r	0	1.6	3	4.2	5.2	5.8	6	5.8	5.2	4.2	3	1.6	0

Table 7.1

Plotting the points gives the curve shown in Figure 7.6. It is a circle as will be confirmed by the other two methods.

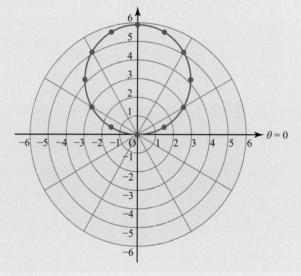

Figure 7.6

Method 2 – Convert to Cartesian form

If $r \neq 0$ then

$r = 6\sin\theta$

$\Leftrightarrow r^2 = 6r\sin\theta$ ← Multiply both sides by r.

$\Leftrightarrow x^2 + y^2 = 6y$ ← Use the conversion equations to substitute, giving the equation of a circle radius 3, centre $(0, 3)$.

Method 3 – Using geometric reasoning

This involves working backwards from the answer.

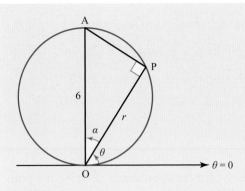

Figure 7.7

In the triangle OAP, $r = 6\cos\alpha$ and $\alpha = \dfrac{\pi}{2} - \theta$.

So $r = 6\cos\left(\dfrac{\pi}{2} - \theta\right) = 6\sin\theta$.

It is known that:

$r = a$ is a circle with centre at the origin and radius a.

$\theta = k$, $(-\pi < \theta \leq \pi)$ is a half line from the origin making an angle of k with the initial line.

Example 7.3

Sketch the curve $r = 5\cos3\theta$ for $0 < \theta \leq \pi$, identifying the range of values of θ that generate each loop.

Note

It is most useful to plot points in this case. Unravelling the multiple angles to convert to Cartesian form and reasoning geometrically are unlikely to produce the required answers.

Solution

Using values of θ from 0 to 2π, in increments of $\dfrac{\pi}{12}$ as this will generate an appropriate number of points, produces Table 7.2.

θ	0	$\dfrac{\pi}{12}$	$\dfrac{\pi}{6}$	$\dfrac{\pi}{4}$	$\dfrac{\pi}{3}$	$\dfrac{5\pi}{12}$	$\dfrac{\pi}{2}$	$\dfrac{7\pi}{12}$	$\dfrac{2\pi}{3}$	$\dfrac{3\pi}{4}$	$\dfrac{5\pi}{6}$	$\dfrac{11\pi}{12}$	π
r	5	3.5	0	−3.5	−5	−3.5	0	3.5	5	3.5	0	−3.5	−5

Table 7.2

There are several negative values for r so care should be taken when plotting them.

Here $\theta = \dfrac{7\pi}{12}, \dfrac{2\pi}{3}$ and $\dfrac{3\pi}{4}$.

Here $\theta = 0, \dfrac{\pi}{12}$ and $\dfrac{11\pi}{12}$.

Here $\theta = \dfrac{\pi}{6}, \dfrac{\pi}{2}$ and $\dfrac{5\pi}{6}$.

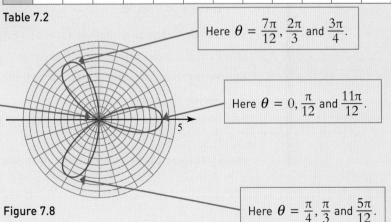

Figure 7.8

Here $\theta = \dfrac{\pi}{4}, \dfrac{\pi}{3}$ and $\dfrac{5\pi}{12}$.

So the right hand loop is generated by $0 < \theta < \frac{\pi}{6}$ and $\frac{5\pi}{6} < \theta < \pi$.

The top loop is generated by $\frac{\pi}{2} < \theta < \frac{5\pi}{6}$.

The bottom loop is generated by $\frac{\pi}{6} < \theta < \frac{\pi}{2}$.

Review exercise

① Which of these coordinates describe the point $\left(2, \frac{2\pi}{3}\right)$?

$\left(2, \frac{4\pi}{3}\right)$ $\quad$ $\left(2, \frac{8\pi}{3}\right)$ $\quad$ $\left(-2, \frac{5\pi}{3}\right)$ $\quad$ $\left(-2, \frac{\pi}{3}\right)$ $\quad$ $\left(2, \frac{14\pi}{3}\right)$ $\quad$ $\left(2, \frac{11\pi}{3}\right)$

② (i) Write the polar coordinate $\left(6, \frac{7\pi}{6}\right)$ in Cartesian form.

$\quad$ (ii) Write the Cartesian coordinate $(5, -5)$ in polar form.

③ What is the polar equation of a circle, centre $\left(10, \frac{\pi}{2}\right)$?

④ The circles $r = 4\sin\theta$ and $r = 3\cos\theta$ intersect at the points A and B. Find the equation of the line AB.

⑤ Sketch the curve $r = 5\cos\theta + 1$ for $-\pi \leqslant \theta \leqslant \pi$.

⑥ The equation of a polar curve is given by $r = 2 + 2\cos 3\theta$ for $-\pi \leqslant \theta \leqslant \pi$.

$\quad$ (i) Verify that the point A with polar coordinates $\left(2, \frac{\pi}{6}\right)$ lies on the curve.

$\quad$ (ii) Find the point B on the curve where $\theta = -\frac{\pi}{6}$.

$\quad$ (iii) Find the angles of the triangle AOB where O is the pole.

$\quad$ (iv) What property does triangle AOB possess?

⑦ The equation of a polar curve is given by $r = 4\sin 3\theta$ for $-\pi \leqslant \theta \leqslant \pi$.

$\quad$ (i) Sketch the curve.

$\quad$ (ii) Find the equation of the normal, in polar form, to the curve at the point where $\theta = \frac{\pi}{3}$.

$\quad$ (iii) Find the equation of the tangent, in polar form, to the curve at the point where $\theta = \frac{\pi}{2}$.

$\quad$ (iv) Find the equation, in polar form, of the smallest circle that encloses the curve $r = 4\sin 3\theta$.

⑧ A circle has Cartesian equation $(x + 1)^2 + (y - 2)^2 = 5$.

$\quad$ (i) Find the polar equation of the circle.

$\quad$ (ii) Verify that the circle passes through the pole, O.

$\quad$ (iii) Sketch the circle.

$\quad$ (iv) Find the polar equation of the point A such that OA is a diameter of the circle.

$\quad$ (v) Find the equation of the tangent to the circle at A in polar form.

1 Finding the area enclosed by a polar curve

Prior knowledge

You need to be able to integrate polynomial functions and trigonometric functions of the form $a \sin bx$ and $a \cos bx$.

Look at the region in Figure 7.9 bounded by the lines OU and OV and the curve UV. To find the area of this region, start by dividing it up into smaller regions OPQ. Let OU and OV have angles $\theta = \alpha$ and $\theta = \beta$ respectively.

If the curve has equation $r = \mathrm{f}(\theta)$, P and Q have coordinates (r, θ) and $(r + \delta r, \theta + \delta \theta)$.

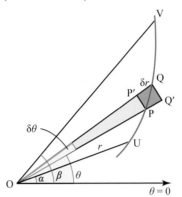

Figure 7.9

Let the area of OUV be A and the area of OPQ be δA.

The area δA lies between the circular sectors OPP′ and OQQ′, so:

$$\tfrac{1}{2} r^2 \delta \theta < \delta A < \tfrac{1}{2}(r + \delta r)^2 \delta \theta$$

> Remember that the area of a sector of a circle is given by $\frac{1}{2} r^2 \theta$, where θ is in radians.

therefore

$$\tfrac{1}{2} r^2 < \frac{\delta A}{\delta \theta} < \tfrac{1}{2}(r + \delta r)^2$$

As $\delta \theta \to 0$, $\delta r \to 0$ and so $\tfrac{1}{2}(r + \delta r)^2 \to \tfrac{1}{2} r^2$. Therefore $\dfrac{\delta A}{\delta \theta}$ must also tend to $\tfrac{1}{2} r^2$ as $\delta \theta \to 0$.

But as $\delta \theta \to 0$, $\dfrac{\delta A}{\delta \theta} \to \dfrac{\mathrm{d}A}{\mathrm{d}\theta}$

Therefore $\dfrac{\mathrm{d}A}{\mathrm{d}\theta} = \tfrac{1}{2} r^2$.

Integrating both sides with respect to θ shows the result for the area of a region bounded by a polar curve and two straight lines $\theta = \alpha$ and $\theta = \beta$ is:

$$A = \int_\alpha^\beta \tfrac{1}{2} r^2 \, \mathrm{d}\theta$$

Example 7.4

Figure 7.10 shows the curve $r = 1 + 2\cos\theta$.

Find the area of the inner loop of the limaçon $r = 1 + 2\cos\theta$.

> **Note**
>
> Pointing key points of the curve helps identify what values of θ generate the inner loop

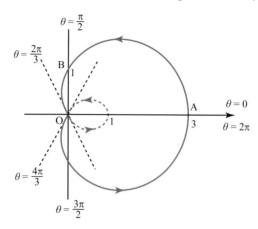

Figure 7.10

Solution

It can be shown that the inner loop is formed as θ varies from $\dfrac{2\pi}{3}$ to $\dfrac{4\pi}{3}$ so its area is given by

$$A = \int_{\frac{2\pi}{3}}^{\frac{4\pi}{3}} \frac{1}{2}r^2\,\mathrm{d}\theta = \int_{\frac{2\pi}{3}}^{\frac{4\pi}{3}} \frac{1}{2}(1 + 2\cos\theta)^2\,\mathrm{d}\theta$$

$$\Rightarrow A = \int_{\frac{2\pi}{3}}^{\frac{4\pi}{3}} \frac{1}{2}\left(1 + 4\cos\theta + 4\cos^2\theta\right)\mathrm{d}\theta \longleftarrow$$

> Using $\cos^2\theta = \frac{1}{2}(1 + \cos 2\theta)$.
>
> Note that even though the value of r is negative between $\dfrac{2\pi}{3}$ and $\dfrac{4\pi}{3}$ the integrand $\frac{1}{2}r^2$ is always positive so there is no issue of 'negative areas' as there is when curves go below the x-axis when using Cartesian coordinates.

$$= \left[\frac{3\theta}{2} + 2\sin\theta + \frac{1}{2}\sin 2\theta\right]_{\frac{2\pi}{3}}^{\frac{4\pi}{3}}$$

$$= \pi - \frac{3\sqrt{3}}{2}$$

Exercise 7.1

① (i) Check that the integral $\int \frac{1}{2}r^2\,\mathrm{d}\theta$ correctly gives the area of the circle $r = 10\cos\theta$ when it is evaluated from $-\dfrac{\pi}{2}$ to $\dfrac{\pi}{2}$.

   (ii) What happens when the integral is evaluated from 0 to 2π?

② A curve has equation $r = 5\cos 4\theta$.

   (i) Sketch the curve for the interval $0 \leqslant \theta \leqslant 2\pi$.

   (ii) Find the area of one loop of the curve.

③ A curve has equation $r = 3 + 3\sin\theta$.

   (i) Sketch the curve for the interval 0 to 2π.

   (ii) Find the area enclosed by the curve.

④ Find the area bounded by the spiral $r = \dfrac{4\theta}{\pi}$ from $\theta = 0$ to $\theta = 2\pi$ and the initial line.

⑤ For the limaçon $r = 1 + 2\cos\theta$ in Example 7.4, find:

   (i)   the total area contained inside the outer loop

   (ii)   the area between the two loops

⑥ Find the exact areas of the two portions into which the line $\theta = \dfrac{\pi}{2}$ divides the upper half of the cardioid $r = 8(1 + \cos\theta)$.

⑦ Sketch the lemniscate $r^2 = a^2 \cos 2\theta$ and find the area of one of its loops.

⑧ The diagram shows the **equiangular spiral** $r = a\mathrm{e}^{k\theta}$ where a and k are positive constants and e is the exponential constant 2.71828…

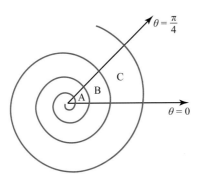

Figure 7.11

Prove that the areas A, B and C formed by the lines $\theta = 0$ and $\theta = \dfrac{\pi}{4}$ and the spiral form a geometric sequence and find its common ratio.

⑨ Find the area enclosed between the curves $r = 3 - 3\cos\theta$ and $r = 4\cos\theta$. Give your final answer to three significant figures.

LEARNING OUTCOMES

Now you have finished this chapter, you should be able to:

➤ understand and use polar coordinates

➤ convert from polar to Cartesian coordinates and vice versa

➤ sketch curves with simple polar equations in the form $r = f(\theta)$

➤ find the area enclosed by a polar curve.

KEY POINTS

1 To convert from polar coordinates to Cartesian coordinates $x = r\cos\theta$, $y = r\sin\theta$.

2 To convert from Cartesian coordinates to polar coordinates $r = \sqrt{x^2 + y^2}$, $\theta = \arctan\dfrac{y}{x}$ ($\pm\pi$ if necessary).

3 The principal polar coordinates $\left(r, \theta\right)$ are those for which $r > 0$ and $-\pi < \theta \leqslant \pi$.

4 The area of a sector is $\displaystyle\int_{\alpha}^{\beta} \frac{1}{2} r^2 \, \mathrm{d}\theta$.

Series and limits

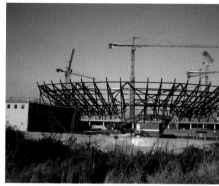

'If I feel unhappy, I do mathematics to become happy. If I feel happy, I do mathematics to keep happy.'

Alfred Renyi, 1921–1970

The four photographs above were taken over a period of time. They build up to the final picture which shows the complete situation.

In this chapter you will meet a similar idea building up a polynomial series to represent a function. Instead of showing how the Olympic Stadium developed over time as new parts were added to it, you will be expressing a function as a series of ever increasing accuracy by adding on successive terms of a polynomial.

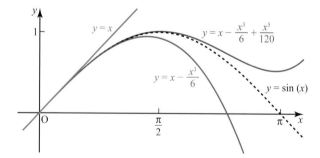

Figure 8.1

Discussion point

➔ Look at the diagram above. Can $\sin x$ be represented as a polynomial for all values of x?

1 Polynomial approximations and Maclaurin series

Since polynomial functions are easy to evaluate, differentiate and integrate (among other things), they are often useful approximations to more complicated functions. Here you will see one way of building these approximations, starting with the example of the exponential function: $f(x) = e^x$.

To build up a polynomial function $p(x)$ which approximates $f(x)$, start by making sure it has the correct value at $x = 0$ (where the graph cuts the y-axis).

Since $e^0 = 1$, the first term of the polynomial approximation is 1, and so $p(x) = 1 + \ldots$ (see Figure 8.2).

You can certainly find a better approximation than this. The next step is to consider the gradient of your approximation. The derivative of e^x is e^x, and so at $x = 0$ the gradient is 1.

The next term of $p(x)$ will be a multiple of x and, since its derivative is 1, this will be x itself. So $p(x) = 1 + x + \ldots$ (Figure 8.3).

This is a better approximation, but, again, can be improved further. The next step is to ensure that the second derivatives of e^x and $p(x)$ have the same values when $x = 0$. The second derivative of $f(x) = e^x$ at $x = 0$ is also 1. The next term of $p(x)$ will be a multiple of x^2. Since the second derivative of x^2 is 2, the quadratic term must be $\frac{1}{2}x^2$ to give a second derivative of 1. So $p(x) = 1 + x + \frac{1}{2}x^2 + \ldots$, and the graph of this function is a much better approximation to $y = f(x)$ (Figure 8.4).

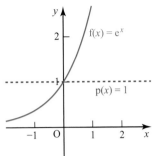

Figure 8.2

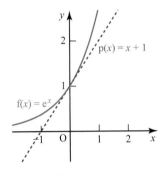

Figure 8.3

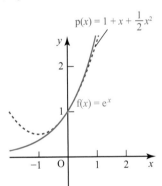

Figure 8.4

ACTIVITY 8.1

(i) Extend this method one more step and show that the cubic approximation to $f(x) = e^x$ is
$$p(x) = 1 + x + \frac{1}{2}x^2 + \frac{1}{6}x^3$$

(ii) Extend the method two further steps to find a degree 5 polynomial approximation for $f(x)$.

You will see from the graph below (Figure 8.5) that, for positive values of x around $x = 0$, the accuracy of the approximation improves as you add more terms. This is also true for negative x-values, although it is not so clear from the diagram.

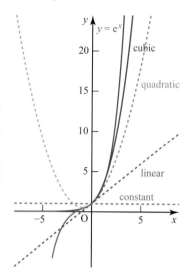

ACTIVITY 8.2

Write down the cubic approximation to e^x and substitute $(-x)$ in place of x to generate a cubic approximation to e^{-x}.

Multiply these two cubics together and comment on your answer.

Figure 8.5

Example 8.1

(i) In the polynomial approximation to e^x, find an expression for the term in x^r.

(ii) Hence express e^x as the sum of an infinite series.

(iii) Investigate whether the terms of the series converge for all values of x.

Solution

(i) Let the term in x^r be kx^r.

Differentiating this r times gives $kr \times (r - 1) \times (r - 2) \times .... \times 2 \times 1 = kr!$
The rth derivative of e^x at $x = 0$ is 1, so $kr! = 1$ and therefore $k = \dfrac{1}{r!}$.
So the term in x^r is $\dfrac{x^r}{r!}$.

(ii) $p(x) = \displaystyle\sum_{r=0}^{\infty} \dfrac{x^r}{r!}$

(iii) For $x = 2$, $p(x) = 1 + 2 + \dfrac{2^2}{2!} + \dfrac{2^3}{3!} + \dfrac{2^4}{4!} + ...$

In the 5th term, 4! is greater than 2^4 so this term is less than 1. After this, the terms continue to get smaller, so it appears that the terms converge.

If you take any value of x, at some point the value of $r!$ will be greater than x^r, so after that, the terms are less than 1 and will continue to get smaller. So it seems that the terms converge for all values of x.

Discussion point

➜ Is it always the case that if the terms of a series converge, the sum of the series also converges?

In Example 8.1 you saw that the terms of the series for e^x converge, whatever the value of x. In fact, the sum of the terms also converges for all values of x. The formal proof of the convergence is beyond the scope of this book.

ACTIVITY 8.3

This spreadsheet shows the start of a method for approximating e^x.

	A	**B**
1	0.5	=SUM(B3:B10)
2		
3	0	1
4	=A3+1	=A1*B3/A4

Copy these cells into your own spreadsheet. Copy and paste the formulae in Row 4 down to Row 10. Check the formula does indeed behave as explained above.

The number in cell A1 is the value of x, try changing it and watch the effect on the other cells. In particular, explain why the value in cell B1 is an approximation to e^x, and state the order of the polynomial approximation this example achieves.

Change your spreadsheet so that it gives a polynomial approximation up to the term in x^{10}. Use it to calculate the value, and the percentage error, of the Maclaurin approximation up to the term in x^{10}, for the calculation e^2.

> **Note**
>
> Notice the $ signs in the formula in cell B4: these indicate an **absolute reference**, meaning this part of the formula will always refer to cell A1, whereas the other parts of the formula are **relative references**, meaning they will change relative to where the formula is copied.

Maclaurin approximations and series

In general, you can find a polynomial $p(x)$, of order n, for any function $f(x)$, for which its first n derivatives at $x = 0$ exist.

To ensure that $p(x)$ takes the same value as $f(x)$ for each derivative, at $x = 0$, then you need:

$$p(0) = f(0)$$
$$p'(0) = f'(0)$$
$$p''(0) = f''(0)$$
$$p^{(3)}(0) = f^{(3)}(0)$$

> A third derivative $f'''(x)$ can be written as $f^{(3)}(x)$, and similarly for higher derivatives.

$$\ldots$$
$$p^{(n)}(0) = f^{(n)}(0)$$

This is a list of $n + 1$ conditions, and you can see the general nth order polynomial has $n + 1$ constants to determine:

$$p(x) = a_0 + a_1 x + a_2 x^2 + a_3 x^3 + \ldots + a_r x^r + \ldots + a_n x^n$$

Substituting in $x = 0$ immediately gives:

$$a_0 = f(0)$$

Doing this with the derivatives of p gives the various values of a_r:

$$p'(x) = a_1 + 2a_2 x + 3a_3 x^2 + \ldots + ra_r x^{r-1} + \ldots + na_n x^{n-1}$$

so

$$a_1 = f'(0)$$

Then

$$p''(x) = 2a_2 + 6a_3 x + \ldots + r(r-1)a_r x^{r-2} + \ldots + n(n-1)a_n x^{n-2}$$
$$2a_2 = f''(0)$$
$$a_2 = \frac{1}{2}f''(0)$$

and

$$p^{(3)}(0) = 6a_3 + \ldots + r(r-1)(r-2)a_r x^{r-3} + \ldots + n(n-1)(n-2)a_n x^{n-3}$$
$$6a_3 = f^{(3)}(0)$$
$$a_3 = \frac{1}{6}f^{(3)}(0)$$

In general, the nth derivative gives the condition

$$n!a_n = f^{(n)}(0)$$

so

$$a_n = \frac{1}{n!}f^{(n)}(0)$$

Putting all these back into the $p(x)$ definition to find the approximation for $f(x)$ gives:

$$f(x) \approx f(0) + f'(0)x + \frac{f''(0)}{2!}x^2 + \frac{f^{(3)}(0)}{3!}x^3 + \ldots + \frac{f^{(r)}(0)}{r!}x^r + \ldots + \frac{f^{(n)}(0)}{n!}x^n.$$

This is the **Maclaurin approximation** or **Maclaurin expansion** for $f(x)$ up to the term in x^n.

The accuracy of the approximation is usually defined as follows:

$$\text{error} = \text{approximate value} - \text{exact value}$$

$$\text{percentage error} = \frac{\text{approximate value} - \text{exact value}}{\text{exact value}} \times 100\%$$

Example 8.2	Find the Maclaurin expansion for $(1-x)^{-1}$, as far as x^n.

Solution

Let $f(x) = (1-x)^{-1}$.

$f(x) = (1-x)^{-1}$	$f(0) = 1$	
$f'(x) = (1-x)^{-2}$	$f'(0) = 1$	
$f''(x) = 2(1-x)^{-3}$	$f''(0) = 2$	
$f^{(3)}(x) = 6(1-x)^{-4}$	$f^{(3)}(0) = 6$	
$f^{(4)}(x) = 24(1-x)^{-5}$	$f^{(4)}(0) = 24$	
$\vdots$	$\vdots$	
$f^{(n)}(x) = n!(1-x)^{-(n+1)}$	$f^{(n)}(0) = n!$	

It is useful to tabulate the function $f(x)$ and its derivatives, then evaluate them at $x = 0$.

Table 8.1

So

$$(1-x)^{-1} \approx 1 + x + \frac{2}{2!}x^2 + \frac{6}{3!}x^3 + \frac{24}{4!}x^4 + \ldots + \frac{n!}{n!}x^n$$

$$= 1 + x + x^2 + x^3 + x^4 + \ldots + x^n$$

ACTIVITY 8.4

Compare the result from Example 8.2 with the binomial expansion for $(1-x)^{-1}$ (see Chapter 7 in AQA A-level Mathematics Year 2) and the sum to infinity of the geometric series $1 + x + x^2 + x^3 + \ldots$

Note

A Maclaurin *expansion* for a function involves a finite number of terms and is an approximation to the function. A Maclaurin *series* is the sum of an infinite number of terms.

If the function and *all* its derivatives exist at $x = 0$ then, of course, this expansion could be continued indefinitely, in which case you would get an infinite order polynomial:

$$f(x) = f(0) + f'(0)x + \frac{f''(0)}{2!}x^2 + \frac{f^{(3)}(0)}{3!}x^3 + \ldots + \frac{f^{(r)}(0)}{r!}x^r + \ldots$$

This is known as the **Maclaurin series** for $f(x)$. Care must be taken with infinite polynomials like this, but if the sum of the series up to and including the term in x^n tends to a limit as n tends to infinity, and this limit is $f(x)$, then you can say that the expansion **converges** to $f(x)$.

Validity of Maclaurin series

The example above showed that the nth Maclaurin expansion for $(1-x)^{-1}$ is the geometric series $1 + x + x^2 + x^3 + \dots + x^n$. If you let n tend to infinity this would become the **sum to infinity**, but you will already know that this sum only converges if $|x| < 1$. This is an example of a Maclaurin series which only converges for a limited range of x values – these are described as the values for which the series is **valid**.

Other Maclaurin series are valid for different ranges of x. As you saw in Example 8.1, the Maclaurin series for e^x

$$e^x \equiv 1 + x + \frac{1}{2!}x^2 + \frac{1}{3!}x^3 + \dots + \frac{1}{r!}x^r + \dots$$

is valid for *all* values of x.

A power series may be regarded as incomplete without a statement of the values of x for which it is valid. However, if there is no such statement it may be taken that it is valid for all x.

Example 8.3

Find the first three non-zero terms and the general term of the Maclaurin series for $\sin x$.

Solution

Let $f(x) = \sin x$.

$f(x) = \sin x$	$f(0) = 0$
$f'(x) = \cos x$	$f'(0) = 1$
$f''(x) = -\sin x$	$f''(0) = 0$
$f^{(3)}(x) = -\cos x$	$f^{(3)}(x) = -1$
$f^{(4)}(x) = \sin x$	$f^{(4)}(0) = 0$
$f^{(5)}(x) = \cos x$	$f^{(5)}(0) = 1$
...	...
$f^{(2r)}(x) = (-1)^r \sin x$	$f^{(2r)}(0) = 0$
$f^{(2r+1)}(x) = (-1)^r \cos x$	$f^{(2r+1)}(0) = (-1)^r$

Table 8.2

> **Note**
>
> All the odd derivatives are zero at $x = 0$. The even derivatives alternate between 1 and -1 at $x = 0$.

> **Note**
>
> Note that for this series, x is measured in radians, as the rules for differentiating $\sin x$ and $\cos x$ require x to be measured in radians.

Using $f(x) = f(0) + f'(0)x + \dfrac{f''(0)}{2!}x^2 + \dfrac{f^{(3)}(0)}{3!}x^3 + \dots + \dfrac{f^{(r)}(0)}{r!}x^r + \dots$

gives

$$\sin x = x - \frac{1}{3!}x^3 + \frac{1}{5!}x^5 - \dots + \frac{(-1)^r x^{2r+1}}{(2r+1)!} + \dots$$

2 Using Maclaurin series for standard functions

You met the Maclaurin series for some common functions in *AQA A-level Further Mathematics Year 1*. Here they are again, listed below, together with the values of x for which they are valid.

Note

It is always good practice to state the values of x for which it is valid when you write down a Maclaurin series, particularly in cases where this is not for all x.

$e^x = 1 + x + \dfrac{x^2}{2!} + \ldots + \dfrac{x^r}{r!} + \ldots$	Valid for all x
$\ln(1 + x) = x - \dfrac{x^2}{2} + \dfrac{x^3}{3} - \ldots + (-1)^{r+1}\dfrac{x^r}{r} + \ldots$	Valid for $-1 < x \leqslant 1$
$\sin x = x - \dfrac{x^3}{3!} + \dfrac{x^5}{5!} - \ldots + (-1)^r \dfrac{x^{2r+1}}{(2r+1)!} + \ldots$	Valid for all x
$\cos x = 1 - \dfrac{x^2}{2!} + \dfrac{x^4}{4!} - \ldots + (-1)^r \dfrac{x^{2r}}{(2r)!} + \ldots$	Valid for all x
$(1 + x)^n = 1 + nx + \dfrac{n(n-1)}{2!}x^2 + \ldots$ $+ \dfrac{n(n-1)\ldots(n-r+1)}{r!}x^r + \ldots$	Valid for $\lvert x \rvert < 1, \ n \in \mathbb{R}$

Table 8.3

These standard series can be used to find Maclaurin series for related functions.

Example 8.4

Find the Maclaurin expansion for $\sin(-2x)$ up to the term in x^5.

Find the general term for the expansion.

For what values of x is the expansion valid?

Solution

$$\sin x = x - \frac{x^3}{3!} + \frac{x^5}{5!} + \cdots$$

Substituting $-2x$ for x:

$$\sin(-2x) = (-2x) - \frac{(-2x)^3}{3!} + \frac{(-2x)^5}{5!} + \cdots$$

$$= -2x + \frac{8x^3}{6} + \frac{-32x^5}{120} + \cdots$$

$$= -2x + \frac{4x^3}{3} - \frac{4x^5}{15} + \cdots$$

$$\text{General term} = (-1)^r \frac{x^{2r+1}}{(2r+1)!} = (-1)^r \frac{(-2x)^{2r+1}}{(2r+1)!} = (-1)^{r+1} \frac{2^{2r+1}}{(2r+1)!} x^{2r+1}$$

Since the expansion for $\sin x$ is valid for all values of x, this expansion is also valid for all values of x.

Sometimes a Maclaurin series can be found by adapting one or more known Maclaurin series. An example you will have already verified is that of differentiating the series for $\sin x$ to obtain the series for $\cos x$. It is important to question whether this is a justifiable method:

■ Is it valid to integrate or differentiate an infinite series term by term?

■ Can you form the product of two infinite series by multiplying terms?

■ Is the series obtained identical to the series that would have been obtained by evaluating the derivatives?

Answering these questions in detail is beyond the scope of this book, but you may take it that the answer to all of them is, 'Yes, subject to certain conditions.' In the work in this book you may safely assume the conditions are met, but strange things can happen with infinite series in other situations.

ACTIVITY 8.5

The first few terms of the Maclaurin series for $\sec x$ can be found from the first three terms of the series for $(1 + y)^{-1}$, when $y = -\dfrac{x^2}{2!} + \dfrac{x^4}{4!}$.

Find these terms and explain why this method works.

Exercise 8.1

① (i) Show that the Maclaurin series for $\cos x$ is

$$\cos x = 1 - \frac{1}{2!}x^2 + \frac{1}{4!}x^4 - \frac{1}{6!}x^6 + \cdots + \frac{(-1)^r x^{2r}}{(2r)!} + \cdots$$

(ii) Use the first three terms of the series to calculate an approximate value for $\cos(0.2)$.

(iii) Use your calculator to find $\cos(0.2)$ and find the percentage error in your answer to (i).

② Write down the Maclaurin series for $e^{-\frac{1}{2}x^2}$ as far as the term in x^8 *and including the general term*. For what values of x is it valid?

③ (i) Use known Maclaurin series to expand the following functions in ascending powers of x, up to and including the term in x^4.

(a) $(1 + x)e^x$

(b) $\sin x + \cos x$

(c) $e^x - \sin x$

(ii) Determine the range of values of x for which each expansion in part (i) is valid.

④ The function $f(x) = \ln\left(1 + \frac{1}{2}x\right) - k(1 - x)^{\frac{1}{3}}$.

When it is expanded as a series of ascending powers of x, $f(x)$ has no term in x^2.

Determine the value of k.

Determine the range of values for x for which the expansion is valid.

⑤ (i) By substituting a suitable value for x in the first three terms of the expansion for $\cos x$, find the value of $\dfrac{1}{\sqrt{2}}$ to 4 decimal places.

(ii) How many terms of the expansion of $\cos x$ do you need to be confident of the value of $\dfrac{1}{\sqrt{2}}$ to 4 decimal places?

⑥ (i) Use known Maclaurin series to expand the following functions in ascending powers of x, up to and including, the term in x^4.

(a) $\cos^2 x$

(b) $\cos 2x$

(ii) Use your results in part (i) to verify the double angle formula for $\cos 2x$, as far as the term in x^4.

⑦ The third Maclaurin approximation to a function is $f(x) = 1 - \dfrac{3}{2}x^2 + \dfrac{5}{2}x^3$.

Write down the values of $f'(0)$, $f''(0)$ and $f^{(3)}(0)$.

Sketch the graph of $y = f(x)$ near $x = 0$.

⑧ (i) Find $\displaystyle\int \dfrac{1}{\sqrt{1 - 4x^2}}\, dx$.

(ii) By expanding $\left(1 - 4x^2\right)^{-\frac{1}{2}}$ and integrating term by term, or otherwise, find the series expansion for $\arcsin 2x$, when $|x| < \dfrac{1}{2}$ as far as the term in x^7.

(iii) Use your expansion to find an approximate value for $\arcsin(0.5)$ and find the percentage error for this approximate value.

⑨ (i) Write down the first four terms of the series for $\dfrac{1}{1 - x}$.

(ii) By comparing with part (i), find an expression for the sum of the series $1 + 2x + 3x^2 + 4x^3 + \ldots$

⑩ (i) By integrating $\dfrac{1}{1 - x^2}$ and its Maclaurin expansion, show that the Maclaurin series for $\arctan x$ is:

$$x - \dfrac{x^3}{3} + \dfrac{x^5}{5} - \dfrac{x^7}{7} + \ldots$$

This is known as Gregory's series, after Scottish mathematician James Gregory. It is valid for $|x| \leqslant 1$.

(ii) By putting $x = 1$ show that:

$$\dfrac{\pi}{4} = 1 - \dfrac{1}{3} + \dfrac{1}{5} - \dfrac{1}{7} + \ldots$$

This is known as Leibniz's series. It converges very slowly.

(iii) Show that:

Known as Euler's formula for π.

(a) $\dfrac{\pi}{4} = \arctan\left(\dfrac{1}{2}\right) + \arctan\left(\dfrac{1}{3}\right)$

Known as Machin's formula.

(b) $\dfrac{\pi}{4} = 4\arctan\left(\dfrac{1}{5}\right) - \arctan\left(\dfrac{1}{239}\right)$.

(iv) Use Machin's formula with Gregory's series to find the value of π to five decimal places.

⑪ In this question y_n and a_n are used to denote $f^{(n)}(x)$ and $f^{(n)}(0)$ respectively.

(i) Let $f(x) = \arcsin x$.

Show that $(1 - x^2)y_1^2 = 1$ and $(1 - x^2)y_2 - xy_1 = 0$.

(ii) Find a_1 and a_2.

(iii) Prove by induction that $(1 - x^2)y_{n+2} - (2n + 1)xy_{n+1} - n^2 y_n = 0$
and deduce that $a_{n+2} = n^2 a_n$.

(iv) Find the Maclaurin expansion of $\arcsin x$, giving the first three non-zero terms and the general term.

3 Evaluating limits

Maclaurin series are examples of infinite series that converge to a particular function. It is also useful to determine the limit of a function (if it exists) as it approaches a particular value of x, usually zero or $\pm\infty$. When differentiating a function from first principles, you found the gradient of a chord between two points $(x, f(x))$ and $(x + h, f(x + h))$. You then used the limit as x approaches zero to work out its value. Maclaurin series can be used to find a limit for a function $f(x)$ as x approaches zero as higher order terms become negligible.

Using Maclaurin series

Example 8.5

Find $\lim\limits_{x \to 0} \dfrac{\sin x}{x}$.

Solution

First check that substituting $x = 0$ does not work:

$\dfrac{\sin x}{x} = \dfrac{\sin 0}{0} = \dfrac{0}{0}$ and that is undefined, so another approach is needed.

Use the Maclaurin series for $\sin x = x - \dfrac{x^3}{3!} + \dfrac{x^5}{5!} + \cdots$ in place of $\sin x$:

$$\lim_{x \to 0} \frac{\sin x}{x} = \lim_{x \to 0} \frac{x - \dfrac{x^3}{3!} + \dfrac{x^5}{5!} + \cdots}{x}$$

Now simplify the expression:

$$\lim_{x \to 0} \frac{\sin x}{x} = \lim_{x \to 0}\left(1 - \frac{x^2}{3!} + \frac{x^4}{5!} + \cdots\right)$$

$$= 1$$

> All the terms in x become zero.

You may have noticed that you could have found the limit in Example 8.5 by using the small angle approximation for $\sin x$, which is introduced in Chapter 2 of *AQA A-level Mathematics Year 2*. The small angle approximation is, in fact, the first term of the Maclaurin series.

The use of Maclaurin's series to find a limit is quick and easy when you are using a standard function for which the Maclaurin series is known. The function for which the limit as x approaches zero is to be found may be a composite function of the standard functions.

Example 8.6

Find $\lim\limits_{x \to 0} \dfrac{4.5x^2 - 1 + \cos 3x}{x^4}$.

Solution

First check that substituting $x = 0$ does not work:

$\dfrac{4.5x^2 - 1 + \cos 3x}{x^4} = \dfrac{0 - 1 + 1}{0} = \dfrac{0}{0}$ and that is undefined.

Use the Maclaurin series for $\cos x = 1 - \dfrac{1}{2!}x^2 + \dfrac{1}{4!}x^4 - \dfrac{1}{6!}x^6 + \dots$ and replace $\cos x$ by $\cos 3x$:

$\cos 3x = 1 - \dfrac{1}{2!}(3x)^2 + \dfrac{1}{4!}(3x)^4 - \dfrac{1}{6!}(3x)^6 + \dots$ ⟵ Use the standard series.

$\qquad = 1 - \dfrac{9}{2}x^2 + \dfrac{81}{24}x^4 - \dfrac{729}{720}x^6 + \dots$

$\lim\limits_{x \to 0} \dfrac{4.5x^2 - 1 + \cos 3x}{x^4} = \lim\limits_{x \to 0} \dfrac{4.5x^2 - 1 + 1 - \dfrac{9}{2}x^2 + \dfrac{81}{24}x^4 - \dfrac{729}{720}x^6 + \dots}{x^4}$

$\qquad = \lim\limits_{x \to 0} \dfrac{\dfrac{81}{24}x^4 - \dfrac{729}{720}x^6 + \dots}{x^4}$

$\qquad = \lim\limits_{x \to 0} \left(\dfrac{81}{24} - \dfrac{729}{720}x^2 + \dots \right)$ ⟵ Allow x to become zero.

$\qquad = \dfrac{81}{24} = \dfrac{27}{8}$

ACTIVITY 8.6

Try to use the Maclaurin series for $\sin x$ to find $\lim\limits_{x \to 0} \dfrac{\sin x}{x^2}$. What goes wrong?

Sometimes a limit cannot be found using a Maclaurin series to substitute for a function. In some of these cases, another method, called L'Hôpital's rule, can be used. It also is used to find limits as x approaches infinity or a constant value.

Using L'Hôpital's rule

L'Hôpital's rule states that:

If $\lim\limits_{x \to c} f(x) = 0$ and $\lim\limits_{x \to c} g(x) = 0$ and $\lim\limits_{x \to c} \dfrac{f'(x)}{g'(x)} = L$

then $\lim\limits_{x \to c} \dfrac{f(x)}{g(x)} = L$

If $\lim\limits_{x \to c} f(x) = \pm\infty$ and $\lim\limits_{x \to c} g(x) = \pm\infty$ and $\lim\limits_{x \to c} \dfrac{f'(x)}{g'(x)} = L$

then $\lim\limits_{x \to c} \dfrac{f(x)}{g(x)} = L$

This is a formal statement of the idea that the limit of $\dfrac{f(x)}{g(x)}$, when $\dfrac{f(x)}{g(x)} = \dfrac{0}{0}$ or $\pm\dfrac{\infty}{\infty}$, equals the limit of $\dfrac{f'(x)}{g'(x)}$ as long as certain conditions are satisfied.

Often the limit required is for when x approaches 0, as in the following example.

Example 8.7

Find $\displaystyle\lim_{x \to 0} \dfrac{e^{2x} - 1}{\ln\left(3 - 2e^{2x}\right)}$.

Solution

Here $f(x) = e^{2x} - 1$ and $g(x) = \ln\left(3 - 2e^{2x}\right)$.

Check $f(0) = e^0 - 1 = 0$, $g(0) = \ln\left(3 - 2e^0\right) = 0$ giving $\dfrac{f(x)}{g(x)} = \dfrac{0}{0}$.

The condition that $\displaystyle\lim_{x \to 0} \dfrac{f'(x)}{g'(x)}$ exists is checked when the two derivatives are calculated as part of the method.

$f(x) = e^{2x} - 1 \Rightarrow f'(x) = 2e^{2x}$

$g(x) = \ln\left(3 - 2e^{2x}\right) \Rightarrow g'(x) = \dfrac{-4e^{2x}}{3 - 2e^{2x}}$

$$\lim_{x \to 0} \dfrac{e^{2x} - 1}{\ln\left(3 - 2e^{2x}\right)} = \lim_{x \to 0} \dfrac{2e^{2x}}{\dfrac{-4e^{2x}}{3 - 2e^{2x}}}$$

$$= \dfrac{2e^0\left(3 - 2e^0\right)}{-4e^0}$$

$$= -\dfrac{1}{2}$$

Discussion point

➜ Try using L'Hôpital's rule to find $\displaystyle\lim_{x \to 0} \dfrac{\sin x}{x}$.

L'Hôpital's rule gives the same result as using the Maclaurin's series for $\displaystyle\lim_{x \to 0} \dfrac{\sin x}{x}$, that it is 1.

The limit for x is not always 0, as shown in the next example.

Example 8.8

Find $\lim\limits_{x \to 1} \dfrac{xe^{3x} - e^3}{2x^2 - 2}$.

Solution

Here $f(x) = xe^{3x} - e^3$ and $g(x) = 2x^2 - 2$.

Check $f(1) = 1e^3 - e^3 = 0$, $g(1) = 2 - 2 = 0$

The condition that $\lim\limits_{x \to 1} \dfrac{f'(x)}{g'(x)}$ exists is checked when the two derivatives are calculated as part of the method.

$f(x) = xe^{3x} - e^3 \implies f'(x) = 3xe^{3x} + e^{3x}$

$g(x) = 2x^2 - 2 \implies g'(x) = 4x$

$$\lim\limits_{x \to 1} \dfrac{xe^{3x} - e^3}{2x^2 - 2} = \lim\limits_{x \to 1} \dfrac{3xe^{3x} + e^{3x}}{4x}$$

$$= \lim\limits_{x \to 1} \dfrac{3e^3 + e^3}{4}$$

$$= e^3$$

The limit as x approaches infinity is another instance where L'Hôpital's rule can be used.

Example 8.9

Investigate $\lim\limits_{x \to \infty} \dfrac{3x^2 - x}{(\ln x)^2}$.

Solution

Here $f(x) = 3x^2 - x$ and $g(x) = (\ln x)^2$.

Check $f(\infty) = \infty$, $g(\infty) = \infty$.

The condition that $\lim\limits_{x \to \infty} \dfrac{f'(x)}{g'(x)}$ exists is checked when the two derivatives are calculated as part of the method.

$f(x) = 3x^2 - x \implies f'(x) = 6x - 1$

$g(x) = (\ln x)^2 \implies g'(x) = 2\ln x \times \dfrac{1}{x}$

$$\lim\limits_{x \to \infty} \dfrac{3x^2 - x}{(\ln x)^2} = \lim\limits_{x \to \infty} \dfrac{6x - 1}{2\ln x \times \dfrac{1}{x}} = \lim\limits_{x \to \infty} \dfrac{6x^2 - x}{2\ln x}$$

But this is still undefined as $\lim\limits_{x \to \infty} \dfrac{6x^2 - x}{2\ln x} = \dfrac{\infty}{\infty}$.

This can be considered as a new starting point and so apply L'Hôpital's rule again.

$f(x) = 6x^2 - x \implies f'(x) = 12x - 1$

$$g(x) = 2\ln x \implies g'(x) = 2 \times \frac{1}{x}$$

$$\lim_{x \to \infty} \frac{6x^2 - x}{2\ln x} = \lim_{x \to \infty} \frac{12x - 1}{\frac{2}{x}} = \lim_{x \to \infty} \frac{12x^2 - x}{2}$$

Now $\dfrac{12x^2 - x}{2}$ is unbounded, so the limit does not exist.

L'Hôpital's rule can be used repeatedly as each quotient can be regarded as a new pair of functions to start the process.

> **ACTIVITY 8.7**
>
> Find $\displaystyle\lim_{x \to 0} \frac{\sin x^2}{1 - \cos 2x}$ using the Maclaurin series and also using L'Hôpital's rule.
>
> Decide which is easier and why.

A proof of L'Hôpital's rule is beyond the scope of this book but you can get a feel for why it works by considering a special case.

Functions $f(x)$ and $g(x)$ have the properties that $f(a) = 0$ and $g(a) = 0$, and also that $f'(a)$ and $g'(a)$ exist.

$$\frac{f'(x)}{g'(x)} = \lim_{x \to a} \frac{\dfrac{f(x) - f(a)}{x - a}}{\dfrac{g(x) - g(a)}{x - a}} \quad \longleftarrow \boxed{\text{Using the definition of a derivative}}$$

$$= \lim_{x \to a} \frac{f(x) - f(a)}{g(x) - g(a)}$$

Now $f(a) = 0$ and $g(a) = 0$

So $\dfrac{f'(x)}{g'(x)} = \displaystyle\lim_{x \to a} \frac{f(x)}{g(x)}$

Exercise 8.2

① Use an appropriate standard Maclaurin series to determine:

(i) $\displaystyle\lim_{x \to 0} \frac{\ln(1 + x)}{x}$

(ii) $\displaystyle\lim_{x \to 0} \frac{e^x - 1}{x}$

(iii) $\displaystyle\lim_{x \to 0} \frac{\cos x - 1}{x^2}$

(iv) $\displaystyle\lim_{x \to 0} \frac{x - \sin x}{x^3}$

② Using L'Hôpital's rule evaluate the following limits.

(i) $\displaystyle\lim_{x \to 0} \frac{e^{2x} - 1}{\ln(3 - 2e^x)}$

(ii) $\displaystyle\lim_{x\to\infty}\frac{1-2x}{3x+5}$

(iii) $\displaystyle\lim_{x\to\frac{1}{2}}\frac{x^2\cos(\pi x)}{3e^x-3\sqrt{e}}$

(iv) $\displaystyle\lim_{x\to4}\frac{x^2-16}{\sin\frac{\pi}{2}x}$

③ Explain why L'Hôpital's rule does not work for $\displaystyle\lim_{x\to\infty}\frac{x+\sin x}{x}$.

④ Use L'Hôpital's rule to evaluate the following limits.

(i) $\displaystyle\lim_{x\to\infty}\frac{e^x}{x^3}$

(ii) $\displaystyle\lim_{x\to0}\frac{\left(x^2+1\right)\sin x}{x\cos x}$

(iii) $\displaystyle\lim_{x\to\infty}\frac{5x^2-x-4}{e^{x^2}}$

⑤ (i) Show that the Maclaurin expansion of $\sqrt{4+x}$ up to the term in x^3 is
$2+\dfrac{x}{4}-\dfrac{x^2}{64}+\dfrac{x^3}{512}$.

(ii) Find $\displaystyle\lim_{x\to0}\frac{\sqrt{4+x}-2}{x}$.

⑥ Find $\displaystyle\lim_{x\to\infty}\frac{2x^2+3x}{5-8x^2}$ using two different methods. Which method is better? Justify your choice.

⑦ (i) Find $\displaystyle\lim_{x\to0}\frac{\ln\left(1+\sin x\right)-x}{e^x-1-x}$ using Maclaurin series.

(ii) Find $\displaystyle\lim_{x\to0}\frac{\ln\left(1+\sin x\right)-x}{e^x-1-x}$ using L'Hôpital's rule.

(iii) Which method is more efficient in this case? Justify your answer.

(iv) Will that always be the case where both can be used? Justify your answer.

⑧ (i) Show that the first three terms of the Maclaurin expansion for $\ln\left(\cos x+\sin x\right)$ are $x-x^2+\dfrac{2}{3}x^3$.

(ii) By substituting a suitable value of x show that $\ln2\approx\dfrac{\pi}{2}-\dfrac{\pi^2}{8}+\dfrac{\pi^3}{48}$.

(iii) Evaluate $\ln2$ and $\dfrac{\pi}{2}-\dfrac{\pi^2}{8}+\dfrac{\pi^3}{48}$.

(iv) Comment on the results from (iii).

⑨ (i) Given that $f(x)=e^{2x}\sin3x$, show that $f''(x)=4f'(x)-13f(x)$.

(ii) Differentiate this result twice to find expressions for $f^{(3)}(x)$ and $f^{(4)}(x)$ in terms of lower derivatives.

(iii) Hence find the Maclaurin expansion for $f(x)$ up to the term in x^4.

(iv) Find $\displaystyle\lim_{x\to0}\frac{e^{2x}\sin3x}{x}$.

⑩ Figure 8.6 illustrates a Maclaurin expansion. Find it. (You may assume that $0 < r < 1$.)

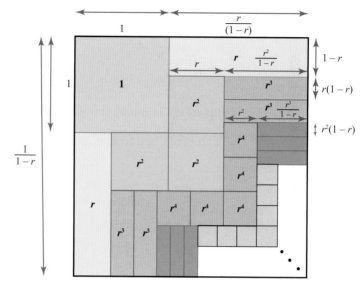

Figure 8.6

LEARNING OUTCOMES

Now you have finished this chapter, you should be able to:

➤ find the Maclaurin series of a function, including the general term

➤ know that a Maclaurin series may converge only for a restricted set of values of x

➤ recognise and use the Maclaurin series of standard functions: e^x, $\ln(1 + x)$, $\sin x$, $\cos x$ and $(1 + x)^n$

➤ demonstrate an understanding of the limiting process used

➤ use Maclaurin's expansion together with any necessary algebraic manipulation before taking limits

➤ understand the circumstances under which L'Hôpital's rule may be applied

➤ apply L'Hôpital's rule once or more than once, together with any necessary algebraic manipulation, to find a limit.

KEY POINTS

1 The general form of the Maclaurin series for $f(x)$ is:

$$f(x) = f(0) + f'(0)x + \frac{f''(0)}{2!}x^2 + \frac{f^{(3)}(0)}{3!}x^3 + \ldots + \frac{f^{(r)}(0)}{r!}x^r + \ldots$$

2 Series which are valid for all x:

$$e^x = 1 + x + \frac{x^2}{2!} + \frac{x^3}{3!} + \ldots + \frac{x^r}{r!} + \ldots$$

$$\sin x = x - \frac{x^3}{3!} + \frac{x^5}{5!} - \frac{x^7}{7!} + \ldots + \frac{(-1)^r x^{2r+1}}{(2r+1)!} + \ldots$$

$$\cos x = 1 - \frac{x^2}{2!} + \frac{x^4}{4!} - \frac{x^6}{6!} + \ldots + \frac{(-1)^r x^{2r}}{(2r)!} + \ldots$$

3 Series valid for $|x| \leqslant 1$:

$$\arctan x = x - \frac{x^3}{3} + \frac{x^5}{5} - \frac{x^7}{7} + \ldots + \frac{(-1)^{r-1} x^{2r-1}}{2r-1} + \ldots$$

4 Series valid for $-1 < x \leqslant 1$:

$$\ln(1 + x) = x - \frac{x^2}{2} + \frac{x^3}{3} - \ldots + \frac{(-1)^{r-1} x^r}{r} + \ldots$$

5 Series where validity depends on n:

$$(1 + x)^n = 1 + nx + \frac{n(n-1)}{2!}x^2 + \ldots + \frac{n(n-1)\ldots(n-r+1)}{r!}x^r + \ldots$$

If n is a positive integer: the series terminates after $n + 1$ terms, and is valid for all x.

If n is not a positive integer: the series is valid for $|x| < 1$;
also for $|x| = 1$ if $n \geqslant 1$;
and for $x = -1$ if $n > 0$.

6 Maclaurin series are used when x approaches zero.

7 L'Hôpital's rule is used to find limits as x approaches c where c can take any value.

8 L'Hôpital's rule states that:

If $\lim\limits_{x \to c} f(x) = 0$ and $\lim\limits_{x \to c} g(x) = 0$ and $\lim\limits_{x \to c} \dfrac{f'(x)}{g'(x)} = L$ then $\lim\limits_{x \to c} \dfrac{f(x)}{g(x)} = L$

If $\lim\limits_{x \to c} f(x) = \pm\infty$ and $\lim\limits_{x \to c} g(x) = \pm\infty$ and $\lim\limits_{x \to c} \dfrac{f'(x)}{g'(x)} = L$ then $\lim\limits_{x \to c} \dfrac{f(x)}{g(x)} = L$

PRACTICE QUESTIONS FURTHER MATHEMATICS 2

① Solve the inequality $\dfrac{5x-6}{x} > x(x-2)$. [5 marks]

MP ② Figure 1 shows the curve $y = \dfrac{1}{x}$. The area below the curve between $x = 1$ and $x = a$ is shaded.

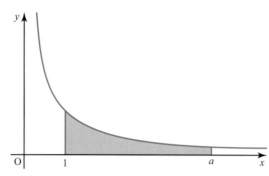

Figure 1

(a) (i) Find the shaded area in terms of a. [2 marks]

(ii) Show that the area becomes infinite as $a \to \infty$. [1 mark]

(b) Show that the volume of revolution about the x-axis of $y = \dfrac{1}{x}$ for $x \geqslant 1$ is π. [3 marks]

PS ③ Do not use your calculator in this question. Give your answer in exact form.

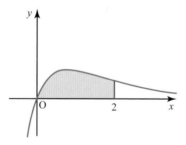

Figure 2

Calculate the area of the region of the plane shaded in the diagram. This region is bounded by the lines $y = 0$ and $x = 2$ and the curve $y = \dfrac{2x(6-x)}{(3x+2)(x^2+4)}$ where $0 \leqslant x \leqslant 2$. [10 marks]

MP ④ The function $f(x)$ is given by $f(x) \equiv \dfrac{x^3 - 9x + 1}{x^2 + 3x + 2}$.

(i) Sketch the graph of $y = f(x)$, stating the equations of the asymptotes. [7 marks]

(ii) Sketch the graph of $y = \dfrac{1}{f(x)}$. [4 marks]

T **PS** ⑤ In this question, the notation y_n denotes the nth derivative of y with respect to x.

(i) Given $y = \ln(\cos x)$, where $0 \leqslant x < \dfrac{\pi}{2}$ show that $y_3 + 2y_2 y_1 = 0$. [3 marks]

(ii) Hence show that $y_4 = -4y_2y_1^2 - 2y_2^2$. Using this result, or otherwise, obtain the Maclaurin expansion of y up to and including the term in x^4. **[5 marks]**

(iii) Taking $x = \frac{\pi}{3}$, deduce the approximation $\ln(2) \approx \frac{\pi^2}{18}\left(1 + \frac{\pi^2}{54}\right)$. **[3 marks]**

(iv) Taking $x = \frac{\pi}{4}$, deduce another approximation for $\ln(2)$.

Which of these approximations is better and why? **[4 marks]**

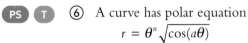

⑥ A curve has polar equation
$$r = \theta^n \sqrt{\cos(a\theta)}$$
where n and a are constants.

(i) Write down an integral that gives the area A enclosed by the curve and the lines $\theta = 0$ and $\theta = \frac{\pi}{6}$, in terms of n and a. **[2 marks]**

(ii) Show that, if a is small enough for a^3 and higher powers of a to be neglected,
$$A = \left(\frac{\pi}{6}\right)^{2n+1}\left[\frac{72(2n+3) - a^2\pi^2(2n+1)}{144(2n+1)(2n+3)}\right].$$ **[6 marks]**

(iii) Evaluate the formula exactly in part (ii) in the case $n = 0$, $a = 0$. **[1 mark]**

(iv) Explain how your answer in part (iii) could have been obtained more easily. **[1 mark]**

(v) In the case $n = 1$, $a = 1$, find the Cartesian equation of the curve. Give your answer in the form $y = x\tan[f(x, y)]$. **[4 marks]**

9

Matrices 2

Discussion points

Are there any points which map to themselves for each of the following transformations?

→ a reflection

→ a rotation

→ an enlargement

Are there any lines which map to themselves?

Prior knowledge

You need to be able to calculate the determinants of matrices with and without a calculator. You should understand and use the fact that a matrix is singular if and only if it has a determinant of zero.

$MM^{-1} = M^{-1}M = I$ for any non-singular matrix M

$$A(BC) = (AB)C$$

$$(\mathbf{AB})^{-1} = \mathbf{B}^{-1}\mathbf{A}^{-1}$$
$$(\mathbf{AB})^{\mathrm{T}} = \mathbf{B}^{\mathrm{T}}\mathbf{A}^{\mathrm{T}}$$

If (x, y) is an invariant point under a transformation represented by the matrix $\mathbf{M}$ then $\mathbf{M}\begin{pmatrix} x \\ y \end{pmatrix} = \begin{pmatrix} x \\ y \end{pmatrix}$.

A line AB is known as an invariant line under a transformation if the image of every point on AB is also on AB.

1 Eigenvalues and eigenvectors

2 × 2 matrices

In a reflection, every point on the mirror line maps to itself. The line may be described as a **line of invariant points**, since every point on the line is itself invariant.

A line of invariant points is a special case of an **invariant line**, where the image of every point on the line is itself on the line but is not necessarily the original point.

As an example, think of the lines through the origin in an enlargement, scale factor 2, centre the origin. Each point on the line in Figure 9.1 maps to another point on the line, but the origin maps to itself.

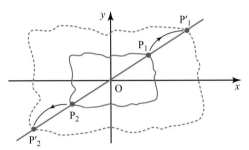

Figure 9.1

This idea is now developed in terms of matrices, but only for lines that pass through the origin.

As an example look at the effect of the transformation with matrix

$$\mathbf{M} = \begin{pmatrix} 4 & 2 \\ 1 & 3 \end{pmatrix}.$$

Since

$$\begin{pmatrix} 4 & 2 \\ 1 & 3 \end{pmatrix}\begin{pmatrix} 1 \\ 1 \end{pmatrix} = \begin{pmatrix} 6 \\ 4 \end{pmatrix}$$

the transformation defined by pre-multiplying position vectors by matrix $\mathbf{M}$ maps the vector $\begin{pmatrix} 1 \\ 1 \end{pmatrix}$ to $\begin{pmatrix} 6 \\ 4 \end{pmatrix}$.

Similarly the image of $\begin{pmatrix} k \\ k \end{pmatrix}$ is $\begin{pmatrix} 6k \\ 4k \end{pmatrix}$. Each point on the line $y = x$ can be represented by the position vector of the form $\begin{pmatrix} k \\ k \end{pmatrix}$, and the points with position vectors $\begin{pmatrix} 6k \\ 4k \end{pmatrix}$ form the line $y = \frac{2}{3}x$. This means that under the transformation represented by the matrix **M**, the image of the line $y = x$ is the line $y = \frac{2}{3}x$.

Similarly since

$$\begin{pmatrix} 4 & 2 \\ 1 & 3 \end{pmatrix} \begin{pmatrix} 1 \\ 2 \end{pmatrix} = \begin{pmatrix} 8 \\ 7 \end{pmatrix}$$

the image of the line $y = 2x$ is the line $y = \frac{7}{8}x$.

ACTIVITY 9.1

(i) Find the images of the following position vectors under the transformation given by $\mathbf{M} = \begin{pmatrix} 4 & 2 \\ 1 & 3 \end{pmatrix}$.

(a) $\begin{pmatrix} 1 \\ 0 \end{pmatrix}$ (b) $\begin{pmatrix} 2 \\ 1 \end{pmatrix}$ (c) $\begin{pmatrix} 0 \\ 1 \end{pmatrix}$

(d) $\begin{pmatrix} -1 \\ 2 \end{pmatrix}$ (e) $\begin{pmatrix} -1 \\ 1 \end{pmatrix}$ (f) $\begin{pmatrix} -2 \\ 1 \end{pmatrix}$

(ii) Use your answer to part **(i)** to find the equations of the images of the following lines.

(a) $y = 0$ (b) $y = \frac{1}{2}x$ (c) $x = 0$

(d) $y = -2x$ (e) $y = -x$ (f) $y = -\frac{1}{2}x$

TECHNOLOGY

Check you are confident using a calculator to multiply matrices.

Using technology, explore the images of each of the vectors in Activity 9.1 under the transformation represented by **M**.

The information you have just gathered may be represented as in Figure 9.2, where the object lines and their images are shown in separate diagrams.

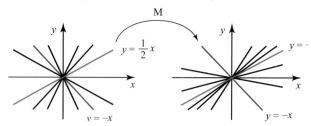

Figure 9.2

However, you can show all the information on one diagram, as in Figure 9.3, where (parts of) the object lines are shown at the centre of the diagram, and (parts of) their image lines are shown in the outer section of the diagram.

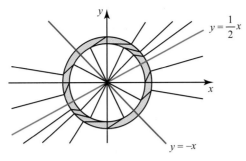

Figure 9.3

The shaded part of Figure 9.3 is not directly relevant but shows lines connecting each object line to its image. You will notice that there are two invariant lines, $y = \frac{1}{2}x$ and $y = -x$, which map to themselves under this transformation. The other lines appear to crowd towards $y = \frac{1}{2}x$, moving away from $y = -x$. This diagram prompts several questions:

- Are there other invariant lines?
- Why does $y = \frac{1}{2}x$ attract and $y = -x$ repel?
- Do all transformations behave like this?
- How can such lines be found efficiently?

Terminology

To answer these questions you need suitable terminology.

If **u** is a non-zero vector such that $\mathbf{Mu} = \lambda\mathbf{u}$, where **M** is a matrix and λ is a scalar, then **u** is called an **eigenvector** of **M**. The scalar λ is known as an **eigenvalue**.

Therefore, since $\begin{pmatrix} 4 & 2 \\ 1 & 3 \end{pmatrix} \begin{pmatrix} 2 \\ 1 \end{pmatrix} = \begin{pmatrix} 10 \\ 5 \end{pmatrix} = 5 \begin{pmatrix} 2 \\ 1 \end{pmatrix}$

and $\begin{pmatrix} 4 & 2 \\ 1 & 3 \end{pmatrix} \begin{pmatrix} -1 \\ 1 \end{pmatrix} = \begin{pmatrix} -2 \\ 2 \end{pmatrix} = 2 \begin{pmatrix} -1 \\ 1 \end{pmatrix}$

$\begin{pmatrix} 2 \\ 1 \end{pmatrix}$ and $\begin{pmatrix} -1 \\ 1 \end{pmatrix}$ are eigenvectors of the matrix $\mathbf{M} = \begin{pmatrix} 4 & 2 \\ 1 & 3 \end{pmatrix}$.

The corresponding eigenvalues are 5 and 2, respectively. It will become evident later that these are the only two eigenvalues.

Properties of eigenvectors

> **Note**
>
> The eigenvectors of a transformation determine the directions of all invariant lines that pass through the origin.

Notice the following properties of eigenvectors:

1. All non-zero scalar multiples of $\begin{pmatrix} 2 \\ 1 \end{pmatrix}$ and $\begin{pmatrix} -1 \\ 1 \end{pmatrix}$ are also eigenvectors of **M**, with (respectively) the same eigenvalues.

2. Under the transformation the eigenvector is enlarged by a scale factor equal to its eigenvalue.

3. The direction of an eigenvector is unchanged by the transformation. (If the eigenvalue is negative the sense of the eigenvector will be reversed.)

When finding eigenvectors you need to solve the equation $\mathbf{Mu} = \lambda\mathbf{u}$.

$$\mathbf{Mu} = \lambda\mathbf{u}$$

$$\Leftrightarrow \mathbf{Mu} - \lambda\mathbf{u} = 0$$

$$\Leftrightarrow \mathbf{Mu} - \lambda\mathbf{Iu} = 0 \quad \longleftarrow$$

| $\mathbf{I}$ is the identity matrix: it is superfluous in this line. |

$$\Leftrightarrow (\mathbf{M} - \lambda\mathbf{I})\mathbf{u} = 0 \quad \longleftarrow$$

| $\mathbf{I}$ is essential here. |

Clearly $\mathbf{u} = 0$ is a solution, but you are seeking a non-zero solution for $\mathbf{u}$. For non-zero solutions you require $\det(\mathbf{M} - \lambda\mathbf{I}) = 0$.

| Since $\mathbf{u}$ is non-zero, this means that the matrix $(\mathbf{M} - \lambda\mathbf{I})$ is singular, which in turn means that its determinant is zero. |

The equation $\det(\mathbf{M} - \lambda\mathbf{I}) = 0$ is known as the **characteristic equation** of $\mathbf{M}$. The left-hand side of the characteristic equation is a polynomial in λ: this polynomial is known as the **characteristic polynomial**.

(The German word for 'characteristic' is *eigen*: eigenvectors are also known as *characteristic* vectors; eigenvalues are also known as *characteristic* values.)

Finding eigenvectors

The following steps are for finding eigenvectors, and they are illustrated in Example 9.1.

1. Form the characteristic equation: $\det(\mathbf{M} - \lambda\mathbf{I}) = 0$.

2. Solve the characteristic equation to find the eigenvalues λ.

3. For each eigenvalue λ find a corresponding eigenvector $\mathbf{u}$ by solving $(\mathbf{M} - \lambda\mathbf{I})\mathbf{u} = 0$.

Example 9.1

Find the eigenvectors of the matrix $\mathbf{M} = \begin{pmatrix} 4 & 2 \\ 1 & 3 \end{pmatrix}$.

Solution

1. Form the characteristic equation, $\det(\mathbf{M} - \lambda\mathbf{I}) = 0$.

$$\mathbf{M} - \lambda\mathbf{I} = \begin{pmatrix} 4 & 2 \\ 1 & 3 \end{pmatrix} - \lambda\begin{pmatrix} 1 & 0 \\ 0 & 1 \end{pmatrix} = \begin{pmatrix} 4 - \lambda & 2 \\ 1 & 3 - \lambda \end{pmatrix}$$

so that $\det(\mathbf{M} - \lambda\mathbf{I}) = 0 \Leftrightarrow \begin{vmatrix} 4 - \lambda & 2 \\ 1 & 3 - \lambda \end{vmatrix} = 0$

$$\Leftrightarrow (4 - \lambda)(3 - \lambda) - 2 = 0$$

$$\Leftrightarrow \lambda^2 - 7\lambda + 10 = 0$$

2. Solve the characteristic equation to find the eigenvalues, λ.

$$\lambda^2 - 7\lambda + 10 = 0 \quad \Leftrightarrow \quad (\lambda - 5)(\lambda - 2) = 0$$

$$\Leftrightarrow \quad \lambda = 2 \text{ or } 5$$

3. For each eigenvalue λ find a corresponding eigenvector $\mathbf{u}$ by solving $(\mathbf{M} - \lambda\mathbf{I})\mathbf{u} = 0$.

When $\lambda = 2$: $(\mathbf{M} - \lambda\mathbf{I})\mathbf{u} = 0 \Leftrightarrow \begin{pmatrix} 2 & 2 \\ 1 & 1 \end{pmatrix}\begin{pmatrix} x \\ y \end{pmatrix} = \begin{pmatrix} 0 \\ 0 \end{pmatrix}$, where $\mathbf{u} = \begin{pmatrix} x \\ y \end{pmatrix}$

$\Leftrightarrow x + y = 0$

> This tells you that if y is any number, k say, then x is $-k$.

$\Leftrightarrow \mathbf{u} = \begin{pmatrix} -k \\ k \end{pmatrix} = k\begin{pmatrix} -1 \\ 1 \end{pmatrix}$

When $\lambda = 5$: $(\mathbf{M} - \lambda\mathbf{I})\mathbf{u} = 0 \Leftrightarrow \begin{pmatrix} -1 & 2 \\ 1 & -2 \end{pmatrix}\begin{pmatrix} x \\ y \end{pmatrix} = \begin{pmatrix} 0 \\ 0 \end{pmatrix}$

> If y is any number, k say, then x is $2k$.

$\Leftrightarrow -x + 2y = 0$

$\Leftrightarrow \mathbf{u} = \begin{pmatrix} 2k \\ k \end{pmatrix} = k\begin{pmatrix} 2 \\ 1 \end{pmatrix}$

Thus the eigenvectors are $\begin{pmatrix} -1 \\ 1 \end{pmatrix}$ and $\begin{pmatrix} 2 \\ 1 \end{pmatrix}$ or any non-zero scalar multiples of these vectors.

ACTIVITY 9.2

By expressing three vectors of your choice in terms of the eigenvectors

$\mathbf{u}_1 = \begin{pmatrix} -1 \\ 1 \end{pmatrix}$ and $\mathbf{u}_2 = \begin{pmatrix} 2 \\ 1 \end{pmatrix}$ explain why the line $y = \frac{1}{2}x$ attracts and the

line $y = -x$ repels under the transformation with matrix $\mathbf{M} = \begin{pmatrix} 4 & 2 \\ 1 & 3 \end{pmatrix}$.

3 × 3 matrices

The definitions of eigenvalue and eigenvector apply to all square matrices. The characteristic equation of matrix $\mathbf{M}$ is $\det(\mathbf{M} - \lambda\mathbf{I}) = 0$. When $\mathbf{M}$ is a 2×2 matrix the characteristic equation is quadratic, and may or may not have real roots. When $\mathbf{M}$ is a 3×3 matrix of real elements the characteristic equation is cubic, with real coefficients; this must have at least one real root. This proves that every real 3×3 matrix has at least one real eigenvector, and so every linear transformation of three-dimensional space has at least one invariant line. You use the same procedure as before for finding the eigenvalues and eigenvectors of a 3×3 matrix, though the work will generally be lengthy.

Example 9.2

Find the eigenvectors of the matrix $M = \begin{pmatrix} 3 & 1 & 1 \\ 1 & 3 & 1 \\ 1 & 1 & 3 \end{pmatrix}$.

Solution

$$\mathbf{M} - \lambda\mathbf{I} = \begin{pmatrix} 3 & 1 & 1 \\ 1 & 3 & 1 \\ 1 & 1 & 3 \end{pmatrix} - \lambda \begin{pmatrix} 1 & 0 & 0 \\ 0 & 1 & 0 \\ 0 & 0 & 1 \end{pmatrix}$$

$$= \begin{pmatrix} 3-\lambda & 1 & 1 \\ 1 & 3-\lambda & 1 \\ 1 & 1 & 3-\lambda \end{pmatrix}$$

$$\det(\mathbf{M} - \lambda\mathbf{I}) = \begin{vmatrix} 3-\lambda & 1 & 1 \\ 1 & 3-\lambda & 1 \\ 1 & 1 & 3-\lambda \end{vmatrix}$$

$$= (3-\lambda)((3-\lambda)(3-\lambda)-1) - ((3-\lambda)-1) + (1-(3-\lambda))$$

$$= -(\lambda^3 - 9\lambda^2 + 24\lambda - 20)$$

$$= -(\lambda - 5)(\lambda - 2)^2$$

so that $\det(\mathbf{M} - \lambda\mathbf{I}) = 0 \iff \lambda = 5$ or 2 (repeated root).

When $\lambda = 5$: $(\mathbf{M} - \lambda\mathbf{I})\mathbf{u} = 0 \iff \begin{pmatrix} -2 & 1 & 1 \\ 1 & -2 & 1 \\ 1 & 1 & -2 \end{pmatrix}\begin{pmatrix} x \\ y \\ z \end{pmatrix} = \begin{pmatrix} 0 \\ 0 \\ 0 \end{pmatrix}$

$$\iff \left.\begin{array}{r} -2x + y + z = 0 \\ x - 2y + z = 0 \\ x + y - 2z = 0 \end{array}\right\} \iff x = y = z = k \text{ say,}$$

so that $\mathbf{u} = \begin{pmatrix} k \\ k \\ k \end{pmatrix} = k\begin{pmatrix} 1 \\ 1 \\ 1 \end{pmatrix}$ is an eigenvector, with eigenvalue 5.

When $\lambda = 2$, a repeated root:

$$(\mathbf{M} - \lambda\mathbf{I})\mathbf{u} = 0 \iff \begin{pmatrix} 1 & 1 & 1 \\ 1 & 1 & 1 \\ 1 & 1 & 1 \end{pmatrix}\begin{pmatrix} x \\ y \\ z \end{pmatrix} = \begin{pmatrix} 0 \\ 0 \\ 0 \end{pmatrix}$$

$$\iff x + y + z = 0$$

i.e. any vector in the plane $x + y + z = 0$ is an eigenvector, with eigenvalue 2.

A general vector in that plane is $\mathbf{u} = \begin{pmatrix} p \\ q \\ -p-q \end{pmatrix}$.

Thus the eigenvectors are $k\begin{pmatrix} 1 \\ 1 \\ 1 \end{pmatrix}$ and $\begin{pmatrix} p \\ q \\ -p-q \end{pmatrix}$ where p and q are not both zero.

Chapter 9 Matrices 2

The ideas above also apply to larger square matrices, but if **M** is $n \times n$, its characteristic equation is of degree n, and solving polynomial equations of higher degree is generally not straightforward. In practice eigenvalues are not usually found by solving characteristic equations! Numerical methods will usually be applied to matrices, using a computer, with consequent problems caused by approximation and rounding errors.

ACTIVITY 9.3

The 3×3 matrix **M** has three eigenvalues $\lambda_1, \lambda_2, \lambda_3$, which are the roots of the polynomial equation $\det(\mathbf{M} - \lambda \mathbf{I}) = 0$.

(i) Imagine factorising the polynomial $\det(\mathbf{M} - \lambda \mathbf{I})$ into linear factors, and hence show that the product of the three eigenvalues is $\det \mathbf{M}$.

(ii) By considering the coefficient of the term in λ^2 in the polynomial $\det(\mathbf{M} - \lambda \mathbf{I})$ show that the sum of the three eigenvalues is the sum of the elements on the leading diagonal of **M**. This sum is known as the **trace** of matrix **M**, $\text{tr}(\mathbf{M})$.

These properties also hold for $n \times n$ matrices.

Exercise 9.1

① The matrix $\mathbf{M} = \begin{pmatrix} 2 & 3 \\ -1 & -2 \end{pmatrix}$ represents a transformation.

(i) Find $\det \mathbf{M}$ and give a geometrical interpretation of this result.

(ii) Show that the characteristic equation of **M** is $\lambda^2 - 1 = 0$, where λ is the eigenvalues of **M**.

(iii) Hence find the eigenvectors of **M**.

(iv) Write down the equations of the lines of invariant points.

② Find the eigenvalues and corresponding eigenvectors of the following 2×2 matrices.

(i) $\begin{pmatrix} 5 & 3 \\ 2 & 4 \end{pmatrix}$ (ii) $\begin{pmatrix} 7 & 2 \\ -12 & -4 \end{pmatrix}$ (iii) $\begin{pmatrix} 1 & 2 \\ 1 & 1 \end{pmatrix}$

(iv) $\begin{pmatrix} 1 & -1 \\ 1 & 3 \end{pmatrix}$ (v) $\begin{pmatrix} 1.1 & -0.4 \\ 0.2 & 0.2 \end{pmatrix}$ (vi) $\begin{pmatrix} p & 0 \\ 0 & q \end{pmatrix}, p \neq q$

③ Find the eigenvalues and corresponding eigenvectors of these 3×3 matrices.

(i) $\begin{pmatrix} 3 & 0 & 0 \\ 0 & 2 & 1 \\ 0 & 0 & -1 \end{pmatrix}$ (ii) $\begin{pmatrix} 1 & 1 & 2 \\ 4 & 2 & -3 \\ 4 & 2 & 3 \end{pmatrix}$ (iii) $\begin{pmatrix} 1 & 1 & 2 \\ 5 & -2 & 1 \\ 1 & 1 & 2 \end{pmatrix}$

(iv) $\begin{pmatrix} 0 & 0 & 2 \\ 0 & 3 & 0 \\ 2 & 0 & 0 \end{pmatrix}$ (v) $\begin{pmatrix} 1 & -3 & -3 \\ -8 & 6 & -3 \\ 8 & -2 & 7 \end{pmatrix}$

④ The matrix $\mathbf{M} = \begin{pmatrix} 0 & 1 & -4 \\ -10 & 7 & -20 \\ -2 & 1 & -2 \end{pmatrix}$ represents a transformation T.

(i) Show that 2 is a repeated eigenvalue of $\mathbf{M}$ and find the other eigenvalue of $\mathbf{M}$.

(ii) For each eigenvalue of $\mathbf{M}$ find a full set of eigenvectors.

(iii) Describe the geometrical significance of the eigenvectors of $\mathbf{M}$ in relation to T.

⑤ Matrix $\mathbf{M}$ is 2×2. Find the real eigenvalues of $\mathbf{M}$ and the corresponding eigenvectors when $\mathbf{M}$ represents

(i) reflection in $y = x \tan\theta$

(ii) a rotation through angle θ about the origin.

⑥ Vector $\mathbf{u}$ is an eigenvector of matrix $\mathbf{A}$, with eigenvalue α, and also an eigenvector of matrix $\mathbf{B}$, with eigenvalue β. Prove that $\mathbf{u}$ is an eigenvector of

(i) $\mathbf{A} + \mathbf{B}$ (ii) $\mathbf{AB}$

and find the corresponding eigenvalues.

[Hint: $\mathbf{u}$ is an eigenvector of $\mathbf{M} \Leftrightarrow \mathbf{Mu} = \lambda\mathbf{u}, \mathbf{u} \neq 0$.]

⑦ Matrix $\mathbf{M}$ has eigenvalue λ with corresponding eigenvector $\mathbf{u}$; k is a non-zero scalar. Prove that the matrix $k\mathbf{M}$ has eigenvalue $k\lambda$ and that $\mathbf{u}$ is a corresponding eigenvector.

⑧ Matrix $\mathbf{M}$ is $n \times n$. For $n = 2$ and for $n = 3$ prove that if the sum of the elements in each row of $\mathbf{M}$ is 1 then 1 is an eigenvalue of $\mathbf{M}$. (This property holds for all values of n.)

⑨ Show that if λ is an eigenvalue of the square matrix $\mathbf{M}$ and the corresponding eigenvector is $\mathbf{u}$, then:

λ^2 is an eigenvalue of $\mathbf{M}^2$

λ^3 is an eigenvalue of $\mathbf{M}^3$

λ^n is an eigenvalue of $\mathbf{M}^n$

λ^{-1} is an eigenvalue of $\mathbf{M}^{-1}$.

Show further that $\mathbf{u}$ is the corresponding eigenvector in all cases.

⑩ Find the eigenvalues of

(i) $\mathbf{M}$ (b) $\mathbf{M}^2$ (c) $\mathbf{M}^5$ (d) $\mathbf{M}^{-1}$ for the following matrices:

(i) $\mathbf{M} = \begin{pmatrix} 4 & -1 \\ 2 & 1 \end{pmatrix}$ (ii) $\mathbf{M} = \begin{pmatrix} 3 & 2 & 2 \\ 1 & 4 & 1 \\ -2 & -4 & -1 \end{pmatrix}$.

⑪ The 2×2 matrix $\mathbf{M}$ has real eigenvalues λ_1, λ_2 and associated eigenvectors $\mathbf{u}_1, \mathbf{u}_2$, where $|\lambda_1| > |\lambda_2|$. By expressing any vector $\mathbf{v}$ in terms of $\mathbf{u}_1$ and $\mathbf{u}_2$, describe the behaviour of $\mathbf{M}^n\mathbf{v}$ as n increases when

(i) $|\lambda_1| < 1$ (ii) $|\lambda_1| = 1$ (iii) $|\lambda_1| > 1$.

⑫ Given $\mathbf{M} = -\mathbf{M}^{\mathrm{T}}$, where $\mathbf{M}$ is a 3×3 matrix, prove that

$\det(\mathbf{M} - k\mathbf{I}) = -\det(\mathbf{M} + k\mathbf{I})$.

Deduce that if λ is a non-zero eigenvalue of $\mathbf{M}$ then $-\lambda$ is also an eigenvalue of $\mathbf{M}$. (Such matrices are called *skew-symmetric*.)

⑬ The self-drive camper-van hire firm DIY has depots at Calgary and Vancouver. The hire period commences on Saturday afternoon and all vans are returned (to either depot) the following Saturday morning. Each week:

- all DIY's vans are hired out
- of the vans hired in Calgary, 50% are returned there, 50% to Vancouver
- of the vans hired in Vancouver, 70% are returned there, 30% to Calgary.

(i) One Saturday the Calgary depot has c vans and the Vancouver depot has v vans. Form matrix $\mathbf{M}$ so that the product $\mathbf{M}\begin{pmatrix} c \\ v \end{pmatrix}$ gives the number of vans in each depot the following Saturday.

(ii) At the start of the season each depot has 100 vans. Use matrix multiplication to find out how many vans will be at each depot two weeks later.

(iii) Solve the equation $\mathbf{M}\mathbf{x} = \mathbf{x}$ where $\mathbf{x} = \begin{pmatrix} c \\ v \end{pmatrix}$ and explain the connection with eigenvalues.

(iv) How many vans should DIY stock at Calgary and Vancouver if they want the number of vans available at those depots to remain constant?

(The process described above is an example of a *Markov process*. The matrix governing it is a *transition* or *stochastic matrix*. Each column of the transition matrix consists of non-negative elements with a sum of 1.)

⑭ At time t, the rabbit and wolf populations (r and w respectively) on a certain island are described by the differential equations:

$$\begin{cases} \dfrac{\mathrm{d}r}{\mathrm{d}t} = 5r - 3w \\ \dfrac{\mathrm{d}w}{\mathrm{d}t} = r + w \end{cases} \quad ①$$

Throughout this question $\mathbf{p}$ represents $\begin{pmatrix} r \\ w \end{pmatrix}$ and $\mathbf{M}$ represents $\begin{pmatrix} 5 & -3 \\ 1 & 1 \end{pmatrix}$.

(i) Show that the differential equations may be written as:

$$\frac{\mathrm{d}\mathbf{p}}{\mathrm{d}t} = \mathbf{M}\mathbf{p} \quad ②$$

(ii) Show that if $\mathbf{p} = \mathbf{p}_1(t)$ and $\mathbf{p} = \mathbf{p}_2(t)$ satisfy ② then $\mathbf{p} = a\mathbf{p}_1(t) + b\mathbf{p}_2(t)$ also satisfies ②, where a and b are constants.

(iii) Show that if $\mathbf{p} = \mathrm{e}^{\lambda t}\mathbf{k}$ satisfies ②, where $\mathbf{k}$ is constant, then $\mathbf{M}\mathbf{k} = \lambda\mathbf{k}$.

(iv) Find the eigenvalues and eigenvectors of $\mathbf{M}$ and hence solve ① given that there are 1000 rabbits and 50 wolves at $t = 0$.

2 Diagonalisation of matrices

A diagonal matrix is a square matrix that only has non-zero elements on the leading diagonal, for example,

$$\mathbf{M} = \begin{pmatrix} 2 & 0 & 0 \\ 0 & 1 & 0 \\ 0 & 0 & -4 \end{pmatrix}$$

The geometrical effect of a diagonal matrix can be seen easily. In the case of matrix **M**,

$\begin{pmatrix} 1 \\ 0 \\ 0 \end{pmatrix}$ is mapped to 2 times itself and so the x-axis is invariant,

$\begin{pmatrix} 0 \\ 1 \\ 0 \end{pmatrix}$ is mapped to itself so the y-axis is a line of invariant points,

$\begin{pmatrix} 0 \\ 0 \\ 1 \end{pmatrix}$ is mapped to -4 times itself so the z-axis is also invariant.

So, the geometrical effect of **M** is that of a one-way stretch scale factor 2 in the x-axis and a one-way stretch scale factor -4 in the z-axis.

Sometimes it is possible to write a transformation matrix as a product of matrices including a diagonal matrix, which makes it easier to recognise the transformation and enables powers of the matrix to be calculated efficiently.

ACTIVITY 9.4

In Example 9.1 you found that the transformation matrix $\mathbf{M} = \begin{pmatrix} 4 & 2 \\ 1 & 3 \end{pmatrix}$

had eigenvalues of $\lambda_1 = 2$ and $\lambda_2 = 5$ with corresponding eigenvectors

$\mathbf{u}_1 = \begin{pmatrix} -1 \\ 1 \end{pmatrix}$ and $\mathbf{u}_2 = \begin{pmatrix} 2 \\ 1 \end{pmatrix}$.

Use your understanding of eigenvalues and vectors to show that

$$\begin{pmatrix} 4 & 2 \\ 1 & 3 \end{pmatrix}\begin{pmatrix} -1 & 2 \\ 1 & 1 \end{pmatrix} = \begin{pmatrix} -1 & 2 \\ 1 & 1 \end{pmatrix}\begin{pmatrix} 2 & 0 \\ 0 & 5 \end{pmatrix}.$$

Show why

$$\begin{pmatrix} 4 & 2 \\ 1 & 3 \end{pmatrix} = \begin{pmatrix} -1 & 2 \\ 1 & 1 \end{pmatrix}\begin{pmatrix} 2 & 0 \\ 0 & 5 \end{pmatrix}\begin{pmatrix} -1 & 2 \\ 1 & 1 \end{pmatrix}^{-1}$$

In general, if any 2×2 matrix has eigenvectors $\mathbf{u}_1$ and $\mathbf{u}_2$ with corresponding eigenvalues λ_1 and λ_2 then $\mathbf{M}\mathbf{u}_1 = \lambda_1\mathbf{u}_1$ and $\mathbf{M}\mathbf{u}_2 = \lambda_2\mathbf{u}_2$ so that $\mathbf{MU} = \mathbf{UD}$

where $\mathbf{U} = (\mathbf{u}_1, \mathbf{u}_2)$, the 2×2 matrix which has the eigenvectors as columns,

and $\mathbf{D} = \begin{pmatrix} \lambda_1 & 0 \\ 0 & \lambda_2 \end{pmatrix}$, a matrix with the corresponding eigenvalues on the

leading diagonal and zeros elsewhere.

If $\mathbf{U}$ is non-singular, $\mathbf{U}^{-1}$ exists and pre-multiplying $\mathbf{MU} = \mathbf{UD}$ by $\mathbf{U}^{-1}$ gives $\mathbf{U}^{-1}\mathbf{MU} = \mathbf{D}$; you then say that $\mathbf{M}$ has been reduced to a diagonal form or that $\mathbf{M}$ has been diagonalised.

> In quantum mechanical and quantum chemical computations, matrix diagonalisation is one of the most frequently applied numerical processes.

Being able to reduce $\mathbf{M}$ to a diagonal form $\mathbf{D}$ helps if you want to raise $\mathbf{M}$ to a power. Post-multiplying $\mathbf{MU} = \mathbf{UD}$ by $\mathbf{U}^{-1}$ gives $\mathbf{M} = \mathbf{UDU}^{-1}$ so that

$$\mathbf{M}^4 = \left(\mathbf{UDU}^{-1}\right)\left(\mathbf{UDU}^{-1}\right)\left(\mathbf{UDU}^{-1}\right)\left(\mathbf{UDU}^{-1}\right)$$

$$= \mathbf{UD}\left(\mathbf{U}^{-1}\mathbf{U}\right)\mathbf{D}\left(\mathbf{U}^{-1}\mathbf{U}\right)\mathbf{D}\left(\mathbf{U}^{-1}\mathbf{U}\right)\mathbf{DU}^{-1}$$

$$= \mathbf{UDIDIDIDU}^{-1}$$

$$= \mathbf{UD}^4\mathbf{U}^{-1}$$

This idea can be generalised to give $\mathbf{M}^n = \mathbf{UD}^n\mathbf{U}^{-1}$. You can prove this using induction in Exercise 9.2.

Note that diagonalisation can only be carried out if $\mathbf{U}$ is non-singular. The matrix cannot be diagonalised if $\mathbf{U}$ is singular as the matrix $\mathbf{U}^{-1}$ would not exist.

Discussion points

→ Consider the case of a repeated eigenvalue: what does this mean for the matrix $\mathbf{U}$?

→ Why does this mean that $\mathbf{U}$ is non-singular?

Example 9.3

Find $\mathbf{M}^n$, where $\mathbf{M} = \begin{pmatrix} 4 & 2 \\ 1 & 3 \end{pmatrix}$.

Solution

You have already seen that $\begin{pmatrix} -1 \\ 1 \end{pmatrix}$ and $\begin{pmatrix} 2 \\ 1 \end{pmatrix}$ are eigenvectors, with

eigenvalues 2 and 5 respectively, so take

$$\mathbf{U} = \begin{pmatrix} -1 & 2 \\ 1 & 1 \end{pmatrix} \text{ and } \mathbf{D} = \begin{pmatrix} 2 & 0 \\ 0 & 5 \end{pmatrix}.$$

> You could use any non-zero multiples of $\begin{pmatrix} -1 \\ 1 \end{pmatrix}, \begin{pmatrix} 2 \\ 1 \end{pmatrix}$ but these are the simplest.

Then $\mathbf{U}^{-1} = \begin{pmatrix} -\dfrac{1}{3} & \dfrac{2}{3} \\ \dfrac{1}{3} & \dfrac{1}{3} \end{pmatrix}$ and $\mathbf{D}^n = \begin{pmatrix} 2^n & 0 \\ 0 & 5^n \end{pmatrix}$.

Therefore $\mathbf{M}^n = \mathbf{U}\,\mathbf{D}^n\mathbf{U}^{-1} = \begin{pmatrix} -1 & 2 \\ 1 & 1 \end{pmatrix}\begin{pmatrix} 2^n & 0 \\ 0 & 5^n \end{pmatrix}\begin{pmatrix} -\dfrac{1}{3} & \dfrac{2}{3} \\ \dfrac{1}{3} & \dfrac{1}{3} \end{pmatrix}$

$$= \frac{1}{3}\begin{pmatrix} 2 \times 5^n + 2^n & 2 \times 5^n - 2^{n+1} \\ 5^n - 2^n & 5^n + 2^{n+1} \end{pmatrix}$$

Again the work with 3×3 and other square matrices follows the same pattern, though the calculations are more complicated.

Use of the characteristic equation

ACTIVITY 9.5

For the familiar matrix $\mathbf{M} = \begin{pmatrix} 4 & 2 \\ 1 & 3 \end{pmatrix}$, calculate $\mathbf{M}^2 - 7\mathbf{M} + 10\mathbf{I}$. What do you notice?

Prove for a general 2×2 matrix $\mathbf{M} = \begin{pmatrix} a & c \\ b & d \end{pmatrix}$ that the matrix will satisfy its own characteristic equation.

 Historical note

The fact that a matrix satisfies its own characteristic equation is called the Cayley-Hamilton theorem and it was announced by Arthur Cayley in 'A Memoir on the Theory of Matrices' in 1858, in which he proved the theorem for 2×2 matrices and checked it for 3×3 matrices. Amazingly, he went on to say, 'I have not thought it necessary to undertake the labour of a formal proof of the theorem in the general case of a matrix of any degree.' Essentially, the same property was contained in Sir William Hamilton's 'Lectures on Quaternions' in 1853, with a proof covering 4×4 matrices. The name 'characteristic equation' is attributed to Augustin Louis Cauchy (1789–1857), and the first general proof of the theorem was supplied in 1878 by Georg Frobenius (1849–1917), complete with modifications to take account of the problems caused by repeated eigenvalues.

Example 9.4

Given that the matrix $\mathbf{M} = \begin{pmatrix} 3 & 1 & 1 \\ 1 & 3 & 1 \\ 1 & 1 & -3 \end{pmatrix}$ satisfies its characteristic

equation $\lambda^3 + 9\lambda^2 - 24\lambda + 20 = 0$, use this to find an expression for $\mathbf{M}^{-1}$ in terms of $\mathbf{M}^2, \mathbf{M}^1,$ and $\mathbf{I}$.

Solution

Since $\mathbf{M}^3 + 9\mathbf{M}^2 - 24\mathbf{M} + 20\mathbf{I} = 0$

$\mathbf{M}^3\mathbf{M}^{-1} + 9\mathbf{M}^2\mathbf{M}^{-1} - 24\mathbf{M}\mathbf{M}^{-1} + 20\mathbf{I}\mathbf{M}^{-1} = 0$ ⟵ Post multiply by $\mathbf{M}^{-1}$.

$\mathbf{M}^2 + 9\mathbf{M}^1 - 24\mathbf{I} + 20\mathbf{M}^{-1} = 0$

$\mathbf{M}^{-1} = \dfrac{1}{20}(-\mathbf{M}^2 - 9\mathbf{M} + 24\mathbf{I}) = 0$

① Find matrices $\mathbf{U}$ and $\mathbf{D}$ such that $\mathbf{M} = \mathbf{UDU}^{-1}$.

(i) $\mathbf{M} = \begin{pmatrix} 5 & 4 \\ 3 & 6 \end{pmatrix}$

(ii) $\mathbf{M} = \begin{pmatrix} 7 & -10 \\ 3 & -4 \end{pmatrix}$

(iii) $\mathbf{M} = \begin{pmatrix} 0.5 & 0.5 \\ 0.3 & 0.7 \end{pmatrix}$

② Express $\mathbf{M} = \begin{pmatrix} 1.9 & -1.5 \\ 0.6 & 0 \end{pmatrix}$ in the form of $\mathbf{UDU}^{-1}$ and hence find $\mathbf{M}^4$.

What can you say about $\mathbf{M}^n$ when n is very large?

③ Calculate the following:

(i) $\begin{pmatrix} 6 & -6 \\ 2 & -1 \end{pmatrix}^5$

(ii) $\begin{pmatrix} 3 & -1 \\ -1 & 3 \end{pmatrix}^{10}$

(iii) $\begin{pmatrix} 0.7 & 0.3 \\ 0.6 & 0.4 \end{pmatrix}^4$.

④ Show that $\begin{pmatrix} 1 \\ 1 \\ 0 \end{pmatrix}$ is an eigenvector of $\mathbf{M} = \begin{pmatrix} 1 & 4 & -1 \\ -1 & 6 & -1 \\ 2 & -2 & 4 \end{pmatrix}$ and state

the corresponding eigenvalue. By finding the other eigenvalues and their eigenvectors express $\mathbf{M}$ in the form $\mathbf{UDU}^{-1}$.

⑤ Find examples of 2×2 matrices to illustrate the following:

(i) $\mathbf{M}$ has repeated eigenvalues and cannot be diagonalised

(ii) $\mathbf{M}$ has repeated eigenvalues and can be diagonalised

(iii) $\mathbf{M}$ has 0 as an eigenvalue and cannot be diagonalised

(iv) $\mathbf{M}$ has 0 as an eigenvalue and can be diagonalised.

⑥ A matrix $\mathbf{M}$ is given by $\mathbf{M} = \begin{pmatrix} -1 & -1 & 1 \\ 6 & 2 & k \\ 0 & -2 & 1 \end{pmatrix}$.

(i) Find, in terms of k,

(a) the determinant of $\mathbf{M}$

(b) the inverse matrix $\mathbf{M}^{-1}$.

One of the eigenvalues of $\mathbf{M}$ is 2.

(ii) Find the value of k, and show that the other two eigenvalues are 1 and -1.

(iii) Find integers p, q and r such that $\mathbf{M}^2 = p\mathbf{M} + q\mathbf{I} + r\mathbf{M}^{-1}$.

(iv) Show that $\mathbf{M}^4 = 10\mathbf{M} + \mathbf{I} - 10\mathbf{M}^{-1}$.

⑦ You are given the matrix $\mathbf{M} = \begin{pmatrix} -2 & -5 \\ 3 & 6 \end{pmatrix}$.

(i) Show that 3 is an eigenvalue of $\mathbf{M}$, and find the other eigenvalue.

(ii) For each eigenvalue, find a corresponding eigenvector.

(iii) Write down a matrix $\mathbf{P}$ such that $\mathbf{P}^{-1}\mathbf{M}\mathbf{P}$ is a diagonal matrix.

(iv) Hence show that $\mathbf{M}^n = \dfrac{1}{2} \begin{pmatrix} 5 - 3^{n+1} & 5 - 5 \times 3^n \\ -3 + 3^{n+1} & -3 + 5 \times 3^n \end{pmatrix}$.

⑧ Given that $\mathbf{M} = \mathbf{U}\mathbf{D}\mathbf{U}^{-1}$ use induction to prove that $\mathbf{M}^n = \mathbf{U}\mathbf{D}^n\mathbf{U}^{-1}$.

⑨ (i) $\mathbf{A}$ and $\mathbf{B}$ are 2×2 matrices with eigenvalues α_1, α_2 and β_1, β_2 respectively. Are the eigenvalues of the product $\mathbf{AB}$ the products of the eigenvalues of $\mathbf{A}$ and $\mathbf{B}$? Justify your answer.

(ii) Find the fallacy in this 'proof ':

$\mathbf{A}$ has eigenvalue λ and $\mathbf{B}$ has eigenvalue μ

$\Rightarrow \mathbf{ABs} = \mathbf{A}\mu\mathbf{s}$

$\Rightarrow \mathbf{ABs} = \mu\mathbf{As}$

$\Rightarrow \mathbf{ABs} = \mu\lambda\mathbf{s}$

$\Rightarrow \mathbf{AB}$ has eigenvalue $\lambda\mu$.

⑩ Matrices $\mathbf{A}$ and $\mathbf{B}$ can be diagonalised. Assuming neither matrix has repeated eigenvalues, show that:

$\mathbf{A}$ and $\mathbf{B}$ share the same eigenvectors $\Leftrightarrow \mathbf{AB} = \mathbf{BA}$.

(This result, useful in quantum mechanics, also holds if eigenvalues are repeated.)

LEARNING OUTCOMES

Now you have finished this chapter, you should be able to:

➤ find eigenvalues and eigenvectors of 2×2 and 3×3 matrices

➤ find and use the characteristic equation

➤ understand the geometrical significance of eigenvalues and eigenvectors

➤ understand the term diagonal matrix

➤ understand how to use the eigenvalues and vectors to form $\mathbf{U}$ and $\mathbf{D}$ in $\mathbf{M} = \mathbf{U}\mathbf{D}\mathbf{U}^{-1}$ and understand that when it is not possible to be written in that form you should use $\mathbf{M} = \mathbf{U}\mathbf{D}\mathbf{U}^{-1}$ to find $\mathbf{M}^n$.

KEY POINTS

1 An eigenvector of a square matrix $\mathbf{M}$ is a non-zero vector $\mathbf{u}$ such that $\mathbf{Mu} = \lambda\mathbf{u}$; the scalar λ is the corresponding eigenvalue.

2 The characteristic equation of $\mathbf{M}$ is $\det(\mathbf{M} - \lambda\mathbf{I}) = 0$.

3 Every square matrix satisfies its own characteristic equation.

4 If $\mathbf{U}$ is the matrix formed of eigenvectors of $\mathbf{M}$, and $\mathbf{D}$ is the diagonal matrix formed of the corresponding eigenvalues, then $\mathbf{M} = \mathbf{U}\mathbf{D}\mathbf{U}^{-1}$.

Review: Complex numbers

1 Working with complex numbers

The different types of numbers in the number system can be represented in a diagram as in Figure R.1.

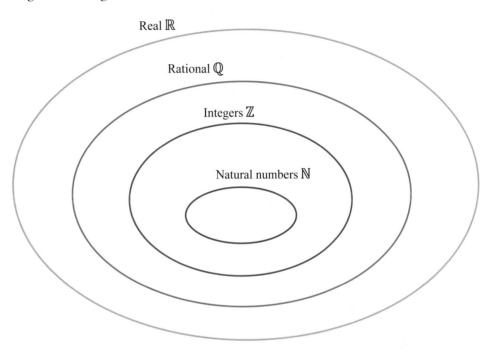

Figure R.1

The number system can be extended to include complex numbers, denoted $\mathbb{C}$, by introducing a new number $i = \sqrt{-1}$. Any number z of the form $x + yi$, where x and y are real, is called a **complex number**. x is called the **real part** of the complex number, denoted by $\mathrm{Re}(z)$, and y is called the **imaginary part**, denoted by $\mathrm{Im}(z)$. If $y = 0$ the number $z = x + yi$ is wholly real, so the set $\mathbb{R}$ is a subset of the set $\mathbb{C}$.

$z^* = x - yi$ is the **complex conjugate** of $z = x + yi$.

Example R.1

Use the quadratic formula to solve the quadratic equation $z^2 - 2z + 10 = 0$, simplifying your answer as far as possible.

Solution

$$z^2 - 2z + 10 = 0$$

$$z = \frac{2 \pm \sqrt{4 - (4 \times 1 \times 10)}}{2 \times 1}$$

> Using the quadratic formula with $a = 1$, $b = -2$ and $c = 10$.

$$= \frac{2 \pm \sqrt{-36}}{2}$$

> $\sqrt{-36} = \sqrt{-1}\sqrt{36} = 6i$.

$$= \frac{2 \pm 6i}{2}$$

$$= 1 \pm 3i$$

The roots have a real part and an imaginary part. The roots form a **conjugate pair** as they have the same real part and the imaginary parts have the opposite signs.

> 1 is the real part of the complex numbers, denoted by $\operatorname{Re}(z)$.

$$1 \pm 3i$$

> 3 is the imaginary part of the complex numbers, denoted by $\operatorname{Im}(z)$.

Example R.2

Calculate:

(i) $(2 - 3i) + (1 + 5i)$

(ii) $(2 - 3i) - (1 + 5i)$

(iii) $(2 - 3i)(1 + 5i)$

(iv) $\dfrac{2 - 3i}{1 + 5i}$

Solution

(i) $(2 - 3i) + (1 + 5i) = (2 + 1) + (-3 + 5)i$

> Add the real parts and add the imaginary parts.

$$= 3 + 2i$$

(ii) $(2 - 3i) - (1 + 5i) = (2 - 1) + (-3 - 5)i$

> Subtract the real parts and subtract the imaginary parts.

$$= 1 - 8i$$

(iii) $(2 - 3i)(1 + 5i) = 2 + 10i - 3i - 15i^2$

> Multiply out the brackets and simplify.

> When simplifying it is important to remember that $i^2 = -1$.

$$= 2 + 7i - 15(-1)$$

$$= 17 + 7i$$

(iv) $\dfrac{2 - 3i}{1 + 5i} \times \dfrac{1 - 5i}{1 - 5i}$

> Multiply the numerator and the denominator by the conjugate of the denominator, then simplify.

$$= \frac{2 - 10i - 3i + 15i^2}{1 - 5i + 5i - 25i^2}$$

$$= \frac{-13 - 13i}{26} = \frac{-1 - i}{2} \text{ or } -\frac{1}{2} - \frac{1}{2}i$$

Do not use a calculator in this exercise.

① (i) Write down the values of:

   (a) i^2 (b) i^3 (c) i^4 (d) i^5 (e) i^6

   (ii) Explain how you would quickly work out the value of i^n for any positive integer value n.

② Find the following:

   (i) $2i(3 - 4i) + i(6 - i)$ (ii) $(2 - 5i)^2$ (iii) $(3 + i)(6 - i)(2 - 5i)$

③ Find the following:

   (i) $\dfrac{5i}{2 - i}$ (ii) $\dfrac{5 + i}{2 - i}$ (iii) $\dfrac{5 - i}{2 + i}$

④ Solve the equation $(7 + 3i)z = (5 + i)(2 - 9i)$, giving your answer in the form $a + bi$.

⑤ Given that the complex numbers

$$z_1 = (2 - a) + 5bi$$
$$z_2 = a^2 + \left(b^2 + 6\right)i$$

are equal, find the possible values of a and b.

Hence list the possible complex numbers z_1 and z_2.

⑥ For the complex number $z = 2 - 5i$ find $\dfrac{1}{z} + \dfrac{1}{z^*}$ in its simplest form.

Write down the value of $\dfrac{1}{z} + \dfrac{1}{z^*}$ for $z = 2 + 5i$.

⑦ For all complex numbers $z = x + yi$ show that $z + z^*$ and zz^* are both real.

⑧ Find the values of the real numbers a and b which satisfy

$$\frac{a}{2 - i} + \frac{2b}{1 + i} = \frac{5i}{2 - i}$$

2 Representing complex numbers geometrically

A complex number $x + yi$ can be represented by the point with Cartesian coordinates (x, y).

For example, in Figure R.2

$5 + 3i$ is represented by $(5, 3)$

$2 - 3i$ is represented by $(2, -3)$

$-6i$ is represented by $(0, -6)$.

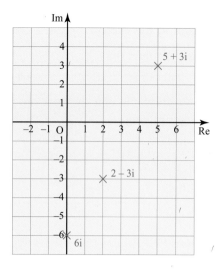

Figure R.2

All real numbers are represented by points on the x-axis, which is therefore called the **real axis**. Purely imaginary numbers which have no real component (of the form $0 + y\mathrm{i}$) give points on the y-axis, which is called the **imaginary axis**.

These axis are labelled as Re and Im.

This geometrical illustration of complex numbers is called the **complex plane** or the **Argand diagram**.

Representing the sum and difference of complex numbers geometrically

Example R.3

Given two complex numbers z_1 and z_2 draw separate Argand diagrams to show geometrically:

(i) $z_1 + z_2$

(ii) $z_1 - z_2$

Solution

(i)

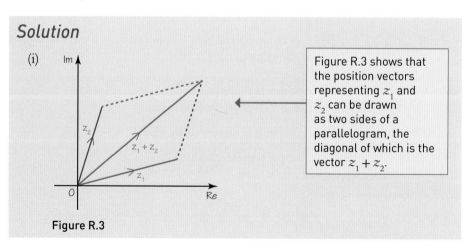

Figure R.3 shows that the position vectors representing z_1 and z_2 can be drawn as two sides of a parallelogram, the diagonal of which is the vector $z_1 + z_2$.

Figure R.3

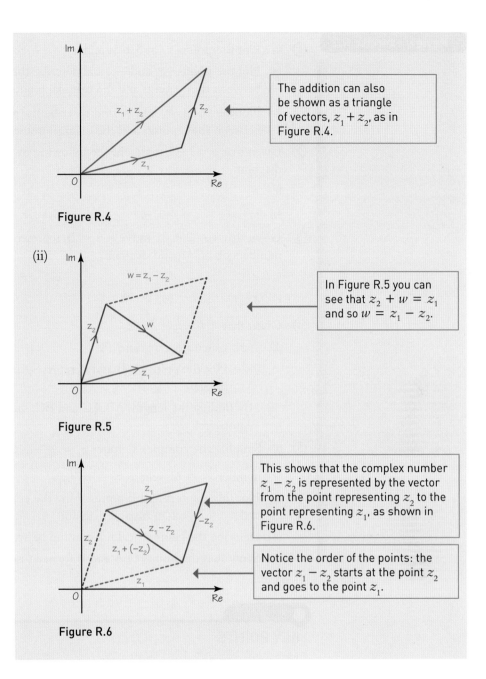

Figure R.4

The addition can also be shown as a triangle of vectors, $z_1 + z_2$, as in Figure R.4.

(ii)

Figure R.5

In Figure R.5 you can see that $z_2 + w = z_1$ and so $w = z_1 - z_2$.

Figure R.6

This shows that the complex number $z_1 - z_2$ is represented by the vector from the point representing z_2 to the point representing z_1, as shown in Figure R.6.

Notice the order of the points: the vector $z_1 - z_2$ starts at the point z_2 and goes to the point z_1.

Review exercise R.2

① (i) Plot the points $z_1 = 3 + 5i$ and $z_2 = -4 + i$ on an Argand diagram.

   (ii) Plot the points $-z_1$ and $-z_2$ and describe the geometrical connection between these points and the original points.

   (iii) Plot the points z_1^* and z_2^* and describe the geometrical connection between these points and the original points.

② Given that $z = 3 - i$, represent the following by points on a single Argand diagram.

   (i) z (ii) $-z$ (iii) z^* (iv) $-z^*$

   (v) iz (vi) $-iz$ (vii) iz^* (viii) $(iz)^*$

③ Given that $z = 2 + 3i$ and $w = 1 - 2i$, represent the following complex numbers on an Argand diagram.

   (i) z (ii) w (iii) $z + w$

   (iv) $z - w$ (v) $w - z$

④ Given that $z = -\dfrac{1}{2} - \dfrac{\sqrt{3}}{2}i$:

   (i) (a) Calculate z^0, z^1 and z^2.

        (b) Plot the points A, B and C representing z^0, z^1 and z^2 on an Argand diagram.

   (ii) By finding the lengths AB, AC and BC, show that triangle ABC is equilateral.

⑤ (i) Simplify the complex number $z_1 = \dfrac{2i}{3 - 7i}$ and find the complex number $z_2 = iz_1$ where $z_1 = \dfrac{2i}{3 - 7i}$.

   (ii) Plot the points A and B representing the complex numbers $z_1 = \dfrac{2i}{3 - 7i}$ and $z_2 = iz_1$ on an Argand diagram.

   (iii) Describe the geometrical relationship between z_1 and z_2.

   (iv) Show that the points O, A and B form an isosceles triangle.

KEY POINTS

1. Complex numbers are of the form $z = x + yi$ with $i^2 = -1$.
   x is called the real part, $\mathrm{Re}(z)$, and y is called the imaginary part, $\mathrm{Im}(z)$.

2. The conjugate of $z = x + yi$ is $z^* = x - yi$.

3. To add or subtract complex numbers, add or subtract the real and imaginary parts separately.
   $$(x_1 + y_1 i) \pm (x_2 + y_2 i) = (x_1 \pm x_2) + (y_1 \pm y_2)i$$

4. To multiply complex numbers, expand the brackets then simplify using the fact that $i^2 = -1$.

5. To divide complex numbers, write as a fraction, then multiply the numerator and the denominator by the conjugate of the denominator and simplify the answer.

6. Two complex numbers $z_1 = x_1 + y_1 i$ and $z_2 = x_2 + y_2 i$ are equal if and only if $x_1 = x_2$ and $y_1 = y_2$.

7. The complex number $z = x + yi$ can be represented geometrically as the point (x, y). The real/imaginary axes system is known as an Argand diagram.

Hyperbolic functions

Everyone knows what a curve is, until he has studied enough mathematics to become confused through the countless number of possible exceptions.

Felix Klein

Discussion point

→ What sort of function do you think you could use to model the arch shown in the photo?

1 Hyperbolic functions

The arch in the photo above looks like a parabola but is actually a catenary. The arch is the reflection in a horizontal axis of a transformation of the graph in Figure 10.1.

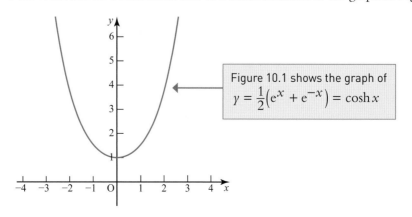

Figure 10.1 shows the graph of
$y = \frac{1}{2}\left(e^x + e^{-x}\right) = \cosh x$

Figure 10.1

> **Note**
>
> This chapter contains some review material that is integrated into the development of new material to give a coherent approach. Examples that are entirely review material are identified as such.

In Chapter 5 of AQA Further Mathematics Book 1 you saw that the functions $x = \cos\theta$ and $y = \sin\theta$ are called circular functions because of their relationship with the unit circle, and how the functions $x = \cosh t$ and $y = \sinh t$ are called hyperbolic functions because of their relationship with the rectangular hyperbola $x^2 - y^2 = 1$.

The functions $\sin x$ and $\cos x$ give rise to $\tan x$, $\sec x$, $\mathrm{cosec}\,x$ and $\cot x$. The hyperbolic functions give rise to a similar family of associated functions. These will be explored in this chapter.

The cosh and sech functions

The graph of $y = \cosh x$ is positive for all real x as can be seen from Figure 10.1. It can also be deduced from $\cosh x = \frac{1}{2}(e^x + e^{-x})$ as both e^x and e^{-x} are always positive.

The function $\cosh x$ is defined for all real values of x but it only takes values from $y = 1$ upwards.

So the function $\cosh x$ has **domain** $x \in \mathbb{R}$ and **range** $y \geqslant 1$.

Example 10.1

Find the exact value of $2\cosh^2 2 - 1$.

Solution

$$\cosh x = \frac{1}{2}(e^x + e^{-x})$$

So $\cosh 2 = \frac{1}{2}(e^2 + e^{-2})$

$$2\cosh^2 2 - 1 = 2 \times \left(\frac{1}{2}\right)^2 (e^2 + e^{-2})^2 - 1$$

$$= 2 \times \frac{1}{4}(e^4 + e^{-4} + 2) - 1$$

$$= \frac{1}{2}(e^4 + e^{-4}) + 1 - 1$$

$$= \frac{1}{2}(e^4 + e^{-4})$$

The function $\mathrm{sech}\,x$ is defined as $\mathrm{sech}\,x = \dfrac{1}{\cosh x}$.

To sketch the graph of $\mathrm{sech}\,x$ from the graph of $\cosh x$ look for key points such as where the graph crosses the y-axis, and the behaviour of the graph as it approaches positive and negative infinity.

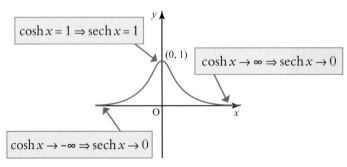

Figure 10.2

The function is defined for all real values of x and takes values between 0 and 1 inclusive.

So $\operatorname{sech} x$ has domain $x \in \mathbb{R}$ and range $0 < y \leqslant 1$.

Since $\cosh x = \frac{1}{2}\left(e^x + e^{-x}\right)$, $\operatorname{sech} x = \dfrac{1}{\frac{1}{2}\left(e^x + e^{-x}\right)} = \dfrac{2}{e^x + e^{-x}}$.

The sinh and cosech functions

The function $\sinh x = \frac{1}{2}\left(e^x - e^{-x}\right)$ is shown in Figure 10.3.

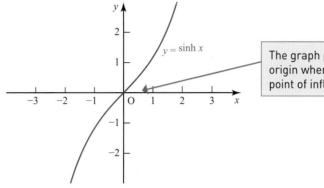

The graph passes through the origin where it has a non-stationary point of inflection.

Figure 10.3

It is defined for all values of x so its domain is $x \in \mathbb{R}$, and takes all real values so its range is $y \in \mathbb{R}$.

The function $y = \operatorname{cosech} x = \dfrac{1}{\sinh x}$ can be sketched by considering the graph of $\sinh x$. This is shown in Figure 10.4.

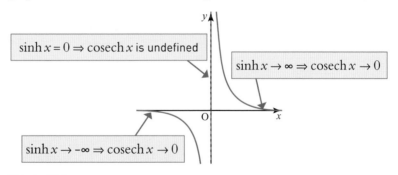

$\sinh x = 0 \Rightarrow \operatorname{cosech} x$ is undefined

$\sinh x \to \infty \Rightarrow \operatorname{cosech} x \to 0$

$\sinh x \to -\infty \Rightarrow \operatorname{cosech} x \to 0$

Figure 10.4

The domain of $\operatorname{cosech} x$ is $x \in \mathbb{R}, \ x \neq 0$ and its range is $y \in \mathbb{R}, \ y \neq 0$. The y-axis is an asymptote of the graph.

Example 10.2

Solve the equation $\cosh x + 2\sinh x = 4$, giving the roots in exact form.

Solution

$$\cosh x + 2\sinh x = 4$$

Substitute using the exponential form.

$$\tfrac{1}{2}\left(e^x + e^{-x}\right) + 2 \times \tfrac{1}{2}\left(e^x - e^{-x}\right) = 4$$

$$e^x + e^{-x} + 2e^x - 2e^{-x} = 8$$

$$3e^x - e^{-x} - 8 = 0$$

$$3e^{2x} - 8e^x - 1 = 0$$

$$e^x = \frac{8 \pm \sqrt{76}}{6} = \tfrac{1}{3}(4 \pm \sqrt{19})$$

Since the negative sign gives a negative value for e^x only the positive value is used.

$$e^x = \tfrac{1}{3}\left(4 + \sqrt{19}\right) \Rightarrow x = \ln\left(\tfrac{1}{3}\left(4 + \sqrt{19}\right)\right)$$

ACTIVITY 10.2

Define $\operatorname{cosech} x$ in terms of e^x.

> **Note**
>
> Pronunciations vary, but usually, 'tanch' or 'tanaitch' or 'than' – with a soft 'th' as in 'thistle', not 'this'.

The tanh and coth functions

In Chapter 5 of *AQA A-level Further Mathematics Year 1* you met $\tanh x = \dfrac{\sinh x}{\cosh x}$.

The equation for $\tanh x$ in terms of e^x is obtained by substituting the equations for $\sinh x$ and $\cosh x$ into $\tanh x = \dfrac{\sinh x}{\cosh x}$:

$$\tanh x = \frac{e^x - e^{-x}}{e^x + e^{-x}} = \frac{1 - e^{-2x}}{1 + e^{-2x}}$$

This makes it easier to sketch $y = \tanh x$ as you can see that it approaches 1 as $x \to \pm\infty$ (see Figure 10.5).

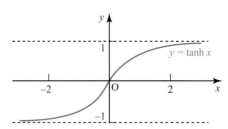

Figure 10.5

The domain of $y = \tanh x$ is $x \in \mathbb{R}$ and the range is $-1 < y < 1$.

The lines $y = 1$ and $y = -1$ are asymptotes of the graph.

The function $\coth x$ is defined as $\coth x = \dfrac{1}{\tanh x}$.

Example 10.3

(i)   Sketch the graph of $y = \coth x$ and list any asymptotes.
(ii) Write down the domain and range of $\coth x$.
(iii) Define $\coth x$ in terms of e^x.

Solution

(i) Asymptotes are $y = 1$ and $y = -1$ and the y-axis.

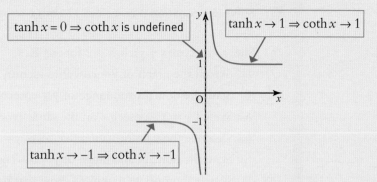

$\tanh x = 0 \Rightarrow \coth x$ is undefined

$\tanh x \to 1 \Rightarrow \coth x \to 1$

$\tanh x \to -1 \Rightarrow \coth x \to -1$

Figure 10.6

(ii) The domain of $y = \coth x$ is $x \in \mathbb{R}, \ x \neq 0$ and the range is $y > 1, \ y < -1$.

(iii) $\coth x = \dfrac{1}{\tanh x} = \dfrac{e^x + e^{-x}}{e^x - e^{-x}}$

Summary

Table 10.1 summarises the information for the six hyperbolic functions.

Function	Definition	Domain	Range
$\cosh x$	$\frac{1}{2}\left(e^x + e^{-x}\right)$	$x \in \mathbb{R}$	$y > 1$
$\sinh x$	$\frac{1}{2}\left(e^x - e^{-x}\right)$	$x \in \mathbb{R}$	$y \in \mathbb{R}$
$\tanh x$	$\dfrac{\sinh x}{\cosh x}$	$x \in \mathbb{R}$	$-1 < y < 1$
$\text{sech}\, x$	$\dfrac{1}{\cosh x}$	$x \in \mathbb{R}$	$0 < y \leqslant 1$
$\text{cosech}\, x$	$\dfrac{1}{\sinh x}$	$x \in \mathbb{R}, \ x \neq 0$	$y \in \mathbb{R}, \ y \neq 0$
$\coth x$	$\dfrac{\cosh x}{\sinh x}$	$x \in \mathbb{R}, \ x \neq 0$	$y > 1, \ y < -1$

Table 10.1

Exercise 10.1

① For each of the following functions:

    (a) $\sinh 2x$ (b) $3\cosh x$ (c) $\cosh x + 1$ (d) $\sinh(x - 3)$

    (i)   write the function in exponential form

    (ii)   sketch the graph

    (iii) state the domain and range of the function.

② (i)   Sketch the graph of $y = 2\tanh x$, identifying any asymptotes.

    (ii)   Describe the transformation that maps $y = \tanh x$ onto $y = 2\tanh x$.

    (iii) State the domain and range of the function $y = 2\tanh x$.

    (iv) Write $2\tanh x$ in exponential form.

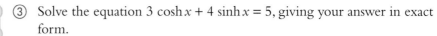

③ Solve the equation $3 \cosh x + 4 \sinh x = 5$, giving your answer in exact form.

④ Solve the equation $4 \tanh x - \operatorname{sech} x = 2$, giving your answer in exact form.

⑤ Solve the equation $3 \operatorname{cosech} x - 2 \coth x = -4$, giving your answer in exact form.

⑥ Show that $\cosh x + \operatorname{sech} x \geqslant 2$ for $x \in \mathbb{R}$.

⑦ (i) Sketch the graph of $y = \operatorname{cosech} x$, identifying any asymptotes.

   (ii) State the domain and range of $y = \operatorname{cosech} x$.

   (iii) Sketch $y = 3 - \operatorname{cosech} x$ on the same axes.

   (iv) Describe the transformation that maps $y = \operatorname{cosech} x$ onto $y = 3 - \operatorname{cosech} x$.

   (v) State the domain and range of $y = 3 - \operatorname{cosech} x$

⑧ (i) Sketch appropriate graphs to show that the equation $\operatorname{sech} x = \tanh x$ has only one root.

   (ii) Find that root in exact form.

⑨ (i) Derive the exponential form of the function $y = 3 - \coth 2x$.

   (ii) Show that $3 - \coth 2x > 0$ for all $x > k$ where k is a positive number to be given in exact form.

⑩ (i) Find exact expressions for p and q, where $\sinh p = \operatorname{sech} p$ and $\cosh q = \coth q$.

   (ii) Arrange $\cosh x$, $\sinh x$, $\tanh x$, $\operatorname{sech} x$, $\operatorname{cosech} x$ and $\coth x$ in ascending order of magnitude when

      (a) $0 < x < p$

      (b) $p < x < q$.

2 Identities

The identities involving hyperbolic functions are similar to those for the trigonometric functions.

The definition of $\tanh x$ is an identity:

$$\tanh x \equiv \frac{\sinh x}{\cosh x}$$

and $\tanh x$ gets its exponential form from this definition. The other identities in this section can be proved using their exponential form.

In Chapter 5 of *AQA A-level Further Mathematics Year 1*, the identity

$$\cosh^2 x - \sinh^2 x \equiv 1$$

was proved by simplifying $\cosh^2 x - \sinh^2 x$.

Example 10.4

Prove that $\cosh 2x \equiv 2\cosh^2 x - 1$.

Solution

$$2\cosh^2 x - 1 = 2 \times \left(\frac{1}{2}\left(e^x + e^{-x}\right)\right)^2 - 1$$

It is usually better to start with the more complicated side.

$$= 2 \times \frac{1}{4}\left(e^{2x} + 2e^x e^{-x} + e^{-2x}\right) - 1$$

Simplify.

$$= \frac{1}{2}\left(e^{2x} + 2 + e^{-2x}\right) - 1$$

Rearrange to get expression for $\cosh 2x$.

$$= \frac{1}{2}\left(e^{2x} + e^{-2x}\right)$$

$$= \cosh 2x$$

There are two more Pythagorean identities involving the hyperbolic functions:

$$\text{sech}^2 x + \tanh^2 x \equiv 1$$
$$\coth^2 x - \text{cosech}^2 x \equiv 1$$

Example 10.5

Prove that $\text{sech}^2 x = 1 - \tanh^2 x$.

Solution

$$1 - \tanh^2 x = 1 - \frac{(e^x - e^{-x})^2}{(e^x + e^{-x})^2}$$

It is usually better to start with the more complicated side.

$$= \frac{(e^x + e^{-x})^2 - (e^x - e^{-x})^2}{(e^x + e^{-x})^2}$$

Expand the brackets.

$$= \frac{e^{2x} + e^{-2x} + 2 - e^{2x} - e^{-2x} + 2}{(e^x + e^{-x})^2}$$

Simplify.

$$= \frac{4}{(e^x + e^{-x})^2}$$

Notice the expression is a perfect square.

$$= \left(\frac{2}{e^x + e^{-x}}\right)^2$$

$$= \text{sech}^2 x$$

Summary

$$\tanh x \equiv \frac{\sinh x}{\cosh x}$$

$$\cosh^2 x - \sinh^2 x \equiv 1$$

$$\text{sech}^2 x + \tanh^2 x \equiv 1$$

$$\coth^2 x - \text{cosech}^2 x \equiv 1$$

$$\sinh 2x \equiv 2\sinh x \cosh x$$

$$\cosh 2x \equiv \cosh^2 x + \sinh^2 x$$

① Prove that $\cosh^2 x - \sinh^2 x \equiv 1$.

② Prove the identity: $\coth^2 x - \operatorname{cosech}^2 x \equiv 1$.

③ Prove the identity: $\sinh 2x \equiv 2 \sinh x \cosh x$.

④ Prove the identity: $\sinh(x + y) \equiv \sinh x \cosh y + \cosh x \sinh y$.

⑤ Prove that $\cosh 2x \equiv 1 + 2 \sinh^2 x$.

⑥ Prove that $\tanh 2x \equiv \dfrac{2 \tanh x}{1 + \tanh^2 x}$.

⑦ Prove that $\sinh 3x \equiv 4 \sinh^3 x + 3 \sinh x$.

3 Inverse hyperbolic functions

Solving equations of the form $\sin x = k$ requires the use of the inverse function $y = \sin^{-1} x$, also known as $y = \arcsin x$.

Similarly, the inverse function of $y = \sinh x$ can be used to solve equations of the form $\sinh x = k$.

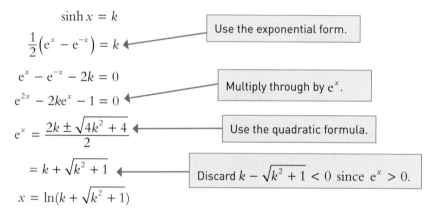

$$\sinh x = k$$

Use the exponential form.

$$\frac{1}{2}\left(e^x - e^{-x}\right) = k$$

$$e^x - e^{-x} - 2k = 0$$

Multiply through by e^x.

$$e^{2x} - 2ke^x - 1 = 0$$

Use the quadratic formula.

$$e^x = \frac{2k \pm \sqrt{4k^2 + 4}}{2}$$

$$= k + \sqrt{k^2 + 1}$$

Discard $k - \sqrt{k^2 + 1} < 0$ since $e^x > 0$.

$$x = \ln(k + \sqrt{k^2 + 1})$$

Note

A function and its inverse are always reflections of each other in $y = x$.

So, in general, $y = \ln(x + \sqrt{x^2 + 1})$ is the inverse function of $y = \sinh x$, and is written $y = \sinh^{-1} x$ or $y = \operatorname{arsinh} x$.

So, for the equation $\sinh x = 4$, $x = \operatorname{arsinh} 4 = \ln\left(4 + \sqrt{17}\right)$.

The graph of $y = \operatorname{arsinh} x$ is the reflection of $y = \sinh x$ in the line $y = x$, as shown in Figure 10.7.

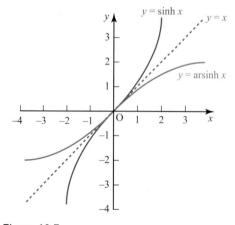

Figure 10.7

It can be seen from the graph that $y = \operatorname{arsinh} x$ is a function as there is only one value of y for each value of x. The domain of $y = \operatorname{arsinh} x$ is $x \in \mathbb{R}$ and its range is $y \in \mathbb{R}$.

The inverse function for $\tanh x$ can be found in a similar way.

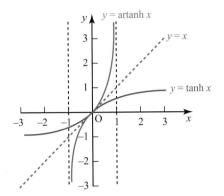

Figure 10.8

The function $y = \tanh x$ is one-to-one with domain $x \in \mathbb{R}$ and range $-1 < y < 1$ and so the inverse function $y = \operatorname{artanh} x$ is defined with domain $-1 < x < 1$ and range $y \in \mathbb{R}$.

The exponential form for $\tanh x$ can be used to find the inverse function, $y = \operatorname{artanh} x$, in logarithmic form.

$$y = \frac{e^x - e^{-x}}{e^x + e^{-x}}$$
Rearrange to make x the subject.

$$y\left(e^x + e^{-x}\right) = e^x - e^{-x}$$
Multiply through by e^x.

$$y(e^{2x} + 1) = e^{2x} - 1$$
Collect terms in e^{2x}.

$$ye^{2x} + e^{2x} = -y - 1$$

$$e^{2x} = \frac{1 + y}{1 - y}$$
Make e^{2x} the subject.

$$x = \frac{1}{2} \ln\left(\frac{1 + y}{1 - y}\right)$$
Take logs of both sides.

So, the logarithmic form of $y = \operatorname{artanh} x = \frac{1}{2} \ln\left(\frac{1 + x}{1 - x}\right)$.

The logarithmic form for $y = \operatorname{arcosh} x$, the inverse function of $y = \cosh x$, is less straightforward. Figure 10.9 shows the graph of $y = \tanh x$ and Figure 10.10 shows its reflection in $y = x$.

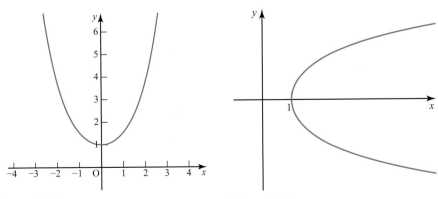

Figure 10.9   Figure 10.10

The reflection is not a function as it is a one-to-many mapping so $\cosh x$ does not have an inverse function as it stands. The domain of $y = \cosh x$ is therefore restricted to non-negative values in order that $y = \operatorname{arcosh} x$ exists.

So, the domain of $y = \cosh x$ is $x \geqslant 0$ and its range is $y \geqslant 1$.

The domain of $y = \operatorname{arcosh} x$ is $x \geqslant 1$ and its range is $y \geqslant 0$.

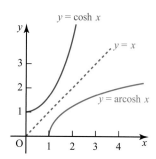

Figure 10.11

This has implications when deriving the logarithmic form for $y = \operatorname{arcosh} x$.

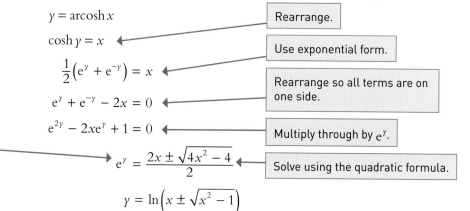

$$y = \operatorname{arcosh} x$$

Rearrange.

$$\cosh y = x$$

Use exponential form.

$$\tfrac{1}{2}\left(e^{y} + e^{-y}\right) = x$$

Rearrange so all terms are on one side.

$$e^{y} + e^{-y} - 2x = 0$$

$$e^{2y} - 2xe^{y} + 1 = 0$$

Multiply through by e^{y}.

Make y the subject.

$$e^{y} = \frac{2x \pm \sqrt{4x^{2} - 4}}{2}$$

Solve using the quadratic formula.

$$y = \ln\left(x \pm \sqrt{x^{2} - 1}\right)$$

However,

$$\ln\left(x + \sqrt{x^{2} - 1}\right) + \ln\left(x - \sqrt{x^{2} - 1}\right) = \ln\left(x + \sqrt{x^{2} - 1}\right)\left(x - \sqrt{x^{2} - 1}\right)$$

$$= \ln\left(x^{2} - \left(x^{2} - 1\right)\right)$$

$$= \ln 1$$

$$= 0$$

So y can be rewritten as $\operatorname{arcosh} x = \pm\ln\left(x + \sqrt{x^{2} - 1}\right)$.

However, $\operatorname{arcosh} x$ is only defined for $x \geqslant 1$ so the logarithmic form for $\operatorname{arcosh} x$ is

$$y = \ln\left(x + \sqrt{x^{2} - 1}\right)$$

However, this does not mean that the solution of, say, $\cosh x = 3$ has only one root, just that the inverse function is only defined for $x \geqslant 1$. The equation has two roots, symmetrically placed either side of the y-axis.

Example 10.6

(i) Sketch the graph of $y = \text{artanh}\, x + 3$, identifying any intersections with the axes and asymptotes.

(ii) Solve the equation $\text{artanh}\, x + 3 = 5$.

Solution

(i)

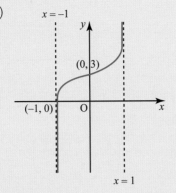

Figure 10.12

(ii) $\text{artanh}\, x + 3 = 5 \Rightarrow \text{artanh}\, x = 2 \Rightarrow x = \tanh 2$

$$x = \frac{e^2 - e^{-2}}{e^2 + e^{-2}}$$

This can be manipulated to give the answer in exact form $x = \dfrac{e^4 - 1}{e^4 + 1}$, or evaluated to give $x = 0.964$ to 3 significant figures.

ACTIVITY 10.3

Use graphing software to explore the graphs of inverses of hyperbolic functions.
Use your software to explore transformations of the inverse hyperbolic functions.

There are three more inverse hyperbolic functions. Their properties and logarithmic forms can be derived in a similar way to those that have already been covered.

Example 10.7

(i) Sketch the graph of $y = \text{sech}^{-1} x$, identifying any asymptotes.

(ii) State the domain and range of $\text{sech}^{-1} x$.

(iii) Express $\text{sech}^{-1} x$ in logarithmic form.

(iv) Hence solve $\text{sech}\, x = 0.2$.

Solution

(i)

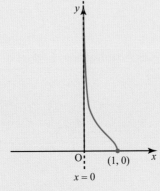

Figure 10.13

(ii) Domain is $0 < x \leqslant 1$, range is $y \geqslant 0$ so that the inverse function is defined.

(iii) $\operatorname{sech}^{-1} x = y \Rightarrow x = \operatorname{sech} y$

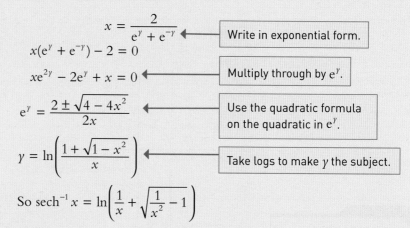

$$x = \frac{2}{e^y + e^{-y}}$$ ← Write in exponential form.

$$x(e^y + e^{-y}) - 2 = 0$$

$$xe^{2y} - 2e^y + x = 0$$ ← Multiply through by e^y.

$$e^y = \frac{2 \pm \sqrt{4 - 4x^2}}{2x}$$ ← Use the quadratic formula on the quadratic in e^y.

$$y = \ln\left(\frac{1 + \sqrt{1 - x^2}}{x}\right)$$ ← Take logs to make y the subject.

So $\operatorname{sech}^{-1} x = \ln\left(\dfrac{1}{x} + \sqrt{\dfrac{1}{x^2} - 1}\right)$

ACTIVITY 10.4

Sketch the graph of $y = \operatorname{arcosech} x$. State its domain and range and derive its logarithmic form.

(iv) $\operatorname{sech} x = 0.2$

$$\Rightarrow x = \ln\left(5 \pm 2\sqrt{6}\right)$$

Since $5 - 2\sqrt{6} > 0$ both roots exist.

Notice that the alternative notation of $\operatorname{sech}^{-1} x$ is used above in place of $\operatorname{arsech} x$. They are interchangeable.

Example 10.8

Solve the equation $4\coth^2 x + 5\operatorname{cosech} x - 10 = 0$.

Solution

$$4\coth^2 x + 5\operatorname{cosech} x - 10 = 0$$ | Use identity $\coth^2 x - \operatorname{cosech}^2 x \equiv 1$.

$$4\operatorname{cosech}^2 x + 4 + 5\operatorname{cosech} x - 10 = 0$$

$$4\operatorname{cosech}^2 x + 5\operatorname{cosech} x - 6 = 0$$ ← Simplify.

$$\left(4\operatorname{cosech} x - 3\right)\left(\operatorname{cosech} x + 2\right) = 0$$ ← Factorise.

$$\operatorname{cosech} x = \frac{3}{4} \Rightarrow \sinh x = \frac{4}{3} \Rightarrow x = \operatorname{arsinh} \frac{4}{3}$$ ← Use logarithmic form of $\operatorname{arsinh} x$.

$$= \ln\left(\frac{4}{3} + \sqrt{\frac{16}{9} + 1}\right) = \ln 3$$

$$\operatorname{cosech} x = -2 \Rightarrow \sinh x = -\frac{1}{2} \Rightarrow x = \operatorname{arsinh}\left(-\frac{1}{2}\right)$$

$$= \ln\left(-\frac{1}{2} + \sqrt{\frac{1}{4} + 1}\right) = \ln\frac{1}{2}\left(\sqrt{5} - 1\right)$$

Table 10.2 summarises the information for the six inverse hyperbolic functions.

Function	Also called	Logarithmic form	Domain	Range
$\operatorname{arcosh} x$	$\cosh^{-1} x$	$\ln\left(x + \sqrt{x^2 - 1}\right)$	$x \geqslant 1$	$y \geqslant 0$
$\operatorname{arsinh} x$	$\sinh^{-1} x$	$\ln\left(x + \sqrt{x^2 + 1}\right)$	$x \in \mathbb{R}$	$y \in \mathbb{R}$
$\operatorname{artanh} x$	$\tanh^{-1} x$	$\frac{1}{2}\ln\left(\frac{1+x}{1-x}\right)$	$-1 < x < 1$	$y \in \mathbb{R}$
$\operatorname{arsech} x$	$\operatorname{sech}^{-1} x$	$\ln\left(\frac{1}{x} + \sqrt{\frac{1}{x^2} - 1}\right)$	$0 < x \leqslant 1$	$y \geqslant 0$
$\operatorname{arcosech} x$	$\operatorname{cosech}^{-1} x$	$\ln\left(\frac{1}{x} + \sqrt{\frac{1}{x^2} + 1}\right)$	$x \in \mathbb{R}$	$y \in \mathbb{R}$
$\operatorname{arcoth} x$	$\coth^{-1} x$	$\frac{1}{2}\ln\left(\frac{1+x}{x-1}\right)$	$x < -1,\ x > 1$	$y \in \mathbb{R}$

Table 10.2

> **Note**
>
> It is not usually necessary to work with the logarithmic forms of $\operatorname{arsech} x$, $\operatorname{arcosech} x$, and $\operatorname{arcoth} x$, since, for example, $\operatorname{arsech} x$ can be written as $\operatorname{arcosh} \frac{1}{x}$.

Exercise 10.3

① Find the exact values of
   (a) $\operatorname{arcosh} 5$
   (b) $\operatorname{artanh} 0.2$
   (c) $\operatorname{arsinh}(-3)$
   (d) $\operatorname{artanh}(-0.5)$
   (e) $\operatorname{arsinh} \frac{3}{4}$
   (f) $\operatorname{arcosh} 2$.

② Find the exact values of
   (a) $\operatorname{arsech} 0.5$
   (b) $\operatorname{arcoth} 5$
   (c) $\operatorname{arcosech}\left(-\frac{1}{3}\right)$.

③ Sketch graphs to show how many solutions there are to the equation $2 - \cosh x = \operatorname{arcosech} x$.

④ Solve the equation $6 \tanh^2 x - 7 \tanh x + 2 = 0$, giving the solutions in exact form.

⑤ (i) Sketch a graph of the function $y = \operatorname{arsinh}(x - 3)$.
   (ii) State the domain and range of $y = \operatorname{arsinh}(x - 3)$.

⑥ Solve the equation $12 \operatorname{cosech}^2 x + 19 \operatorname{cosech} x - 10 = 0$. Give the solutions in exact form.

⑦ Derive the logarithmic form for $\operatorname{arcoth} x$.

⑧ Find all the real solutions of the following equations.
   (i) $4 \tanh x = \coth x$
   (ii) $3 \tanh x = 4(1 - \operatorname{sech} x)$
   (iii) $3 \operatorname{sech}^2 x + \tanh x = 3$

⑨ (i) Sketch the graph of $y = \operatorname{arcoth}(x - 1)$.
   (ii) State the domain and range.

⑩ Solve the equation $\operatorname{arsech}\left(\frac{1}{x^2}\right) = \ln 2$, giving your solution in exact form.

⑪ Express $\operatorname{arcosech} \frac{x}{x - 1}$ in terms of natural logarithms.

4 Calculus and hyperbolic functions

The derivatives of the hyperbolic functions

Differentiating shows that

$$\frac{d}{dx}(\cosh x) = \frac{d}{dx}\left(\frac{e^x + e^{-x}}{2}\right) = \frac{e^x - e^{-x}}{2} = \sinh x$$

and

$$\frac{d}{dx}(\sinh x) = \frac{d}{dx}\left(\frac{e^x - e^{-x}}{2}\right) = \frac{e^x + e^{-x}}{2} = \cosh x$$

The other hyperbolic functions can also be differentiated. As with any differentiation it is worth considering what methods are available and aiming to choose the most efficient.

Example 10.9

Find $\dfrac{d(\tanh x)}{dx}$.

Solution

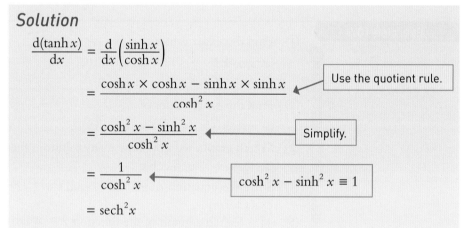

$$\frac{d(\tanh x)}{dx} = \frac{d}{dx}\left(\frac{\sinh x}{\cosh x}\right)$$

$$= \frac{\cosh x \times \cosh x - \sinh x \times \sinh x}{\cosh^2 x} \quad \text{Use the quotient rule.}$$

$$= \frac{\cosh^2 x - \sinh^2 x}{\cosh^2 x} \quad \text{Simplify.}$$

$$= \frac{1}{\cosh^2 x} \quad \cosh^2 x - \sinh^2 x \equiv 1$$

$$= \operatorname{sech}^2 x$$

ACTIVITY 10.5

Differentiate $\operatorname{sech} x$ with respect to x.

Integrating hyperbolic functions

ACTIVITY 10.6

By writing $\tanh x$ as $\dfrac{\sinh x}{\cosh x}$, find $\int \tanh x \, dx$.

It is easy to write down the integrals of some of the hyperbolic functions as they are derivatives of known functions.

$$\int \sinh x \, dx = \cosh x + c \quad \text{and} \quad \int \cosh x \, dx = \sinh x + c$$

Others require the use of a variety of integration techniques.

Differentiating the inverse hyperbolic functions

It is possible to differentiate the logarithmic versions of arcosh x and arsinh x, but it is easier to work with the hyperbolic functions themselves and their identities:

$$y = \operatorname{arcosh} x$$

$$\cosh y = x$$

$$\sinh y \frac{dy}{dx} = 1 \quad \text{Differentiating implicitly.}$$

ACTIVITY 10.7

Use a similar method to show that
$$\frac{d}{dx}\left(\text{arsinh}\, x\right) = \frac{1}{\sqrt{x^2 + 1}}.$$
You should recognise a similarity with the inverse trigonometric results in Chapter 6.

ACTIVITY 10.8

Find $\dfrac{d}{dx}\text{artanh}\, x.$

$$\frac{dy}{dx} = \frac{1}{\sinh y}$$

$$= \frac{1}{\pm\sqrt{\cosh^2 y - 1}} \quad \longleftarrow \boxed{\text{Using the identity } \cosh^2 y - \sinh^2 y \equiv 1.}$$

$$= \frac{1}{\pm\sqrt{x^2 - 1}}$$

Since the gradient of $y = \text{arcosh}\, x$ is always positive you must take the positive square root, so

$$\frac{d}{dx}\left(\text{arcosh}\, x\right) = \frac{1}{\sqrt{x^2 - 1}}.$$

Example 10.10

Find $\dfrac{d}{dx}\text{arsech}\, x.$

Solution

$$y = \text{arsech}\, x \Rightarrow \text{sech}\, y = x \quad \longleftarrow \boxed{\text{Differentiate implicitly.}}$$

$$-\text{sech}\, y \tanh y \frac{dy}{dx} = 1$$

$$\frac{dy}{dx} = -\frac{1}{\text{sech}\, y \tanh y}$$

$$= -\frac{\cosh^2 y}{\sinh y}$$

$$= -\frac{\dfrac{1}{x^2}}{\sqrt{\dfrac{1}{x^2} - 1}}$$

So $\quad \dfrac{d}{dx}\text{arsech}\, x = -\dfrac{1}{x\sqrt{1 - x^2}}.$

Integration using inverse hyperbolic functions

From the derivatives of $\text{arcosh}\, x$ and $\text{arsinh}\, x$, given previously, it is clear by integrating that

$$\int \frac{1}{\sqrt{x^2 - 1}}\, dx = \text{arcosh}\, x + c$$

and

$$\int \frac{1}{\sqrt{x^2 + 1}}\, dx = \text{arsinh}\, x + c.$$

ACTIVITY 10.9

Use the chain rule and the derivatives for $\text{arsinh}\, x$ and $\text{arcosh}\, x$ given above, to show that

- $\dfrac{d}{dx}\left(\text{arsinh}\, \dfrac{x}{a}\right) = \dfrac{1}{\sqrt{x^2 + a^2}}$

- $\dfrac{d}{dx}\left(\text{arcosh}\, \dfrac{x}{a}\right) = \dfrac{1}{\sqrt{x^2 - a^2}}.$

Note

When evaluating a definite integral, it is usually easier to work with the logarithmic form.

The results in Activity 10.9 lead to the following results:

$$\int \frac{1}{\sqrt{x^2 + a^2}}\, dx = \operatorname{arsinh}\frac{x}{a} + c \text{ or } \ln\left(x + \sqrt{x^2 + a^2}\right) + c$$

$$\int \frac{1}{\sqrt{x^2 - a^2}}\, dx = \operatorname{arcosh}\frac{x}{a} + c \text{ or } \ln\left(x - \sqrt{x^2 + a^2}\right) + c.$$

By comparing these results with those from Chapter 6 you should see that the range of functions that you can integrate has been extended. Just as before, more complicated examples use techniques such as taking out constant factors or completing the square. You can often choose an appropriate substitution, even if the integral is not quite a standard one.

Example 10.11

Find $\int \dfrac{1}{\sqrt{9x^2 - 25}}\, dx$.

To use the standard integral, the coefficient of x^2 must be 1, so you need to take out a factor of 9, which becomes 3 when it leaves the square root.

Solution

$$\int \frac{1}{\sqrt{9x^2 - 25}}\, dx = \frac{1}{3}\int \frac{1}{\sqrt{x^2 - \frac{25}{9}}}\, dx$$

This is now a standard integral with $a = \dfrac{5}{3}$.

$$= \frac{1}{3}\operatorname{arcosh}\left(\frac{x}{\left(\frac{5}{3}\right)}\right) + c$$

$$= \frac{1}{3}\operatorname{arcosh}\left(\frac{3x}{5}\right) + c$$

Exercise 10.4

① Use the chain rule or product rule (as appropriate) to differentiate the following functions, with respect to x.

(i) $\sinh 4x$ (ii) $\cosh x^2$

(iii) $\cosh^2 x$ (iv) $\cosh x \sinh x$

② Differentiate the following with respect to x.

(i) $\operatorname{cosech} x$ (ii) $\coth x$

③ Find the following integrals, using the suggested methods:

(i) $\int \sinh 3x\, dx$ (use a substitution of $u = 3x$, if necessary)

(ii) $\int x \sinh x\, dx$ (use integration by parts)

(iii) $\int x \cosh\left(1 + x^2\right) dx$ (use a substitution of $u = 1 + x^2$, if necessary).

④ Differentiate the following.

(i) $\operatorname{arsinh} 3x$ (ii) $\operatorname{arcosh} 4x$

(iii) $\operatorname{arsinh}(2x^2)$ (iv) $\operatorname{arcosh}(2x + 1)$

⑤ Use implicit differentiation to show that

(i) $\dfrac{d}{dx}\left(\operatorname{arsinh}\dfrac{x}{a}\right) = \dfrac{1}{\sqrt{x^2 + a^2}}$

(ii) $\dfrac{d}{dx}\left(\operatorname{arcosh}\dfrac{x}{a}\right) = \dfrac{1}{\sqrt{x^2 - a^2}}.$

⑥ Integrate the following with respect to x.

(i) $\operatorname{arcosh} x$

(ii) $\operatorname{arsinh} x$

(iii) $\operatorname{artanh} x$

Hint: write $\operatorname{arcosh} x = 1 \times \operatorname{arcosh} x$ and integrate by parts.

⑦ Use the standard results to find the following indefinite integrals.

(i) $\displaystyle\int \frac{1}{\sqrt{x^2 - 4}}\,dx$ (ii) $\displaystyle\int \frac{1}{\sqrt{x^2 + 4}}\,dx$ (iii) $\displaystyle\int \frac{1}{\sqrt{9x^2 - 1}}\,dx$

(iv) $\displaystyle\int \frac{1}{\sqrt{9x^2 + 1}}\,dx$ (v) $\displaystyle\int \frac{1}{\sqrt{4x^2 + 9}}\,dx$ (vi) $\displaystyle\int \frac{1}{\sqrt{4x^2 - 9}}\,dx$

⑧ Use the standard results to find the following definite integrals, giving your answers in terms of logarithms.

(i) $\displaystyle\int_3^6 \frac{1}{\sqrt{x^2 - 9}}\,dx$ (ii) $\displaystyle\int_3^6 \frac{1}{\sqrt{x^2 + 9}}\,dx$

(iii) $\displaystyle\int_1^2 \frac{1}{\sqrt{25x^2 - 16}}\,dx$ (iv) $\displaystyle\int_0^2 \frac{1}{\sqrt{9x^2 - 4}}\,dx$

⑨ (i) Use the chain rule to differentiate
$$\ln\left(x + \sqrt{x^2 - 1}\right).$$

(ii) Show, by multiplying the numerator and denominator by $\left(x^2 - 1\right)^{\frac{1}{2}}$, that your answer simplifies to $\dfrac{1}{\sqrt{x^2 - 1}}$.

(iii) Differentiate $\ln\left(x + \sqrt{x^2 + 1}\right)$ to show that the derivative of $\operatorname{arsinh} x$ is $\dfrac{1}{\sqrt{x^2 + 1}}$.

⑩ (i) Prove that $\dfrac{d}{dx}\left(\operatorname{artanh} x\right) = \dfrac{1}{1 - x^2}$.

(ii) By using partial fractions and integrating, deduce the logarithmic form of $\operatorname{artanh} x$.

⑪ (i) Prove that $\sinh^2 x = \frac{1}{2}\left(\cosh 2x - 1\right)$.

(ii) Use a suitable substitution and the result from (i) to find $\displaystyle\int \sqrt{x^2 - 9}\,dx$.

⑫ Find the exact coordinate of the turning point of the graph $y = e^{2x}\sinh 5x$.

⑬ Find the exact area between the curve of $y = 5 - 4\cosh x$ and the x-axis.

⑭ (i) Write $4x^2 + 12x - 40$ in completed square form.

(ii) Evaluate the definite integral
$$\int_{10}^{20} \frac{1}{\sqrt{4x^2 + 12x - 40}}\,dx.$$

⑮ (i) Find $\displaystyle\int x\sinh\left(x^2\right)\,dx$.

(ii) By writing $x^3\sinh x^2$ as $x^2(x\sinh x^2)$, or otherwise, find $\displaystyle\int x^3\sinh\left(x^2\right)\,dx$.

⑯ Prove that the curves $y = \operatorname{arsinh} x$ and $y = \operatorname{arcosh} 2x$ intersect where
$$x = \frac{1}{\sqrt{3}}.$$

Sketch the curves on the same axes and shade the region bounded by the x-axis and the curves.

Find the area of the shaded region.

LEARNING OUTCOMES

Now you have finished this chapter, you should be able to:

➤ know the definitions of the hyperbolic functions and their domains and ranges, and be able to sketch their graphs

➤ understand and use the identities:
  - $\cosh^2 x - \sinh^2 x \equiv 1$
  - $\operatorname{sech}^2 x \equiv 1 - \tanh^2 x$
  - $\operatorname{cosech}^2 x \equiv \coth^2 x - 1$
  - $\cosh 2x \equiv \cosh^2 x + \sinh^2 x$
  - $\sinh 2x \equiv 2 \sinh x \cosh x$

➤ differentiate and integrate hyperbolic functions

➤ understand and use the definitions of the inverse hyperbolic functions and know their domains and ranges

➤ derive and use the logarithmic forms of the inverse hyperbolic functions

➤ recognise integrals of functions of the form $\dfrac{1}{\sqrt{x^2 + a^2}}$ and $\dfrac{1}{\sqrt{x^2 - a^2}}$ and be able to integrate related functions by using substitutions.

KEY POINTS

1

Function	Definition	Domain	Range	Graph
$\cosh x$	$\frac{1}{2}\left(e^x + e^{-x}\right)$	$x \in \mathbb{R}$	$0 < y$	
$\sinh x$	$\frac{1}{2}\left(e^x - e^{-x}\right)$	$x \in \mathbb{R}$	$y \in \mathbb{R}$	
$\tanh x$	$\dfrac{\sinh x}{\cosh x}$	$x \in \mathbb{R}$	$y > 1, y < -1$	

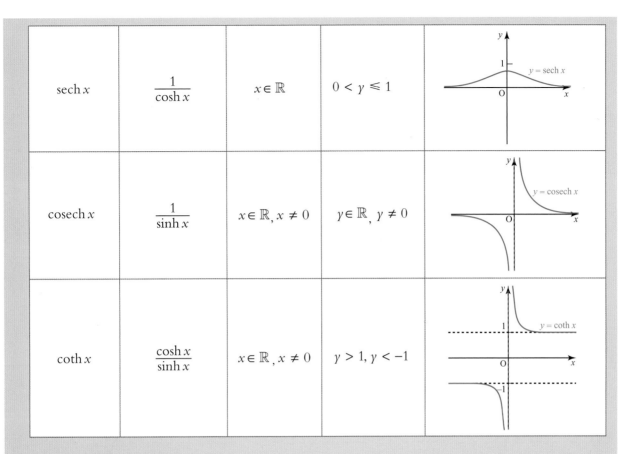

sech x	$\dfrac{1}{\cosh x}$	$x \in \mathbb{R}$	$0 < y \leqslant 1$	
cosech x	$\dfrac{1}{\sinh x}$	$x \in \mathbb{R},\, x \neq 0$	$y \in \mathbb{R},\, y \neq 0$	
coth x	$\dfrac{\cosh x}{\sinh x}$	$x \in \mathbb{R},\, x \neq 0$	$y > 1,\, y < -1$	

2 $\cosh^2 x - \sinh^2 x \equiv 1$

$\operatorname{sech}^2 x \equiv 1 - \tanh^2 x$

$\operatorname{cosech}^2 x \equiv \coth^2 x - 1$

$\cosh 2x \equiv \cosh^2 x + \sinh^2 x$

$\sinh 2x \equiv 2 \sinh x \cosh x$

3 $\dfrac{\mathrm{d}}{\mathrm{d}x}(\cosh x) + \sinh x \qquad \int \cosh x \,\mathrm{d}x = \sinh x + c$

$\dfrac{\mathrm{d}}{\mathrm{d}x}(\sinh x) + \cosh x \qquad \int \sinh x \,\mathrm{d}x = \cosh x + c$

4

Function	Logarithmic form	Domain	Range	Graph
arcosh x	$\ln(x + \sqrt{x^2 - 1})$	$x \geqslant 1$	$y \geqslant 0$	
arsinh x	$\ln(x + \sqrt{x^2 + 1})$	$x \in \mathbb{R}$	$y \in \mathbb{R}$	

artanh x	$\frac{1}{2}\ln\left(\frac{1+x}{1-x}\right)$	$-1 < x < 1$	$y \in \mathbb{R}$	
arsech x	$\ln\left(\frac{1}{x} + \sqrt{\frac{1}{x^2} - 1}\right)$	$0 < x \leqslant 1$	$y \geqslant 0$	
arcosech x	$\ln\left(\frac{1}{x} + \sqrt{\frac{1}{x^2} + 1}\right)$	$x \in \mathbb{R}$	$y \in \mathbb{R}$	
arcoth x	$\frac{1}{2}\ln\left(\frac{1+x}{x-1}\right)$	$x < -1,\ x > 1$	$y \in \mathbb{R}$	

5
$$\int \frac{1}{\sqrt{x^2 - a^2}}\, dx = \operatorname{arcosh}\frac{x}{a} + c \quad \text{or} \quad \ln\left(x + \sqrt{x^2 - a^2}\right) + c$$

$$\int \frac{1}{\sqrt{x^2 + a^2}}\, dx = \operatorname{arsinh}\frac{x}{a} + c \quad \text{or} \quad \ln\left(x + \sqrt{x^2 + a^2}\right) + c$$

11

Applications of integration

> *Mathematics is not only real, but it is the only reality.*
>
> Martin Gardner

Discussion point

→ What plane shape would need to be rotated through 360° to produce a solid in the shape of the above vase?

Review: Volumes of revolution and mean values

Volumes of revolution

When the region between a curve and an axis is rotated around that axis it generates a solid shape. The volume of that shape can be found using calculus.

Figure R.1 shows a curve with the region beneath it shaded and Figure R.2 shows the volume generated when that shaded region is rotated through 360° about the x-axis.

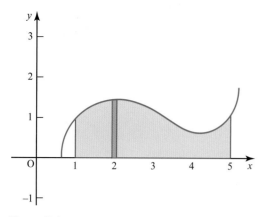

Figure R.1

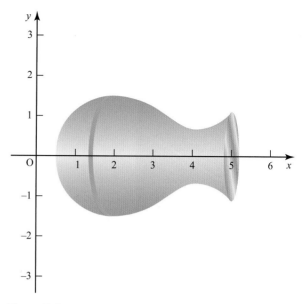

Figure R.2

When the green strip in Figure R.1 is rotated about the x-axis it forms a disc that is approximately a cylinder.

The cylinder has radius y and height δx, so its volume is given by:

$$\delta V = \pi y^2 \delta x$$

Adding all the discs gives an approximation to the volume of the complete solid. Using thinner discs gives a better approximation

$$V \approx \sum \delta V$$

$$V \approx \sum_{x=a}^{x=b} \pi y^2 \delta x$$

The limit of this sum, as $\delta x \to 0$, becomes an integral that enables the exact value of the volume to be calculated:

$$V = \int_a^b \pi y^2 \, dx$$

where a and b are the values of x at either end of the shaded region.

Similarly, the volume generated when the region from $y = a$ to $y = b$, between a curve and the y-axis is rotated $360°$ about the y-axis is given by:

$$V = \int_a^b \pi x^2 \, dy$$

Example 11.1

The region between the curve $y = x^2 + 1$, the y-axis and the lines $y = 2$ and $y = 4$, as shown in Figure R.3, is rotated through $360°$ about the y-axis.

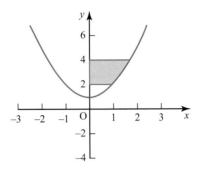

Figure R.3

Find the exact volume of the solid that is formed.

Solution

$y = x^2 + 1 \Rightarrow x^2 = y - 1$

$V = \int_2^4 \pi x^2 \, dy$

$V = \pi \int_2^4 (y - 1) \, dy$ ← Substitute for x^2.

$V = \pi \left[\dfrac{y^2}{2} - y \right]_2^4$ ← Integrate with respect to y.

$V = \pi \big((8 - 4) - (2 - 2) \big)$ ← Substitute both limits.

$\quad = 4\pi$

The mean value of a function

It is possible to find the mean value of a function over a given domain.

Figure R.4 shows a curve, $y = f(x)$. The region under the curve between $x = a$ and $x = b$ is shaded blue.

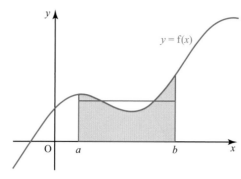

Figure R.4

The area of the green rectangle has the same area as the shaded region. The height of the rectangle is the mean value for $f(x)$ for the domain $a \leqslant x \leqslant b$.

The area is given by $\int_a^b f(x)\,dx$ and the width of the rectangle is $b - a$.

So, the mean value of the function $f(x)$ on the interval $[a, b]$ is given by:

$$\frac{1}{b-a} \int_a^b f(x)\,dx$$

Example 11.2

Find the mean value of the function $y = \cosh x$ for $0 \leqslant x \leqslant 3$.

Solution

$$\text{Mean value} = \frac{1}{3-0} \int_0^3 \cosh x\,dx$$

$$= \frac{1}{3}\left[\sinh x\right]_0^3$$

$$= \frac{1}{3}(\sinh 3 - \sinh 0)$$

$$= \frac{1}{6}\left(e^3 - \frac{1}{e^3}\right)$$

Review exercise

① The region bounded by the curve $y = \sin x$ and the x-axis from 0 to π is rotated through 2π radians about the x-axis to form a solid. Find the exact value of the solid generated.

② A solid is formed when the region between the curve $y = x^3$, the y-axis and the line $y = 8$ is rotated through $360°$ about the y-axis. Find its volume in exact form.

③ Find the mean value of the function $f(x) = \dfrac{1}{\sqrt{9 - x^2}}$ for $0 \leqslant x \leqslant 3$.

④ The mean value of the function $f(x) = x^2$ over the interval $-a \leqslant x \leqslant a$ is $\dfrac{4}{3}$. Find a.

⑤ Figure R.5 shows the function $y = xe^{-x}$. The region, R, is rotated about the x-axis through $360°$. Find the volume of the solid formed to 4 decimal places.

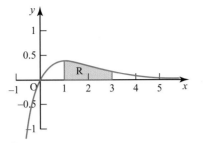

Figure R.5

⑥ The region between the x-axis and the graph of $x^2 + y^2 = 4$ is rotated through $360°$ about the y-axis. Find the volume of the solid generated.

Explain how you could check your answer geometrically.

⑦ Find the mean value of the function $f(x) = x\sinh x$ over the interval $[-1,1]$.

⑧ Figure R.6 shows the curve $y^2 = 8x$. The region, marked R, is the region between the curve, the x-axis and the lines $x = 1$ and $x = 3$.

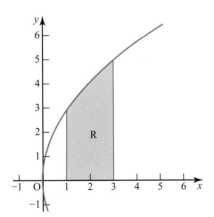

Figure R.6

The region, R, is rotated through $360°$ about the x-axis, forming a solid.

(i) By considering the volume of discs generated by the rotation of rectangular strips, show that the volume of the solid is related to the integral $V = \int_1^3 x \, dx$.

(ii) Find the exact volume of the solid.

⑨ (i) Find the mean value of the function $f(x) = x\ln x$ over the interval $1 \leqslant x \leqslant 5$.

(ii) Deduce the mean value of the function $g(x) = f(x) + a$ over the same interval.

⑩ The region bounded by the curve $y = x^2 - 4$, the x-axis and the line $x = 3$ is rotated through $360°$ about the y-axis. Calculate the volume of the solid formed in exact form.

1 General integration

Hyperbolic and circular substitutions

In Chapter 6 you used trigonometric substitutions (circular functions) to integrate certain functions, and in Chapter 10 you used hyperbolic substitutions to integrate similar looking functions. You should now be in a position to distinguish which method is appropriate, alongside all the other techniques of integration you have learned.

The standard integrals and their connections to the identities are clues to which technique to use:

Function	Underlying relationships
$\displaystyle \int \frac{1}{\sqrt{a^2 - x^2}}\, dx = \arcsin\left(\frac{x}{a}\right) + c$	This works because $\dfrac{d}{dx}(\sin(x)) = \cos(x)$ and $\cos(x) = \sqrt{1 - \sin^2(x)}$.
$\displaystyle \int \frac{1}{a^2 + x^2}\, dx = \frac{1}{a}\arctan\left(\frac{x}{a}\right) + c$	This works because $\dfrac{d}{dx}(\tan(x)) = \sec^2(x)$ and $\sec^2(x) = 1 + \tan^2(x)$.
$\displaystyle \int \frac{1}{\sqrt{x^2 - a^2}}\, dx = \operatorname{arcosh}\left(\frac{x}{a}\right) + c$	This works because $\dfrac{d}{dx}(\cosh(x)) = \sinh(x)$ and $\sinh(x) = \sqrt{\cosh^2(x) - 1}$.
$\displaystyle \int \frac{1}{\sqrt{a^2 + x^2}}\, dx = \operatorname{arsinh}\left(\frac{x}{a}\right) + c$	This works because $\dfrac{d}{dx}(\sinh(x)) = \cosh(x)$ and $\cosh(x) = \sqrt{1 + \sinh^2(x)}$.

Table 11.1

It is helpful to recognise how the trigonometric and hyperbolic identities suggest a useful substitution. While the above results are standard and can be quoted, you may often find integrals that have a slightly different form, but similar terms, where making an appropriate trigonometric or hyperbolic substitution will simplify the calculations.

Example 11.3

Find the following integrals:

(i) $\displaystyle \int \frac{1}{x^2 + 4}\, dx$

(ii) $\displaystyle \int \frac{1}{x^2 - 4}\, dx$

(iii) $\displaystyle \int \frac{1}{\sqrt{x^2 - 4}}\, dx$

(iv) $\displaystyle \int \sqrt{x^2 - 4}\, dx$

(v) $\displaystyle \int \frac{1}{\sqrt{x^2 + 4x + 13}}\, dx$

(vi) $\displaystyle \int \frac{13}{(x - 3)(x^2 + 4)}\, dx$

Solution

(i) Use the standard result (or a $x = 2\tan u$ substitution) to get

$$\int \frac{1}{x^2 + 4}\,dx = \frac{1}{2}\arctan\left(\frac{x}{2}\right) + c$$

(ii) Although it looks like the standard examples in the table above, the denominator is the difference of two squares, and so should be factorised and then partial fractions used:

$$\frac{1}{4}\int \frac{1}{x-2} - \frac{1}{x+2}\,dx = \frac{1}{4}\left(\ln|x-2| - \ln|x+2|\right) + c$$
$$= \frac{1}{4}\ln\left|\frac{x-2}{x+2}\right| + c$$

(iii) Use the standard result (or a $x = 2\cosh u$ substitution) to get

$$\int \frac{1}{\sqrt{x^2 - 4}}\,dx = \operatorname{arcosh}\left(\frac{x}{2}\right) + c$$

(iv) This is not a standard result, but it looks like the form of a $\cosh u$ identity. Try the substitution $x = 2\cosh u$ then $\frac{dx}{du} = 2\sinh u$ and so

$$\int \sqrt{x^2 - 4}\,dx = \int \sqrt{4\cosh^2 u - 4} \times 2\sinh u\,du$$
$$= 4\int \sinh^2 u\,du$$
$$= \int (e^{2u} - 2 + e^{-2u})\,du$$
$$= \frac{1}{2}e^{2u} - 2u - \frac{1}{2}e^{-2u} + c$$
$$= \sinh 2u - 2\operatorname{arcosh}\left(\frac{x}{2}\right) + c$$
$$= 2\sinh u\cosh u - 2\operatorname{arcosh}\left(\frac{x}{2}\right) + c$$
$$= \frac{1}{2}x\sqrt{x^2 - 4} - 2\operatorname{arcosh}\left(\frac{x}{2}\right) + c$$
$$= \frac{1}{2}x\sqrt{x^2 - 4} - 2\ln\left|\frac{x}{2} + \frac{1}{2}\sqrt{x^2 - 4}\right| + c$$

(v) The quadratic in the denominator has no real roots since $b^2 - 4ac = 16 - 4 \times 13 < 0$.
Complete the square:

$$\int \frac{1}{\sqrt{(x^2 + 4x + 13)}}\,dx = \int \frac{1}{\sqrt{(x+2)^2 + 9}}\,dx$$

> It looks like a $\sinh u$ substitution will help: so let $x + 2 = 3\sinh u$, then $dx = 3\cosh u\,du$.

$$= \int \frac{1}{\sqrt{9\sinh^2 u + 9}} \times 3\cosh u\,du$$
$$= \int 1\,du$$
$$= u + c$$
$$= \operatorname{arsinh}\left(\frac{x+2}{3}\right) + c$$

(vi) Using partial fractions gives

$$\int \frac{13}{(x-3)(x^2+4)}\,dx = \int \frac{1}{x-3} - \frac{x+3}{x^2+4}\,dx$$

$$= \ln|x-3| - \int \frac{x}{x^2+4}\,dx - \int \frac{3}{x^2+4}\,dx + c$$

$$= \ln|x-3| - \frac{1}{2}\ln|x^2+4| - \frac{3}{2}\arctan\left(\frac{x}{2}\right) + c$$

Exercise 11.1

① Find the following integrals:

(i) $\displaystyle\int \frac{1}{\sqrt{9-x^2}}\,dx$ (ii) $\displaystyle\int \frac{1}{\sqrt{x^2-9}}\,dx$ (iii) $\displaystyle\int \frac{1}{\sqrt{9+x^2}}\,dx$

(iv) $\displaystyle\int \frac{1}{9+x^2}\,dx$ (v) $\displaystyle\int \frac{1}{9-x^2}\,dx$

② Evaluate the following, giving your answers in exact form:

(i) $\displaystyle\int_0^5 \frac{1}{x^2+25}\,dx$ (ii) $\displaystyle\int_0^4 \frac{1}{25-x^2}\,dx$ (iii) $\displaystyle\int_0^5 \frac{1}{\sqrt{25+x^2}}\,dx$

(iv) $\displaystyle\int_5^{10} \frac{1}{\sqrt{x^2-25}}\,dx$ (v) $\displaystyle\int_0^5 \frac{1}{\sqrt{25-x^2}}\,dx$

③ Find:

(i) $\displaystyle\int \frac{1}{\sqrt{4-9x^2}}\,dx$ (ii) $\displaystyle\int \frac{1}{\sqrt{4+9x^2}}\,dx$ (iii) $\displaystyle\int \frac{1}{\sqrt{9x^2-4}}\,dx$

④ Evaluate the following, giving your answers in exact form:

(i) $\displaystyle\int_0^1 \frac{1}{3x^2+1}\,dx$ (ii) $\displaystyle\int_{\frac{2}{\sqrt 3}}^{\frac{3}{\sqrt 3}} \frac{1}{\sqrt{3x^2-4}}\,dx$ (iii) $\displaystyle\int_0^1 \frac{1}{\sqrt{4-3x^2}}\,dx$

⑤ Show that the volume of the solid of revolution formed by rotating the function $f(x) = \dfrac{1}{\sqrt{4+x^2}}$ between $x = 0$ and $x = 5$ about the x-axis, is 1.87 units³ to 3 d.p.

⑥ Calculate the mean value of the function $f(x) = \dfrac{3x}{\sqrt{x^2-9}}$ on the interval $[5, 6]$.

⑦ Use a suitable trigonometric or hyperbolic substitution to find $\displaystyle\int_0^6 \frac{1}{\left(x^2+64\right)^{\frac{3}{2}}}\,dx$, giving your answer in exact form.

⑧ Use partial fractions to find $\displaystyle\int \frac{4x^2-6x+5}{(x-1)^2\left(x^2+2\right)}\,dx$.

⑨ Find $\displaystyle\int_0^{\sqrt 2} \frac{3x-1}{x^2+2}\,dx$, giving your answer in exact form.

⑩ Figure 11.1 shows the curve $y = f(x)$, where $f(x) = \dfrac{10}{(x+1)(x+2)(x^2+1)}$.

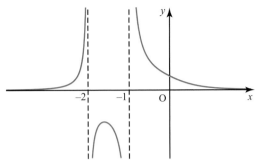

Figure 11.1

(i) Find the area under the graph between $x = 0$ and $x = 1$.

(ii) Why is it not possible to find the area under the graph for all values of x?

(iii) For which of the following intervals does the area under the graph converge? Give the area in any cases which do converge.

　(a)　$[-3, -2]$　　　(b)　$[-2, -1]$　　　(c)　$[-1, \infty]$　　　(d)　$[0, \infty]$

⑪ Figure 11.2 shows the graph of $y = f(x)$, where $f(x) = x^2(2 - x)$.

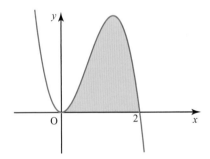

Figure 11.2

(i) Find the area of the shaded region.

(ii) The line $y = a$ is drawn so that the area under the line between $x = 0$ and $x = 2$ is the same as the area in (i). Find the value of a.

(iii) Find the volume of the solid formed by rotating the shaded region through $360°$ about the x-axis.

(iv) The line $y = b$ is drawn so that when the region under the line between $x = 0$ and $x = 2$ is rotated through $360°$ about the x-axis, the cylinder formed has the same volume as the solid in (iii). Find the value of b.

⑫ Evaluate $\displaystyle\int_0^1 \frac{2}{\sqrt{x^2 + 6x + 10}}\, dx$, giving your answer in exact form.

2 Two limits

In Chapter 6 the limits you worked with were obvious but this is not always the case.

Think about $\lim\limits_{x \to \infty} x^k e^{-x}$. As x approaches infinity x^k also approaches infinity but e^{-x} approaches zero. So it is not clear what the limit of the expression is, and therefore it is necessary to look more closely at what is happening.

First, look at the case when $k = 2$: $\lim\limits_{x \to \infty} x^2 e^{-x}$.

$$\lim_{x \to \infty} x^2 e^{-x} = \lim_{x \to \infty} \frac{x^2}{e^x}$$

$$= \lim_{x \to \infty} \frac{x^2}{1 + x + \dfrac{x^2}{2} + \dfrac{x^3}{6} + \ldots + \dfrac{x^k}{k!} + \ldots}$$

Substitute for e^x using its Maclaurin series expansion

$$= \lim_{x \to \infty} \frac{1}{\dfrac{1}{x^2} + \dfrac{1}{x} + \dfrac{1}{2} + \dfrac{x}{6} + \ldots + \dfrac{x^{k-2}}{k!} + \ldots}$$

Divide top and bottom by x^2.

$\dfrac{1}{x^2} \to 0$ and $\dfrac{1}{x} \to 0$ as $x \to \infty$

and $\dfrac{x}{6} + \ldots + \dfrac{x^{k-2}}{k!} + \ldots \to \infty$ as $x \to \infty$

So $\lim\limits_{x \to \infty} \dfrac{1}{\dfrac{1}{x^2} + \dfrac{1}{x} + \dfrac{1}{2} + \dfrac{x}{6} + \ldots + \dfrac{x^{k-2}}{k!} + \ldots} = 0 \Rightarrow \lim\limits_{x \to \infty} x^2 e^{-x} = 0$

So, in the general case:

$$\lim_{x \to \infty} x^k e^{-x} = \lim_{x \to \infty} \frac{x^k}{e^x}$$

$$= \lim_{x \to \infty} \frac{x^k}{1 + x + \dfrac{x^2}{2} + \dfrac{x^3}{6} + \ldots + \dfrac{x^k}{k!} + \ldots}$$

Substitute for e^x using its Maclaurin series expansion.

$$= \lim_{x \to \infty} \frac{1}{\dfrac{1}{x^k} + \dfrac{1}{x^{k-1}} + \dfrac{1}{2x^{k-2}} + \ldots \dfrac{1}{k!} + \ldots + \dfrac{x^{n-k}}{n!} + \ldots}$$

Divide top and bottom by x^k.

As x approaches infinity the first k terms in the denominator approach zero.

Beyond the $(k + 1)$th term, the terms in the denominator approach infinity.

So $\lim\limits_{x \to \infty} x^k e^{-x} = 0$.

Example 11.4

Show that $\lim_{x \to \infty} x^3 e^{-3x} = 0$.

Solution

Let $y = 3x$ as $x \to \infty$, $y \to \infty$

> To use the result the expression must be in the appropriate form.

$$\lim_{x \to \infty} x^3 e^{-3x} = \lim_{y \to \infty} \frac{1}{27} y^3 e^{-y}$$

> Here $k = 3$ so the result may be applied.

$$= \frac{1}{27} \lim_{y \to \infty} y^3 e^{-y}$$

$$= 0$$

Another important limit is $\lim_{x \to 0} x^k \ln x$. Here x must be positive in order that $\ln x$ is defined.

The problem here is that x^k approaches zero as x approaches zero but $\ln x$ approaches negative infinity so it is difficult to tell what the limit of the product is.

In fact,

$$\lim_{x \to 0} x^k \ln x = 0 \text{ for all } k > 0$$

This can be proved using the previous result. The proof is shown in Example 11.5.

Example 11.5

Prove that $\lim_{x \to 0} x^k \ln x = 0$ for all $k > 0$.

Solution

Let $x = e^{-\frac{y}{k}}$

As $x \to 0$, $y \to \infty$ and the limit becomes

$$\lim_{y \to \infty} \left(e^{-\frac{y}{k}} \right)^k \ln e^{-\frac{y}{k}} = \lim_{y \to \infty} \left(e^{-y} \times -\frac{y}{k} \right)$$

$$= -\frac{1}{k} \lim_{y \to \infty} y e^{-y}$$

$$= 0$$

So $\lim_{x \to 0} x^k \ln x = 0$ for all $k > 0$

Example 11.6

Determine $\lim_{x \to 0}\left((1-x)^2 - 1\right)\ln x$.

Solution

Simplify the expression first:

$$\lim_{x \to 0}\left((1-x)^2 - 1\right)\ln x = \lim_{x \to 0}\left(x^2 - 2x\right)\ln x$$

$$= \lim_{x \to 0} x^2 \ln x - 2 \times \lim_{x \to 0} x \ln x$$

$$= 0$$

> Use the result
> $$\lim_{x \to 0} x^k \ln x = 0$$
> with $k = 2$, and $k = 1$.

Exercise 11.2

① Find $\lim_{x \to \infty} \dfrac{x^6}{e^x}$.

② Find $\lim_{x \to \infty} \dfrac{\ln x}{x^3}$, showing the limiting process involved.

③ Find $\lim_{x \to a}(x - a)\ln(x - a)$. Justify your answer.

④ Show $\lim_{x \to 0} x \ln\left(x^3 + x^2\right) = 0.$

⑤ (i) Express x^x in the form e^y.

　 (ii) Show $\lim_{x \to 0} x^x = 1.$

⑥ Prove $\lim_{x \to \infty} x^3 e^{-x} = 0.$

⑦ (i) Explain why $\displaystyle\int_4^\infty (x - 4)e^{-3x}\,\mathrm{d}x$ is an improper integral.

　 (ii) Evaluate $\displaystyle\int_4^\infty (x - 4)e^{-3x}\,\mathrm{d}x$, clearly showing your method.

⑧ (i) Explain why $\displaystyle\int_0^2 x^3 \ln x\,\mathrm{d}x$ is an improper integral.

　 (ii) Work out $\displaystyle\int_0^2 x^3 \ln x\,\mathrm{d}x$, showing the limiting process used.

3 Reduction formulae

Prior knowledge

You need to be able to integrate trigonometric functions and be fluent using integration by parts from *AQA A-level Mathematics Year 2*. You also need to be able to integrate hyperbolic functions from Chapter 10.

Integrating a function such as xe^x can be done using integration by parts and takes just one application of the technique. When integrating the function x^2e^x it is necessary to apply integration by parts a second time.

For the more general function, x^ne^x, it is necessary to integrate by parts repeatedly.

In this section, you will see how to derive a formula for I_n in terms of I_{n-1} or perhaps I_{n-2} or some other similar term.

Example 11.7

These formulae are **reduction formulae**.

(i) Find an expression for $I_n = \int_0^1 x^ne^x\,dx$ in terms of I_{n-1}.

(ii) Find I_3.

Solution

(i) $I_n = \int_0^1 x^ne^x\,dx$ Integrate by parts with x^n as u.

$= \left[x^ne^x\right]_0^1 - \int_0^1 nx^{n-1}e^x\,dx$ Notice the new integrand has the same structure as the original.

$= (1 \times e - 0) - n\int_0^1 x^{n-1}e^x\,dx$ This integrand is I_{n-1}.

$= e - nI_{n-1}$

So, the required reduction formula is $I_n = e - nI_{n-1}$

(ii) Begin by working out I_0.

$$I_0 = \int_0^1 e^x\,dx = \left[e^x\right]_0^1 = e - 1$$

$I_1 = e - 1I_0 = e - 1(e - 1) = 1$

$I_2 = e - 2I_1 = e - 2$

$I_3 = e - 3I_2 = e - 3(e - 2) = 6 - 2e$

In Example 11.5, part (ii), an alternative approach is to do one calculation.

$$I_3 = e - 3I_2$$

$$= e - 3\left[e - 2I_1\right] \longleftarrow \boxed{\text{Substitute for } I_2.}$$

$$= e - 3\left[e - 2\left[e - 1I_0\right]\right] \longleftarrow \boxed{\text{Substitute for } I_1.}$$

$$= e - 3\left[e - 2\left[e - \int_0^1 e^x \, dx\right]\right] \longleftarrow \boxed{\text{Substitute for } I_0.}$$

$$= e - 3\left[e - 2\left[e - \left[e^x\right]_0^1\right]\right] \longleftarrow \boxed{\text{Integrate with respect to } x.}$$

$$= e - 3\left[e - 2\left[e - (e - 1)\right]\right]$$

$$= e - 3\left[e - 2\right]$$

$$= 6 - 2e$$

Discussion point

➜ Which of the two methods for working out I_3 do you prefer? Justify your answer.

Discussion point

➜ Why do you think that reduction formulae involving trigonometric functions are likely to use I_{n-2}?

Reduction formulae can also be used to evaluate integrals involving trigonometric functions. In these cases, the reduction formula for I_n may use I_{n-2} rather than I_{n-1}.

Example 11.8

(i) Find a reduction formula for $I_n = \int \cos^n x \, dx$.

(ii) Hence find $\int \cos^4 x \, dx$.

Solution

(i) $\quad I_n = \int \cos^n x \, dx \longleftarrow \boxed{\text{Split the integrand.}}$

$$= \int \cos^{n-1} x \cos x \, dx \longleftarrow \boxed{\text{Integrate by parts.}}$$

$$= \left[\cos^{n-1} x \sin x\right] + \int (n-1) \cos^{n-2} x \sin x \sin x \, dx$$

$$= \left[\cos^{n-1} x \sin x\right] + (n-1) \int \cos^{n-2} x \sin^2 x \, dx$$

$$= \left[\cos^{n-1} x \sin x\right] + (n-1) \int \cos^{n-2} x (1 - \cos^2 x) \, dx$$

$$= \left[\cos^{n-1} x \sin x\right] + (n-1) \int (\cos^{n-2} x - \cos^n x) \, dx$$

$$= \left[\cos^{n-1} x \sin x\right] + (n-1) \int \cos^{n-2} x \, dx - (n-1) \int \cos^n x \, dx$$

$$= \left[\cos^{n-1} x \sin x\right] + (n-1) I_{n-2} - (n-1) I_n$$

Rearrange to give I_n:

$$I_n \equiv \frac{1}{n}\cos^{n-1} x \sin x + \frac{n-1}{n} I_{n-2}$$

(ii) $\displaystyle\int \cos^4 x \,\mathrm{d}x = \frac{1}{4}\cos^3 x \sin x + \frac{3}{4} I_2$

$$I_2 = \frac{1}{2}\cos x \sin x + \frac{1}{2} I_0$$

$$I_2 = \frac{1}{2}\cos x \sin x + \frac{1}{2}\int 1 \,\mathrm{d}x$$

$$= \frac{1}{2}\cos x \sin x + \frac{1}{2}x \quad (+ \text{ integrating constant})$$

So $\displaystyle\int \cos^4 x \,\mathrm{d}x = \frac{1}{4}\cos^3 x \sin x + \frac{3}{4}\left(\frac{1}{2}\cos x \sin x + \frac{1}{2}x\right) + c$

$$= \frac{1}{4}\cos^3 x \sin x + \frac{3}{8}\cos x \sin x + \frac{3}{8}x + c$$

Discussion point

→ Try finding I_2 using the double angle formula. Is it a preferable approach?"

ACTIVITY 11.1

Integrate by parts twice to find $\displaystyle\int \mathrm{e}^x \sin x \,\mathrm{d}x$.

Exercise 11.3

① (i) Given that $I_n = \displaystyle\int_0^\pi x^n \cos x \,\mathrm{d}x$, show that $I_n = -n\pi^{n-1} - n(n-1)I_{n-2}$, for $n > 2$.

Find: (ii) I_0 (iii) I_1 (iv) I_2 .

② (i) Given that $I_n = \displaystyle\int (\ln x)^n \,\mathrm{d}x$, show that $I_n = x (\ln x)^n - nI_{n-1}$ for $n > 1$ and $x > 0$.

(ii) Find I_3 .

③ $I_n = \displaystyle\int_{-\pi}^\pi x^n \cos\frac{1}{2}x \,\mathrm{d}x$

(i) Show that $I_n = 2x^n \sin\frac{1}{2}x + 4nx^{n-1}\cos\frac{1}{2}x - 4n(n-1)I_{n-2}$ for $n > 2$.

(ii) Hence find $\displaystyle\int_{-\pi}^\pi x^4 \cos\frac{1}{2}x \,\mathrm{d}x$ in exact form.

④ Find $\displaystyle\int \mathrm{e}^x \cos x \,\mathrm{d}x$.

⑤ Given that $I_n = \displaystyle\int \sin^n x \,\mathrm{d}x$, show that $nI_n = -\sin^{n-1} x \cos x + (n-1)I_{n-2}$, for $n > 2$.

⑥ Find a reduction formula for $I_n = \displaystyle\int x^n \mathrm{e}^{3x} \,\mathrm{d}x$ and hence find I_5.

⑦ Use a reduction formula to evaluate $\displaystyle\int_0^{\frac{\pi}{6}} \cos^9 3x \,\mathrm{d}x$ in exact form.

⑧ (i) Show that, for $I_n = \displaystyle\int x^n \cosh x \,\mathrm{d}x$,
$I_n = x^n \sinh x - nx^{n-1}\cosh x + n(n-1)I_{n-2}$ for $n > 2$.

(ii) Hence evaluate $\displaystyle\int_0^1 x^4 \cosh x \,\mathrm{d}x$, giving your answer to 4 decimal places.

4 Curved lengths and surface areas

Arc length

One practical way to find the length of a curve between two of its points (for example, to find the distance along the road between two towns from a map) is to mark several intermediate points along the arc, join them successively with straight lines to form a polygonal line, and measure the total length of this open polygon to give an approximation to the length of the arc. If you start with a polygon P1 and construct a new polygon P2 by inserting extra points along the arc then P2 will fit better than P1, and the length of P2 will be greater than the length of P1 (see Figure 11.3).

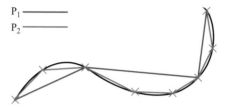

Figure 11.3

In this way, you can form more and more polygons with successively greater lengths. But since the shortest route between two points is the straight line joining them, the length of any such polygon does not exceed the length of the curve. So you would expect that the lengths of successive approximations would be bounded above (by the length of the curve), and that by putting the intermediate points sufficiently close together you could get an approximation as close as you like to the arc length.

For most curves that occur in practice this approach works, and leads to the calculus method of finding arc length.

Arc length with Cartesian coordinates

To see how to find the length of an arc, look at part of a general curve, as shown in Figure 11.4. There, C is a fixed point on the curve, P is the point with parameter p, and s is the arc length from C to P, where s is positive if and only if the motion from C to P is in the positive sense along the curve.

Let P and Q have coordinates (x, y) and $(x + \delta x, y + \delta y)$, corresponding to parametric values p and $p + \delta p$ respectively, and let the length of arc PQ $= \delta s$.

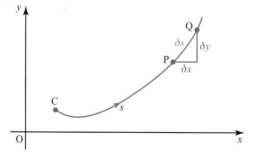

Figure 11.4

The chord length PQ is $|PQ| = \sqrt{(\delta x)^2 + (\delta y)^2}$.

Therefore $\dfrac{\delta s}{\delta p} = \dfrac{\delta s}{|PQ|} \times \dfrac{|PQ|}{\delta p}$

$$= \dfrac{\delta s}{|PQ|} \times \sqrt{\left(\dfrac{\delta x}{\delta p}\right)^2 + \left(\dfrac{\delta y}{\delta p}\right)^2}.$$

As $\delta p \to 0$, $\dfrac{\delta s}{\delta p}, \dfrac{\delta x}{\delta p}, \dfrac{\delta y}{\delta p}$ tend to $\dfrac{ds}{dp}, \dfrac{dx}{dp}, \dfrac{dy}{dp}$ respectively, and by the assumption

stated above $\dfrac{\delta s}{|PQ|} \to 1$. So taking limits as $\delta p \to 0$ gives the basic result:

$$\dfrac{ds}{dp} = \sqrt{\left(\dfrac{dx}{dp}\right)^2 + \left(\dfrac{dy}{dp}\right)^2}.$$

From this s can be found by integrating with respect to p.

If the independent variable is x (i.e. the equation of the curve is given in the form $y = f(x)$), then you put $p = x$ in the basic result. Then

$$\dfrac{dx}{dp} = \dfrac{dx}{dx} = 1 \text{ and } \dfrac{dy}{dp} = \dfrac{dy}{dx}$$

so that

$$\dfrac{ds}{dx} = \sqrt{1 + \left(\dfrac{dy}{dx}\right)^2}.$$

Similarly, when the independent variable is y,

$$\dfrac{ds}{dy} = \sqrt{\left(\dfrac{dx}{dy}\right)^2 + 1}$$

All these are easy to remember from the right-angled 'triangle' (see Figure 11.5) in which $(\delta s)^2 \approx (\delta x)^2 + (\delta y)^2$ by Pythagoras' theorem. The three results follow in the limit from dividing by $(\delta p)^2$, $(\delta x)^2$ or $(\delta y)^2$ as appropriate, and then taking the positive square root of each side. Notice that the *positive* root is needed in each case, since by definition s increases with the independent variable.

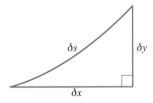

Figure 11.5

Example 11.9

Find the length of the astroid $x = a\cos^3\theta$, $y = a\sin^3\theta$.

Solution

$$\frac{dx}{d\theta} = -3a\cos^2\theta\sin\theta \qquad \frac{dy}{d\theta} = 3a\sin^2\theta\cos\theta$$

$$\Rightarrow \frac{ds}{d\theta} = \sqrt{9a^2\cos^4\theta\sin^2\theta + 9a^2\sin^4\theta\cos^2\theta}$$

$$= 3a\sqrt{\cos^2\theta\sin^2\theta\left(\cos^2\theta + \sin^2\theta\right)} \qquad (1)$$

$$= 3a\cos\theta\sin\theta$$

The values of θ at the four cusps of the curve are shown in Figure 11.6, so the length of arc in the first quadrant is:

$$\int_0^{\pi/2} 3a\cos\theta\sin\theta \, d\theta = \left[\frac{3a}{2}\sin^2\theta\right]_0^{\frac{\pi}{2}} = \frac{3a}{2}$$

and the length of the complete astroid is

$$4 \times \frac{3a}{2} = 6a$$

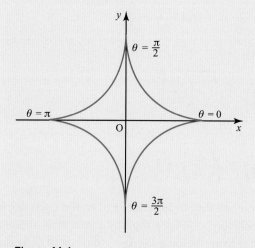

Figure 11.6

> ### Note
> ---
> If you try to find the whole length in a single integration you get
> $$\int_0^{2\pi} 3a\cos\theta\sin\theta \, d\theta = \left[\frac{3a}{2}\sin^2\theta\right]_0^{2\pi} = 0. \text{ This is because } \cos\theta\sin\theta < 0 \text{ in the}$$
> second and fourth quadrants, and the positive and negative contributions in the four quadrants have cancelled. When taking the square root (and in similar cases) it is essential to check that you use an expression which is never negative throughout the range of integration.

The curved surface area of a solid of revolution

First consider a simple case: the curved surface area of a right circular cone. If the cone has base radius r and slant height l its curved surface, flattened out, is a sector of a circle with radius l and arc length $2\pi r$, as in Figure 11.7.

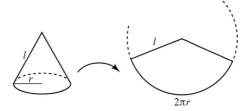

Figure 11.7

The angle at the centre of this sector is $\dfrac{2\pi r}{l}$ radians, and so the curved surface area of the cone is given by

$$\frac{1}{2}l^2 \times \frac{2\pi r}{l} = \pi r l.$$

Now consider the solid of revolution formed by rotating about the x-axis the line segment joining the points (x_1, y_1) and (x_2, y_2), where the distance between these points is δs. This solid, which is called a **frustum**, is the difference between two cones. Let the slant heights of these cones be l_1 and l_1 as shown in Figure 11.8. Then $\delta s = l_2 - l_1$.

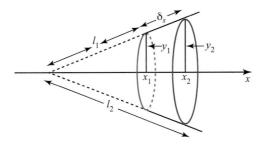

Figure 11.8

The curved surface area of the frustum, δS, is given by

$$\delta S = \pi y_2 l_2 - \pi y_1 l_1 = \pi\left(y_2 l_2 - y_1 l_1\right).$$

But by similar triangles $\dfrac{l_1}{y_1} = \dfrac{l_2}{y_2}$ so that $y_2 l_1 = y_1 l_2$. Therefore

$$\delta S = \pi\left(y_2 l_2 - y_2 l_1 + y_1 l_2 - y_1 l_1\right) \quad \longleftarrow \boxed{\text{Inserting extra terms that cancel.}}$$

$$= \pi\left(y_2 + y_1\right)\left(l_2 - l_1\right)$$

$$= 2\pi y\,\delta s$$

where $y = \dfrac{1}{2}\left(y_1 + y_2\right)$, the average radius of the frustum.

It is now easy to see how to find the curved surface area of a solid of revolution: divide the arc AB which is rotated into elements of arc δs; each of which generates a surface with area approximately $2\pi y \delta s$. Then the total curved surface area is:

$$S = \lim_{\delta s \to 0} \sum_{A}^{B} 2\pi y\,\delta s = \int_A^B 2\pi y\,\mathrm{d}s$$

Now $\dfrac{ds}{dx} dx$ can be substituted for ds to give

$S = \displaystyle\int_a^b 2\pi y \dfrac{ds}{dx} dx$, where $x = a$ at A and $x = b$ at B.

You saw, in the previous section on arc length, that $\dfrac{ds}{dx} = \sqrt{1 + \left(\dfrac{dy}{dx}\right)^2}$.

So $S = \displaystyle\int_a^b 2\pi y \sqrt{1 + \left(\dfrac{dy}{dx}\right)^2}\, dx$ when working with a function of x and y.

When x and y are given in terms of a parameter, t, with $t = t_1$ at A and $t = t_2$ at B, $\dfrac{ds}{dt} dt$ is substituted for ds to give

$$S = \int_{t_1}^{t_2} 2\pi y \dfrac{ds}{dt}\, dt$$

Again, from the previous section,

$$\dfrac{ds}{dt} = \sqrt{\left(\dfrac{dx}{dt}\right)^2 + \left(\dfrac{dy}{dt}\right)^2}$$

so

$$S = \int_{t_1}^{t_2} 2\pi y \sqrt{\left(\dfrac{dx}{dt}\right)^2 + \left(\dfrac{dy}{dt}\right)^2}\, dt.$$

Example 11.10

Find the curved surface area of the *paraboloid* formed by rotating the parabola $y^2 = 4ax$ about the x-axis from $x = 0$ to $x = a$.

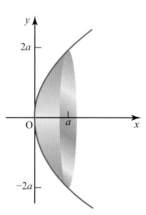

Figure 11.9

Solution

$$y^2 = 4ax \implies 2y\frac{dy}{dx} = 4a \implies \frac{dy}{dx} = \frac{2a}{y}$$

$$\implies \left(\frac{dy}{dx}\right)^2 = \frac{4a^2}{y^2} \implies \left(\frac{dy}{dx}\right)^2 = \frac{4a^2}{4ax} = \frac{a}{x}$$

So that $S = \displaystyle\int_a^b 2\pi y\sqrt{1 + \left(\frac{dy}{dx}\right)^2}\,dx$

$$= \int_0^a 2\pi\sqrt{4ax}\sqrt{1 + \frac{a}{x}}\,dx \quad\longleftarrow\quad \boxed{\text{Substitute for } \left(\frac{dy}{dx}\right)^2.}$$

$$= \int_0^a 2\pi\sqrt{4ax}\sqrt{\frac{x+a}{x}}\,dx$$

$$= \int_0^a 4\pi\sqrt{a}\sqrt{x+a}\,dx$$

$$= \int_0^a 4\pi\sqrt{a}\,(x+a)^{\frac{1}{2}}\,dx$$

$$= 4\pi\sqrt{a}\left[\frac{2}{3}(x+a)^{\frac{3}{2}}\right]_0^a$$

$$= 4\pi\sqrt{a} \times \frac{2}{3}\left((2a)^{\frac{3}{2}} - a^{\frac{3}{2}}\right)$$

$$= \frac{8}{3}\pi a^2 \times \left(2^{\frac{3}{2}} - 1\right)$$

Exercise 11.4

① Find the length of the semi-cubical parabola $y^2 = x^3$ from $(0,0)$ to $(4,8)$.

(This curve was the first for which the length was found by calculus methods by the Dutchman Heinrich van Heuraet, the Englishman William Neil and the Frenchman Pierre de Fermat independently, all between 1658 and 1660.)

② Find the curved surface area of the solid generated by rotating the following curves about the x-axis. Leave your answers is terms of π.

   (i)   The line $4y = 3x$ from $x = 4$ to $x = 8$

   (ii)   The circle $x^2 + y^2 = a^2$ from $x = -a$ to $x = a$

   (iii)   The catenary $y = c\cosh\dfrac{x}{c}$ from $x = -a$ to $x = a$.

③ Find the length of the curve $x = \dfrac{1}{1 + p^2}$, $y = \dfrac{p}{1 + p^2}$ from $p = 0$ to $p = 1$, and draw a sketch of the curve to explain your answer.

④ Find the length of the catenary $y = c\cosh\dfrac{x}{c}$ from $x = 0$ to $x = X$.

⑤ Find the curved surface area of the solid generated by rotating the following curves about the x-axis. Leave your answers in terms of π.

(i) The parabola $x = ap^2$, $y = 2ap$ from $p = 1$ to $p = 2$

(ii) One arch of the cycloid $x = a(\theta - \sin\theta)$, $y = a(1 - \cos\theta)$

(iii) The astroid $x = a\cos^3 p$, $y = a\sin^3 p$.

⑥ Show that $x = a\sinh^2 p$, $y = 2a\sinh p$ are parametric equations of the parabola $y^2 = 4ax$, and that the arc length from $p = 0$ to $p = P$ is

$$a\left(P + \frac{1}{2}\sinh 2P\right).$$

⑦ Prove that the length of one complete arch of the cycloid $x = a(\theta - \sin\theta)$, $y = a(1 - \cos\theta)$ is $8a$.

(This result was first given by Christopher Wren in 1659.)

⑧ Find the length of the nephroid $x = 3a\cos\theta - a\cos 3\theta$, $y = 3a\sin\theta - a\sin 3\theta$.

⑨ (i) For the ellipse $x = a\cos\theta$, $y = b\sin\theta$ prove that $\dfrac{ds}{d\theta} = a\sqrt{1 - e^2\cos^2\theta}$, where $b^2 = a^2\left(1 - e^2\right)$.

(ii) Prove that the perimeter of this ellipse is exactly the same as the length of one complete wave of the curve $y = ae\cos\dfrac{x}{b}$.

(iii) Prove that, if e is small, then the perimeter of this ellipse is approximately $2\pi a\left(1 - \dfrac{1}{4}e^2\right)$.

⑩ A solid of revolution is generated by rotating the curve $y = \dfrac{1}{x}$ about the x-axis from $x = 1$ to $x = k$, where $k > 1$.

(i) Prove that the volume, V, of this solid is $\pi\left(1 - \dfrac{1}{k}\right)$.

(ii) Prove that the curved surface area, S, is given by $S = \displaystyle\int_1^k \frac{2\pi}{x}\sqrt{1 + \frac{1}{x^4}}\,dx$.

(iii) Given that $1 + \dfrac{1}{x^4} > 1$, deduce that $S > 2\pi\ln k$.

LEARNING OUTCOMES

Now you have finished this chapter, you should be able to:

➤ recognise integrals of functions of the form $\left(a^2 - x^2\right)^{-\frac{1}{2}}$, $\left(a^2 + x^2\right)^{-1}$, $\left(x^2 - a^2\right)^{-\frac{1}{2}}$ and $\left(x^2 + a^2\right)^{-\frac{1}{2}}$, and be able to integrate related functions by using trigonometric or hyperbolic substitutions

➤ understand, prove and use the result $\lim\limits_{x \to \infty} x^k e^{-x} = 0$

➤ understand and use the result $\lim\limits_{x \to 0} x^k \ln x = 0$ for all $k > 0$

➤ understand how reduction formulae can be used repeatedly to evaluate integrals

➤ understand how to produce reduction formulae using integration by parts

➤ understand and use formulae for the arc length of curves in Cartesian form, and those defined parametrically

➤ understand and use formulae for the area of the surface of revolution for curves in Cartesian form, and those defined parametrically.

KEY POINTS

1 $\displaystyle\int \frac{1}{\sqrt{a^2 - x^2}}\, \mathrm{d}x = \arcsin \frac{x}{a} + c$

$\displaystyle\int \frac{1}{a^2 + x^2}\, \mathrm{d}x = \frac{1}{a}\arctan \frac{x}{a} + c$

$\displaystyle\int \frac{1}{\sqrt{x^2 - a^2}}\, \mathrm{d}x = \operatorname{arcosh} \frac{x}{a} + c$

$\displaystyle\int \frac{1}{\sqrt{x^2 + a^2}}\, \mathrm{d}x = \operatorname{arsinh} \frac{x}{a} + c$

2 $\displaystyle\lim_{x\to\infty} x^k \mathrm{e}^{-x} = 0$ and $\displaystyle\lim_{x\to 0} x^k \ln x = 0$ for all $k > 0$.

3 $\displaystyle s = \int_a^b \sqrt{1 + \left(\frac{\mathrm{d}y}{\mathrm{d}x}\right)^2}\, \mathrm{d}x$ is the arc length from $x = a$ to $x = b$ for a curve in Cartesian form.

$\displaystyle s = \int_{t_1}^{t_2} \sqrt{\left(\frac{\mathrm{d}x}{\mathrm{d}t}\right)^2 + \left(\frac{\mathrm{d}y}{\mathrm{d}t}\right)^2}\, \mathrm{d}t$ is the arc length from $t = t_1$ to $t = t_2$ for a curve in parametric form.

$\displaystyle S = \int_a^b 2\pi y \sqrt{1 + \left(\frac{\mathrm{d}y}{\mathrm{d}x}\right)^2}\, \mathrm{d}x$ is the area of the surface of revolution from $x = a$ to $x = b$ for a curve in Cartesian form.

$\displaystyle S = \int_{t_1}^{t_2} 2\pi y \sqrt{\left(\frac{\mathrm{d}x}{\mathrm{d}t}\right)^2 + \left(\frac{\mathrm{d}y}{\mathrm{d}t}\right)^2}\, \mathrm{d}t$ is the area of the surface of revolution from $t = t_1$ to $t = t_2$ for a curve in parametric form.

R Review: Roots of polynomials

1 Roots and coefficients

There are relationships between the roots and coefficients of polynomial equations. These relationships hold for complex roots as well as for real roots.

Roots and coefficients of quadratic equations

For the general quadratic equation $az^2 + bz + c = 0$, with roots α and β:

- $\alpha + \beta = -\dfrac{b}{a}$

- $\alpha\beta = \dfrac{c}{a}$

For example: the roots α and β of the quadratic equation $3z^2 + 2z - 7 = 0$ satisfy $\alpha + \beta = -\dfrac{2}{3}$ and $\alpha\beta = -\dfrac{7}{3}$.

 Note

You can derive these results by comparing the factorised form $(z - \alpha)(z - \beta) = 0$ with the equation $z^2 + \dfrac{b}{a}z + \dfrac{c}{a} = 0$.

Roots and coefficients of cubic equations

For the general cubic equation $az^3 + bz^2 + cz + d = 0$, with roots α, β and γ:

- $\alpha + \beta + \gamma = -\dfrac{b}{a}$

- $\alpha\beta + \beta\gamma + \gamma\alpha = \dfrac{c}{a}$

- $\alpha\beta\gamma = -\dfrac{d}{a}$

 Note

As for quadratic equations, you can derive these results by comparing the factorised form $(z - \alpha)(z - \beta)(z - \gamma) = 0$ with the equation $z^3 + \dfrac{b}{a}z^2 + \dfrac{c}{a}z + \dfrac{d}{a} = 0$.

Roots and coefficients of quartic equations

For the general quartic equation $az^4 + bz^3 + cz^2 + dz + e = 0$, with roots α, β, γ and δ:

- $\sum \alpha = \alpha + \beta + \gamma + \delta = -\dfrac{b}{a}$

- $\sum \alpha\beta = \alpha\beta + \alpha\gamma + \alpha\delta + \beta\gamma + \beta\delta + \gamma\delta = \dfrac{c}{a}$

- $\sum \alpha\beta\gamma = \alpha\beta\gamma + \beta\gamma\delta + \gamma\delta\alpha + \delta\alpha\beta = -\dfrac{d}{a}$

- $\alpha\beta\gamma\delta = \dfrac{e}{a}$

 Note

Shorthand notation can be used to save writing out all the combinations. So $\sum \alpha$ means the sum of the individual roots, $\sum \alpha\beta$ means the sum of all possible products of pairs of roots, and so on.

Forming new equations

The relationships between roots and coefficients can often be used to form a new equation with roots that are related to the roots of the original equation. Two different methods are shown in the following example.

| **Example R.1** | The quadratic equation $2z^2 - 5z + 8 = 0$ has roots α and β. Find a quadratic equation with roots $\alpha + 2$ and $\beta + 2$. |

Solution 1

From the original equation.

$$\alpha + \beta = \frac{5}{2}$$

$$\alpha\beta = \frac{8}{2} = 4$$

For the new equation: $-\dfrac{b}{a} = (\alpha + 2) + (\beta + 2)$

$$= \alpha + \beta + 4$$

$$= \frac{5}{2} + 4$$

$$= \frac{13}{2}$$

$$\frac{c}{a} = (\alpha + 2)(\beta + 2)$$

$$= \alpha\beta + 2(\alpha + \beta) + 4$$

$$= 4 + 2 \times \frac{5}{2} + 4$$

$$= 13$$

Taking $a = 2$ makes the coefficients integers. Any other value for a would give a multiple of the same equation.

Taking $a = 2$ gives $b = -13$ and $c = 26$.

So the new equation is $2z^2 - 13z + 26 = 0$.

Solution 2 (substitution method)

Think of the graph $y = f(x)$. The new graph is found by translating the original graph 2 units to the right, so the new graph has equation $y = f(x - 2)$.

In the new equation, $w = z + 2$ so $z = w - 2$.

Replacing z in the original equation:

$$2(w - 2)^2 - 5(w - 2) + 8 = 0$$

$$2w^2 - 8w + 8 - 5w + 10 + 8 = 0$$

$$2w^2 - 13w + 26 = 0$$

The substitution method shown above is often more efficient when dealing with higher order equations.

The cubic equation $z^3 - 3z^2 + z + 4 = 0$ has roots α, β and γ. Find a cubic equation with roots $2\alpha - 1, 2\beta - 1$ and $2\gamma - 1$.

Solution

In the new equation, $w = 2z - 1$ so $z = \dfrac{w+1}{2}$.

Replacing z in the original equation:

$$\left(\frac{w+1}{2}\right)^3 - 3\left(\frac{w+1}{2}\right)^2 + \left(\frac{w+1}{2}\right) + 4 = 0$$

$$\frac{(w+1)^3}{8} - \frac{3(w+1)^2}{4} + \frac{w+1}{2} + 4 = 0$$

$$(w+1)^3 - 6(w+1)^2 + 4(w+1) + 32 = 0$$

$$w^3 + 3w^2 + 3w + 1 - 6w^2 - 12w - 6 + 4w + 4 + 32 = 0$$

$$w^3 - 3w^2 - 5w + 29 = 0$$

Multiply through by 8. It is usually easier to eliminate the fractions before expanding the brackets.

① The quadratic equation $2z^2 - 3z - 5 = 0$ has roots α and β.

Write down the values of:

(i) $\alpha + \beta$ (ii) $\alpha\beta$

② The cubic equation $3z^3 + z^2 + 4z - 7 = 0$ has roots α, β and γ.

Write down the values of:

(i) $\alpha + \beta + \gamma$ (ii) $\alpha\beta + \beta\gamma + \gamma\alpha$ (iii) $\alpha\beta\gamma$

③ The quartic equation $z^4 - 2z^3 - 5z^2 - 3 = 0$ has roots α, β, γ and δ.

Write down the values of:

(i) $\alpha + \beta + \gamma + \delta$ (ii) $\alpha\beta + \alpha\gamma + \alpha\delta + \beta\gamma + \beta\delta + \gamma\delta$

(iii) $\alpha\beta\gamma + \beta\gamma\delta + \gamma\delta\alpha + \delta\alpha\beta$ (iv) $\alpha\beta\gamma\delta$

④ Find a cubic equation with roots 2, −4 and 5.

⑤ The quadratic equation $3z^2 + 2z + 4 = 0$ has roots α and β.

Find quadratic equations with roots:

(i) $\alpha - 1$ and $\beta - 1$ (ii) 2α and 2β (iii) $2\alpha + 3$ and $2\beta + 3$

⑥ The cubic equation $z^3 - 3z^2 - z + 1 = 0$ has roots α, β and γ.

Find cubic equations with roots:

(i) $\alpha + 2, \beta + 2$ and $\gamma + 2$

(ii) $\frac{1}{2}\alpha, \frac{1}{2}\beta$ and $\frac{1}{2}\gamma$

(iii) $1 - 3\alpha, 1 - 3\beta$ and $1 - 3\gamma$

⑦ The cubic equation $f(x) = 0$, where $f(x) = x^3 - 4x^2 - 3x + k$, has a pair of equal roots.

(i) Find the two possible values of the roots of the equation and the corresponding values of k.

(ii) Sketch the graph of $y = f(x)$ for each of the possible values of k.

⑧ One of the roots of the equation $x^2 + px + q = 0$ is three times the other. Prove that $3p^2 = 16q$.

⑨ One of the roots of the equation $z^3 - 7z^2 + k = 0$ is twice one of the other roots.

Solve the equation and find the value of k.

⑩ The equation $x^4 + px^3 - qx - 4 = 0$ has roots $\alpha, 2\alpha, \beta$ and $-\beta$.

Find the two possible sets of values for p and q, and give the roots of the equation in each case.

2 Complex roots of polynomial equations

All polynomial equations of degree n have exactly n roots, some of which may be complex and some of which may be repeated.

You often need to use the factor theorem to help you solve polynomial equations.

Example R.3

(i) Show that $z = 2$ is a root of the equation $z^3 + 2z^2 - 3z - 10 = 0$.

(ii) Solve the equation.

Solution

$f(z) = z^3 + 2z^2 - 3z - 10$

$f(2) = 2^3 + 2 \times 2^2 - 3 \times 2 - 10$

$= 8 + 8 - 6 - 10$

$= 0$

The factor theorem states $f(a) = 0 \Leftrightarrow (x - a)$ is a factor of $f(x)$.

$f(2) = 0$ so $(z - 2)$ is a factor of $f(z)$.

$z^3 + 2z^2 - 3z - 10 = 0$

$(z - 2)(z^2 + 4z + 5) = 0$ ← Factorise using inspection or polynomial division.

$z = 2$ or $z^2 + 4z + 5 = 0$

$z = \dfrac{-4 \pm \sqrt{16 - 4 \times 1 \times 5}}{2} = \dfrac{-4 \pm \sqrt{-4}}{2} = -2 \pm i$

So the solution of the equation is $z = 2, z = 2 + i, z = 2 - i$.

Notice that in the example above, the two complex roots are conjugates of each other.

If a polynomial equation has real coefficients, any complex roots always occur in conjugate pairs.

Sometimes the relationships between roots and coefficients of polynomial equations can be useful in solving problems involving polynomial equations with complex roots.

Example R.4

Find a cubic equation with roots $1, 2 + i$ and $2 - i$.

Note

It would be possible to solve this problem by writing the equation in factorised form and multiplying out. However, the method shown here is more efficient.

Solution

$$\sum \alpha = 1 + 2 + i + 2 - i = 5$$

$$\sum \alpha\beta = 1(2 + i) + 1(2 - i) + (2 + i)(2 - i)$$

$$= 2 + i + 2 - i + 4 + 1 = 9$$

$$\alpha\beta\gamma = 1(2 + i)(2 - i)$$

$$= 4 + 1 = 5$$

So $-\dfrac{b}{a} = 5, \dfrac{c}{a} = 9, -\dfrac{d}{a} = 5$.

Taking $a = 1$ gives $b = -5, c = 9, d = -5$. The equation is $z^3 - 5z^2 + 9z - 5 = 0$.

> If you use a different value for a, you will get a multiple of the same equation.

The next example shows two methods of solving the same problem.

Example R.5

One of the roots of the equation $2z^3 - 5z^2 + 12z - 5 = 0$ is $1 + 2i$.

Solve the equation and show the roots on an Argand diagram.

Solution 1

Since one of the roots is $1 + 2i$, another root is $1 - 2i$.
So a quadratic factor is $(z - 1 - 2i)(z - 1 + 2i)$

> Writing the expression in this form shows that it is the difference of two squares.

$$= \big((z - 1) - 2i\big)\big((z - 1) + 2i\big)$$

$$= (z - 1)^2 + 4$$

$$= z^2 - 2z + 5$$

$$2z^3 - 5z^2 + 12z - 5 = 0$$

> Find the linear factor by inspection.

$$(z^2 - 2z + 5)(2z - 1) = 0$$

The roots are $1 + 2i, 1 - 2i$ and $\dfrac{1}{2}$.

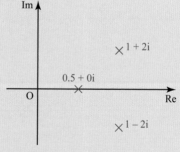

Figure R.1

Solution 2

Since one of the roots is $1 + 2i$, another root is $1 - 2i$.

The sum of the three roots is $\dfrac{5}{2}$.

So $1 + 2i + 1 - 2i + \gamma = \dfrac{5}{2} \Rightarrow \gamma = \dfrac{1}{2}$.

> Using $\alpha + \beta + \gamma = \dfrac{b}{a}$

The roots are $1 + 2i, 1 - 2i$ and $\dfrac{1}{2}$.

① One of the roots of the quadratic equation $x^2 + px + q = 0$ is $2 - 5i$.

   (i)   Write down the other root of the equation.

   (ii) Find the values of p and q.

② (i)   Show that $z = -1$ is a root of the equation $z^3 - 5z^2 + 19z + 25 = 0$.

   (ii) Find the other two roots.

   (iii) Show the three roots on an Argand diagram.

③ Find a quartic equation with roots $2, 5, 3 + 2i$ and $3 - 2i$.

④ The cubic equation $x^3 + px^2 - 2x + 4 = 0$ has a root of -2.
Find the value of p and solve the equation.

⑤ One of the roots of the cubic equation $3z^3 - 14z^2 + 47z - 26 = 0$
is $2 + 3i$. Solve the equation.

⑥ The four roots of a quartic equation are shown on the Argand diagram in
Figure R.2.

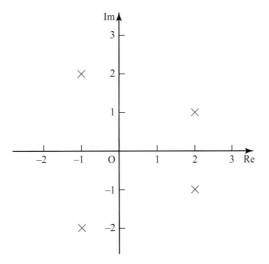

Figure R.2

Find the equation.

⑦ The cubic equation $z^3 + 4z^2 + 14z + 20 = 0$ has an integer root. Solve the
equation and show the roots on an Argand diagram.

⑧ (i)   Given that $2 - i$ is a root of the equation $2z^3 + az^2 + 14z + b = 0$, find
the values of a and b.

   (ii) Solve the equation $2z^3 + az^2 + 14z + b = 0$.

⑨ One of the roots of the equation $4z^4 - 8z^3 + 9z^2 - 2z + 2 = 0$ is $1 - i$.
Solve the equation.

⑩ One of the roots of the equation $z^4 - 4z^3 + pz^2 + qz + 50 = 0$ is $-1 + 2i$.
Find the values of p and q and solve the equation.

KEY POINTS

1 If α and β are the roots of the quadratic equation $az^2 + bz + c = 0$, then

$$\alpha + \beta = -\frac{b}{a}$$

$$\alpha\beta = \frac{c}{a}.$$

2 If α, β and γ are the roots of the cubic equation $az^3 + bz^2 + cz + d = 0$, then

$$\sum \alpha = \alpha + \beta + \gamma = -\frac{b}{a}$$

$$\sum \alpha\beta = \alpha\beta + \beta\gamma + \gamma\alpha = \frac{c}{a}$$

$$\alpha\beta\gamma = -\frac{d}{a}.$$

3 If α, β, γ and δ are the roots of the quartic equation
$az^4 + bz^3 + cz^2 + dz + e = 0$, then

$$\sum \alpha = \alpha + \beta + \gamma + \delta = -\frac{b}{a}$$

$$\sum \alpha\beta = \alpha\beta + \alpha\gamma + \alpha\delta + \beta\gamma + \beta\delta + \gamma\delta = \frac{c}{a}$$

$$\sum \alpha\beta\gamma = \alpha\beta\gamma + \beta\gamma\delta + \gamma\delta\alpha + \delta\alpha\beta = -\frac{d}{a}$$

$$\alpha\beta\gamma\delta = \frac{e}{a}.$$

4 All of these formulae may be summarised using the shorthand sigma
notation for elementary symmetric functions as follows:

$$\sum \alpha = -\frac{b}{a}$$

$$\sum \alpha\beta = \frac{c}{a}$$

$$\sum \alpha\beta\gamma = -\frac{d}{a}$$

$$\sum \alpha\beta\gamma\delta = \frac{e}{a}$$

etc. (using the convention that polynomials of degree n are labelled
$az^n + bz^{n-1} + \ldots = 0$ and have roots α, β, γ, $\ldots$).

5 A polynomial equation of degree n has n roots, taking into account complex
roots and repeated roots. In the case of polynomial equations with real
coefficients, complex roots always occur in conjugate pairs.

12 First order differential equations

It isn't that they can't see the solution.

It's that they can't see the problem.

G.K. Chesterton

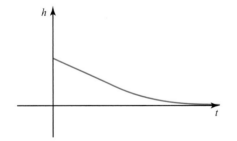

Figure 12.1

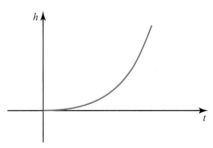

Figure 12.2

Discussion point

One of these graphs applies to an aeroplane landing and the other to one taking off. Which is which? What do these graphs tell you about $\dfrac{\mathrm{d}h}{\mathrm{d}t}$ in each of these circumstances?

1 Modelling rates of change

Modelling

Modelling is the process of representing situations in the real world in mathematical form. It is an important skill. Sometimes rates of change are involved and in those cases the models involve differential equations.

Forming differential equations

If you are given sufficient information about the rate of change of a quantity, such as the caffeine level in the body or the height of water in a harbour, you can form a differential equation to model the situation.

It is important to look carefully at the problem before writing down an equivalent mathematical statement. You have to decide whether you need a model for a rate of change with respect to time or with respect to another variable such as distance or height. You need to be familiar with the language used in these different cases.

If the altitude, h, of an aircraft is being considered, the phrase **rate of change of altitude** might be used. This actually means the **rate of change of altitude with respect to time**. You could write it as $\dfrac{dh}{dt}$ where t stands for time.

However, you might be more interested in how the altitude of the aircraft changes with the horizontal distance it has travelled. In this case you would talk about the **rate of change of altitude with respect to horizontal distance**, and you could write it as $\dfrac{dh}{dx}$ where x stands for the horizontal distance travelled.

Notation

- Any equation which involves a derivative such as $\dfrac{dq}{dt}$, $\dfrac{dy}{dt}$, or $\dfrac{d^2x}{dt^2}$, is called a differential equation.

- A differential equation which involves a first derivative such as $\dfrac{dq}{dt}$ is called a **first order differential equation**.

> This is just like the convention of naming polynomials e.g. cubics which contain x^3 terms, but might also contain x^2 and x terms too.

- One which involves a second order derivative such as $\dfrac{d^2x}{dt^2}$ is called a **second order differential equation**. A second order differential equation may also involve first derivatives as well – it is the *highest* derivative that matters.

- A third order differential equation involves a third order derivative (e.g. $\dfrac{d^3y}{dx^3}$), and so on.

You should be aware of two shorthand notations.

- Differentiation with respect to time is often indicated by writing a dot above the variable.

  For example $\dfrac{dx}{dt}$ may be written as $\dot{x}$ | You would say this as 'x dot'. |

  and $\dfrac{d^2y}{dt^2}$ may be written as $\ddot{y}$. | You would say this as 'y double dot'. |

- Differentiation with respect to x may be denoted by the use of the symbol $'$.

For example y' means $\dfrac{\mathrm{d}y}{\mathrm{d}x}$

> You would say this as 'y dash'.

and f'' means $\dfrac{\mathrm{d}^2 f}{\mathrm{d}x^2}$.

> You would say this as 'f double dash'.

The following examples show how differential equations can be formed from descriptions of a wide variety of situations.

Example 12.1

 Note

ρ is the Greek letter 'Rho'; be careful as it looks very similar to the letter p.

A simple model of the atmosphere above the Earth's surface is as a perfect gas at constant temperature. This means that:

- the pressure, p, at any point is proportional to the density, ρ
- the rate at which the pressure decreases with respect to height, z, is proportional to the density.

Use this information to find an expression for $\dfrac{\mathrm{d}p}{\mathrm{d}z}$ in terms of p.

Solution

Since the pressure, p, at any point is proportional to the density at that point, you can write:

$$p = c_1 \rho$$

where c_1 is a positive constant.

Since the rate of change of pressure, p, with height, z, is proportional to the density, ρ, you can write:

$$\frac{\mathrm{d}p}{\mathrm{d}z} = -c_2 \rho$$

> The negative sign is needed since pressure *decreases* with height (assume $c_2 > 0$).

where c_2 is a positive constant.

Substituting for ρ from the first equation into the second gives

$$\frac{\mathrm{d}p}{\mathrm{d}z} = -\frac{c_2}{c_1} p$$

> c is a constant formed from c_1 and c_2.

$$= -cp$$

Example 12.2

The volume, V, of a spherical raindrop of radius r is decreasing (due to evaporation) at a rate proportional to its surface area, S. Find an expression for $\dfrac{\mathrm{d}r}{\mathrm{d}t}$.

Solution

Note that the problem description contains four variables, V, S, r and t, but that V and S are themselves functions of the radius r:

$$V = \frac{4}{3}\pi r^3 \text{ and } S = 4\pi r^2.$$

You can use this to write the problem more simply in terms of the two variables, r and t.

The wording of the problem gives:

$$\frac{dV}{dt} = -kS$$

> The next step is to write this equation in terms of r.

$$\frac{dV}{dt} = -k\left(4\pi r^2\right)$$

where k is a positive constant.

The left-hand side of the differential equation may be rewritten using the chain rule in the form:

$$\frac{dV}{dt} = \frac{dV}{dr} \times \frac{dr}{dt}$$

Since $V = \frac{4}{3}\pi r^3$ so $\frac{dV}{dr} = 4\pi r^2$, and then $\frac{dV}{dt} = 4\pi r^2 \times \frac{dr}{dt}$.

The differential equation is now written:

$$4\pi r^2 \times \frac{dr}{dt} = -4k\pi r^2$$

$$\frac{dr}{dt} = -k$$

> This very simple differential equation tells you that the radius decreases at a constant rate. You can solve this equation directly, by integration.

For a model to be useful you need to not only set up a differential equation but also solve it.

You are familiar with the idea that integration is the inverse of differentiation. In taking this step you often solve a differential equation.

So, for example, if

$$\frac{dy}{dx} = 2x - 1$$

> This is a differential equation.

then

$$y = x^2 - x + c$$

> This is the solution (c is an arbitrary constant).

Note that the solution contains a constant c. This is because for this **first order** differential equation you carried out *one* integration (which always introduces the possibility of an arbitrary constant). If the original equation was a **second order** differential equation, you would have introduced *two* constants. This is true for solutions of all differential equations. Whatever method is used to solve them, the solution will contain a number of constants equal to the order of the differential equation.

A solution which contains constants like this is called the **general solution** of the differential equation. A general solution forms a family of solutions and can be thought of, and represented as, a set of curves on a graph.

For example, the differential equation

$$\frac{dy}{dx} = 2x - 1,$$

with general solution

$$y(x) = x^2 - x + c,$$

will produce particular curves for different values of c. e.g. $y = x^2 - x + 3$ or $y = x^2 - x - 1$.

Figure 12.3 shows some of these possibilities.

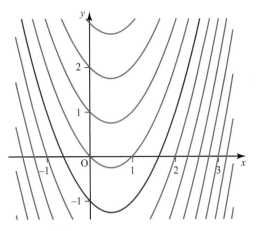

Figure 12.3

If you are investigating the outputs from graphing software you may come across a diagram like Figure 12.3. It is called a tangent field and consists of many small line segments that show the gradients of potential solutions at a large number of points. It can help you to see what the family of solution curves is likely to look like.

Figure 12.4 is a common diagram in the study of differential equations. The tangents are plotted locally for points on the graph to indicate the general shape of the particular solutions for given initial or boundary conditions.

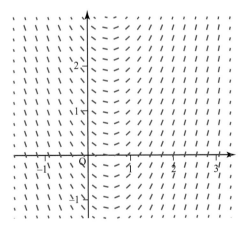

Figure 12.4

Suppose that you are now given one extra piece of information, such as $y = -1$ when $x = 1$. You can now use this information to find the particular value of c, and therefore the particular curve that fits your information.

Substituting $y = -1$ and $x = 1$ into the general solution gives

$$-1 = 1^2 - 1 + c$$

$$c = -1$$

The solution which fits your extra information is $y = x^2 - x - 1$. This is the **particular solution** of the differential equation. It is highlighted in blue in Figure 12.3 and passes through $(1, -1)$. Once a particular solution is found you can use it to find any value for y given a value of x.

The extra piece of information that allows you to identify the particular solution for your problem is called a **condition**. It effectively gives you the coordinates of one of the points through which the required solution curve passes. In the case of a **second order differential equation**, the general solution contains *two constants of integration,* and so *two conditions* are needed to uniquely specify a particular solution. For a **third order differential equation**, *three conditions* are needed, and so on.

Verification of solutions

You will meet situations in which you think you know the solution of a differential equation without actually having to solve it. It may be that the situation, or one like it, is familiar, or you may have experimental data that suggest a solution. In such situations you need to check or **verify** that your solution does indeed satisfy the differential equation. You do this by substituting the solution into the differential equation, a process called **verification**. You would also have to verify that the proposed solution satisfies any conditions that are given.

Verification is also helpful in cases where you think you have found a solution, but want to check that it fits the original problem.

Example 12.3

Verify that $y = Ae^{-3x}$ is the general solution of the differential equation $\dfrac{\mathrm{d}y}{\mathrm{d}x} = -3y$.

Solution

Differentiating $y = Ae^{-3x}$ gives

$$\frac{\mathrm{d}y}{\mathrm{d}x} = -3Ae^{-3x}$$

$$= -3\left(Ae^{-3x}\right)$$

$$= -3y$$

so $y = Ae^{-3x}$ does satisfy the original differential equation, and has one arbitrary constant, so it is the general solution.

Some possible solution curves (red) are shown over the tangent field (blue) for this differential equation in Figure 12.5.

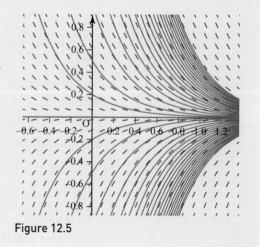

Figure 12.5

Example 12.4

Verify that

$$v = 20e^{-2t} + 5$$

is the particular solution of the differential equation

$$\frac{\mathrm{d}v}{\mathrm{d}t} = 10 - 2v$$

that satisfies the condition $v = 25$ when $t = 0$.

Solution

Differentiating $v = 20e^{-2t} + 5$ gives:

$$\frac{\mathrm{d}v}{\mathrm{d}t} = -40e^{-2t}$$

Start with the right-hand side of the differential equation:

$$10 - 2v = 10 - 2\left(20e^{-2t} + 5\right)$$

$$= 10 - 40e^{-2t} - 10$$

$$= -40e^{-2t}$$

$$= \frac{\mathrm{d}v}{\mathrm{d}t}$$

So the solution $v = 20e^{-2t} + 5$ does satisfy the differential equation $\frac{\mathrm{d}v}{\mathrm{d}t} = 10 - 2v$.

Checking the condition, substitute $t = 0$ into the solution:

$$v = 20e^{-2 \times 0} + 5$$

$$= 20 + 5$$

$$= 25$$

So the solution satisfies the condition too, and is therefore the particular solution required.

Exercise 12.1

> It should be inferred that the rate of mass lost is with respect to time if no other variable is mentioned.

① During the decay of a radioactive substance, the rate at which mass is lost is proportional to the mass present at that instant. Write down a differential equation to describe this relationship.

② The rate at which the population of a particular country increases is proportional to its population. Currently the population is 68 million and it is increasing at a rate of 2 million per year. Form a differential equation for P, the population of the country in millions.

③ The rate at which water leaves a tank is modelled as being proportional to the square root of the height of water in the tank. Initially the height of water is $100\,\mathrm{cm}$ and water is leaving at the rate of $20\,\mathrm{cm}^3$ per second. Form a differential equation that describes this model.

④ Verify that $P = 40e^{0.2t}$ is a solution of $\frac{\mathrm{d}P}{\mathrm{d}t} = \frac{P}{5}$.

⑤ The temperature, T, of an object changes such that

$$\frac{\mathrm{d}T}{\mathrm{d}t} = 2 - 0.1T$$

(i)   Verify that $T = 20 + 60e^{-0.1t}$ is a solution of this differential equation.

(ii)   Verify that $T = 20 - 18e^{-0.1t}$ is also a solution.

(iii)   One of these solutions models the temperature change of a cup of coffee placed in a room, and the other models that of a cold carton of juice just taken from a fridge. Which solution corresponds to which object?

⑥   A radioactive substance decays so that its mass, m mg, decreases according to the equation

$$\frac{dm}{dt} = -\frac{m}{100}$$

where time (t) is measured in hours.

(i)   Verify that $m = 20e^{-\frac{t}{100}}$ is a solution of this differential equation for a particular sample of this substance.

(ii)   Find the half-life (the time taken for the mass to halve) for this substance.

(iii)   What is the initial mass of this sample?

(iv)   Write down the solution that would apply if the initial mass were 50 mg.

⑦   Air is escaping from a spherical balloon at a rate proportional to its surface area. Given that the air is escaping at $4\,\text{cm}^3\,\text{s}^{-1}$ when the radius is 10 cm, find an expression for the rate of change of the radius, $\frac{dr}{dt}$.

⑧   The water pressure in the sea increases with depth. At depth h below the surface, the rate of pressure increase with respect to depth is proportional to the density of sea water. The constant of proportionality is the acceleration due to gravity, g ($9.8\,\text{m}\,\text{s}^{-2}$).

The density ρ (in kg m^{-3}) of sea water in part of an ocean is modelled by:

$$\rho = 1000(1 + 0.001h) \qquad 0 \leqslant h \leqslant 100$$

$$\rho = 1100 \qquad\qquad\quad h > 100$$

Find a differential equation to model this situation.

⑨   A volume V of water is held in a tank that has a square base, side 2 m, and height 4 m. Initially the tank is full. Water leaves the tank through a hole at the bottom at a rate of $\sqrt{20h}$ litres per second, where h is the depth of water in the tank. Find expressions for $\frac{dV}{dt}$ and $\frac{dh}{dt}$ in terms of h.

⑩   Verify that $y = -\dfrac{2}{x^2 + 2}$ is the solution of the differential equation

$$\frac{dy}{dx} = xy^2$$

that satisfies the initial condition $y = -1$ when $x = 0$.

⑪   By integrating both sides find the general solution of the differential equation

$$\frac{dy}{dx} = 3x^2 - x + 1$$

Find the particular solution that satisfies the condition $y = 4$ when $x = 1$. Sketch a number of curves to illustrate the family of solutions including the particular solution that passes through the point $(1, 4)$.

⑫ Which of the following are possible solutions of the differential equation $\dfrac{dy}{dx} = -8y$?

(i) $y = 4e^{-8x}$ (ii) $y = 8e^{-4x}$ (iii) $y = 4e^{-8x} + 2$

(iv) $y = 4e^{-8x} + 8$ (v) $y = 8e^{-8x}$

⑬ (i) Verify that $T = \alpha + Ae^{-kt}$ is a solution of the differential equation

$$\frac{dT}{dt} = -k(T - \alpha), \quad (k > 0)$$

(which models Newton's law of cooling).

(ii) Find, in terms of A and α, the initial and final temperatures of the object.

(iii) An object at a temperature of $90\,°C$ is placed in a room at $25\,°C$. State the values of α and A in this case.

(iv) Sketch this family of curves including cases where A is both positive and negative.

⑭ The range of an aircraft equipped with an inflight refuelling probe can be vastly extended if it can rendezvous with a tanker aircraft and take on board more fuel.

To take on board more fuel, an aircraft is gaining altitude to join the tanker. The tanker is at a height b relative to the initial position of the aircraft that needs refuelling.

The tanker is flying horizontally with constant speed v. At the instant the aircraft begins to ascend, the tanker aircraft's initial position has coordinates (a, b) relative to the aircraft. At time t, the aircraft has position (x, y).

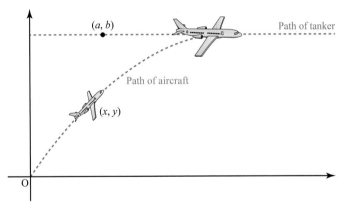

Figure 12.6

Formulate a differential equation for $\dfrac{dy}{dx}$, in terms of x, y and t, to model the path of the aircraft as it travels to intercept the tanker.

⑮ Water is pumped into a conical tank at a rate of $0.3\,m^3$ per second. The tank has the dimensions shown in Figure 12.7, and the depth of the water is h metres at time t seconds.

Find an expression for $\dfrac{dh}{dt}$.

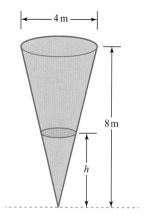

Figure 12.7

⑯ In a model (which applies up to heights of about 10 km) for the pressure p in the lower atmosphere at a height z metres above sea level, it is assumed that gravity is constant and that the air behaves as a perfect gas. This gives the differential equation

$\dfrac{\mathrm{d}p}{\mathrm{d}z} = -\dfrac{pg}{RT}$ where g is the acceleration due to gravity, R is the universal gas constant and T is the absolute temperature of the air in Kelvin (K). The value of R is $8.314\,\mathrm{JK^{-1}\,mol^{-1}}$.

The temperature distribution in the lower atmosphere is given by $T = 300 - 0.006z$.

(i) Taking $g = 10\,\mathrm{m\,s^{-2}}$, show that this differential equation may, to a good approximation, be written as:

$$\frac{1}{p}\frac{\mathrm{d}p}{\mathrm{d}z} = -\frac{200}{50\,000 - z}$$

(ii) Show that $p = p_0\left(1 - \dfrac{z}{50\,000}\right)^{200}$ satisfies this differential equation, given $p = p_0$, when $z = 0$.

2 Separation of variables

If a first order differential equation is written in the form:

$$\frac{\mathrm{d}y}{\mathrm{d}x} = \mathrm{f}(x)$$

(i.e. the RHS is a function of x only)

then you can solve it by directly integrating – assuming you can integrate $\mathrm{f}(x)$.

If, however, the differential equation is in the form:

$$\frac{dy}{dx} = f(x)g(y)$$

then you cannot directly integrate because of the y variable on the RHS. One method of proceeding is to use instead a technique covered in A-level Mathematics, called **separating the variables**. For example, look at the differential equation:

$$\frac{dy}{dx} = xy^2$$

Integrating directly is a problem because you do not know what y is in terms of x (that's precisely the final solution to the differential equation). However, since the RHS is just a product of a function of x and a function of y, you can separate them by writing:

$$\frac{1}{y^2}\frac{dy}{dx} = x$$

This is the process called **separating the variables**. Notice that the RHS is a function of x only, and the LHS is $\frac{dy}{dx}$ multiplied by a function of y only.

Integrating both sides with respect to x gives:

$$\int \frac{1}{y^2}\frac{dy}{dx}dx = \int x\,dx$$

It can be shown that the LHS simplifies, giving:

$$\int \frac{1}{y^2}dy = \int x\,dx$$

Since the LHS involves integrating a function involving y only, with respect to y, and the RHS involves integrating a function of x only, with respect to x, you can proceed by integrating each side separately.

This results in:

$$-\frac{1}{y} + c_1 = \frac{1}{2}x^2 + c_2$$

where c_1 and c_2 are the constants from the two integrations. It is simpler to combine these constants into one $c = c_2 - c_1$, giving:

$$-\frac{1}{y} = \frac{1}{2}x^2 + c$$

Tidying up the solution to make y the subject:

$$y = -\frac{1}{\frac{1}{2}x^2 + c}$$

The new constant $k = 2c$, but in practice it doesn't matter what you call it, as long as you keep track of where it fits.

which can be simplified further by writing:

$$y = -\frac{2}{x^2 + k}$$

This is the general solution of the differential equation $\frac{dy}{dx} = xy^2$.

You can use this method of separating the variables on any first order differential equation which can be written in the form

$$\frac{dy}{dx} = f(x)\,g(y)$$

by rewriting it as

$$\int \frac{1}{g(y)}\,dy = \int f(x)\,dx.$$

The same approach can be used even if the equation is of the form

$$\frac{dy}{dx} = g(y)$$

as in the following example.

Example 12.5

The population, P (in millions), of a country grows so that $\dfrac{dP}{dt} = \dfrac{P}{50}$.

(i) Find the general solution of this differential equation.

(ii) Find the particular solution if $P = 100$ when $t = 0$.

Solution

(i) Dividing both sides by P gives

$$\frac{1}{P}\frac{dP}{dt} = \frac{1}{50}$$

so

$$\int \frac{1}{P}\,dP = \int \frac{1}{50}\,dt$$

$$\ln\left|P\right| = \frac{t}{50} + c$$

Since the population is always positive, $\left|P\right| = P$.

Rearranging gives

$$P = e^{\left(\frac{t}{50} + c\right)}$$

which can be written as:

$$P = Ae^{\frac{t}{50}}$$

> This is a very common occurrence, where an additive constant becomes a multiplicative one.

where $A = e^c$.

The general solution is

$$P = Ae^{\frac{t}{50}}$$

(ii) The particular solution needs $P = 100$ when $t = 0$, so substitute these in:

$$100 = Ae^0$$

$$A = 100$$

so the particular solution is $P = 100e^{\frac{t}{50}}$.

In the previous example the modulus sign that occurs when you integrated $\dfrac{1}{P}$ did not matter, but the following example shows that it is important not to neglect it.

Example 12.6

A drink is placed in a room where the ambient temperature is $20\,°\text{C}$. A model for the subsequent temperature T of the drink at time t, in hours, is given by Newton's law of cooling. This leads to the differential equation

$$\frac{\mathrm{d}T}{\mathrm{d}t} = -5(T - 20).$$

Find the solution of this differential equation, in the cases where

(i) $T = 80$ when $t = 0$

(ii) $T = 0$ when $t = 0$

Solution

Dividing both sides of the differential equation by $(T - 20)$, and rewriting it as two integrals gives

$$\int \frac{1}{T - 20}\,\mathrm{d}T = \int -5\,\mathrm{d}t$$

$$\ln|T - 20| = -5t + c$$

$$|T - 20| = e^{(-5t+c)}$$

$$|T - 20| = A e^{-5t}$$

> This is the general solution of the differential equation, and since you don't know that $T - 20$ is necessarily positive you should not remove the modulus signs.

(i) $T = 80$ when $t = 0$. Substituting these gives

$$|80 - 20| = A, \quad A = 60$$

> The RHS clearly always remains greater than zero, so the LHS must also never reach zero. Since $T - 20$ starts positive, it must always remain positive, so you can remove the modulus signs.

so:

$$|T - 20| = 60 e^{-5t}$$

$$T = 20 + 60 e^{-5t} \text{ as the particular solution.}$$

(ii) $T = 0$ when $t = 0$. Substituting gives

$$|0 - 20| = A \Rightarrow A = 20$$

> The drink is cooling (the e^{-5t} term gets smaller as t gets bigger) towards the ambient room temperature of $20\,°\text{C}$. See the red curve in Figure 12.8.

so:

$$|T - 20| = 20 e^{-5t}$$

$$-(T - 20) = 20 e^{-5t} \quad -T + 20 = 20 e^{-5t}$$

so:

$$T = 20 - 20 e^{-5t}$$

is the particular solution.

> The drink is warming up (the e^{-5t} term gets smaller again, but it is being subtracted from the ambient temperature this time) to the ambient room temperature of $20\,°\text{C}$. See the blue curve in Figure 12.8.

> Again, the RHS is always greater than zero, so the LHS can never equal zero. Since $T - 20$ is negative to start with, it must remain negative. This means the modulus sign will always reverse the sign of the left-hand side and can be replaced with a negative sign.

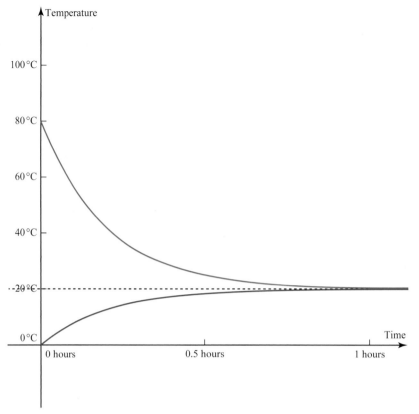

Figure 12.8 The red curve is the drink cooling down, starting at 80 °C, while the blue curve is the drink warming up having started at 0 °C. The ambient temperature can be seen as the horizontal asymptote at 20 °C which both curves approach.

 Historical note

The discovery of calculus is usually attributed to Isaac Newton with the invention of the **method of fluxions** in 1665, when he was 23 years old, although it is also claimed that Gottfried Leibniz discovered it first but did not publicise his work. The study of differential equations followed on naturally. Newton is known to have solved a differential equation in 1676 but he only published details of it in 1693, the same year in which a differential equation featured in Leibniz's work.

The **method of separation of variables** was developed by Johann Bernoulli between 1694 and 1697. He was a famous teacher and wrote the first calculus textbook in 1696; one of his students was Leonhard Euler. Johann Bernoulli was one of the older members of a quite remarkable family of Swiss mathematicians (three of them called Johann) spread over three generations, at least eight of whom may be regarded as famous.

① Which of the following differential equations can be solved by the method of separation of variables? Give the general solution of those which can be solved in this way.

(i) $\dfrac{dy}{dx} = y$ (ii) $\dfrac{dy}{dx} = xy$ (iii) $\dfrac{dy}{dx} = x^3 y$

(iv) $\dfrac{dy}{dx} = 3x^2 e^{-y}$ (v) $\dfrac{dy}{dx} = x + yx$ (vi) $\dfrac{dy}{dx} = x + y^2$

(vii) $\dfrac{dy}{dx} = e^{x+y}$ (viii) $\dfrac{dy}{dx} = \dfrac{y}{x(x-1)}$ (ix) $\dfrac{dy}{dx} = x^2 - y^2$

(x) $\dfrac{dy}{dx} = \dfrac{\sin x}{y^2}$ (xi) $\dfrac{dy}{dx} = \dfrac{y+2}{x-2}$ (xii) $\dfrac{dx}{dt} = x^2 - 8x$

② Find the particular solution of each of the differential equations below for the given conditions.

(i) $x\dfrac{dy}{dx} = y^2$ $y = 10$ when $x = 1$

(ii) $\dfrac{dz}{dt} = \dfrac{z^2}{t}$ $z = 2$ when $t = 1$

(iii) $\dfrac{dy}{dx} = \dfrac{x^2}{y}$ $y = 10$ when $x = 1$

(iv) $\dfrac{dp}{ds} = e^{-p} \sin 2s$ $p = 0$ when $s = 0$

(v) $\dfrac{dy}{dx} = x^2 e^{-y}$ $y = 10$ when $x = 0$

(vi) $\dfrac{dy}{dt} = \dfrac{e^y}{y}$ $y = 2$ when $t = 0$

③ The rate of radioactive decay of a chemical is given by:

$$\dfrac{dm}{dt} = -5m$$

(i) Find the general solution of this equation.

(ii) Given that $m = 10$ when $t = 0$, find the particular solution.

④ A bacterium reproduces such that the rate of increase (in bacteria per minute) of the population is given by:

$$\dfrac{dP}{dt} = 0.7P$$

(i) Find the general solution of this differential equation.

(ii) Find the particular solution if $P = 100$ when $t = 0$.

(iii) How long does it take for the population to double?

⑤ An object is projected horizontally on a surface with an initial speed of $20\,\text{ms}^{-1}$. Its speed is modelled by the differential equation:

$$\dfrac{dv}{dt} = -\dfrac{v^2}{10}$$

(i) Find the particular solution of this differential equation that corresponds to the initial conditions given.

(ii) How long does it take for the speed to drop to 10% of its original value?

⑥ Water evaporates from a conical tank such that the rate of change of the height of water is modelled by the differential equation

$$\frac{dh}{dt} = -\frac{\pi h^2}{4}.$$

(i) Find the general solution of this differential equation.

(ii) Initially the height of water is H. Find, in terms of H, how long it takes for the height of the water to decrease by 10%.

⑦ Water is leaving a tank through a pipe at the bottom, such that change in the height of water is modelled by the differential equation

$$4\frac{dh}{dt} = -\sqrt{20h}, \text{ where } t \text{ is in minutes and } h \text{ in centimetres.}$$

(i) Find the particular solution of this differential equation given that the initial height is 4 cm.

(ii) How long does the tank take to empty?

⑧ The differential equation below models the motion of a body falling vertically subject to air resistance:

$$\frac{dv}{dt} = 10 - 0.2v, \text{ where } v \text{ is the downward vertical speed of the body in m s}^{-1}$$

and time t is measured in seconds.

(i) Find the general solution of the differential equation.

(ii) Find particular solutions when

(a) the body starts from rest

(b) the body has initial downward vertical speed of 80 m s^{-1}.

In each case state the terminal velocity.

⑨ The temperature of a hot body is initially 100 °C. It is placed in a tank of water at a temperature of 20 °C. The rate of change of temperature is modelled by the differential equation $\frac{dT}{dt} = -0.5(T - 20)$, where t is the time in minutes.

(i) Find the particular solution of this differential equation that fits the initial condition given.

(ii) After what time interval has the temperature of the body fallen to 50 °C?

⑩ A small particle moving in a fluid satisfies the differential equation $\frac{dv}{dt} = -0.2(v + v^2)$. Find the particular solution of this differential equation given that $v = 40$ when $t = 0$.

⑪ For the electrical circuit shown below, the current, I amperes, flowing once the switch is closed is given by

$$L\frac{dI}{dt} + RI = V$$

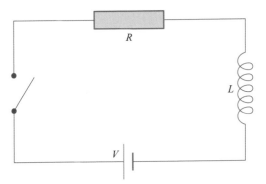

Figure 12.9

L is the inductance, measured in henries.

R is the resistance, measured in ohms.

V is the applied voltage, measured in volts.

(i) If $L = 0.04$, $V = 20$ and $R = 100$, find the general solution of the differential equation.

(ii) If $I = 0$ when $t = 0$, find the particular solution.

(iii) Find the general solution of the original differential equation in terms of R, V and L.

⑫ At 0300 in the morning the police were called to a house where the body of a murder victim had been found. The police doctor arrived at 0345 and took the temperature of the body, which was $34.5\,°C$. One hour later he took the temperature again and measured it to be $33.9\,°C$. The temperature of the room was fairly constant at $15.5\,°C$.

Take the normal body temperature of a human being as $37.0\,°C$, and use Newton's law of cooling

$$\frac{dT}{dt} = -k\left(T - T_{room}\right)$$

as a model to estimate the time of death of the victim.

⑬ A tank contains 2000 litres of salt solution with an initial concentration of $0.3\,\text{kg}\,l^{-1}$. It is necessary to reduce the concentration to $0.2\,\text{kg}\,l^{-1}$. In order to do this, pure water is pumped into the tank at a rate of 8 litres per minute, and the solution is pumped out of the tank at the same rate. The liquid in the tank is stirred so that perfect mixing may be assumed. At time t minutes the mass of salt in the tank is denoted by $M\,\text{kg}$ and the concentration of salt in the solution by $C\,\text{kg}\,l^{-1}$.

(i) By considering the rate of change of mass of salt in the tank show that $\frac{dM}{dt} = -8C$ and also state the relationship between M and C.

(ii) Hence form a differential equation for C (with respect to time).

(iii) Find the time it will take to reduce the concentration to the required level (to the nearest minute).

3 Integrating factors and integration by substitution

When a first order differential equation cannot be written in the form $\frac{dy}{dx} = f(x)\,g(y)$, it cannot be solved using the method of separation of variables.

If, however, it is a **linear** equation, you can multiply it by a special function called an **integrating factor** which converts it to a form which can be integrated.

Linear equations

Strictly, it is linear in y.

A linear differential equation is one in which the independent variable (y in these examples) only appears to the power of 1. So the differential equation $\frac{dy}{dx} = x^2 - xy$ is linear because the only terms that involve y are $\frac{dy}{dx}$ and $-xy$. There are no terms in y^2, y, $\sin y$, $\left(\frac{dy}{dx}\right)^2$, etc.

Any linear first order differential equation may be written in the form:

$$\frac{dy}{dx} + P(x)y = Q(x)$$

where P and Q are functions of x only.

For example, the equation

$$\frac{dy}{dx} = x^2 - xy$$

can be rewritten as

$$\frac{dy}{dx} + xy = x^2$$

This is in the form

$$\frac{dy}{dx} + Py = Q$$

where the functions P and Q are given by $P = x$ and $Q = x^2$.

ACTIVITY 12.1

Which of the following differential equations

(i) can be written in the form $\frac{dy}{dx} = f(x)\,g(y)$ (i.e. which can be solved by separating variables)?

(ii) can be written in the form $\frac{dy}{dx} + P(x)y = Q(x)$ (i.e. which are **linear**)? Identify P and Q if so.

(a) $\frac{dy}{dx} = x^2 + x^2 y$

(b) $\frac{dy}{dx} = x^2 - xy$

(c) $\frac{dy}{dx} = x^2 + x + xy + y$

(d) $\frac{dy}{dx} = x + y^2$

Example 12.7

Find the general solution of the differential equation:

$$\cos x \frac{dy}{dx} - y \sin x = x^2$$

Solution

The equation looks forbidding until you notice that the left-hand side is a perfect derivative.

Since $\frac{d}{dx}(\cos x) = -\sin x$, it follows (using the product rule) that

$$\frac{d}{dx}(y \cos x) = \cos x \frac{dy}{dx} - y \sin x.$$

So you can rewrite the differential equation as

$$\frac{d}{dx}(y \cos x) = x^2.$$

You may now integrate both sides to obtain

$$y \cos x = \frac{x^3}{3} + c.$$

Dividing both sides by $\cos x$, the general solution is

$$y = \frac{x^3}{3 \cos x} + \frac{c}{\cos x}.$$

In the example above the left-hand side was already a perfect derivative. That is not the case in the next example but it is a simple matter to convert it into one.

Example 12.8

Find the general solution of the differential equation

$$\frac{dy}{dx} + \frac{2}{x} y = \frac{4}{x^2} \text{ for } x \neq 0.$$

Solution

First note that the equation is linear, because it can be written in the form

$$\frac{dy}{dx} + Py = Q$$

where $P = \frac{2}{x}$ and $Q = \frac{4}{x^2}$.

> There are no terms in y^2, $\frac{1}{y}$, $\sqrt{y}$, etc.

If you now multiply through by x^2, the equation becomes

$$x^2 \frac{dy}{dx} + 2xy = 4.$$

The left-hand side of this equation is now the expression you obtain when you differentiate $x^2 y$ with respect to x, using the product rule:

$$\frac{d}{dx}(x^2 y) = x^2 \frac{dy}{dx} + 2xy$$

So the differential equation can be rewritten as

$$\frac{d}{dx}(x^2 y) = 4.$$

Now integrating both sides with respect to x gives

$$x^2 y = \int 4 \, dx = 4x + c.$$

The general solution is

$$y = \frac{4}{x} + \frac{c}{x^2}.$$

In the previous two examples the differential equations could be rewritten in the form:

$$\frac{d}{dx}(Ry) = \text{function of } x$$

where R was some function of x. In Example 12.7 R $= y \cos x$, and in Example 12.8, R $= x^2 y$. Once the differential equation was written in this form, it was a straightforward task to solve it, since all that remained was to integrate the function of x on the right-hand side.

However, in the Example 12.8 we had to multiply each term in the equation by a factor x^2 to bring the left-hand side into the required form. This factor of x^2 is an example of an **integrating factor**; multiplying by it made the left-hand side a perfect derivative.

To see how to calculate the integrating factor from a differential equation in the standard form, think about the general case:

$$\frac{dy}{dx} + Py = Q$$

You need a function R to multiply everything by:

$$R\frac{dy}{dx} + RPy = RQ$$

so that the LHS can be written as $\frac{d}{dx}(Ry)$.

This means you need:

$$\frac{d}{dx}(Ry) = R\frac{dy}{dx} + RPy$$

Differentiating the LHS (using the product rule and chain rule), remembering that R and y are functions of x) gives:

$$R\frac{dy}{dx} + \frac{dR}{dx}y = R\frac{dy}{dx} + RPy$$

Comparing the two sides, and realising that $y \neq 0$, you should be able to see that this is only true if:

$$\frac{dR}{dx} = RP$$

This is a first order differential equation, in R and x, but the variables are separable:

$$\frac{1}{R}\frac{dR}{dx} = P$$

$$\ln|R| = \int P \, dx$$

so

$$R = e^{\int P\,dx}$$

This means that *any* first order linear differential equation written in the standard form can be multiplied by an **integrating factor** $R = e^{\int P\,dx}$ to convert it to the compact form: $\dfrac{d}{dx}(Ry) = RQ$

You can then solve the equation by integrating the right-hand side, which is a function of x only.

Example 12.9

Solve the differential equation:

$$x\frac{dy}{dx} + 2y = \frac{4}{x} \text{ for } x \neq 0$$

Solution

Dividing through by x gives the standard form:

$$\frac{dy}{dx} + \frac{2y}{x} = \frac{4}{x^2}$$

and with the standard notation $P = \dfrac{2}{x}$ and $Q = \dfrac{4}{x^2}$.

The integrating factor is:

$$R = e^{\int \frac{2}{x}\,dx} = e^{2\ln|x|+c} = Ae^{\ln|x|^2} = Ax^2$$

> The constant of integration c becomes $A = e^c$.

You will multiply the standard form equation by this. But the constant of integration will always become a multiplier, and since it cannot be zero (otherwise R would be zero) you can immediately divide by the constant again, eliminating it. In practice this means you can safely ignore this constant, so:

$$R = x^2$$

> One of the few times you can safely ignore a constant of integration is when calculating the integrating factor!

Multiplying the standard form of the differential equation by R gives:

$$x^2\frac{dy}{dx} + 2yx = 4$$

The left-hand side is now the derivative of a product, and can be written as $\dfrac{d}{dx}(Ry)$. In this case:

$$\frac{d}{dx}\left(x^2 y\right) = 4$$

$$x^2 y = 4x + c$$

Dividing by x^2 gives the general solution:

$$y = \frac{4}{x} + \frac{c}{x^2}$$

This is, of course, the same solution as you obtained in Example 12.8, but this time you didn't need to 'spot' a convenient form for the LHS, but used an explicit method to find an integrating factor.

Check you can follow the method in the following example, which also requests a particular solution to satisfy a condition.

Example 12.10

Find the particular solution of:

$$x^2 \frac{dy}{dx} + xy = \frac{2}{x}$$

that satisfies the condition $y = 1$ when $x = 2$.

Solution

Write the equation in the standard form:

$$\frac{dy}{dx} + \frac{y}{x} = \frac{2}{x^3}$$

Find the integrating factor R:

$$R = e^{\int \frac{1}{x} dx} = e^{\ln|x|} = x$$

Multiply through by R:

$$x \frac{dy}{dx} + y = \frac{2}{x^2}$$

$$\frac{d}{dx}(xy) = \frac{2}{x^2}$$

Integrate with respect to x:

$$xy = \int \frac{2}{x^2} dx$$

$$xy = -\frac{2}{x} + c$$

$$y = -\frac{2}{x^2} + \frac{c}{x}$$

which is the general solution.

The condition states that $y = 1$ when $x = 2$, so we need:

$$1 = -\frac{2}{2^2} + \frac{c}{2}$$

$$c = 3$$

Therefore, the particular solution is:

$$y = \frac{3}{x} - \frac{2}{x^2}$$

Using substitutions

Differential equations that are not linear in y cannot be immediately solved using the integrating factor method. However, sometimes it is possible to use a substitution $y = f(u)$ to transform the differential equation into a form that is linear in u. The new equation in u and x can then be solved using the techniques in this chapter.

Example 12.11

(i) Use the substitution $y = \dfrac{1}{u}$ to transform the differential equation
$x^2 \dfrac{dy}{dx} - 2xy = x^2 y^2$ into a linear differential equation in u.

(ii) By solving this equation find the general solution to the original equation where $y = f(x)$.

Solution

(i) If $y = \dfrac{1}{u}$ then $\dfrac{dy}{dx} = -\dfrac{1}{u^2} \dfrac{du}{dx}$

Substituting these into the differential equation gives

$$x^2 \left(-\frac{1}{u^2} \frac{du}{dx} \right) - 2x \frac{1}{u} = x^2 \frac{1}{u^2}$$

$$x^2 \frac{du}{dx} + 2xu = -x^2 \qquad \boxed{\text{Multiplying throughout by } -u^2.}$$

(ii) The left-hand side of this equation is now a perfect integral and can be written as

$$\frac{d(x^2 u)}{dx} = -x^2$$

Integrating both sides, with respect to x.

$$\int \frac{d(x^2 u)}{dx}\, dx = \int -x^2\, dx$$

$$x^2 u = -\frac{x^3}{3} + C$$

> Remember from the earlier section on separation of variables, that there is actually a constant of integration on both sides. It is simpler to combine these as a single constant on one side of the equation.

$y = \dfrac{1}{u}$ can be substituted, to find an equation where $y = f(x)$, giving

$$x^2 \frac{1}{y} = -\frac{x^3}{3} + C$$

or

$$x^2 \frac{1}{y} = \frac{3C - x^3}{3}$$

> Dividing throughout by x^2 and then taking the reciprocal of both sides will give your final answer.

$$y = \frac{3x^2}{3C - x^3}$$

You will sometimes need to use the **integrating factor** method to solve equations of this type.

Exercise 12.3

① Find the integrating factor for each of the following differential equations:

(i) $\dfrac{dy}{dx} + x^2 y = x$

(ii) $\dfrac{dy}{dx} + y\sin(x) = x$

(iii) $4x\dfrac{dy}{dx} - y = x^2$

(iv) $x^2\dfrac{dy}{dx} + xy = 2$

(v) $\dfrac{dy}{dx} + 7y = 1$

(vi) $\cos(x)\dfrac{dy}{dx} + y\sin(x) = e^{-2x}$

② Consider the differential equation

$x^2\dfrac{dy}{dx} + xy = 1$, with condition $y = 0$ when $x = 1$.

(i) Rewrite the differential equation into the form $\dfrac{dy}{dx} + P(x)\,y = Q(x)$.

(ii) Find the integrating factor by calculating $e^{\int P\,dx}$.

(iii) Multiply your answer from part (i) by the integrating factor from part (ii).

(iv) Rewrite your answer from part (iii) in the form $\dfrac{d}{dx}(Ry) = RQ$.

(v) Integrate both sides with respect to x.

(vi) Rearrange your answer from part (v) to give the general solution.

(vii) Substitute the condition into the general solution to find the value of the constant.

(viii) Write down the particular solution.

③ For each of the following use the substitution $u = \dfrac{y}{x}$ find the general solution to the differential equation.

(i) $\dfrac{dy}{dx} = \dfrac{y}{x} + \dfrac{2x}{y}$ for $x \ne 0$ and $y \ne 0$, leaving your answer in the form $y^2 = f(x)$.

(ii) $\dfrac{dy}{dx} = \dfrac{x}{2y} + \dfrac{y}{2x}$ for $x \ne 0$ and $y \ne 0$, leaving your answer in the form $y = f(x)$.

(iii) $\dfrac{dy}{dx} = \dfrac{3x^2 + 6y^2}{4xy}$ for $x \ne 0$ and $y \ne 0$, leaving your answer in the form $y^2 = f(x)$.

④ (i) Using the substitution $u = x + y$ show that differential equation $(x + y + n)\dfrac{dy}{dx} = x + y + m$ can be written in the form $\dfrac{du}{dx} = \dfrac{m - n}{n + u}$ for $x \ne 0$, $y \ne 0$ and $m \ne n$.

(ii) Show that the general solution can be written in the form $x^2 + y^2 + 2xy + (4n - 2m) + 2ny = 2(m - n)C$.

⑤ Find the particular solution of each of the following differential equations.

(i) $x\dfrac{dy}{dx} + 2y = x^2$ $y = 0$ when $x = 1$

(ii) $\dfrac{dy}{dx} + xy = 4x$ $y = 2$ when $x = 0$

(iii) $6xy + \dfrac{dy}{dx} = 0$ $\qquad y = 3$ when $x = 1$

(iv) $\dfrac{dx}{dt} - 2tx = t$ $\qquad x = 1$ when $t = 0$

(v) $\dfrac{dy}{dx} - \dfrac{3}{x}y = x$ $\qquad y = 0$ when $x = 1$

(vi) $\dfrac{dv}{dt} + 3t^2 v = t^2$ $\qquad v = -1$ when $t = 0$

⑥ An object falling vertically experiences air resistance so that the velocity satisfies the differential equation

$$\dfrac{dv}{dt} = 10 - 0.4v$$

(i) Use the integrating factor method to find the general solution of this differential equation.

(ii) Find the particular solution if, initially, $v = 0$.

(iii) Solve the original differential equation, with the condition, using the method of separation of variables.

(iv) Compare your solutions and comment on which method you would prefer to use, assuming that both were available (i.e. it is a first order, linear, differential equation).

⑦ A parachutist has a terminal speed of $30\,\text{ms}^{-1}$. The magnitude of the air resistance acting when the parachute is open is modelled by $F = kmv$ newtons, where k is a constant, v is the speed and m is the mass of the parachutist.

(i) By considering the forces acting at terminal velocity, find the value for k, taking the acceleration due to gravity to be $10\,\text{ms}^{-2}$.

(ii) Formulate a first order differential equation for the velocity, v, at a given time t, after the parachute opens.

(iii) Use the integrating factor method to find the general solution of the differential equation.

(iv) Find the particular solution if the parachutist is moving at $60\,\text{ms}^{-1}$ when the parachute opens.

⑧ A solution is sought to the differential equation:

$$\dfrac{dy}{dx} + \dfrac{y}{x} = \cos x \qquad (x > 0)$$

(i) Find the general solution for y in terms of x.

(ii) Given that $y = 0$ when $x = \pi$, find the particular solution.

(iii) Write down the function which approximates the solution as x gets very large.

(iv) State the behaviour of y as $x \to 0$ and sketch the shape of graph of y against x, focusing on the behaviour at large and small x.

⑨ Using the substitution $y = \dfrac{1}{u}$ find the general solution to the following differential equations.

(i) $\dfrac{dy}{dx} + \dfrac{2y}{x} = x^3 y^2$ for $x \neq 0$, leaving your answer in the form $y = \text{f}(x)$.

(ii) $\dfrac{dy}{dx} + y \cot x = 2y^2 \sin^2 x$, leaving your answer in the form $y = \text{f}(x)$.

(iii) $\dfrac{dy}{dx} - 2xy = e^{x^2}y^2$, leaving your answer in the form $y = f(x)$.

(iv) $\dfrac{dy}{dx} - \dfrac{1}{x}y = -(1 + x^2)^4 y^2$ for $x \neq 0$, leaving your answer in the form $y = f(x)$.

Hint: in this question it is useful to write your constant of integration as $\ln k$ instead of the usual C.

⑩ The radioactive isotope uranium-238 decays into thorium-234, which in turn decays into protactinium-234. This can be summarised as

$$^{238}_{92}U \xrightarrow{k_1} {}^{234}_{90}Th \xrightarrow{k_2} {}^{234}_{91}Pa$$ where k_1 and k_2 are reaction constants (i.e. the constants of proportionality by which the rates of decay occur).

The amounts of U-238, Th-234 and Pa-234 at time t are denoted by x, y and z, respectively.

You may assume that the rate of decay of an isotope is proportional to the amount present, and that the constant of proportionality is the relevant k-value.

An experiment begins with an amount a of U-238, but no Th-234 or Pa-234. The amount y of Th-234 present at time t satisfies the differential equation

$$\dfrac{dy}{dt} + k_2 y = k_1 a e^{-k_1 t}.$$

(i) Find the integrating factor for the differential equation, and hence its general solution.

(ii) Find the particular solution that satisfies the initial conditions.

(iii) Write down a differential equation for the variable x.

(iv) Solve the model you suggested in part (iii) and explain how its particular solution has been incorporated into the suggested differential equation for y.

(v) Find the particular solution of the differential equation for y, in the case where $k_1 = k_2 = k$.

(vi) Write down a differential equation for the variable z (still assuming $k_1 = k_2 = k$). By substituting in the solution you found in part (v) for y, solve this differential equation to find a particular solution for the variable z at any time t.

⑪ The differential equation

$$(1 - x)\dfrac{dy}{dx} + \dfrac{2}{1 + x}y = (1 - x)^2 \text{ is to be solved for } |x| < 1.$$

(i) Solve the differential equation.

(ii) Find particular solutions when

(a) $y = 0$ when $x = 0$

(b) $y = 1$ when $x = 0$

in each case stating the behaviour of the solution as $x \to -1$.

(iii) Find a particular solution which tends to a finite limit as $x \to -1$. Sketch the graph of this solution.

⑫ The function $y = f(x)$ satisfies the differential equation

$$x\dfrac{dy}{dx} + 2y = \sqrt{1 + x^2}.$$

(i) Using the integrating factor method, or otherwise, find the general solution of this differential equation.

(ii) Find the particular solution which satisfies the condition that $y = 1$ when $x = 1$. How does y behave when x becomes very small?

(iii) Write down the first three non-zero terms in the expansion of $\left(1 + x^2\right)^{\frac{3}{2}}$ in ascending powers of x.

(iv) Using the expansion in part (iii) and the general solution found in part (i), write down the power series expansion of the general solution y for small values of x up to and including the term in x^2. Hence find the particular solution of the differential equation which crosses the y-axis from the region $x > 0$ into the region $x < 0$.

⑬ (i) Using the substitution $u = \dfrac{dy}{dx}$ write the differential equation $x\dfrac{d^2y}{dx^2} + \dfrac{dy}{dx} = 3x^2$ as a first order differential equation, linear in u.

(ii) Find the general solution to this differential equation, where $\dfrac{dy}{dx} = f(x)$.

(iii) By further integration, find the general solution to the original differential equation, leaving your answer in the form $y = f(x)$.

⑭ Using the substitution $u = \dfrac{dy}{dx}$ and by integration find the general solution to the differential equation $x^2\dfrac{d^2y}{dx^2} + x\dfrac{dy}{dx} = 2x^4 + 6x^3$ which can be written as a first order differential equation in the form $\dfrac{du}{dx} + \dfrac{1}{x}u = 2x^2 + 6x$, leaving your answer in the form $y = f(x)$.

LEARNING OUTCOMES

Now you have finished this chapter, you should be able to:

➤ formulate a differential equation from verbal descriptions involving rates of change

➤ use differential equations in modelling

➤ know the difference between a general solution and a particular solution

➤ find a particular solution to a differential equation by using initial or boundary conditions

➤ recognise differential equations which can be solved by separation of variables

➤ use separation of variables to solve a first order separable differential equation, to find both general and particular solutions

➤ recognise differential equations which can be solved using an integrating factor

➤ find an integrating factor and use it to solve a first order linear differential equation, to find both general and particular solutions

➤ by using a given substitution transform a differential equation into a linear differential equation to be solved.

KEY POINTS

1 A differential equation is an equation involving derivatives such as $\dfrac{\mathrm{d}y}{\mathrm{d}x}$, $\dfrac{\mathrm{d}V}{\mathrm{d}t}$ or $\dfrac{\mathrm{d}^2 x}{\mathrm{d}t^2}$.

2 The order of a differential equation is the order of its highest derivative.

3 Differential equations are used to model situations which involve rates of change.

4 The solution of a differential equation gives the relationship between the variables themselves, not their derivatives.

5 The general solution of a first order differential equation satisfies the differential equation and has one constant of integration in the solution.

6 A particular solution of a differential equation is one in which additional information has been used to calculate the constant of integration.

7 The general solution may be represented by a family of curves and a particular solution is one member of that family.

8 The method of separation of variables can be used to solve first order differential equations of the form $\dfrac{\mathrm{d}y}{\mathrm{d}x} = \mathrm{f}(x)\mathrm{g}(y)$. Separating the variables gives $\displaystyle\int \dfrac{1}{\mathrm{g}(y)}\,\mathrm{d}y = \int \mathrm{f}(x)\,\mathrm{d}x$.

9 Any first order linear differential equation can be written in the form:

$$\dfrac{\mathrm{d}y}{\mathrm{d}x} + \mathrm{P}y = \mathrm{Q}$$ where P and Q are functions of x only.

10 To solve the equation $\dfrac{\mathrm{d}y}{\mathrm{d}x} + \mathrm{P}y = \mathrm{Q}$ you multiply throughout by the integrating factor $\mathrm{e}^{\int \mathrm{P}\,\mathrm{d}x}$.

11 Differential equations that are non-linear in y can be transformed by a given substitution and solved using the above techniques.

① Software has been used to plot the locus of a complex number z in an Argand diagram with the following result.

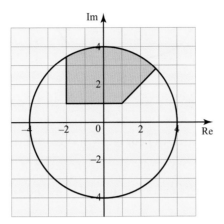

Figure 1

Write down the four inequalities, one for $\mathrm{Re}(z)$, one for $\mathrm{Im}(z)$, one for $\arg(z)$ and one for $|z|$, satisfied by the points in the region shaded. (Note: the boundaries are included in the region.) [4 marks]

② Find the particular solution of the differential equation

$$\frac{\mathrm{d}y}{\mathrm{d}x} - 2xy = \mathrm{e}^{2x+x^2}$$

for which $y = 1$ when $x = 0$. [5 marks]

PS ③ The following equation has $1 + \mathrm{i}$ as one of its roots. m is a real constant.

$$(m + 2)x^3 + (m^2 - 8)x^2 + (m + 3)x - 2 = 0$$

Find the other roots of the equation. [6 marks]

④ **M** is the matrix $\begin{pmatrix} \frac{1}{2} & \frac{1}{4} & 0 \\ \frac{1}{2} & \frac{1}{2} & \frac{1}{2} \\ 0 & \frac{1}{4} & \frac{1}{2} \end{pmatrix}$.

(i) Find the eigenvalues and the corresponding eigenvectors of **M**. [8 marks]

(ii) Write **M** in diagonal form. [3 marks]

⑤ In this question $I_n = \displaystyle\int_0^1 \frac{x^n}{\sqrt{1 + x^2}}\,\mathrm{d}x$.

(i) Show that $nI_n = \sqrt{2} - (n - 1)I_{n-2}$. [6 marks]

(ii) Hence evaluate I_4 in terms of $s = \sinh^{-1}(1)$. [4 marks]

⑥ A curve is given parametrically by $x = 3t^2$, $y = t^3 - 3t$.

(i) Find the length of one arc of the curve between the points $(0, 0)$ and $(9, 0)$. [6 marks]

(ii) The curve is rotated through 2π about the x-axis. Find the surface area of the solid of revolution between $(0, 0)$ and $(9, 0)$. [4 marks]

MP ⑦ (i) Prove that $\cosh 2x = 2\sinh^2 x + 1$. [3 marks]

Figure 2 shows the curve $y = f(x)$, where $f(x) = 2\sinh^2 x - 3\cosh x$.

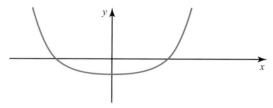

Figure 2

(ii) Show that the curve $y = f(x)$ crosses the x-axis at $x = \pm\ln(2 + \sqrt{3})$. [3 marks]

(iii) Find the exact area bounded by the curve $y = f(x)$ and the x-axis. [8 marks]

13 Numerical methods

Science is the art of the appropriate approximation. While the flat earth model is usually spoken of with derision it is still widely used. Flat maps, either in atlases or road maps, use the flat earth model as an approximation to the more complicated shape.

Byron K. Jennings

Discussion point

In a sunny garden, a patio is to be constructed in a sheltered corner by the house. An architect produces the design shown in Figure 13.1.

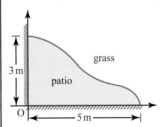

Figure 13.1

The patio is to be covered in concrete up to a depth of 80 mm. The concrete is supplied already mixed. The problem is to estimate the volume of concrete which should be ordered.

Since the volume is equal to the product of the area of the patio and the depth of the concrete, the problem reduces to one of finding the area of the patio.

→ How might you find the exact area of the patio?

→ How might you approximate the area of the patio?

→ In each case what information do you need?

The area is given by $\int_0^5 f(x)\,dx$, where the curve is given by $y = f(x)$. The problem is that often you do not know $f(x)$, or you know it, but cannot do the integration.

In the design of buildings and manufactured goods, irregular shapes are often used for aesthetic or practical reasons. Since properties of these shapes such as the length of a curve, an area or a volume, may be required, mathematical techniques are available to approximate them. You will meet some of these here.

1 Numerical integration

Mid-ordinate rule

The mid-ordinate rule uses rectangles to approximate the area underneath a curve.

In Figure 13.2 four rectangles, each with the same width, are used to approximate the area under the graph of a function $f(x)$ between $x = a$ and $x = b$.

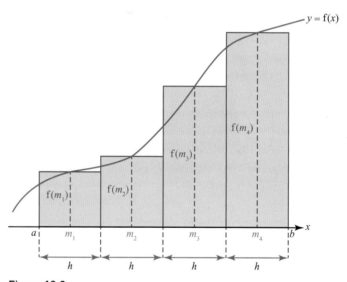

Figure 13.2

Since there are four rectangles of equal width, the width of each rectangle is $\frac{b-a}{4}$. This value is equal to h in Figure 13.2. The height of the first rectangle is the value of the function at the midpoint of the interval from a to $a + h$, i.e. at $a + \frac{h}{2}$.

This has been labelled m_1 in Figure 13.2.

The height of this rectangle is $f(m_1)$, its width is h and so its area is $h f(m_1)$.

The midpoints of the four rectangles are

$$m_1 = a + \frac{h}{2}, \quad m_2 = a + \frac{3h}{2}, \quad m_3 = a + \frac{5h}{2} \quad \text{and} \quad m_4 = a + \frac{7h}{2}.$$

The heights of the corresponding rectangles are

$$f(m_1), \quad f(m_2), \quad f(m_3) \quad \text{and} \quad f(m_4).$$

You can see that the total area of the four rectangles in Figure 13.2 is

$$h\,f\left(m_1\right) + h\,f\left(m_2\right) + h\,f\left(m_3\right) + h\,f\left(m_4\right) = h\left[f\left(m_1\right) + f\left(m_2\right) + f\left(m_3\right) + f\left(m_4\right)\right].$$

This particular approximation to $\int_a^b f(x)\,dx$ is called M_4. This means the mid-ordinate rule with four 'strips', or five ordinates.

You can carry out this procedure with any number of rectangles. Intuitively, you would expect that the more rectangles you use, the closer the value obtained will be to the exact area beneath the curve.

The general form of the mid-ordinate rule, using n strips, each of width h gives the following approximation of $\int_a^b f(x)\,dx$.

$$M_n = h\left(f\left(m_1\right) + f\left(m_2\right) + \ldots + f\left(m_n\right)\right)$$

Discussion point

➡ Why must $h = \dfrac{b-a}{n}$?

where $m_1, m_2, \ldots, m_n$ are the values of x at the midpoints of the strips, and

$$h = \frac{b-a}{n}.$$

You will sometimes see this formula stated as

$$M_n = h\left(y_{\frac{1}{2}} + y_{\frac{3}{2}} + \ldots + y_{n-\frac{3}{2}} + y_{n-\frac{1}{2}}\right).$$

Here $y_{\frac{1}{2}}$ means the value of the function half-way along the first strip, $y_{\frac{3}{2}}$ means the function value half-way along the second strip and so on.

As an example, think about calculating the mid-ordinate rule approximations, M_1, M_2, M_4 and M_8, to the area beneath the graph of $f(x) = \cos x$ (with x in radians) between $a = 0$ and $b = \dfrac{\pi}{2}$.

The value M_1 uses just one rectangle with a base that stretches from $a = 0$ to $b = \dfrac{\pi}{2}$ (see Figure 13.3). This has its midpoint at $\dfrac{\pi}{4}$.

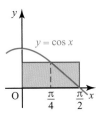

Figure 13.3

Therefore the height of the single rectangle used to estimate the area is

$$f\left(\frac{\pi}{4}\right) = \cos\frac{\pi}{4} = \frac{1}{\sqrt{2}}$$

and its width is $\dfrac{\pi}{2}$.

So $M_1 = \dfrac{1}{\sqrt{2}} \times \dfrac{\pi}{2}$

$\qquad = 1.110\,720\,735$ (to 9 d.p.)

For M_2 you divide the interval into two strips of equal width, $h = \frac{\pi}{4}$. The first of these goes from $a = 0$ to $a + h = \frac{\pi}{4}$ and the second from $\frac{\pi}{4}$ to $b = \frac{\pi}{2}$.

The respective midpoints are at $\frac{\pi}{8}$ and $\frac{3\pi}{8}$ (see Figure 13.4).

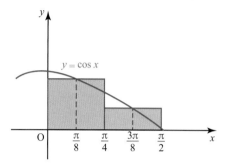

Figure 13.4

Therefore the two rectangles have width $\frac{\pi}{4}$ and have heights of $\cos\frac{\pi}{8}$ and $\cos\frac{3\pi}{8}$ respectively. This gives

$$M_2 = \frac{\pi}{4}\left(\cos\frac{\pi}{8} + \cos\frac{3\pi}{8}\right)$$
$$= 1.026\,172\,153 \text{ (to 9 d.p.)}$$

With four strips, each rectangle has a width of $\frac{\pi}{8}$.

The midpoints of the strips are at $x = \frac{\pi}{16}, \frac{3\pi}{16}, \frac{5\pi}{16}$ and $\frac{7\pi}{16}$.

Therefore,

$$M_4 = \frac{\pi}{8}\left(\cos\frac{\pi}{16} + \cos\frac{3\pi}{16} + \cos\frac{5\pi}{16} + \cos\frac{7\pi}{16}\right)$$
$$= 1.006\,454\,543 \text{ (to 9 d.p.)}$$

Finally

$$M_8 = \frac{\pi}{16}\left(\cos\frac{\pi}{32} + \cos\frac{3\pi}{32} + \dots + \cos\frac{13\pi}{32} + \cos\frac{15\pi}{32}\right)$$
$$= 1.001\,608\,189 \text{ (to 9 d.p.)}$$

Below the same values are obtained using a spreadsheet.

Column A contains the values of x at which the value of $\cos x$ is needed (the midpoints of the rectangles). In this case it contains some values that are not used in the approximations for M_1 to M_4. This is to make the spreadsheet formulae easier and so that it's simpler to keep track of where the values are on the x-axis.

Column B contains the values of $\cos x$.

'0' is entered into cell A2.

The formula '=A2+PI()/32' is in cell A3 and is copied down the column. 'PI()' gives a good approximation to π. Why is it impossible for this to be an exact value?

The formula '=COS(A2)' is in cell B2 and is copied down the column.

The formula in this cell is '=(A18–A2)*B10' which is how to calculate M_1. Make sure you understand this.

The formula in this cell is '=((A18–A2)/2)*(B6+B14)' which is how to calculate M_2.

The formula in this cell, for M_4, is '=((A18–A2)/4)* (B4+B8+B12+B16)'.

Try to work out what the formula in this cell is.

This value is very close to $\frac{\pi}{2}$. Why is the value in cell B18 -3.28×10^{-16} and not 0?

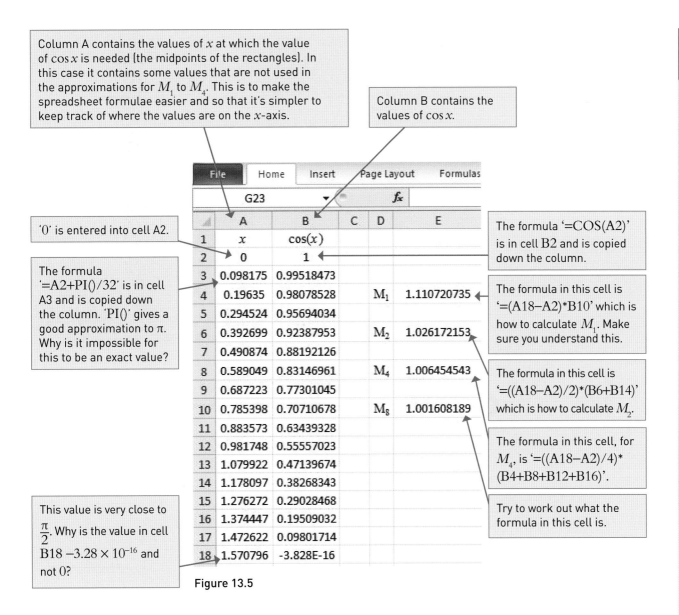

	A	B	C	D	E
1	x	$\cos(x)$			
2	0	1			
3	0.098175	0.99518473			
4	0.19635	0.98078528		M_1	1.110720735
5	0.294524	0.95694034			
6	0.392699	0.92387953		M_2	1.026172153
7	0.490874	0.88192126			
8	0.589049	0.83146961		M_4	1.006454543
9	0.687223	0.77301045			
10	0.785398	0.70710678		M_8	1.001608189
11	0.883573	0.63439328			
12	0.981748	0.55557023			
13	1.079922	0.47139674			
14	1.178097	0.38268343			
15	1.276272	0.29028468			
16	1.374447	0.19509032			
17	1.472622	0.09801714			
18	1.570796	-3.828E-16			

Figure 13.5

It is useful to determine whether the value obtained is an overestimate or an underestimate.

Example 13.1

(i) Use the mid-ordinate rule with 1 step to find an estimate for
$$\int_{0.5}^{1.5} \operatorname{cosech} x \, dx .$$

(ii) Determine whether the estimate is an overestimate or an underestimate.

Solution

(i) There is only one strip so only one mid-ordinate is used, $y_1 = f(1)$.

So, $\int_{0.5}^{1.5} \operatorname{cosech} x \, dx \approx 1 \times f(1) \approx 0.850918128$ to 9 d.p.

(ii) Figure 13.6 shows the area beneath the curve, together with the estimate using the mid-ordinate rule with one strip.

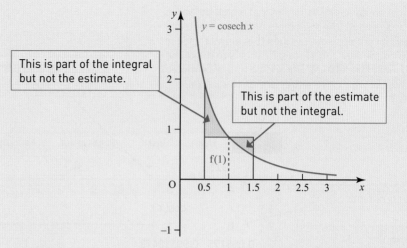

Figure 13.6

Comparing the shaded areas shows that the area missed by the estimate is greater than the area 'wrongly' included by the estimate. Thus, the estimate is an underestimate of the value of the integral.

ACTIVITY 13.1

(i) Decide whether each of the following mid-ordinate estimates show overestimates or underestimates of the areas between the curves and the x-axis.

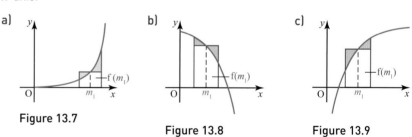

Figure 13.7

Figure 13.8   **Figure 13.9**

(ii) What property of the curve is associated with an overestimate rather than an underestimate when the mid-ordinate rule is used?

Note

A curve is concave upwards when the second derivative is positive, and concave downwards, or convex, when the second derivative is negative.

The four cases show that an underestimate occurs when the curve is concave upwards and an overestimate occurs when the curve is concave downwards, or convex.

Whether the estimate is an overestimate or an underestimate the accuracy can be increased by using more strips. The rectangles leave smaller gaps as the number of strips increases.

Simpson's rule

Simpson's rule involves fitting quadratic functions to points on a curve and using the area under these quadratic functions to estimate the area under the curve.

The idea is shown in the Figure 13.10. The solid line is part of the curve $y = f(x)$, so the shaded area is $\int_a^b f(x)\,dx$.

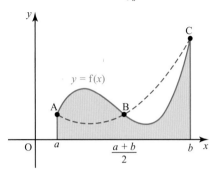

Figure 13.10

The points A, B and C are on the curve $y = f(x)$ and have equally spaced x-coordinates, so the x-coordinate of B is $\dfrac{a+b}{2}$, the midpoint of a and b. The dashed line is a quadratic function whose graph passes through A, B and C. The area under that quadratic (which can be calculated by integration) is used as an approximation to $\int_a^b f(x)\,dx$.

More than one quadratic can be used to improve this estimate. In Figure 13.11 two quadratics are used, one which fits the points A, D and B and one which fits the points B, E and C. The points A, D, B, E and C have evenly spaced x-coordinates.

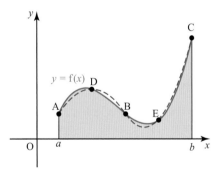

Figure 13.11

There is no universally agreed notation for Simpson's rule. In this book the approximation using one quadratic shown in Figure 13.10 is called S_2. The approximation using two quadratics shown in Figure 13.11 is called S_4. These can both be expressed in terms of values of the function.

In general, the Simpson's rule approximation S_{2n} to the integral $\int_a^b f(x)\,dx$ can be calculated directly from function values as follows.

Divide the interval from a to b into $2n$ pieces, each of width $h = \dfrac{b-a}{2n}$.

As usual, let f_0 be the value of the function at a, f_1, the value at the right-hand end of the first strip, and so on, so that f_{2n} is the value of the function at b.

Then

$$S_{2n} = \frac{h}{3}\left[f_0 + f_{2n} + 4(f_1 + f_3 + f_5 + \ldots + f_{2n-1}) + 2(f_2 + f_4 + f_6 + \ldots + f_{2n-2})\right].$$

As for the other formulae in this chapter, you might see this written as

$$S_{2n} = \frac{h}{3}\left[y_0 + y_{2n} + 4(y_1 + y_3 + y_5 + \ldots + y_{2n-1}) + 2(y_2 + y_4 + y_6 + \ldots + y_{2n-2})\right].$$

This is **Simpson's rule**.

ACTIVITY 13.2

This activity is to derive the formula for Simpson's rule.

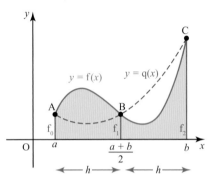

Figure 13.12

Points A, B and C on a curve $y = f(x)$ are shown in Figure 13.12.

A has coordinates (a, f_0), B has coordinates $\left(\frac{a+b}{2}, f_1\right)$ and C has coordinates

(b, f_2), so $f_0 = f(a)$, $f_1 = f\left(\frac{a+b}{2}\right)$ and $f_2 = f(b)$. h is equal to $\frac{b-a}{2}$.

The dashed line is part of the curve $y = q(x)$, where $q(x)$ is a quadratic which passes through the points A, B and C.

$\int_a^b q(x)\,dx$ is to be used as an approximation to $\int_a^b f(x)\,dx$, giving the Simpson's

rule approximation S_2. Follow steps (i) to (iv) to find $S_2 = \int_a^b q(x)\,dx$ in terms of f_0, f_1, f_2 and h.

(i)   Imagine translating the curve $y = q(x)$ horizontally so that a moves to $-h$,

   $\frac{a+b}{2}$ moves to 0 and b moves to h. Call this new curve $y = r(x)$.

   Why is $\int_{-h}^{h} r(x)\,dx = \int_a^b q(x)\,dx$?

   What are $r(-h)$, $r(0)$ and $r(h)$ in terms of values already introduced?

(ii) Suppose $r(x) = ax^2 + bx + c$.

   Show that $\int_{-h}^{h} r(x)\,dx = \frac{2ah^3}{3} + 2ch$, $f_0 + f_2 = 2ah^2 + 2c$ and $f_1 = c$.

(iii) Hence show that the Simpson's rule estimate S_2 to $\int_a^b f(x)\,dx$ is

   $\frac{h}{3}\left(f_0 + 4f_1 + f_2\right)$.

(iv) Explain the general Simpson's rule formula

   $$S_{2n} = \frac{h}{3}\Big[f_0 + f_{2n} + 4\left(f_1 + f_3 + f_5 + \ldots + f_{2n-1}\right)$$
   $$+ 2\left(f_2 + f_4 + f_6 + \ldots + f_{2n-2}\right)\Big],$$

   where $h = \frac{b-a}{2n}$, in terms of this.

Historical note

Thomas Simpson (1710–61) was a weaver from Spitalfields who taught himself mathematics and, as a break from working at his loom, taught mathematics to others. A textbook which he wrote in 1745 ran to eight editions, the last of which was published in 1809. He became Professor of Mathematics at Woolwich College and was noted for his work on trigonometric proofs and for the derivation of formulae for use in the computation of tables of values of trigonometric functions. The result with which his name is associated had been published in draft form by the Scottish mathematician James Gregory in 1668 and was published in complete form by Simpson in his 'Mathematical Dissertation on Physical and Analytical Subjects' in 1743.

Example 13.2

Use Simpson's rule with

(i) 2 strips   (ii) 4 strips

to calculate an estimate of $\int_{0.5}^{1.5} \sin\left(x^2\right) dx$.

Solution

(i)

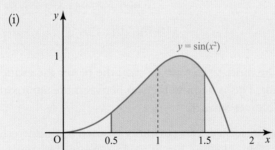

Figure 13.13

$$\int_a^b y\, dx \approx \frac{h}{3}\left(y_0 + y_n + 4\left(y_1 + y_3 + \ldots + y_{n-1}\right) + 2\left(y_2 + y_4 + \ldots + y_{n-2}\right)\right),$$

where $h = \dfrac{b-a}{n}$, and y is even.

Table 13.1 shows the values of $y = \sin\left(x^2\right)$ required for two strips, to 9 d.p.

x	0.5	1	1.5
$y = \sin\left(x^2\right)$	0.247 403 959	0.841 470 985	0.778 073 197

Table 13.1

$$\int_{0.5}^{1.5} \sin\left(x^2\right) dx \approx \frac{0.5}{3}\left(y_0 + y_n + 4y_1\right)$$

> The strip width is 0.5 and there are 3 ordinates.

$$\approx \frac{0.5}{3}\left(0.247\,403\,959 + 0.778\,073\,197 + 4 \times 0.841\,470\,985\right)$$

$$\approx 0.731\,893\,516$$

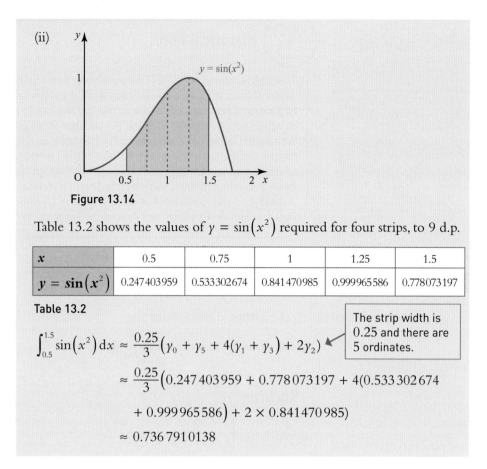

Figure 13.14

Table 13.2 shows the values of $y = \sin(x^2)$ required for four strips, to 9 d.p.

x	0.5	0.75	1	1.25	1.5
$y = \sin(x^2)$	0.247403959	0.533302674	0.841470985	0.999965586	0.778073197

Table 13.2

> The strip width is 0.25 and there are 5 ordinates.

$$\int_{0.5}^{1.5} \sin(x^2)\, dx \approx \frac{0.25}{3}\left(y_0 + y_5 + 4(y_1 + y_3) + 2y_2\right)$$

$$\approx \frac{0.25}{3}\left(0.247403959 + 0.778073197 + 4(0.533302674\right.$$

$$\left. + 0.999965586\right) + 2 \times 0.841470985)$$

$$\approx 0.7367910138$$

The more strips that are used the better the estimate obtained using Simpson's rule. This is because the gap between the approximating quadratics and the curve becomes smaller.

A spreadsheet can obviously be used to do the calculations, and is particularly useful as the number of strips increases and, with that, the accuracy of the estimate.

Example 13.3

Use Simpson's rule with eight strips to estimate $\int_{1}^{2} \sqrt{1 + \cos x}\, dx$.

Solution

> The interval is from 1 to 2, and there are to be 8 strips so the formula here is $= \dfrac{2 - 1}{8}$.

> The value of h is added on each time. What formula is in cell B3 in order that, when it is dragged down, the values shown appear?

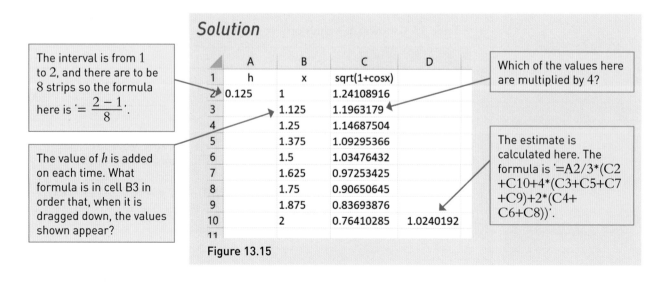

	A	B	C	D
1	h	x	sqrt(1+cosx)	
2	0.125	1	1.24108916	
3		1.125	1.1963179	
4		1.25	1.14687504	
5		1.375	1.09295366	
6		1.5	1.03476432	
7		1.625	0.97253425	
8		1.75	0.90650645	
9		1.875	0.83693876	
10		2	0.76410285	1.0240192
11				

> Which of the values here are multiplied by 4?

> The estimate is calculated here. The formula is '=A2/3*(C2 +C10+4*(C3+C5+C7 +C9)+2*(C4+ C6+C8))'.

Figure 13.15

ACTIVITY 13.3

Use a spreadsheet to calculate $\int_1^2 \sqrt{1 + \cos x}\, \mathrm{d}x$ using Simpson's rule with 20 strips.

Exercise 13.1

① Figure 13.16 shows the curve $y = \sqrt{x^3 + 1}$.

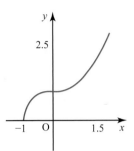

Figure 13.16

Use the mid-ordinate rule with four strips of equal width to find an estimate for $\int_0^1 \sqrt{x^3 + 1}\, \mathrm{d}x$ correct to 4 significant figures.

② Figure 13.17 shows the curve $y = x^x$.

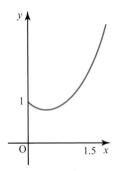

Figure 13.17

(i) Use the mid-ordinate rule with five strips of equal width to find an estimate for $\int_{0.5}^{1.5} x^x$ correct to 3 significant figures.

(ii) Is your answer an underestimate or an overestimate?

(iii) Justify your answer to part (ii).

③ Use Simpson's rule to find an approximation to $\int_0^2 \sqrt{1 + \sin x + \cos x}\, \mathrm{d}x$ using

(i) 2 strips   (ii) 4 strips   (iii) 8 strips.

Comment on the accuracy of your answers.

④ The values of a function $f(x)$ are known only for the values of x shown in Table 13.3.

x	0	0.1	0.2	0.4
$f(x)$	1.270	1.662	2.138	4.535

Table 13.3

(i) An approximation for $\int_0^{0.4} f(x)\,dx$ is made using Simpson's rule. What is the largest number of strips that can be used with the information in Table 13.3?

(ii) The value of $f(0.3)$ now becomes available. What is the largest number of strips now?

⑤ Each of the following definite integrals can be obtained using calculus.

(a) $\int_1^3 \frac{1}{x}\,dx$ (b) $\int_0^2 x^3\,dx$ (c) $\int_0^1 \sin \pi x\,dx$ (d) $\int_0^{2\pi} x \sin x\,dx$

(i) Determine the integrals.

(ii) Approximate the integrals using the mid-ordinate rule with 4 strips.

(iii) Approximate the integrals using the mid-ordinate rule with 8 strips.

(iv) Compare your answers to (i), (ii) and (iii), explaining what you find.

⑥ For each of the integrals in question 5, determine, with the aid of a graph, whether the approximations using the mid-ordinate rule are overestimates or underestimates.

Explain why it was difficult to answer the question for the integral in (d).

⑦ Use Simpson's rule to approximate the following integrals using 2, 4 and 8 strips. State the value of the integral to the maximum number of decimal places you consider to be correct. Explain how you could improve the accuracy of your approximation.

(i) $\int_{0.5}^1 x^3 e^x\,dx$ (ii) $\int_1^3 \sqrt[x]{1+x^2}\,dx$

2 Euler's method

You have seen in Chapter 12 that many differential equations of the form $\frac{dy}{dx} = f(x)$ can be solved analytically, that is, by finding the function itself.

When an analytical method is not possible a numerical method can be used to find an approximate solution.

ACTIVITY 13.4

Which of the following differential equations cannot be solved analytically?

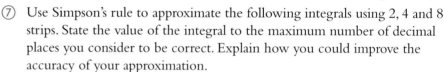

(a) $\frac{dy}{dx} = 5e^{3x}$ (b) $\frac{dy}{dx} = 4e^{x^3}$ (c) $\frac{dy}{dx} = x^2 y^2$ (d) $\frac{dy}{dx} = x^2 + y^2$

Euler's method is one of the simplest methods that can be used to solve first order differential equations numerically. It uses the idea that, if you start from a known point on the curve and move along the tangent at that point for a short distance, you can use that line segment to approximate the solution curve. Repeating the process from the new point obtained produces a sequence of straight line segments approximating the curve, as shown in Figure 13.18.

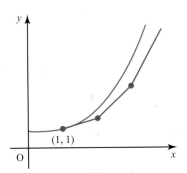

Figure 13.18

Consider the differential equation $\frac{dy}{dx} = x^2$ with initial condition that $y = 1$ when $x = 1$.

Although this equation can be solved analytically, it serves to illustrate the method.

At the point $(1, 1)$, $\frac{dy}{dx} = 1^2 = 1$.

The increment in x, $h = 0.1 \Rightarrow x_1 = 1.1$.

The increment in $y = h \times$ gradient at $(1, 1) = 0.1 \times 1 \Rightarrow y_1 = 1.1$.

See Figure 13.19.

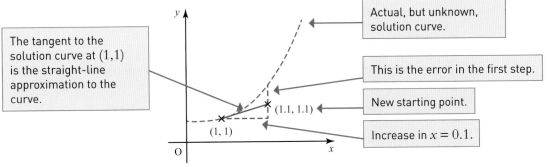

The tangent to the solution curve at $(1,1)$ is the straight-line approximation to the curve.

Actual, but unknown, solution curve.

This is the error in the first step.

New starting point.

Increase in $x = 0.1$.

Figure 13.19

The new starting point is $(1.1, 1.1)$ and the process is repeated.

At the point $(1.1, 1.1)$, $\frac{dy}{dx} = 1.1^2 = 1.21$.

The increment in x, $h = 0.1 \Rightarrow x_2 = 1.1 + 0.1 = 1.2$.

The increment in $y = h \times$ gradient at $(1.1, 1.1) = 0.1 \times 1.21 = 0.121$

$\Rightarrow y_2 = 1.1 + 0.121 = 1.221$.

See Figure 13.20.

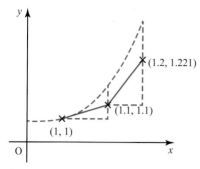

Figure 13.20

(1.21, 1.221) is now the new starting point. You can see that the error is now greater than after the first step. The approximating function consists of the red straight line segments and can be continued as needed.

The spreadsheet in Figure 13.21 shows the steps until $x = 2$.

> *x* increases in steps of 0.1.

> What is the formula in cell B4?

> The gradient is x^2 so the formula in C3 is '=A3^2'.

	A	B	C
1	x	y	y'
2	1	1	1
3	1.1	1.1	1.21
4	1.2	1.221	1.44
5	1.3	1.365	1.69
6	1.4	1.534	1.96
7	1.5	1.73	2.25
8	1.6	1.955	2.56
9	1.7	2.211	2.89
10	1.8	2.5	3.24
11	1.9	2.824	3.61
12	2	3.185	4

> What is the formula in cell C9?

Figure 13.21

Discussion point

→ Cell B3 contains the formula '=B2+0.1*C2'. Explain why it takes this form.

You may wish to work through the calculations for the remainder of the values until $x = 2$.

ACTIVITY 13.5

(i) Solve the differential equation $\dfrac{dy}{dx} = x^2$ with initial condition that $y = 1$ when $x = 1$ analytically and compare the true values of the solution function with those of the approximating function.

(ii) By creating a spreadsheet compare the accuracy of the approximating function for both smaller and larger values of h, the increase in x at each step.

You should now have a feeling for the process involved and it is appropriate to formalise it.

In general, the differential equation $\frac{dy}{dx} = f(x, y)$ with initial condition of (x_0, y_0) is to be solved using Euler's method. After n steps the most recent point is shown in Figure 13.22 as point A with coordinates (x_n, y_n) and so the gradient of the line AB = f (x_n, y_n).

So $f(x_n, y_n) = \frac{A'B}{h} \Rightarrow A'B = h\,f(x_n, y_n)$

The increase in x is h and the increase in y is $h\,f(x_n, y_n)$ so the new starting point is $(x_n + h, \ y_n + h\,f(x_n, y_n))$.

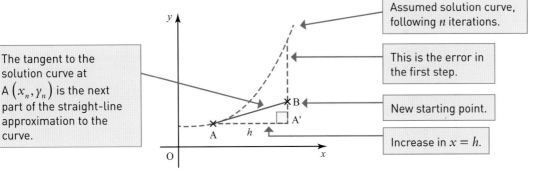

The tangent to the solution curve at $A\left(x_n, y_n\right)$ is the next part of the straight-line approximation to the curve.

Assumed solution curve, following n iterations.

This is the error in the first step.

New starting point.

Increase in $x = h$.

Figure 13.22

The new starting point is B, (x_{n+1}, y_{n+1}), and the process is repeated.

So Euler's formula is $y_{r+1} = y_r + hf\left(x_r, y_r\right)$ for equations of the form $\frac{dy}{dx} = f(x)$ and $\frac{dy}{dx} = f(x, y)$.

The values of x are called $x_1, x_2, \ldots$ and similarly the values of y are called $y_1, y_2, \ldots$

$f\left(x_r, y_r\right)$ is the value of the derivative at the point $\left(x_r, y_r\right)$ and h is the step length between successive values of x.

Euler's method can also be used with equations of the form $\frac{dy}{dx} = f(x, y)$, as included in the formal statement of the method.

The only difference is that the value of the gradient involves y as well as x.

Example 13.4

The function $y(x)$ satisfies the differential equation $\frac{dy}{dx} = f(x, y)$, where $f(x, y) = x^2 + y^2$, and $y(0) = 0.5$. Use the Euler formula with $h = 0.25$ to obtain an approximation to $y(1)$.

Solution

It is a good idea to lay out the solution in a table that shows the calculations when you are doing it without a spreadsheet (see Table 13.4).

n	x_n	y_n	$f(x_n, y_n)$	$hf(x_n, y_n)$	$y_{n+1} = y_n + hf(x_n, y_n)$
0	0	0.5	0.25	0.0625	0.5625
1	0.25	0.5625	0.3789	0.0947	0.6572
2	0.5	0.6572	0.6819	0.1705	0.8277
3	0.75	0.8277	1.2476	0.3119	1.1396
4	1.0	1.1396			

Table 13.4

Or, alternatively, with a spreadsheet (see Figure 13.23):

	A	B	C
1	x	y	dy/dx=f(x,y)
2	0	0.5	0.25
3	0.25	0.5625	0.37890625
4	0.5	0.657226563	0.681946754
5	0.75	0.827713251	1.247609226
6	1	1.139615558	2.298723619
7			

The formula in cell C2 is '= A2^2+B2^2' and this is copied down to C6.

Figure 13.23

The formula in cell B3 is '=B2+0.25*C2' and this is copied down to B6.

In both cases you can see that $y(1) = 1.1396$ to 4 decimal places.

ACTIVITY 13.6

Use a spreadsheet to find the value of y when $x = 1$ for $\dfrac{\mathrm{d}y}{\mathrm{d}x} = x^2 + y^2$, given that $y = 0.5$ when $x = 0$ with steps of $h = 0.1$.

The improved Euler method

One way to improve the accuracy of an approximation using Euler's formula is to reduce the step size.

Another method, sometimes called the improved Euler method, uses the average of the gradients from the differential equation at each end of the line segment.

So, for the first order differential equation $\dfrac{\mathrm{d}y}{\mathrm{d}x} = f(x, y)$ the formula is:

$$y_{n+1} = y_n + \frac{1}{2}(k_1 + k_2)$$

where $k_1 = hf(x_n, y_n)$ and $k_2 = hf(x_n + h, y_n + k_1)$.

Example 13.5

The function $y(x)$ satisfies the differential equation $\dfrac{dy}{dx} = f(x, y)$ where $f(x, y) = 2(x + y)\cos x$ and $y(0) = 1$.

Use the improved Euler formula $y_{n+1} = y_n + \dfrac{1}{2}(k_1 + k_2)$, where $k_1 = hf(x_n, y_n)$ and $k_2 = hf(x_n + h, y_n + k_1)$ and $h = 0.2$ to obtain an approximation to $y(0.2)$ to 4 decimal places.

Solution

$(x_0, y_0) = (0, 1)$ ← Substitute into $\dfrac{dy}{dx}$.

$\Rightarrow f(x_0, y_0) = f(0, 1) = 2 \times (0 + 1) \times \cos 0 = 2$

$\Rightarrow k_1 = hf(x_n, y_n) = 0.2 \times 2 = 0.4$

$\Rightarrow k_2 = hf(x_n + h, y_n + k_1) = 0.2 \times 2 \times (0.2 + 1 + 0.4) \times \cos 0.2 = 0.627243$

And so $y(0.2) = y_0 + \dfrac{1}{2}(k_1 + k_2)$

$= 1 + \dfrac{1}{2}(0.4 + 0.627243) = 1.5136$ to 4 d.p.

There are other methods that can improve the accuracy of Euler's approximation, for example:

$y_{r+1} = y_{r-1} + 2hf(x_r, y_r)$ with the usual notation.

Exercise 13.2

You may wish to use a spreadsheet for questions 1 and 2.

① $\dfrac{dy}{dx} = x - y^2; y = 1$ when $x = 2$

Using $h = 0.2$, estimate the value of $y(3)$.

② $\dfrac{dy}{dx} = \sqrt{x + y}; y = 2$ when $x = 1$

Using $h = 0.1$, estimate the value of $y(1.4)$.

③ $\dfrac{dy}{dx} = x^2 - y^2; y = 2.1$ when $x = 1.5$

Using $h = 0.2$, estimate the value of $y(2.5)$.

How could you improve the accuracy of this approximation?

④ The function $y(x)$ satisfies the differential equation $\dfrac{dy}{dx} = f(x, y)$, where $f(x, y) = 2xy + 1$ and $y(0) = 0$.

Use the Euler formula $y_{r+1} = y_r + hf(x_r, y_r)$ with $h = 0.1$ to obtain an approximate solution to $y(1)$.

⑤ The function $y(x)$ satisfies the differential equation $\dfrac{dy}{dx} = f(x, y)$, where $f(x, y) = 1 + x\sin y$ and $y(0) = 0.5$.

Use the formula $y_{n+1} = y_n + \dfrac{1}{2}(k_1 + k_2)$ where $k_1 = hf(x_n, y_n)$ and $k_2 = hf(x_n + h, y_n + k_1)$ with a step length of 0.2 to approximate $y(0.2)$.

⑥ It is given that $y(x)$ satisfies the differential equation $\dfrac{dy}{dx} = f(x, y)$, where $f(x, y) = 4x - 0.05y$ and $y(5) = 80$.

(i) Use the improved Euler formula $y_{n+1} = y_n + \dfrac{1}{2}(k_1 + k_2)$, where $k_1 = hf(x_n, y_n)$ and $k_2 = hf(x_n + h, y_n + k_1)$ with a step length of 0.5, to approximate $y(6)$.

(ii) Find an analytical solution to the differential equation.

(iii) Hence find an exact value of $y(6)$.

(iv) Suggest a way of improving the approximation in part (i).

⑦ The function $y(x)$ satisfies the equation $\dfrac{dy}{dx} = f(x, y)$, where $f(x, y) = xy^2$ and $y(1) = 0.5$.

(i) Use Euler's method to approximate the value of $y(2)$ using a step length of $h = 0.2$.

(ii) Find an analytical solution to the differential equation and hence find an exact value of $y(2)$.

(iii) How could the approximation in part (i) be improved?

⑧ The function $y(x)$ satisfies the differential equation $\dfrac{dy}{dx} = f(x, y)$, where $f(x, y) = y + e^{-x}\sin x$ and $y(0) = 1$.

(i) Use the Euler formula $y_{r+1} = y_r + hf(x_r, y_r)$ with $h = 0.25$ to obtain an approximate solution to $y(0.25)$.

(ii) Use the formula $y_{r+1} = y_{r-1} + 2hf(x_r, y_r)$ with your answer to part (i) to obtain $y(0.5)$ giving your answer to four decimal places.

⑨ (i) Use Euler's method with step length 0.1 to estimate the value of $y(1.2)$ given the differential equation $\dfrac{dy}{dx} = x^2 - y^2 \; (x \geqslant 0)$ with initial condition $y = 0$ when $x = 1$.

(ii) The table below shows approximate values of $y(1.2)$ for various step lengths.

Step length k	$y(1.2)$
0.01	0.237287
0.005	0.238220
0.002	0.238778
0.001	0.238964

Draw a graph of these approximations to $y(1.2)$ against step length h.

What does your graph suggest to you about the relationship between the Euler approximation to $y(1.2)$ and the step length h for small h?

Use the graph to estimate the value of $y(1.2)$. How many decimal places can you justify?

⑩ The application of Euler's method to solve the differential equation $\dfrac{dy}{dx} = y^2 - x + 1$, with the initial condition $y = 0$ when $x = 0$, results in the incomplete table below, where the values shown have been rounded to 4 decimal places.

n	x_n	y_n	y'_n	y_{n+1}
0	0	0	1	0.1
1	0.1	0.1	0.91	0.191
2	0.2	0.191	0.8365	0.2746
3	0.3	0.2746	0.7754	0.3522
4	0.4	0.3522	0.7240	0.4246
5	0.5	0.4246	0.6803	
6	0.6			
7	0.7			
8	0.8		0.5818	0.6761
9	0.9	0.6761	0.5571	0.7318
10	1.0	0.7318	0.5355	0.7853

(i) State the value of the step length, h, used in this case.

(ii) Complete the table by finding values for the remaining entries, giving your answers to 4 decimal places.

(iii) Sketch the graph of y against x, for $0 \leqslant x \leqslant 1$.

(iv) Why does your graph suggest that $y'' < 0$ in the range $0 < x < 1$? If the solution is continued beyond $x = 1$, do you think that $y'' < 0$ will continue to be true? Give reasons for your answer.

(v) If a step length of $h = 0.2$ were used throughout, instead, would the solution be more accurate or less accurate, and why?

(iv) Is the true value of y at $x = 1.0$ greater or less than the value 0.7318 given in the table? Would your conclusion remain the same if a step length $h = 0.001$ were used instead in a new calculation? Explain your answers briefly.

LEARNING OUTCOMES

Now you have finished this chapter, you should be able to:

➤ know how to use the mid-ordinate rule and Simpson's rule to calculate approximations to a definite integral

➤ know how to apply these rules using a spreadsheet

➤ know that by looking at a graph of a function it is possible to work out when you will obtain underestimates or overestimates from the mid-ordinate rule

➤ know that increasing the number of strips increases the accuracy of the approximation

➤ know how to apply Euler's formula to find approximate solutions to first order differential equations of the form $\dfrac{dy}{dx} = f(x)$ or $\dfrac{dy}{dx} = f(x, y)$

➤ understand the geometrical derivation of Euler's formula

➤ understand how the accuracy of these methods is increased by reducing the step size

➤ understand ways in which Euler's formula may be improved

➤ be able to use given improved methods to find approximate solutions to first order differential equations of the form $\dfrac{dy}{dx} = f(x)$ or $\dfrac{dy}{dx} = f(x, y)$.

KEY POINTS

1 The value of an integral can be approximated by the mid-ordinate rule using the formula

$$\int_a^b f(x)\,dx \approx h\big(f(m_1) + f(m_2) + \ldots + f(m_n)\big)$$

where $m_1, m_2, \ldots, m_n$ are the values of x at the midpoints of n strips, each of width $h = \dfrac{b-a}{n}$.

2 The value of an integral can be approximated by Simpson's rule using the formula

$$\int_a^b y\,dx \approx \frac{h}{3}\big(y_0 + y_n + 4(y_1 + y_3 + \ldots + y_{n-1}) + 2(y_2 + y_4 + \ldots + y_{n-2})\big)$$

where $h = \dfrac{b-a}{2n}$.

3 Euler's numerical step-by-step method for solving the differential equation $\dfrac{dy}{dx} = f(x, y)$ with initial condition $y(a) = b$ and a step length of h is given by:

$$y_{r+1} = y_r + hf(x_r, y_r) \text{ with } y_0 = b,\ x_n = a + nh$$

4 There are other methods that improve on Euler's formula.

Complex numbers

The shortest path between two truths in the real domain passes through the complex domain.

Jacques Hadamard, 1865–1963

Review: The modulus and argument of a complex number

Figure 14.1 shows the point representing $z = x + y\mathrm{i}$ on an Argand diagram.

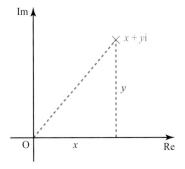

Figure 14.1

> **Prior knowledge**
>
> You need to be able to work with complex numbers and be familiar with using an Argand diagram to represent complex numbers. This is covered in Review: Complex numbers

The distance of this point from the origin is $\sqrt{x^2 + y^2}$. ← Using Pythagoras' theorem.

This distance is called the **modulus** of z, and is denoted by $|z|$.

So, for the complex number $z = x + y\mathrm{i}$, $|z| = \sqrt{x^2 + y^2}$.

Notice that since $zz^* = (x + \mathrm{i}y)(x - \mathrm{i}y) = x^2 + y^2$, then $|z|^2 = zz^*$.

Figure 14.2 shows the complex number z on an Argand diagram. The length r represents the modulus of the complex number and is denoted $|z|$. The angle θ is called the **argument** of the complex number.

> The argument of $z = 0$ is undefined.

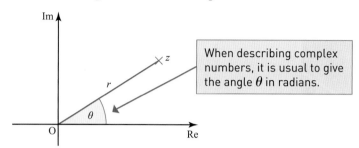

> When describing complex numbers, it is usual to give the angle θ in radians.

Figure 14.2

The argument is measured anticlockwise from the positive real axis.

This angle is not uniquely defined since adding any multiple of 2π to θ gives the same direction. To avoid confusion, it is usual to choose that value of θ for which $-\pi < \theta \leqslant \pi$. This is called the **principal argument** of z and is denoted by arg z. Every complex number except zero has a unique principal argument.

Always draw a diagram when finding the argument of a complex number. This tells you which quadrant the complex number lies in.

In Figure 14.3, you can see the relationship between the components of a complex number and its modulus and argument.

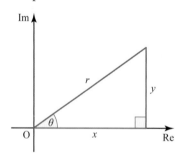

> **Discussion point**
> → Give one similarity and one difference between modulus-argument form and polar coordinates.

Figure 14.3

Using trigonometry, you can see that $\sin\theta = \dfrac{y}{r}$ and so $y = r\sin\theta$.

Similarly, $\cos\theta = \dfrac{x}{r}$ so $x = r\cos\theta$.

Therefore, the complex number $z = x + y\mathrm{i}$ can be written

$$z = r\cos\theta + r\sin\theta\,\mathrm{i}$$

> This is called the **modulus–argument** or polar form of a complex number and is denoted (r, θ).

or $\qquad z = r(\cos\theta + \mathrm{i}\sin\theta)$

Example 14.1

For each of the following:

(a) Show the complex number on Argand diagram.

(b) Write the complex number in modulus-argument form.

    (i) $z_1 = 3\mathrm{i}$ (ii) $z_2 = \sqrt{3} + \mathrm{i}$

    (iii) $z_3 = -3 - 3\mathrm{i}$ (iv) $z_4 = -1 + \sqrt{3}\mathrm{i}$

Solution

(i) (a)

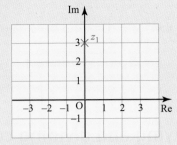

Figure 14.4

(b) z_1 has modulus 3 and argument $\frac{\pi}{2}$ so $z_1 = 3\left(\cos\frac{\pi}{2} + i\sin\frac{\pi}{2}\right)$

(ii) (a)

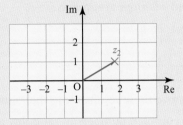

Figure 14.5

(b) z_2 has modulus $\sqrt{\left(\sqrt{3}\right)^2 + 1^2} = 2$ and argument $\tan^{-1}\left(\frac{1}{\sqrt{3}}\right) = \frac{\pi}{6}$
so

$$z_2 = 2\left(\cos\frac{\pi}{6} + i\sin\frac{\pi}{6}\right).$$

(iii) (a)

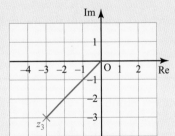

Figure 14.6

(b) z_3 has modulus $\sqrt{3^2 + 3^2} = 3\sqrt{2}$

$\tan^{-1}\left(\frac{3}{3}\right) = \frac{\pi}{4}$ so the argument of z_3 is $-\frac{3\pi}{4}$

$$z_3 = 3\sqrt{2}\left(\cos\left(-\frac{3\pi}{4}\right) + i\sin\left(-\frac{3\pi}{4}\right)\right)$$

(iv) (a)

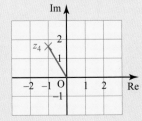

Figure 14.7

(b) z_4 has modulus $\sqrt{(-1)^2 + \left(\sqrt{3}\right)^2} = 2$

$\tan^{-1}\left(\dfrac{\sqrt{3}}{1}\right) = \dfrac{\pi}{3}$ so the argument of z_4 is $\dfrac{2\pi}{3}$.

$z_4 = 2\left(\cos\dfrac{2\pi}{3} + i\sin\dfrac{2\pi}{3}\right)$

Multiplying and dividing complex numbers in modulus–argument form

To multiply complex numbers in modulus–argument form, you *multiply* their moduli and *add* their arguments.

> You may need to add or subtract 2π to give the principal argument. For example, if
> $$\arg(z_1) + \arg(z_2) = \dfrac{7\pi}{3}$$
> then $\arg(z_1 z_2) = \dfrac{\pi}{3}$.

$$\left|z_1 z_2\right| = \left|z_1\right|\left|z_2\right|$$

$$\arg(z_1 z_2) = \arg z_1 + \arg z_2$$

You can prove these results using the compound angle formulae.

To divide complex numbers in modulus–argument form, you *divide* their moduli and *subtract* their arguments.

$$\left|\dfrac{z_1}{z_2}\right| = \dfrac{\left|z_1\right|}{\left|z_2\right|}$$

$$\arg\left(\dfrac{z_1}{z_2}\right) = \arg z_1 - \arg z_2$$

You can prove this easily from the multiplication results by letting $\dfrac{z_1}{z_2} = w$, so that $z_1 = w z_2$.

Example 14.2

The complex numbers w and z are given by $w = 3\left(\cos\dfrac{\pi}{3} + i\sin\dfrac{\pi}{3}\right)$ and

$z = 6\left(\cos\left(-\dfrac{\pi}{6}\right) + i\sin\left(-\dfrac{\pi}{6}\right)\right)$.

Find (i) wz and (ii) $\dfrac{w}{z}$ in modulus–argument form. Illustrate each of these on a separate Argand diagram.

Solution

$\left|w\right| = 3$ $\qquad \arg w = \dfrac{\pi}{3}$

$\left|z\right| = 6$ $\qquad \arg z = -\dfrac{\pi}{6}$

(i) $\qquad \left|wz\right| = \left|w\right|\left|z\right| = 3 \times 6 = 18$

$\qquad\qquad \arg wz = \arg w + \arg z = \dfrac{\pi}{3} + \left(-\dfrac{\pi}{6}\right) = \dfrac{\pi}{6}$

So $wz = 18\left(\cos\dfrac{\pi}{6} + i\sin\dfrac{\pi}{6}\right).$

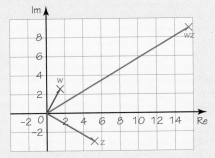

Figure 14.8

(ii) $\quad\left|\dfrac{w}{z}\right| = \dfrac{|w|}{|z|} = \dfrac{1}{2}$

$\arg\dfrac{w}{z} = \arg w - \arg z = \dfrac{\pi}{3} - \left(-\dfrac{\pi}{6}\right) = \dfrac{\pi}{2}$

$\dfrac{w}{z} = \dfrac{1}{2}\left(\cos\dfrac{\pi}{2} + i\sin\dfrac{\pi}{2}\right)$

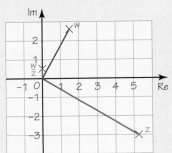

Figure 14.9

The effect of multiplication by a complex number in an Argand diagram

Much of the geometrical power of complex numbers comes from the result about multiplication of complex numbers in polar form: 'multiply the moduli, add the arguments'.

ACTIVITY 14.1

(i) Write the numbers i and −2 in modulus-argument form.

(ii) Using the result about multiplication of complex numbers in modulus–argument form investigate:

   (a)   multiplication of a complex number z by i

   (b)   multiplication of a complex number z by −2.

You will have found in Activity 14.1 that multiplication of complex numbers in modulus-argument form gives rise to a simple geometrical interpretation of multiplication.

This combination of an enlargement followed by a rotation is called a **spiral dilation**.

To obtain the line representing $z_1 z_2$ enlarge the line representing z_2 by the scale factor $|z_1|$ and rotate it through arg z_1 anticlockwise about O (see Figure 14.10).

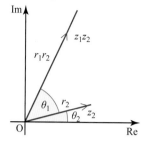

Figure 14.10

In modulus–argument form, you can say that multiplication by $r(\cos\theta + i\sin\theta)$ corresponds to enlargement with scale factor r with (anticlockwise) rotation through θ about the origin.

Example 14.3

Explain the geometrical effect of multiplying a complex number z by

(i) $-3i$

(ii) $-3\sqrt{3} + 3i$.

Solution

(i) $-3i$ has modulus 3 and argument $-\dfrac{\pi}{2}$.

Multiplying by $-3i$ would enlarge the vector z by scale factor 3 and rotate it through $\dfrac{\pi}{2}$ radians clockwise (or $\dfrac{3\pi}{2}$ radians anticlockwise).

(ii) $-3\sqrt{3} + 3i$ has modulus 6 and argument $\dfrac{5\pi}{6}$.

Multiplying by $-3\sqrt{3} + 3i$ would enlarge the vector z by scale factor 6 and rotate it through $\dfrac{5\pi}{6}$ radians anticlockwise.

Loci in the Argand diagram

The complex number $z_2 - z_1$ can be represented on an Argand diagram by the vector from the point representing z_1 to the point representing z_2, as shown in Figure 14.11.

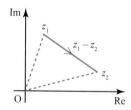

Figure 14.11

This means that:

■ **a locus of the form $|z - a| = r$ is a circle, centre a and radius r.**
   For example, the locus $|z - 3i| = 4$ represents a circle, centre 3i, radius 4 (see Figure 14.12).

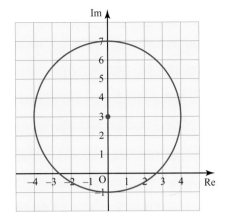

Figure 14.12

- **a locus of the form arg$(z - a) = \theta$ is a half line of points from the point a and with angle θ from the positive horizontal axis.**

  For example, the locus arg$(z - 4) = \dfrac{\pi}{4}$ is the half line of points shown in Figure 14.13.

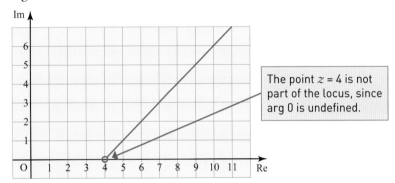

The point $z = 4$ is not part of the locus, since arg 0 is undefined.

Figure 14.13

- **a locus of the form $|z - a| = |z - b|$ represents the locus of all points which lie on the perpendicular bisector between the points represented by the complex numbers a and b.**

  For example, the locus $|z - (2 + 3i)| = |z - (1 - i)|$ is the perpendicular bisector shown in Figure 14.14.

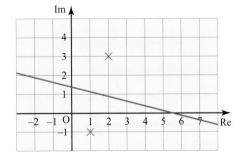

Figure 14.14

Example 14.4

Draw Argand diagrams showing the following sets of points z for which:

(i) $|z - 3i| \leqslant 3$

(ii) $0 < \arg(z - (2 - 3i)) < -\dfrac{2\pi}{3}$

(iii) $|z - 4| < |z + 4i|$ and $|z - (4 - 4i)| \leqslant 4$

Solution

(i)

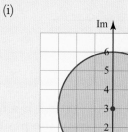

$|z - 3i| = 3$ is a circle centre 3i, radius 3. The points for which $|z - 3i| \leqslant 3$ are the points inside the circle.

The circle is shown as a solid line to indicate that it is included as part of the locus.

Figure 14.15

(ii)

$\arg(z - (2 - 3i)) = 0$ and $\arg(z - (2 - 3i)) = -\dfrac{2\pi}{3}$ are represented by the two half lines. The points for which $0 < \arg(z - (2 - 3i)) < -\dfrac{2\pi}{3}$ lie in the shaded region between the two half lines.

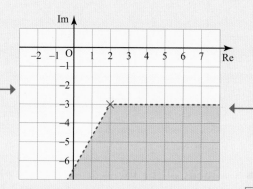

The lines are shown dotted to indicate that they are not included as part of the locus

Figure 14.16

(iii)

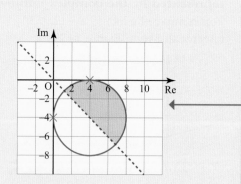

The locus $|z - 4| = |z + 4i|$ is represented by the perpendicular bisector of line segment joining the points $z = 4$ and $z = -4i$. The locus $|z - (4 - 4i)| = 4$ is represented by the circle pictured.

So the locus where both $|z - 4| < |z + 4i|$ and $|z - (4 - 4i)| \leqslant 4$ lies inside the circle and above the perpendicular bisector.

Figure 14.17

① Write each of the following complex numbers in the form $x + y\mathrm{i}$, giving surds in your answer where appropriate.

(i) $6\left(\cos\left(-\dfrac{\pi}{4}\right) + \mathrm{i}\sin\left(-\dfrac{\pi}{4}\right)\right)$

(ii) $3\left(\cos\left(-\dfrac{\pi}{6}\right) + \mathrm{i}\sin\left(-\dfrac{\pi}{6}\right)\right)$

(iii) $2\left(\cos\dfrac{3\pi}{4} + \mathrm{i}\sin\dfrac{3\pi}{4}\right)$

(iv) $7\left(\cos\left(-\dfrac{5\pi}{6}\right) + \mathrm{i}\sin\left(-\dfrac{5\pi}{6}\right)\right)$

② For each complex number, find the modulus and principal argument, and hence write the complex number in modulus-argument form. Give the argument in radians, as a multiple of π.

(i) $2\sqrt{3} + 2\mathrm{i}$

(ii) $-2\sqrt{3} + 2\mathrm{i}$

(iii) $2\sqrt{3} - 2\mathrm{i}$

(iv) $-2\sqrt{3} - 2\mathrm{i}$

③ Represent each of the following complex numbers on a separate Argand diagram, and write it in the form $x + y\mathrm{i}$, giving your answer in surd form where appropriate.

(i) $|z| = 3$, $\arg z = \dfrac{\pi}{4}$

(ii) $|z| = 5$, $\arg z = \dfrac{\pi}{2}$

(iii) $|z| = 4$, $\arg z = -\dfrac{5\pi}{6}$

(iv) $|z| = 6$, $\arg z = -\dfrac{\pi}{4}$

④ Given that $z = 5\left(\cos\left(-\dfrac{\pi}{3}\right) + \mathrm{i}\sin\left(-\dfrac{\pi}{3}\right)\right)$ and $w = 3\left(\cos\dfrac{5\pi}{6} + \mathrm{i}\sin\dfrac{5\pi}{6}\right)$,

find the following complex numbers in modulus-argument form.

(i) wz

(ii) $\dfrac{w}{z}$

(iii) $\dfrac{z}{w}$

(iv) $\dfrac{1}{z}$

⑤ Explain the geometrical effect of multiplying a complex number z by

(i) $5 - 5\mathrm{i}$

(ii) $-\dfrac{1}{4} - \dfrac{1}{4}\sqrt{3}\mathrm{i}$.

⑥ Write down the complex number w such that the product wz represents the following transformations of z:

(i) an enlargement by scale factor 2 and a rotation of $\dfrac{\pi}{3}$ radians anticlockwise

(ii) an enlargement by scale factor $\dfrac{1}{3}$ and a rotation of $\dfrac{2\pi}{3}$ radians clockwise.

⑦ For each of the parts (i) to (iii), draw an Argand diagram showing the set of points z for which the given condition is true.

(i) $|z - 5\mathrm{i}| \geqslant 3$

(ii) $|z - 5| < 2$

(iii) $\left|z + \sqrt{5} + \sqrt{5}\mathrm{i}\right| \leqslant 5$

⑧ For each of the parts (i) to (iii), draw an Argand diagram showing the set of points z for which the given condition is true.

(i) $\arg z = \dfrac{\pi}{2}$

(ii) $0 < \arg(z + 2 + \mathrm{i}) < \dfrac{\pi}{3}$

(iii) $-\dfrac{2\pi}{3} \leqslant \arg(z + 2 + \mathrm{i}) \leqslant \dfrac{2\pi}{3}$

⑨ For each of the parts (i) to (iii), draw an Argand diagram showing the set of points z for which the given condition is true.

(i) $|z - 2| = |z - 4\mathrm{i}|$

(ii) $|z - 1 - \mathrm{i}| < |z - 3 + 4\mathrm{i}|$

(iii) $|z + 3 + 2\mathrm{i}| \geqslant |z + 3\mathrm{i}|$

⑩ Write down, in terms of z, the loci for the regions that are represented in each of the Argand diagrams pictured below.

(i)

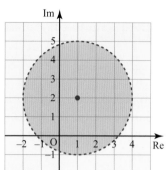

Figure 14.18

(ii)

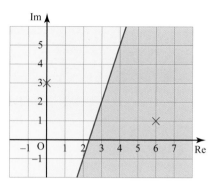

Figure 14.19

(iii)

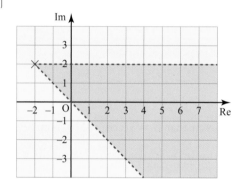

Figure 14.20

⑪ Sketch on the same Argand diagram:

(i)   The locus of points $|z - 4\mathrm{i}| = 1$

(ii)   The locus of points $\arg(z - 4\mathrm{i}) = -\dfrac{\pi}{4}$

(iii)   The locus of points $\arg(z - 4\mathrm{i}) = -\dfrac{\pi}{2}$

(iv)   The locus of points $|z - 4\mathrm{i}| = |z|$. Shade the region defined by the inequalities $|z - 4\mathrm{i}| \geqslant 1$, $-\dfrac{\pi}{2} \leqslant \arg(z - 4\mathrm{i}) \leqslant -\dfrac{\pi}{4}$ and $|z - 4\mathrm{i}| < |z|$.

⑫ The complex number z is multiplied by the complex number $w*w$, where $w = a + b\mathrm{i}$ and $a, b \in \mathbb{Z}$. Find the possible values of a and b if the geometrical effect of the multiplication is to enlarge the vector z by a scale factor of 13.

⑬ (i)   Given the point representing a complex number z on an Argand diagram, explain how to find the following points geometrically.

(a) $3z$ (b) $2\mathrm{i}z$ (c) $(3 + 2\mathrm{i})z$

(ii)   Sketch an Argand diagram to represent the points O, $3z$, $2\mathrm{i}z$ and $(3 + 2\mathrm{i})z$ and explain the geometrical connection between the points.

1 De Moivre's theorem

On page 300 you saw that when you multiply two complex numbers in modulus-argument form you multiply their moduli and add their arguments.

ACTIVITY 14.2

For the complex number $z = 2(\cos 0.1 + i \sin 0.1)$ write down z^2, z^3, z^4 and z^5.

Use your answers to write down an expression for z^n.

For the general complex number $z = r(\cos\theta + i\sin\theta)$ write down an expression for z^n.

The product of:

$$z_1 = r_1\left(\cos\theta_1 + i\sin\theta_1\right) \text{ and } z_2 = r_2\left(\cos\theta_2 + i\sin\theta_2\right)$$

is: $\quad z_1 z_2 = r_1 r_2\left(\cos\left(\theta_1 + \theta_2\right) + i\sin\left(\theta_1 + \theta_2\right)\right)$

Using this result repeatedly with a single complex number z with modulus 1 allows you to concentrate on what happens to the argument.

$\quad$ If $z = \cos\theta + i\sin\theta$

$\quad$ then $z^2 = \cos\left(\theta + \theta\right) + i\sin\left(\theta + \theta\right) = \cos 2\theta + i\sin 2\theta$

$\quad z^3 = z^2 z = \cos\left(2\theta + \theta\right) + i\sin\left(2\theta + \theta\right) = \cos 3\theta + i\sin 3\theta$

and so on.

The diagram below shows $(\cos\theta + i\sin\theta)^n$ for $n = 1, 2, 3, \ldots, 6$.

In this example $0 < \theta < \dfrac{\pi}{2}$.

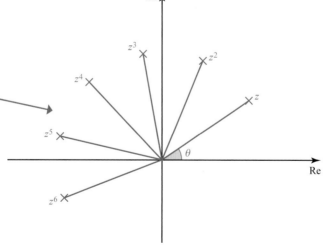

> **Note**
>
> Notice that this shows that multiplying by a complex number with magnitude 1 results in a rotation in the Argand diagram.

Figure 14.21

Discussion point

➜ How does the pattern on the Argand diagram continue when higher powers of n are added?

This suggests the following result which is called **de Moivre's theorem**.

If n is an integer then $(\cos\theta + i\sin\theta)^n = \cos n\theta + i\sin n\theta$

The proof of this result is in three parts, in which n is **(i)** positive **(ii)** zero **(iii)** negative.

(i) When n is a positive integer de Moivre's theorem can be proved by induction.

The theorem is obviously true if $n = 1$.

Assume the result is true for $n = k$, so

$$\left(\cos\theta + i\sin\theta\right)^k = \cos k\theta + i\sin k\theta$$

You want to prove that the result is true for $n = k + 1$ (if the assumption is true).

$$\begin{aligned}
\left(\cos\theta + i\sin\theta\right)^{k+1} &= (\cos k\theta + i\sin k\theta)(\cos\theta + i\sin\theta) \\
&= \cos(k\theta + \theta) + i\sin(k\theta + \theta) \\
&= \cos((k+1)\theta) + i\sin((k+1)\theta)
\end{aligned}$$

If the result is true for $n = k$, then it is true for $n = k + 1$. Since it is true for $n = 1$, it is true for all positive integer values of n.

(ii) By definition, $z^0 = 1$ for all complex numbers $z \neq 0$.

Therefore $\left(\cos\theta + i\sin\theta\right)^0 = 1 = \cos 0 + i\sin 0$.

(iii) For negative n the proof starts with the case $n = -1$.

If $a \times b = 1$ it follows that b is the reciprocal of a.

As $(\cos\theta + i\sin\theta)(\cos(-\theta) + i\sin(-\theta)) = \cos(\theta - \theta) + i\sin(\theta - \theta) = 1$ it follows that

$$\left(\cos\theta + i\sin\theta\right)^{-1} = \cos(-\theta) + i\sin(-\theta). \quad \dagger$$

If n is another negative integer, let $n = -m$ where m is a positive integer.

Then $\left(\cos\theta + i\sin\theta\right)^n = \left(\cos\theta + i\sin\theta\right)^{-m}$

As m is a positive integer de Moivre's theorem holds using part (i).

$$= \left[\left(\cos\theta + i\sin\theta\right)^m\right]^{-1}$$

$$= \left[\cos m\theta + i\sin m\theta\right]^{-1}$$

Using † with $m\theta$ in place of θ.

$$= \cos(-m\theta) + i\sin(-m\theta)$$

$$= \cos n\theta + i\sin n\theta$$

Therefore de Moivre's theorem holds for all integers n.

De Moivre's theorem can also be used for simplifying powers of complex numbers when the modulus is not 1, as in Figure 14.22. If $z = r\left(\cos\theta + i\sin\theta\right)$ then

$$z^n = \left[r(\cos\theta + i\sin\theta)\right]^n = r^n\left(\cos n\theta + i\sin n\theta\right)$$

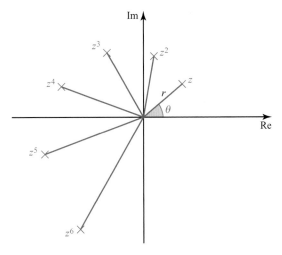

Figure 14.22

Example 14.5

Use de Moivre's theorem to simplify each of the following:

(i) $\left(\cos \dfrac{\pi}{12} + \mathrm{i} \sin \dfrac{\pi}{12} \right)^{18}$

(ii) $\left(-1 + \sqrt{3}\mathrm{i} \right)^{5}$.

Solution

(i) $\left(\cos \dfrac{\pi}{12} + \mathrm{i} \sin \dfrac{\pi}{12} \right)^{18} = \cos\left(18 \times \dfrac{\pi}{12} \right) + \mathrm{i} \sin\left(18 \times \dfrac{\pi}{12} \right)$

$$= \cos \dfrac{3\pi}{2} + \mathrm{i} \sin \dfrac{3\pi}{2}$$

$$= -\mathrm{i}$$

(ii) First convert to modulus-argument form:

$$z = -1 + \sqrt{3}\mathrm{i} \Rightarrow |z| = \sqrt{(-1)^2 + \left(\sqrt{3} \right)^2} = 2$$

$\arctan\left(\dfrac{\sqrt{3}}{1} \right) = \dfrac{\pi}{3}$ so $\arg z = \dfrac{2\pi}{3}$. ⟵ | z is in the second quadrant.

So $\left(-1 + \sqrt{3}\mathrm{i} \right)^{5} = 2^5 \left(\cos \dfrac{2\pi}{3} + \mathrm{i} \sin \dfrac{2\pi}{3} \right)^5 = 2^5 \left(\cos \dfrac{10\pi}{3} + \mathrm{i} \sin \dfrac{10\pi}{3} \right)$

$$= 32 \left(-\dfrac{1}{2} - \dfrac{\sqrt{3}}{2}\mathrm{i} \right)$$

$$= -16 - 16\sqrt{3}\mathrm{i}$$

Example 14.6

Simplify the expression $\dfrac{\left[4\left(\cos 5\theta + \mathrm{i} \sin 5\theta \right) \right]^5}{\left[3\left(\cos 4\theta + \mathrm{i} \sin 4\theta \right) \right]^4}$.

De Moivre's theorem

ACTIVITY 14.3

Use de Moivre's theorem with $\theta = -\phi$ and the facts:
$$\cos(-\theta) = \cos\theta$$
$$\sin(-\theta) = -\sin\theta$$
to show that:
$$(\cos\phi - i\sin\phi)^n = \cos n\phi - i\sin n\phi$$

Solution

$$\frac{\left[4(\cos 5\theta + i\sin 5\theta)\right]^5}{\left[3(\cos 4\theta + i\sin 4\theta)\right]^4} = \frac{1024(\cos 25\theta + i\sin 25\theta)}{81(\cos 16\theta + i\sin 16\theta)}$$

$$= \frac{1024}{81}(\cos 9\theta + i\sin 9\theta)$$

In Activity 14.3 you will prove the useful result

$$(\cos\phi - i\sin\phi)^n = \cos n\phi - i\sin n\phi$$

Exercise 14.1

① Use de Moivre's theorem to simplify each of the following:
   (a) in the form $\cos\alpha + i\sin\alpha$
   (b) in the form $a + ib$.

   (i) $\left(\cos\frac{\pi}{6} + i\sin\frac{\pi}{6}\right)^4$ (ii) $\left(\cos\frac{\pi}{3} + i\sin\frac{\pi}{3}\right)^{-8}$

   (iii) $\left(\cos\left(-\frac{\pi}{12}\right) + i\sin\left(-\frac{\pi}{12}\right)\right)^{10}$

② Given that $w = \cos\frac{\pi}{4} + i\sin\frac{\pi}{4}$, write each of the following complex numbers as a power of w.

   (i) $z_1 = \cos\frac{3\pi}{4} + i\sin\frac{3\pi}{4}$ (ii) $z_2 = \cos\frac{\pi}{2} + i\sin\frac{\pi}{2}$

   (iii) $z_3 = \cos\pi + i\sin\pi$

   Illustrate w, z_1, z_2 and z_3 on an Argand diagram.

③ Use de Moivre's theorem to simplify each of the following:

   (i) $\dfrac{(\cos 3\theta + i\sin 3\theta)^4}{(\cos 5\theta + i\sin 5\theta)^3}$

   (ii) $\dfrac{\left(\cos\frac{\pi}{4} + i\sin\frac{\pi}{4}\right)^3}{\left(\cos\frac{\pi}{6} + i\sin\frac{\pi}{6}\right)^2}$

   (iii) $\left(\cos\frac{\pi}{3} + i\sin\frac{\pi}{3}\right)^5 \left(\cos\frac{\pi}{6} + i\sin\frac{\pi}{6}\right)^{-4}$.

④ Given that $w = \cos\frac{\pi}{6} + i\sin\frac{\pi}{6}$, write each of the following complex numbers as a power of w.

   (i) $z_1 = \cos\frac{\pi}{6} - i\sin\frac{\pi}{6}$ (ii) $z_2 = \cos\frac{\pi}{2} - i\sin\frac{\pi}{2}$

   Illustrate w, z_1 and z_2 on an Argand diagram.

⑤ By converting to modulus-argument form and using de Moivre's theorem, find the following in the form $x + y$i giving x and y as exact expressions or correct to 3 significant figures.

(i) $\left(1 - \sqrt{3}\text{i}\right)^4$ 　　　　(ii) $(-2 + 2\text{i})^7$ 　　　　(iii) $\left(\sqrt{27} + 3\text{i}\right)^6$

⑥ Without using a calculator write $\left(-\sqrt{3} - \text{i}\right)^7$ in the form $x + y$i, where x and y are exact values.

⑦ Simplify the following expressions as far as possible.

(i) $\left[3(\cos 2\theta + \text{i}\sin 2\theta)\right]^4$ 　　　　(ii) $\left[\text{i}\left(\cos 3\theta + \text{i}\sin 3\theta\right)\right]^5$

(iii) $\left[2\text{i}\left(\cos 7\theta + \text{i}\sin 7\theta\right)\right]^{-3}$

⑧ Show that

$$\frac{\left[3(\cos 2\theta - \text{i}\sin 2\theta)\right]^4 \left[2(\cos\theta + \text{i}\sin\theta)\right]^5}{\left[4(\cos 3\theta + \text{i}\sin 3\theta)\right]^2 \left[\frac{1}{2}(\cos\theta - \text{i}\sin\theta)\right]^8}$$

can be expressed in the form $k(\cos\theta - \text{i}\sin\theta)$ where k is a constant to be found.

⑨ The three complex numbers in Figure 14.23 below each have modulus 1. They form an equilateral triangle centred on the origin.

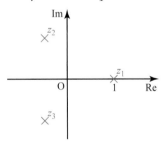

Figure 14.23

(i) Write down each of z_1, z_2 and z_3 in the form $\cos\theta + \text{i}\sin\theta$.

(ii) Use de Moivre's theorem to show that the cube of each of these complex numbers is the real number 1.

(iii) Draw an Argand diagram to show five complex numbers for which $z^5 = 1$, and write down these complex numbers in the form $\cos\theta + \text{i}\sin\theta$.

2 The nth roots of a complex number

The nth roots of unity

You already know that all quadratic equations have two roots (which may be a repeated root, or they may be complex roots). You have also solved cubic equations to find the three roots, some of which may be complex.

As early as 1629 Albert Girard stated that every polynomial equation of degree n has exactly n roots, including repeated roots. Some of these roots may be complex numbers. This was first proved by the 18-year-old Carl Freidrich Gauss 170 years later.

Therefore even the simple equation $z^n = 1$ has n roots. One of these roots is $z = 1$ and, if n is even, then $z = -1$ is another. All the other roots are complex numbers.

In this section you will look at methods for finding the other roots of the equation $z^n = 1$, and the relationship between them.

Example 14.7

(i) Write down the two roots of the equation $z^2 = 1$ and show them on an Argand diagram.

(ii) Use $z^3 - 1 = (z - 1)(z^2 + z + 1)$ to find the three roots of $z^3 = 1$. Show them on an Argand diagram.

(iii) Find the four roots of $z^4 = 1$ and show them on an Argand diagram.

Solution

(i) Using properties of real numbers $z = \pm 1$. These numbers are shown in Figure 14.24.

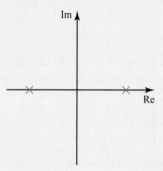

Figure 14.24

(ii) The equation $z^3 = 1$ can be rewritten as $z^3 - 1 = 0$.

So $(z - 1)(z^2 + z + 1) = 0$ ← This result is given in the question.

One of the roots is $z = 1$.

The equation $z^2 + z + 1$ has roots $z = \dfrac{-1 \pm \sqrt{3}i}{2}$.

The three roots

$$z = 1, \quad z = \frac{-1 \pm \sqrt{3}i}{2}$$

Using the quadratic formula.

are shown on an Argand diagram in Figure 14.25.

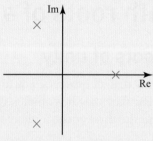

Figure 14.25

$$z = 1, \quad z = \frac{-1 \pm \sqrt{3}i}{2}$$

(iii) $z^4 = 1$ can be written in the form $z^4 - 1 = (z^2 - 1)(z^2 + 1)$ which has the four roots $z = \pm 1$, $z = \pm i$.

These roots are shown on the Argand diagram in Figure 14.26.

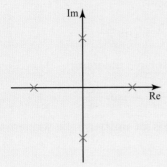

Figure 14.26

In the previous example you may have noticed that the roots of the equations $z^2 = 1$, $z^3 = 1$ and $z^4 = 1$ all lie on a unit circle, centred on the origin, with one root at the point 1.

In fact, every root of the equation $z^n = 1$ must have unit modulus, as otherwise the modulus of z^n would not be 1. So every root is of the form $z = \cos\theta + i\sin\theta$, and

> Using de Moivre's theorem.

$$z^n = 1 \Rightarrow \left(\cos\theta + i\sin\theta\right)^n$$
$$\Rightarrow \quad \cos n\theta + i\sin n\theta = 1$$
$$\Rightarrow \quad n\theta = 2k\pi \text{ where } k \text{ is any integer}$$

> Since in modulus–argument form 1 can be written $(1, 0)$ or $(1, 2\pi)$ or $(1, 4\pi)$ etc.

So, for example, in the case when $n = 3$ the result $n\theta = 2k\pi$ gives roots as follows:

when $k = 0$ $3\theta = 0 \Rightarrow \theta = 0$ so $z = \cos 0 + i\sin 0 = 1$

> Note that these are the same roots of $z^3 = 1$ as those obtained in Example 14.7.

when $k = 1$ $3\theta = 2\pi \Rightarrow \theta = \dfrac{2\pi}{3}$ so $z = \cos\dfrac{2\pi}{3} + i\sin\dfrac{2\pi}{3} = -\dfrac{1}{2} + \dfrac{\sqrt{3}}{2}i$

when $k = 2$ $3\theta = 4\pi \Rightarrow \theta = \dfrac{4\pi}{3}$ so $z = \cos\dfrac{4\pi}{3} + i\sin\dfrac{4\pi}{3} = -\dfrac{1}{2} - \dfrac{\sqrt{3}}{2}i$

For larger values of k the same roots are obtained, so $k = 3$ gives the root $z = 1$, $k = 4$ gives the root $z = -\dfrac{1}{2} + \dfrac{\sqrt{3}}{2}i$, and so on.

So, when $n = 3$, the values of k that need to be considered are $k = 0, 1$ and 2.

Generally, as k takes values $0, 1, 2, 3, \ldots, (n - 1)$ the corresponding values of θ are

$$0, \frac{2\pi}{n}, \frac{4\pi}{n}, \frac{6\pi}{n}, \ldots, \frac{2(n-1)\pi}{n}$$

giving n distinct values of z.

When $k = n$ then $\theta = 2\pi$, which gives the same z as $\theta = 0$. Similarly, any integer value of k larger than n will differ from one of $0, 1, 2, 3, \ldots, (n - 1)$ by a multiple of n, and so gives a value of θ differing by a multiple of 2π from one already listed; the same applies when k is any negative integer.

Therefore, the equation $z^n = 1$ has exactly n roots. These are

$$z = \cos\frac{2k\pi}{n} + i\sin\frac{2k\pi}{n}, \quad k = 0, 1, 2, 3, \ldots, (n - 1)$$

These n complex numbers are called the **nth roots of unity**. They include $z = 1$ when $k = 0$ and, if n is even, $z = -1$ when $k = \dfrac{n}{2}$.

It is usual to use ω (the Greek letter omega) for the root with the smallest positive argument:

$$\omega = \cos\frac{2\pi}{n} + i\sin\frac{2\pi}{n}$$

Then, by de Moivre's theorem,

$$\omega^k = \cos\frac{2k\pi}{n} + i\sin\frac{2k\pi}{n}$$

so the **nth roots of unity** can be written as

$$1, \omega, \omega^2, \ldots, \omega^{n-1}.$$

These complex numbers can be represented on an Argand diagram by the vertices of a regular *n*-sided polygon inscribed in the unit circle, with one vertex at the point 1. Figure 14.27 shows the *n*th roots of unity when $n = 9$.

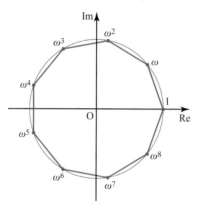

Figure 14.27

The sum of all of the *n*th roots of unity is a geometric series with common ratio ω:

$$1 + \omega + \omega^2 + \ldots + \omega^{n-1}$$

Using the formula
$$S_n = \frac{a(1 - r^n)}{1 - r}.$$

$$= \frac{1 - \omega^n}{1 - \omega}$$

Since $\omega^n = 1$.

$$= 0$$

So the sum of the *n*th roots of unity is always zero.

ACTIVITY 14.4

Verify that the sum of the *n*th roots of unity is equal to zero in the cases where $n = 2$, $n = 3$ and $n = 4$.

Example 14.8

Solve the equation $z^6 = 1$. Show the roots on an Argand diagram.

Solution

The sixth roots of unity are given by

$$z = \omega^k = \cos\frac{2k\pi}{6} + i\sin\frac{2k\pi}{6},$$

where $k = 0, 1, 2, 3, 4, 5$.

This gives the following roots z:

$k = 0$ $z = 1$

$k = 1$ $z = \omega = \cos\dfrac{2\pi}{6} + i\sin\dfrac{2\pi}{6} = \dfrac{1}{2} + \dfrac{\sqrt{3}}{2}i$

$k = 2$ $z = \omega^2 = \cos\dfrac{4\pi}{6} + i\sin\dfrac{4\pi}{6} = -\dfrac{1}{2} + \dfrac{\sqrt{3}}{2}i$

$k = 3$ $z = \omega^3 = \cos\dfrac{6\pi}{6} + i\sin\dfrac{6\pi}{6} = -1$

$k = 4$ $z = \omega^4 = \cos\dfrac{8\pi}{6} + i\sin\dfrac{8\pi}{6} = -\dfrac{1}{2} - \dfrac{\sqrt{3}}{2}i$

$k = 5$ $z = \omega^5 = \cos\dfrac{10\pi}{6} + i\sin\dfrac{10\pi}{6} = \dfrac{1}{2} - \dfrac{\sqrt{3}}{2}i$

The roots form a hexagon inscribed within a unit circle, as in Figure 14.28.

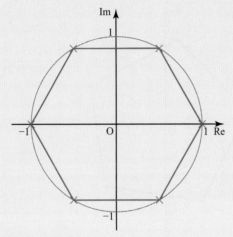

Figure 14.28

The nth roots of any complex number

To find the nth roots of any given non-zero complex number a you have to find z such that $z^n = a$. The method to find the nth roots is the same as that in the previous section on nth roots of unity, adjusted to take account of the modulus s and argument ϕ of a.

Suppose, for example, you wanted to find the fifth roots of the complex number $a = -1 + i$.

You want to find the complex number z such that $z^5 = -1 + i$.

Let $z = r(\cos\theta + i\sin\theta)$, then:

Writing $-1 + i$ in modulus–argument form.

$$\left[r\left(\cos\theta + i\sin\theta\right)\right]^5 = \sqrt{2}\left(\cos\dfrac{3\pi}{4} + i\sin\dfrac{3\pi}{4}\right)$$

Using de Moivre's theorem.

$$\Leftrightarrow r^5\left(\cos 5\theta + i\sin 5\theta\right) = \sqrt{2}\left(\cos\dfrac{3\pi}{4} + i\sin\dfrac{3\pi}{4}\right)$$

Two complex numbers in modulus-argument form are equal only if they have the same moduli and their arguments are equal or differ by a multiple of 2π. Therefore:

$$r = 2^{\frac{1}{10}} \text{ and } 5\theta = \frac{3\pi}{4} + 2k\pi, \text{ where } k \text{ is an integer.}$$

Each of the roots will have the same modulus and so will lie on the circle $\left| z \right| = 2^{\frac{1}{10}}$.

The argument of z is $\theta = \dfrac{\dfrac{3\pi}{4} + 2k\pi}{5}$.

As k takes the values 0, 1, 2, 3 and 4 the arguments obtained are $\dfrac{3\pi}{20}$, $\dfrac{11\pi}{20}$, $\dfrac{19\pi}{20}$, $\dfrac{27\pi}{20}$ and $\dfrac{7\pi}{4}$.

> Larger values of k generate the same set of arguments so, for example, $k = 5$ gives $\dfrac{43\pi}{20}$ which is equivalent to $\dfrac{3\pi}{20}$.

The five roots are shown in Figure 14.29 and form the five vertices of a regular pentagon.

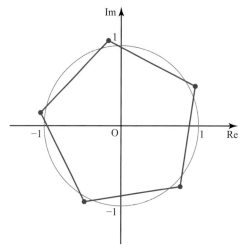

Note

Notice that multiplying by the appropriate complex root of unity results in a rotation in the Argand diagram that produces another of the roots of the original equation.

Figure 14.29

In a general case, suppose $z = r(\cos\theta + \mathrm{i}\sin\theta)$ and $a = s(\cos\phi + \mathrm{i}\sin\phi)$. Then:

$$z^n = a \Leftrightarrow r^n(\cos\theta + \mathrm{i}\sin\theta)^n = s(\cos\phi + \mathrm{i}\sin\phi)$$

$$\Leftrightarrow r^n(\cos n\theta + \mathrm{i}\sin n\theta) = s(\cos\phi + \mathrm{i}\sin\phi)$$

$$\Leftrightarrow r^n = s \quad \text{and} \quad n\theta = \phi + 2k\pi, \text{ where } k \text{ is an integer}$$

Since r and s are positive real numbers the equation $r^n = s$ gives the *unique* value $r = s^{\frac{1}{n}}$ so all the roots lie on the circle $\left| z \right| = s^{\frac{1}{n}}$.

The argument of z is $\theta = \dfrac{\phi + 2k\pi}{n}$. As k can take the values 0, 1, 2, ..., $n - 1$, this gives n distinct complex numbers z and, by the same argument as for the roots of unity, there are no other roots.

Therefore the non-zero complex number $s(\cos\phi + \mathrm{i}\sin\phi)$ has precisely n different nth roots, which are given by

$$s^{\frac{1}{n}}\left(\cos\left(\frac{\phi + 2k\pi}{n}\right) + \mathrm{i}\sin\left(\frac{\phi + 2k\pi}{n}\right) \right),$$

where $k = 0, 1, 2, ..., n - 1$.

You may also express these n roots as α, $\alpha\omega$, $\alpha\omega^2$, ..., $\alpha\omega^{n-1}$, where

$$\alpha = s^{\frac{1}{n}}\left(\cos\frac{\phi}{n} + i\sin\frac{\phi}{n}\right) \quad \text{and} \quad \omega = \cos\frac{2\pi}{n} + i\sin\frac{2\pi}{n}.$$

The sum of these nth roots of w is

Since $\omega^n = 1$. $\longrightarrow$ $\alpha + \alpha\omega + \alpha\omega^2 + \dots + \alpha\omega^{n-1} = \dfrac{\alpha\left(1 - \omega^n\right)}{1 - \omega} = 0.$

Example 14.9

Represent the five roots of the equation $z^5 = 32$ on an Argand diagram.

Hence, represent the five roots of the equation $(z - 3i)^5 = 32$ on an Argand diagram.

Solution

You know mod $32 = 32$ and $\arg(32) = 0$ so the fifth roots of 32 are given by:

$$32^{\frac{1}{5}}\left(\cos\left(\frac{0 + 2\pi k}{5}\right) + i\sin\left(\frac{0 + 2\pi k}{5}\right)\right)$$

Each root has modulus 2 and the arguments of the roots are $0, \dfrac{2\pi}{5}, \dfrac{4\pi}{5}, -\dfrac{4\pi}{5}$ and $-\dfrac{2\pi}{5}$. The roots form a regular pentagon inscribed in a circle, centre the origin and radius 2, as shown by the blue points in Figure 14.30.

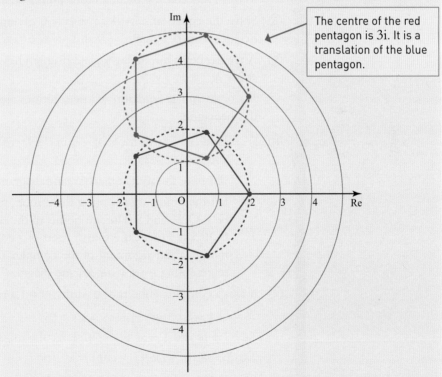

> The centre of the red pentagon is 3i. It is a translation of the blue pentagon.

Figure 14.30

The roots of $(z - 3i) = 32$ are therefore represented by the same pentagon inscribed in a circle, centre $3i$, radius 2 as shown by the red points in Figure 14.30.

Exercise 14.2

① Find the roots of the equation $z^5 = 1$ giving your answers to 3 significant figures.

Show these roots on an Argand diagram.

Describe the polygon formed by the points representing the roots.

② Find the roots of the equation $z^8 = 1$ and show these on an Argand diagram.

Describe the polygon formed by the points representing the roots.

③ Find both square roots of $-7 + 5i$, giving your answers in the form $x + yi$ with x and y correct to 2 decimal places.

④ Find the fourth roots of -4, giving your answers in the form $x + yi$.

Show the roots on an Argand diagram.

⑤ In this question, give answers as exact values or to two decimal places where appropriate.

(i) Find the cube roots of $1 - i$.

(ii) Find the fourth roots of $2 + 3i$.

(iii) Find the fifth roots of $-3 + 4i$.

⑥ Explain geometrically why the set of tenth roots of unity is the same as the set of fifth roots of unity together with their negatives.

⑦ One fourth root of the complex number w is $2 + 3i$. Find w and its other fourth roots and represent all four points on an Argand diagram.

⑧ Solve the equation $z^3 - 4\sqrt{3} - 4i = 0$ giving your solutions in the form $r(\cos\theta + i\sin\theta)$, where $r > 0$ and $-\pi < \theta \leq \pi$.

⑨ A regular heptagon on an Argand diagram has centre $-1 + 3i$ and one vertex at $2 + 3i$.

Write down the equation whose solutions are represented by the vertices of this heptagon.

⑩ (i) Find the fourth roots of $-9i$ in the form $r(\cos\theta + i\sin\theta)$ where $r > 0$ and $0 < \theta < 2\pi$.

Illustrate the roots on an Argand diagram.

(ii) Let the points representing these roots, taken in order of θ increasing, be P, Q, R and S. The midpoints of the sides of the quadrilateral PQRS represent the fourth roots of a complex number w. Find the modulus and argument of the complex number w and mark the point representing w on your Argand diagram.

⑪ If ω is a complex cube root of unity, $\omega \neq 1$, prove that:

(i) $(1 + \omega)(1 + \omega^2) = 1$

(ii) $1 + \omega$ and $1 + \omega^2$ are complex cube roots of -1

(iii) $(a + b)(a + \omega b)(a + \omega^2 b) = a^3 + b^3$

(iv) $(a + b + c)(a + \omega b + \omega^2 c)(a + \omega^2 b + \omega c) = a^3 + b^3 + c^3 - 3abc$

3 Finding multiple angle identities using de Moivre's theorem

When they were first introduced, complex numbers were not generally accepted by mathematicians. However, during the eighteenth century, their usefulness in producing results involving only real numbers was recognised. The results could also be obtained without using complex numbers, but often only with considerably more effort. One of these results is finding expressions for the sine or cosine of multiple angles using de Moivre's theorem, as shown in the following example.

Example 14.10

Express $\cos 5\theta$ in terms of $\cos\theta$.

Solution

By de Moivre's theorem,

$$\cos 5\theta + i\sin 5\theta = (\cos\theta + i\sin\theta)^5$$

Where c and s are used as abbreviations for $\cos\theta$ and $\sin\theta$ respectively.

$$= c^5 + 5ic^4s - 10c^3s^2 - 10ic^2s^3 + 5cs^4 + is^5$$

Equating the real parts gives

$$\cos 5\theta = c^5 - 10c^3s^2 + 5cs^4$$

$$= c^5 - 10c^3(1 - c^2) + 5c(1 - c^2)^2 \quad \boxed{\text{Using } \sin^2\theta + \cos^2\theta \equiv 1.}$$

$$= c^5 - 10c^3 + 10c^5 + 5c - 10c^3 + 5c^5$$

Therefore $\cos 5\theta = 16\cos^5\theta - 20\cos^3\theta + 5\cos\theta$.

ACTIVITY 14.5

By equating the imaginary parts in Example 14.10, find $\sin 5\theta$ in terms of $\sin\theta$.

Example 14.10 expressed $\cos 5\theta$ in terms of powers of $\cos\theta$. Sometimes it is useful to do the reverse, for example if you wanted to integrate $\cos^5\theta$. The next example shows how this can be expressed in the form $a\cos 5\theta + b\cos 3\theta + c\cos\theta$, which is much easier to integrate.

To do this, you need expressions for $\cos n\theta$ and $\sin n\theta$ in terms of z^n and z^{-n}. You can deduce these expressions from de Moivre's theorem as follows.

If

$$z = \cos\theta + i\sin\theta$$

then

$$z^n = \cos n\theta + i\sin n\theta$$

and

As $\cos(-\theta) = \cos\theta$ and $\sin(-\theta) = -\sin\theta$.

$$z^{-n} = \cos(-n\theta) + i\sin(-n\theta) = \cos n\theta - i\sin n\theta$$

Adding these two expressions gives $z^n + z^{-n} = 2\cos n\theta$

so

$$\cos n\theta = \frac{z^n + z^{-n}}{2}.$$

Subtracting the two expressions gives $z^n - z^{-n} = 2i\sin n\theta$

so

$$\sin n\theta = \frac{z^n - z^{-n}}{2i}.$$

Example 14.11

Express $\cos^5\theta$ in terms of multiple angles.

Solution

Let $z = \cos\theta + i\sin\theta$.

Then $2\cos\theta = z + z^{-1}$.

> A rearrangement of $\cos n\theta = \dfrac{z^n + z^{-n}}{2}$ with $n = 1$.

$\Rightarrow \quad (2\cos\theta)^5 = (z + z^{-1})^5$

$\Rightarrow \quad 32\cos^5\theta = z^5 + 5z^3 + 10z + 10z^{-1} + 5z^{-3} + z^{-5}$

> Expanding the right-hand side using the binomial expansion.

$\quad\quad\quad = (z^5 + z^{-5}) + 5(z^3 + z^{-3}) + 10(z + z^{-1})$

$\quad\quad\quad = 2\cos 5\theta + 10\cos 3\theta + 20\cos\theta$

$\Rightarrow \quad \cos^5\theta = \dfrac{\cos 5\theta + 5\cos 3\theta + 10\cos\theta}{16}$

> Using $\cos n\theta = \dfrac{z^n + z^{-n}}{2}$ three times, with $n = 1$, $n = 3$ and $n = 5$.

ACTIVITY 14.6

Use a similar method to that used in Example 14.11 to express $\sin^5\theta$ in terms of multiple angles.

Exercise 14.3

① (i) Using the result $(\cos\theta + i\sin\theta)^3 = \cos 3\theta + i\sin 3\theta$, compare real and imaginary parts to show that
$$\cos 3\theta = 4\cos^3\theta - 3\cos\theta$$
and
$$\sin 3\theta = 3\sin\theta - 4\sin^3\theta$$

(ii) Hence express $\tan 3\theta$ in terms of $\tan\theta$.

② Let $z = \cos\theta + i\sin\theta$.

(i) Write down expressions for z^3 and z^{-3}.

(ii) Use your expressions from (i) to show that $\cos 3\theta = \dfrac{z^3 + z^{-3}}{2}$ and $\sin 3\theta = \dfrac{z^3 - z^{-3}}{2i}$.

③ Let $z = \cos\theta + i\sin\theta$.

(i) Write down an expression for z^{-1}.

(ii) (a) Use your answer to (i) to show that $2\cos\theta = z + z^{-1}$.

(b) Using the result in part (a), express $\cos^4\theta$ in terms of multiple angles.

(iii) (a) Use your answer to (i) to show that $2i\sin\theta = z - z^{-1}$.

(b) Use the result in part (a), express $\sin^5\theta$ in terms of multiple angles.

④ Find $\cos 6\theta$ and $\dfrac{\sin 6\theta}{\sin \theta}$ in terms of $\cos \theta$.

⑤ Find an expression for $\sin^6 \theta$ in terms of multiple angles.

Hence evaluate $\displaystyle\int \sin^6 \theta\, d\theta$.

⑥ Express $\cos^4 \theta \sin^3 \theta$ in terms of multiple angles and hence find $\displaystyle\int \cos^4 \theta \sin^3 \theta\, d\theta$.

⑦ By first using de Moivre's theorem, evaluate

$$\int_0^{\frac{\pi}{6}} \cos^5 \theta\, d\theta .$$

⑧ (i) Use de Moivre's theorem to show that
$$\cos 5\theta = 16\cos^5 \theta - 20\cos^3 \theta + 5\cos \theta$$

(ii) Given $\cos 5\theta = 0$ but $\cos \theta \neq 0$, use your answer from (i) to find two possible values for $\cos^2 \theta$. Give your answers in surd form.

(iii) Use (ii) to show that
$$\cos 18° = \left(\frac{5 + \sqrt{5}}{8}\right)^{\frac{1}{2}}$$

and find, in a similar form, an expression for $\sin 18°$.

⑨ Use $\cos n\theta = \dfrac{z^n + z^{-n}}{2}$ to express
$$\cos \theta + \cos 3\theta + \cos 5\theta + \ldots + \cos(2n-1)\theta$$
as a geometric series in terms of z.

Show that the sum of the geometric series can be expressed in the form $\dfrac{\sin(2n\theta)}{2\sin \theta}$.

⑩ (i) Given that $z = \cos \theta + i\sin \theta$, write down z^n and $\dfrac{1}{z^n}$ in the form $a + ib$.

Simplify $z^n + \dfrac{1}{z^n}$ and $z^n - \dfrac{1}{z^n}$.

(ii) Expand $\left(z^n + \dfrac{1}{z^n}\right)^2 \left(z^n - \dfrac{1}{z^n}\right)^4$ and hence find the constants p, q, r and s such that
$$\sin^4 \theta \, \cos^2 \theta = p + q\cos 2\theta + r\cos 4\theta + s\cos 6\theta$$

(iii) Using a suitable substitution and your answer to part (ii), show that
$$\int_1^2 x^4 \sqrt{4 - x^2}\, dx = \frac{4\pi}{3} + \sqrt{3}.$$

4 The form $z = re^{i\theta}$

In Chapter 8 you saw that the series expansions of $\sin \theta$, $\cos \theta$ and e^x can be written as

$$\sin \theta = \theta - \frac{\theta^3}{3!} + \frac{\theta^5}{5!} - \frac{\theta^7}{7!} + \ldots \frac{(-1)^r \theta^{2r+1}}{(2r+1)!}$$

$$\cos \theta = 1 - \frac{\theta^2}{2!} + \frac{\theta^4}{4!} - \frac{\theta^6}{6!} + \ldots \frac{(-1)^r \theta^{2r}}{(2r)!}$$

$$e^x = 1 + x + \frac{x^2}{2!} + \frac{x^3}{3!} + \frac{x^4}{4!} + \ldots + \frac{x^r}{r!} + \ldots$$

It can be shown that this series expansion is also true for complex powers.

Replacing x by $i\theta$ in the expansion e^x gives

$$e^{i\theta} = 1 + i\theta + \frac{(i\theta)^2}{2!} + \frac{(i\theta)^3}{3!} + \frac{(i\theta)^4}{4!} + \frac{(i\theta)^5}{5!} + \frac{(i\theta)^6}{6!} + \ldots$$

$$= 1 + i\theta + \frac{i^2\theta^2}{2!} + \frac{i^3\theta^3}{3!} + \frac{i^4\theta^4}{4!} + \frac{i^5\theta^5}{5!} + \frac{i^6\theta^6}{6!} + \ldots$$

$$= 1 + i\theta - \frac{\theta^2}{2!} - \frac{i\theta^3}{3!} + \frac{\theta^4}{4!} + \frac{i\theta^5}{5!} - \frac{\theta^6}{6!} + \ldots$$

Collecting together real and imaginary terms.

$$= \left(1 - \frac{\theta^2}{2!} + \frac{\theta^4}{4!} - \frac{\theta^6}{6!} + \ldots\right) + i\left(\theta - \frac{\theta^3}{3!} + \frac{\theta^5}{5!} - \ldots\right)$$

Using the series expansions for $\cos\theta$ and $\sin\theta$.

Therefore
$$e^{i\theta} = \cos\theta + i\sin\theta$$

so

$$z = r(\cos\theta + i\sin\theta)$$

can be rewritten as

$$z = re^{i\theta}$$

This is called the **exponential form** of a complex number with modulus r and argument θ.

Discussion point

→ How would the result
$z = r(\cos\theta + i\sin\theta)$
$= re^{i\theta}$
be adapted for a complex number of the form
$r(\cos\theta - i\sin\theta)$
where $-\pi < \theta \leqslant \pi$?

This format is simply a more compact way of writing familiar expressions, as the modulus-argument form $z = r(\cos\theta + i\sin\theta)$ can now be abbreviated to $z = re^{i\theta}$.

This form allows you to derive de Moivre's theorem very easily for all rational n by using the laws of indices:

$$\cos\theta + i\sin\theta^n = (e^{i\theta})^n$$

$$= e^{in\theta}$$

$$= \cos n\theta + i\sin n\theta$$

Example 14.12

(i) Write $z = 6\left(\cos\frac{\pi}{6} + i\sin\frac{\pi}{6}\right)$ in the form $re^{i\theta}$.

(ii) Write $z = -1 + \sqrt{3}i$ in the form $re^{i\theta}$.

Solution

(i) $z = 6\left(\cos\frac{\pi}{6} + i\sin\frac{\pi}{6}\right)$ has modulus 6 and argument $\frac{\pi}{6}$.

Therefore $z = 6e^{\frac{i\pi}{6}}$.

(ii) $z = -1 + \sqrt{3}i$ has modulus 2 and argument $\frac{2\pi}{3}$.

Therefore $z = 2e^{\frac{2i\pi}{3}}$.

In the discussion point above, you should have noticed that since

$$\cos(-\theta) = \cos\theta \quad \text{and} \quad \sin(-\theta) = -\sin\theta$$

then

$$r(\cos\theta - i\sin\theta) = r(\cos(-\theta) + i\sin(-\theta)).$$

Therefore

$$r(\cos\theta - \mathrm{i}\sin\theta) = r\mathrm{e}^{-\mathrm{i}\theta}.$$

ACTIVITY 14.7

For a complex number $z = x + \mathrm{i}y$, show that:

(i) $\mathrm{e}^z = \mathrm{e}^x(\cos y + \mathrm{i}\sin y)$

(ii) $\mathrm{e}^{z+2\pi n \mathrm{i}} = \mathrm{e}^z$

(iii) $\mathrm{e}^{\mathrm{i}\pi} = -1$

The results in Activity 14.7 are useful when simplifying results involving exponential functions with complex exponents. Part (iii) is often written in the form $\mathrm{e}^{\mathrm{i}\pi} + 1 = 0$ which is a remarkable result that links the five numbers $0, 1, \mathrm{i}, \mathrm{e}$ and π.

The results from Activity 14.7 also give rise to two very interesting mathematical results that are useful when working with complex numbers:

$$\cos\theta = \frac{1}{2}\left(\mathrm{e}^{\mathrm{i}\theta} + \mathrm{e}^{-\mathrm{i}\theta}\right)$$

$$\sin\theta = \frac{1}{2\mathrm{i}}\left(\mathrm{e}^{\mathrm{i}\theta} - \mathrm{e}^{-\mathrm{i}\theta}\right).$$

> Notice that these expressions for $\cos\theta$ and $\sin\theta$ are very similar to the definitions of the hyperbolic functions $\cosh\theta$ and $\sinh\theta$.

To prove these results, you can use

$$\mathrm{e}^{\mathrm{i}\theta} = \cos\theta + \mathrm{i}\sin\theta \qquad ①$$

and

$$\mathrm{e}^{-\mathrm{i}\theta} = \cos\theta - \mathrm{i}\sin\theta. \qquad ②$$

Finding ① + ② gives

$$\mathrm{e}^{\mathrm{i}\theta} + \mathrm{e}^{-\mathrm{i}\theta} = 2\cos\theta$$

so

$$\cos\theta = \frac{1}{2}\left(\mathrm{e}^{\mathrm{i}\theta} + \mathrm{e}^{-\mathrm{i}\theta}\right).$$

Similarly, finding ① − ② gives

$$\mathrm{e}^{\mathrm{i}\theta} - \mathrm{e}^{-\mathrm{i}\theta} = 2\mathrm{i}\sin\theta$$

so

$$\sin\theta = \frac{1}{2\mathrm{i}}\left(\mathrm{e}^{\mathrm{i}\theta} - \mathrm{e}^{-\mathrm{i}\theta}\right).$$

You need to learn the proofs of these results.

Discussion point

➜ These results are essentially the same as the ones given on pages 319 and 320. Explain why this is the case.

Prior knowledge

You need to be familiar with geometric sequences and series, including finding the sum to n terms and the sum to infinity of a geometric series.

Summing series using de Moivre's theorem

This section shows how complex numbers can be used to evaluate certain sums of real quantities. It may be possible to do these summations without using complex numbers, for example by induction if you already know the answer, but this is a lot more difficult.

Sometimes it is worth setting out to do more than is actually required, as shown in the next example.

Example 14.13

(i) Prove that

$$1 + e^{i\theta} = 2\cos\frac{\theta}{2} e^{\frac{i\theta}{2}}$$

and

$$1 - e^{i\theta} = -2i\sin\frac{\theta}{2} e^{\frac{i\theta}{2}}.$$

(ii) Show that the sum of the series

$$1 + {}^nC_1\cos\theta + {}^nC_2\cos 2\theta + {}^nC_3\cos 3\theta + \ldots + \cos n\theta$$

is

$$2_n\cos_n\frac{\theta}{2}\cos\frac{n\theta}{2}.$$

Solution

(i) The factor $e^{\frac{i\theta}{2}}$ on the right-hand side suggests writing each term on the left-hand side as a multiple of $e^{\frac{i\theta}{2}}$.

$$1 = e^{\frac{i\theta}{2}} \times e^{\frac{-i\theta}{2}}$$

$$e^{i\theta} = e^{\frac{i\theta}{2}} \times e^{\frac{i\theta}{2}}$$

So $1 + e^{i\theta} = \left(e^{\frac{i\theta}{2}} \times e^{\frac{-i\theta}{2}}\right) + \left(e^{\frac{i\theta}{2}} \times e^{\frac{i\theta}{2}}\right)$

$$= e^{\frac{i\theta}{2}}\left(e^{\frac{-i\theta}{2}} + e^{\frac{i\theta}{2}}\right) \qquad \longleftarrow \boxed{\text{Taking out a factor of } e^{\frac{i\theta}{2}}.}$$

$$= e^{\frac{i\theta}{2}} \times 2\cos\frac{\theta}{2} \qquad \longleftarrow \boxed{\begin{array}{l}\text{Using the earlier result}\\ \cos\theta = \frac{1}{2}\left(e^{i\theta} + e^{-i\theta}\right).\end{array}}$$

Similarly,

$$1 - e^{i\theta} = \left(e^{\frac{i\theta}{2}} \times e^{\frac{-i\theta}{2}}\right) - \left(e^{\frac{i\theta}{2}} \times e^{\frac{i\theta}{2}}\right)$$

$$= e^{\frac{i\theta}{2}}\left(e^{\frac{-i\theta}{2}} - e^{\frac{i\theta}{2}}\right)$$

$$= e^{\frac{i\theta}{2}} \times -2i\sin\frac{\theta}{2} \qquad \longleftarrow \boxed{\begin{array}{l}\text{Using the earlier result}\\ \sin\theta = \frac{1}{2i}\left(e^{i\theta} - e^{-i\theta}\right).\end{array}}$$

(ii) At first sight this series seems to suggest the binomial expansion $(1 + \cos\theta)^n$. The binomial coefficients $1, {}^nC_1, {}^nC_2, \ldots, 1$ are correct, but there are multiple angles, $\cos r\theta$, instead of powers of cosines, $\cos^r\theta$. This suggests that de Moivre's theorem can be used.

The method involves introducing a corresponding sine series too.

Let $\quad C = 1 + {}^nC_1\cos\theta + {}^nC_2\cos 2\theta + {}^nC_3\cos 3\theta + \ldots + \cos n\theta$

and $\quad S = {}^nC_1\sin\theta + {}^nC_2\sin 2\theta + {}^nC_3\sin 3\theta + \ldots + \sin n\theta$

Then

$$C + iS = 1 + {}^nC_1(\cos\theta + i\sin\theta) + {}^nC_2(\cos 2\theta + i\sin 2\theta) + \ldots$$

$$+ (\cos n\theta + i\sin n\theta)$$

$$= 1 + {}^nC_1 e^{i\theta} + {}^nC_2 e^{i2\theta} + \ldots + e^{in\theta}$$

$$= 1 + {}^nC_1 e^{i\theta} + {}^nC_2 \left(e^{i\theta}\right)^2 + \ldots + \left(e^{i\theta}\right)^n$$

> Using de Moivre's theorem and the fact that $e^{ir\theta} = \left(e^{i\theta}\right)^r$.

This is now recognisable as a binomial expansion, so that

$$C + iS = \left(1 + e^{i\theta}\right)^n$$

To find C you need to find the real part of $\left(1 + e^{i\theta}\right)^n$ and here the results from part(i) are useful. Using the result:

$$1 + e^{i\theta} = 2\cos\frac{\theta}{2} \, e^{\frac{i\theta}{2}}$$

gives

$$C + iS = \left(1 + e^{i\theta}\right)^n = \left(2\cos\frac{\theta}{2} \, e^{\frac{i\theta}{2}}\right)^n$$

$$= 2^n \cos^n\frac{\theta}{2} \, e^{\frac{in\theta}{2}}$$

$$= 2^n \cos^n\frac{\theta}{2} \left(\cos\frac{n\theta}{2} + i\sin\frac{n\theta}{2}\right)$$

Taking the real part:

$$C = 2^n \cos^n\frac{\theta}{2}\cos\frac{n\theta}{2}$$

ACTIVITY 14.8

For Example 14.13, state the corresponding result obtained by equating the imaginary parts.

Exercise 14.4

① Write the following complex numbers in the form $z = re^{i\theta}$, where $-\pi < \theta \leqslant \pi$.

(i) $4\left(\cos\frac{\pi}{3} + i\sin\frac{\pi}{3}\right)$

(ii) $\sqrt{3}\left(\cos\left(-\frac{5\pi}{6}\right) + i\sin\left(-\frac{5\pi}{6}\right)\right)$

(iii) $-5i$

(iv) $-3 - 3i$

(v) $\sqrt{3} - i$

② Write the following complex numbers in the form $x + yi$.

(i) $5e^{i\pi}$ (ii) $\sqrt{2}e^{\frac{3\pi}{4}i}$ (iii) $\sqrt{2}e^{-\frac{3\pi}{4}i}$ (iv) $5e^{\frac{23\pi}{4}i}$

③ Two complex numbers are given by $z = 2e^{\frac{3\pi}{4}i}$ and $w = 3e^{\frac{\pi}{3}i}$.

Find zw and $\frac{z}{w}$, giving your answers in the form $z = re^{i\theta}$, where $r > 0$ and $-\pi < \theta \leqslant \pi$.

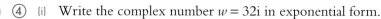

④ (i) Write the complex number $w = 32i$ in exponential form.

(ii) Find the five fifth roots of w, giving your answers in exponential form.

⑤ Let
$$C = 1 + \cos\theta + \cos 2\theta + \ldots + \cos(n-1)\theta$$
and
$$S = \sin\theta + \sin 2\theta + \ldots + \sin(n-1)\theta$$

(i) Find $C + iS$ and show that this forms a geometric series with common ratio $e^{i\theta}$

(ii) Show that the sum of the series in part (i) is $\dfrac{1 - e^{in\theta}}{1 - e^{i\theta}}$.

(iii) By multiplying the numerator and denominator of this sum by $1 - e^{-i\theta}$, show that
$$C = \frac{1 - \cos\theta + \cos(n-1)\theta - \cos n\theta}{2 - 2\cos\theta}$$
and find S.

⑥ (i) Show that $1 + e^{i2\theta} = 2\cos\theta(\cos\theta + i\sin\theta)$.

(ii) The series C and S are defined as follows.
$$C = 1 + \binom{n}{1}\cos 2\theta + \binom{n}{2}\cos 4\theta + \ldots \cos 2n\theta$$

$$S = \binom{n}{1}\sin 2\theta + \binom{n}{2}\sin 4\theta + \ldots \sin 2n\theta$$

By considering $C + iS$, show that
$$C = 2^n \cos^n\theta \cos n\theta$$
and find a corresponding expression for S.

⑦ (i) Use de Moivre's theorem to find the constants a, b, c in the identity
$$\cos 5\theta \equiv a\cos^5\theta + b\cos^3\theta + c\cos\theta.$$

(ii) Let
$$C = \cos\theta + \cos\left(\theta + \frac{2\pi}{n}\right) + \cos\left(\theta + \frac{4\pi}{n}\right) + \ldots\cos\left(\theta + \frac{(2n-2)\pi}{n}\right)$$
and
$$S = \sin\theta + \sin\left(\theta + \frac{2\pi}{n}\right) + \sin\left(\theta + \frac{4\pi}{n}\right) + \ldots\sin\left(\theta + \frac{(2n-2)\pi}{n}\right)$$
where n is an integer greater than 1.

Show that $C + iS$ forms a geometric series and hence show that $C = 0$, $S = 0$.

⑧ Use the result $e^{i\theta} = \cos\theta + i\sin\theta$ to prove that
$$e^{z^*} = \left(e^z\right)^*$$
for all complex numbers $z = a + ib$.

⑨ Let $C = \displaystyle\int e^{3x}\cos 2x \,\mathrm{d}x$ and $S = \displaystyle\int e^{3x}\sin 2x \,\mathrm{d}x$.

(i) Find C and S by using integration by parts twice.

(ii) (a) Show that $C + iS = \dfrac{e^{(3+2i)x}}{3 + 2i} + A$ where A is a constant.

(b) Hence verify your answers for C and S from part (i).

⑩ The infinite series C and S are defined as follows:

$$C = \frac{\cos\theta}{2} - \frac{\cos 2\theta}{4} + \frac{\cos 3\theta}{8} - \frac{\cos 4\theta}{16} + \ldots$$

$$S = \frac{\sin\theta}{2} - \frac{\sin 2\theta}{4} + \frac{\sin 3\theta}{8} - \frac{\sin 4\theta}{16} + \ldots .$$

Show that $C + \mathrm{i}S = \dfrac{2\mathrm{e}^{\mathrm{i}\theta} + 1}{5 + 4\cos\theta}$ and hence find expressions for C and S in terms of $\cos\theta$ and $\sin\theta$.

LEARNING OUTCOMES

Now you have finished this chapter, you should be able to:

➤ find the modulus and argument of a complex number

➤ multiply and divide complex numbers in modulus–argument form

➤ understand the effect of multiplication by a complex number in an Argand diagram

➤ sketch loci of the form $|z - a| = r$, $\arg(z - a) = \theta$ and $|z - a| = |z - b|$ in an Argand diagram

➤ use de Moivre's theorem to simplify expressions involving powers of complex numbers

➤ find the nth roots of a complex number

➤ use de Moivre's theorem to find multiple angle identities

➤ use the exponential form of a complex number $z = r\mathrm{e}^{\mathrm{i}\theta}$

➤ sum series using de Moivre's theorem.

KEY POINTS

1 The modulus r of $z = x + y\mathrm{i}$ is $|z| = \sqrt{x^2 + y^2}$. This is the distance of the point z from the origin on the Argand diagram.

2 The argument of z is the angle θ, measured in radians, between the line connecting the origin and the point z and the positive real axis.

3 The principal argument of z, $\arg z$, is the angle θ, measured in radians, for which $-\pi < \theta \leqslant \pi$, between the line connecting the origin and the point z and the positive real axis.

4 For a complex number z, $zz^* = |z|^2$.

5 The modulus–argument form of z is $z = r(\cos\theta + \mathrm{i}\sin\theta)$, where $r = |z|$ and $\theta = \arg z$. This is often written as (r, θ).

6 For two complex numbers z_1 and z_2:

$$|z_1 z_2| = |z_1||z_2| \qquad \arg(z_1 z_2) = \arg z_1 + \arg z_2$$

$$\left|\frac{z_1}{z_2}\right| = \frac{|z_1|}{|z_2|} \qquad \arg\left(\frac{z_1}{z_2}\right) = \arg z_1 - \arg z_2$$

7 Geometrically, to obtain the vector $z_1 z_2$ enlarge the vector z_2 by the scale factor $|z_1|$ and rotate it through $\arg z_1$ anticlockwise about 0. In polar form, multiplication by $r\mathrm{e}^{\mathrm{i}\theta}$ corresponds to enlargement with scale factor r with (anticlockwise) rotation through θ about the origin.

8 The distance between the points z_1 and z_2 in an Argand diagram is $|z_2 - z_1|$.

9 $|z - a| = r$ represents a circle, centre a and radius r. $|z - a| < r$ represents the interior of the circle, and $|z - a| > r$ represents the exterior of the circle.

10 $\arg(z - a) = \theta$ represents a half line starting at $z = a$ at an angle of θ from the positive real direction joining between the complex numbers $z = a$ and $z = b$

11 $|z - a| = |z - b|$ represents the perpendicular bisector of the points a and b.

12 De Moivre's theorem: If $z = r(\cos\theta + i\sin\theta)$ then
$$z^n = \left[r(\cos\theta + i\sin\theta)\right]^n = r^n(\cos n\theta + i\sin n\theta)$$

13 The nth roots of unity can be written as
$$1, \omega, \omega^2, \ldots, \omega^{n-1},$$
where
$$\omega = \cos\frac{2\pi}{n} + i\sin\frac{2\pi}{n}$$

14 The sum of all of the nth roots of unity is zero.

15 The non-zero complex number $r(\cos\theta + i\sin\theta)$ has precisely n different nth roots, which are
$$r^{\frac{1}{n}}\left(\cos\left(\frac{\theta + 2k\pi}{n}\right) + i\sin\left(\frac{\theta + 2k\pi}{n}\right)\right),$$
where $k = 0, 1, 2, \ldots, n - 1$.

These roots can also be written as $\alpha, \alpha\omega, \alpha\omega^2, \ldots, \alpha\omega^{n-1}$, where
$$\alpha = r^{\frac{1}{n}}\left(\cos\frac{\theta}{n} + i\sin\frac{\theta}{n}\right) \text{ and } \omega = \cos\frac{2\pi}{n} + i\sin\frac{2\pi}{n}.$$

16 The sum of the nth roots is zero.

17 If $z = \cos\theta + i\sin\theta$ then
$$\cos n\theta = \frac{z^n + z^{-n}}{2}$$
and
$$\sin n\theta = \frac{z^n - z^{-n}}{2i}$$

18 The exponential form of a complex number is
$$z = r(\cos\theta + i\sin\theta) = re^{i\theta}$$
For a complex number $z = x + iy$ this can be written as
$$e^z = e^x(\cos y + i\sin y)$$

19 For a complex number in exponential form
$$e^{z + 2\pi n} = e^z$$

20 For the complex number $z = r(\cos\theta + i\sin\theta)$
$$\cos\theta = \frac{1}{2}\left(e^{i\theta} + e^{-i\theta}\right)$$
$$\sin\theta = \frac{1}{2i}\left(e^{i\theta} - e^{-i\theta}\right)$$

FUTURE USES

- Complex numbers will be needed for work on differential equations in A-level Further Mathematics, in particular in modelling oscillations (simple harmonic motion).

15 Vectors 2

> *How can it be that mathematics, being after all a product of human thought which is independent of experience, is so admirably appropriate to the objects of reality?*
>
> Albert Einstein

Discussion point

→ The picture shows the ceiling of King's Cross Station in London. What mathematics might be involved in designing and building a structure like this?

Review: Distance between point and line and between two lines

Finding the distance between a point and a line

Suppose you have a line, *l*, given by $\mathbf{r} = \mathbf{a} + \lambda\mathbf{d}$ and a point P not on the line whose position vector is $\mathbf{p}$. You need to be able to find the shortest distance from P to the line.

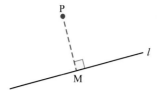

Figure 15.1

> Be careful here. It is the direction vector of the line NOT the position vector of a point on the line, which is perpendicular to $\overrightarrow{PM}$.

M is on l so its position vector $\mathbf{m}$ is given by $\mathbf{m} = \mathbf{a} + \lambda\mathbf{d}$ for some value of λ. But what distinguishes M from every other point on the line is that PM is perpendicular to the line so the vector $\overrightarrow{PM} = \mathbf{m} - \mathbf{p}$ is perpendicular to the vector $\mathbf{d}$.

If two vectors are perpendicular then their scalar product must be 0.

$$(\mathbf{m} - \mathbf{p}) \cdot \mathbf{d} = 0$$

$$(\mathbf{a} + \lambda\mathbf{d} - \mathbf{p}) \cdot \mathbf{d} = 0$$

$$(\mathbf{a} - \mathbf{p}) \cdot \mathbf{d} + \lambda\mathbf{d} \cdot \mathbf{d} = 0$$

$$\lambda = \frac{(\mathbf{p} - \mathbf{a}) \cdot \mathbf{d}}{|\mathbf{d}|^2}$$

> It is best not to try to remember this formula but to remember the technique.

So, the required distance is given by

$$\left|\overrightarrow{PM}\right| = \left|\mathbf{m} - \mathbf{p}\right| = \left|\mathbf{a} + \lambda\mathbf{d} - \mathbf{p}\right| = \left|\mathbf{a} - \mathbf{p} + \frac{(\mathbf{p} - \mathbf{a}) \cdot \mathbf{d}}{|\mathbf{d}|^2}\mathbf{d}\right|$$

Example 15.1

Find the shortest distance between the point P(6, 3, 2) and the line

$$l : \mathbf{r} = \begin{pmatrix} -2 \\ 1 \\ 3 \end{pmatrix} + \lambda\begin{pmatrix} -1 \\ 0 \\ 2 \end{pmatrix}.$$

Solution

$$\mathbf{p} = \overrightarrow{OP} = \begin{pmatrix} 6 \\ 3 \\ 2 \end{pmatrix}$$

If M is the foot of the perpendicular from P to l then

> Remember, M lies on the line.

$$\mathbf{m} = \begin{pmatrix} -2 \\ 1 \\ 3 \end{pmatrix} + \lambda\begin{pmatrix} -1 \\ 0 \\ 2 \end{pmatrix} \quad \text{for some value of } \lambda.$$

$$\overrightarrow{PM} = \mathbf{m} - \mathbf{p} = \begin{pmatrix} -2 \\ 1 \\ 3 \end{pmatrix} + \lambda\begin{pmatrix} -1 \\ 0 \\ 2 \end{pmatrix} - \begin{pmatrix} 6 \\ 3 \\ 2 \end{pmatrix} = \begin{pmatrix} -8 \\ -2 \\ 1 \end{pmatrix} + \lambda\begin{pmatrix} -1 \\ 0 \\ 2 \end{pmatrix}$$

But $\overrightarrow{PM}$ must be perpendicular to the line.

> Remember to use **d** the direction vector of the line here.

$$\overrightarrow{PM} \cdot \begin{pmatrix} -1 \\ 0 \\ 2 \end{pmatrix} = 0$$

$$\left(\begin{pmatrix} -8 \\ -2 \\ 1 \end{pmatrix} + \lambda \begin{pmatrix} -1 \\ 0 \\ 2 \end{pmatrix} \right) \cdot \begin{pmatrix} -1 \\ 0 \\ 2 \end{pmatrix} = 0$$

$$\begin{pmatrix} -8 \\ -2 \\ 1 \end{pmatrix} \cdot \begin{pmatrix} -1 \\ 0 \\ 2 \end{pmatrix} + \lambda \begin{pmatrix} -1 \\ 0 \\ 2 \end{pmatrix} \cdot \begin{pmatrix} -1 \\ 0 \\ 2 \end{pmatrix} = 0$$

$$10 + 5\lambda = 0$$

> Note that you could now find the coordinates of point M if required.

$$\lambda = -2$$

> Note that this is perpendicular to $\begin{pmatrix} -1 \\ 0 \\ 2 \end{pmatrix}$ as required.

$$\overrightarrow{PM} = \begin{pmatrix} -8 \\ -2 \\ 1 \end{pmatrix} + -2 \begin{pmatrix} -1 \\ 0 \\ 2 \end{pmatrix} = \begin{pmatrix} -6 \\ -2 \\ -3 \end{pmatrix}$$

So $PM = \left| \overrightarrow{PM} \right| = \sqrt{(-6)^2 + (-2)^2 + (-3)^2} = \sqrt{49} = 7$

So, the shortest distance from the point to the line is 7.

Finding the distance between two parallel lines

The distance between two parallel lines can be found by choosing any point on one line, and then finding the distance between this point and the second line, using the technique shown in the previous section.

Finding the distance between skew lines

Two lines are skew if they do not intersect and are not parallel.

Figure 15.2 shows the lines $l_1 : \mathbf{r} = \mathbf{a}_1 + \lambda \mathbf{d}_1$ and $l_2 : \mathbf{r} = \mathbf{a}_2 + \mu \mathbf{d}_2$ and the two parallel planes π_1 and π_2 that contain them.

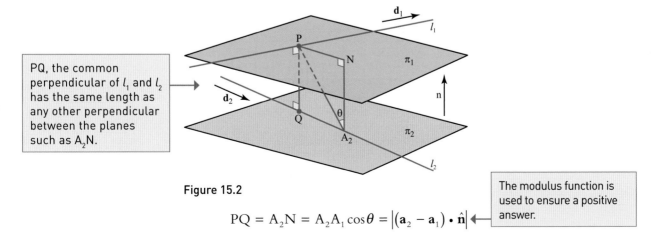

> PQ, the common perpendicular of l_1 and l_2 has the same length as any other perpendicular between the planes such as A_2N.

Figure 15.2

> The modulus function is used to ensure a positive answer.

$$PQ = A_2N = A_2A_1 \cos\theta = \left| (\mathbf{a}_2 - \mathbf{a}_1) \cdot \hat{\mathbf{n}} \right|$$

where $\hat{\mathbf{n}}$ is a unit vector parallel to A_2N, i.e. perpendicular to both planes.

You can find the vector $\hat{\mathbf{n}}$ by first solving the equations $\mathbf{n} \cdot \mathbf{d}_1 = 0$ and $\mathbf{n} \cdot \mathbf{d}_2 = 0$ to find a vector $\mathbf{n}$ that is perpendicular to both planes, and then find the unit vector in the direction of $\mathbf{n}$ by dividing by $|\mathbf{n}|$.

Example 15.2

Find the shortest distance between the skew lines $l_1: \mathbf{r} = \begin{pmatrix} 2 \\ 1 \\ -2 \end{pmatrix} + \lambda \begin{pmatrix} 2 \\ 0 \\ 5 \end{pmatrix}$

and $l_2: \mathbf{r} = \begin{pmatrix} 3 \\ 5 \\ 0 \end{pmatrix} + \mu \begin{pmatrix} -5 \\ 3 \\ 4 \end{pmatrix}$. Give your answer in simplified surd form.

Solution

Line l_1 contains the point $A_1 = (2, 1, -2)$ and is parallel to the vector $\begin{pmatrix} 2 \\ 0 \\ 5 \end{pmatrix}$.

Line l_2 contains the point $A_2 = (3, 5, 0)$ and is parallel to the vector $\begin{pmatrix} -5 \\ 3 \\ 4 \end{pmatrix}$.

Let $\mathbf{n} = \begin{pmatrix} n_1 \\ n_2 \\ n_3 \end{pmatrix}$ be perpendicular to both planes.

So $\begin{pmatrix} n_1 \\ n_2 \\ n_3 \end{pmatrix} \cdot \begin{pmatrix} 2 \\ 0 \\ 5 \end{pmatrix} = 0 \Rightarrow 2n_1 + 5n_3 = 0 \qquad \Rightarrow 2n_1 = -5n_3$

and $\begin{pmatrix} n_1 \\ n_2 \\ n_3 \end{pmatrix} \cdot \begin{pmatrix} -5 \\ 3 \\ 4 \end{pmatrix} = 0 \Rightarrow -5n_1 + 3n_2 + 4n_3 = 0$

Taking $n_1 = 5$ gives $n_3 = -2$ and $n_2 = 11$.

So $\mathbf{n} = \begin{pmatrix} 5 \\ 11 \\ -2 \end{pmatrix}$ and $\hat{\mathbf{n}} = \dfrac{1}{\sqrt{150}} \begin{pmatrix} 5 \\ 11 \\ -2 \end{pmatrix}$

$$a_1 - a_2 = \begin{pmatrix} 2 \\ 1 \\ -2 \end{pmatrix} - \begin{pmatrix} 3 \\ 5 \\ 0 \end{pmatrix} = \begin{pmatrix} -1 \\ -4 \\ -2 \end{pmatrix}$$

So the shortest distance is

$$= \dfrac{1}{\sqrt{150}} \begin{pmatrix} -1 \\ -4 \\ -2 \end{pmatrix} \cdot \begin{pmatrix} 5 \\ 11 \\ -2 \end{pmatrix} = \dfrac{-45}{\sqrt{150}} = \dfrac{-3\sqrt{6}}{2} \approx 3.67 \text{ units}$$

① Two straight lines in three dimensions are given by the equations

$$l_1 = \begin{pmatrix} 2 \\ -3 \\ 0 \end{pmatrix} + \lambda \begin{pmatrix} 1 \\ -3 \\ 2 \end{pmatrix} \text{ and } l_2 = \begin{pmatrix} 4 \\ 2 \\ 1 \end{pmatrix} + \mu \begin{pmatrix} -2 \\ 6 \\ -4 \end{pmatrix}$$

(i)   Show that the two lines are parallel.

(ii)   Write down a point P on l_2 by setting $\mu = 0$.

(iii)   Let Q be a point on l_1. What is the vector $\overrightarrow{PQ}$ in terms of λ?

(iv)   Given that $\overrightarrow{PQ}$ is perpendicular to $\begin{pmatrix} 1 \\ -3 \\ 2 \end{pmatrix}$ deduce the value of λ.

(v)   Hence find the shortest distance between the two lines.

② Two skew lines in three dimensions are given by the equations

$$l_1 = \begin{pmatrix} 2 \\ -1 \\ 0 \end{pmatrix} + \lambda \begin{pmatrix} 1 \\ 3 \\ 2 \end{pmatrix} \text{ and } l_2 = \begin{pmatrix} 4 \\ -5 \\ 1 \end{pmatrix} + \mu \begin{pmatrix} 2 \\ 4 \\ 5 \end{pmatrix}$$

(i)   Write down the direction vectors of each line.

(ii)   Use the scalar product to find a vector $\mathbf{n} = \begin{pmatrix} n_1 \\ n_2 \\ n_3 \end{pmatrix}$ so that $\mathbf{n}$ is

perpendicular to both the parallel planes containing the lines.

(iii)   Hence find the shortest distance between the two lines.

③ Find the shortest distance between the given point P and the given straight line.

(i)   P(5, 2) and $l: \mathbf{r} = \begin{pmatrix} 1 \\ 2 \end{pmatrix} + \lambda \begin{pmatrix} 1 \\ 1 \end{pmatrix}$.

(ii)   P(3, 1, 0) and $l: \mathbf{r} = \begin{pmatrix} 3 \\ 1 \\ 1 \end{pmatrix} + \lambda \begin{pmatrix} 1 \\ 3 \\ -1 \end{pmatrix}$.

(iii)   P(−1, −2) and $l: y = 3x + 2$.

(iv)   P(3, −1, 5) and $l: \dfrac{x-1}{2} = \dfrac{z-2}{3}, y = 1$.

④ Find the shortest distance between the two given lines, in each case stating whether the lines intersect, are parallel or are skew.

(i) $l_1: \mathbf{r} = \begin{pmatrix} 3 \\ 2 \\ 3 \end{pmatrix} + \lambda \begin{pmatrix} 0 \\ 1 \\ 3 \end{pmatrix} \text{ and } l_2: \mathbf{r} = \begin{pmatrix} 7 \\ 5 \\ 8 \end{pmatrix} + \mu \begin{pmatrix} -2 \\ 5 \\ 6 \end{pmatrix}$

(ii) $l_1: \mathbf{r} = \begin{pmatrix} 4 \\ 1 \\ 6 \end{pmatrix} + \lambda \begin{pmatrix} 2 \\ 3 \\ -1 \end{pmatrix}$ and $l_2: \mathbf{r} = \begin{pmatrix} 9 \\ 6 \\ 11 \end{pmatrix} + \mu \begin{pmatrix} 1 \\ 1 \\ 1 \end{pmatrix}$

(iii) $l_1: \mathbf{r} = \begin{pmatrix} 5 \\ -3 \\ -3 \end{pmatrix} + \lambda \begin{pmatrix} -4 \\ 2 \\ -6 \end{pmatrix}$ and $l_2: \mathbf{r} = \begin{pmatrix} 3 \\ 0 \\ -4 \end{pmatrix} + \mu \begin{pmatrix} 2 \\ -1 \\ 3 \end{pmatrix}$

1 Lines and planes

The intersection of a line and a plane

In three dimensions, there are three possibilities for the arrangement of a line and a plane as shown in Figure 15.3.

No intersection   Point   Line contained in plane

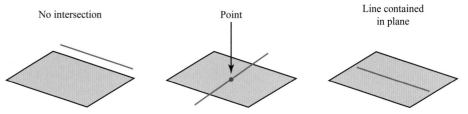

Figure 15.3

The next example shows how you can find the point at which a line intersects a plane.

Example 15.3

Find the point of intersection of the line $\mathbf{r} = \begin{pmatrix} 2 \\ 3 \\ 4 \end{pmatrix} + \lambda \begin{pmatrix} 1 \\ 2 \\ -1 \end{pmatrix}$ and the

plane $5x + y - z = 1$.

Solution

The line is $\mathbf{r} = \begin{pmatrix} x \\ y \\ z \end{pmatrix} = \begin{pmatrix} 2 \\ 3 \\ 4 \end{pmatrix} + \lambda \begin{pmatrix} 1 \\ 2 \\ -1 \end{pmatrix}$ and so for any point on the line

$$x = 2 + \lambda \quad y = 3 + 2\lambda \quad z = 4 - \lambda$$

Substituting these into the equation of the plane $5x + y - z = 1$ gives

$$5(2 + \lambda) + (3 + 2\lambda) - (4 - \lambda) = 1$$
$$8\lambda = -8$$
$$\lambda = -1$$

Substituting $\lambda = -1$ into the equation of the line gives

$$\mathbf{r} = \begin{pmatrix} x \\ y \\ z \end{pmatrix} = \begin{pmatrix} 2 \\ 3 \\ 4 \end{pmatrix} - \begin{pmatrix} 1 \\ 2 \\ -1 \end{pmatrix} = \begin{pmatrix} 1 \\ 1 \\ 5 \end{pmatrix}.$$

So the point of intersection is $(1, 1, 5)$.

ACTIVITY 15.1

Example 15.3 shows the situation where a line intersects a plane at a point. Try to find the point of intersection of

(i)　　the line $\mathbf{r} = \begin{pmatrix} 2 \\ 3 \\ 4 \end{pmatrix} + \lambda \begin{pmatrix} 1 \\ 2 \\ 7 \end{pmatrix}$ and the plane $5x + y - z = 1$

(ii)　　the line $\mathbf{r} = \begin{pmatrix} 1 \\ 1 \\ 5 \end{pmatrix} + \lambda \begin{pmatrix} 1 \\ 2 \\ 7 \end{pmatrix}$ and the plane $5x + y - z = 1$.

Explain what has happened. By considering the direction vector of the line and the normal vector of the plane how can you identify if there is a unique solution or not?

The angle between a line and a plane

If they are not perpendicular, the acute angle between a line and a plane is the acute angle θ between the line and its **orthogonal projection** onto the plane, shown by the dotted line AB in Figure 15.4.

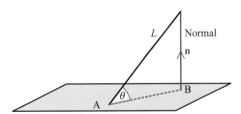

Figure 15.4

You can find the angle θ by first finding the angle α between the direction vector $\mathbf{d}$ of the straight line L and a normal vector $\mathbf{n}$ to the plane, as shown in Figure 15.5.

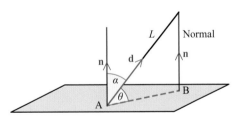

Figure 15.5

The angle θ can then be found by calculating $90 - \alpha$.

This method is illustrated in the following example.

Example 15.4

Find the angle between the line $\mathbf{r} = \begin{pmatrix} 2 \\ 0 \\ -3 \end{pmatrix} + \lambda \begin{pmatrix} 1 \\ 3 \\ 2 \end{pmatrix}$ and the plane

$3x - y + z = 4$.

Solution

For this line, the direction vector $\mathbf{d} = \begin{pmatrix} 1 \\ 3 \\ 2 \end{pmatrix}$ and a normal to the plane is

$$\mathbf{n} = \begin{pmatrix} 3 \\ -1 \\ 1 \end{pmatrix}.$$

The angle α between the normal vector and the direction vector satisfies

$$\mathbf{d}.\mathbf{n} = |\mathbf{d}||\mathbf{n}|\cos\alpha$$

$$\Rightarrow \begin{pmatrix} 1 \\ 3 \\ 2 \end{pmatrix}.\begin{pmatrix} 3 \\ -1 \\ 1 \end{pmatrix} = \sqrt{14}\sqrt{11}\cos\theta$$

$$\Rightarrow \cos\alpha = \frac{2}{\sqrt{14}\sqrt{11}}$$

$$\Rightarrow \alpha = 80.7°$$

So the angle between the line and the plane $\theta = 90° - 80.7° = 9.3°$.

The distance from a point to a plane

In Figure 15.6, the point M is the closest point on the plane to the point P. The line PM is perpendicular to the plane. The distance from the point P to the plane can be found by finding point M and then finding the distance PM.

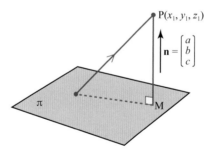

Figure 15.6

Example 15.5

Find the shortest distance from the point P, with coordinates $(2, 4, -2)$, to the plane $6x - y - 3z + 1 = 0$.

Solution

The equation of the line from P perpendicular to the plane to M is given

$$\text{by} = \begin{pmatrix} 2 \\ 4 \\ -2 \end{pmatrix} + \lambda \begin{pmatrix} 6 \\ -1 \\ -3 \end{pmatrix}.$$

This normal to the plane $6x - y - 3z + 1 = 0$ will be the direction vector for this line.

This line intersects the plane when

$$6(2 + 6\lambda) - (4 - \lambda) - 3(-2 - 3\lambda) + 1 = 0$$

$$12 + 36\lambda - 4 + \lambda + 6 + 9\lambda + 1 = 0$$

$$15 + 46\lambda = 0$$

$$\lambda = \frac{15}{46}$$

M is found by substituting the value for λ into the equation for the line.

So the point of intersection M is $\left(\frac{91}{23}, \frac{169}{46}, \frac{-137}{46}\right)$.

The required distance from P to the intersection point M can then be found.

$$\left|\overrightarrow{PM}\right| = \left|\mathbf{m} - \mathbf{p}\right| = \left|\sqrt{\left(2 - \frac{91}{23}\right)^2 + \left(4 - \frac{169}{46}\right)^2 + \left(-2 + \frac{137}{46}\right)^2}\right| = \frac{15}{\sqrt{46}} \approx 2.21$$

ACTIVITY 15.2

Given the plane in Figure 15.6 has equation $\mathbf{r} \cdot \mathbf{n} + d = 0$ and points P and M have position vectors $\mathbf{p}$ and $\mathbf{m}$ respectively, write down the equation of the line PM in terms of λ. Then find λ by substituting into the equation of the plane.

It is best not to try to remember this formula but to remember the technique.

Hence show that the distance PM is given by $\dfrac{\left|\mathbf{p} \cdot \mathbf{n} + d\right|}{\left|\mathbf{n}\right|}$.

Exercise 15.1

① Show that the point of intersection of the line $\mathbf{r} = \begin{pmatrix} 1 \\ 3 \\ 0 \end{pmatrix} + \lambda \begin{pmatrix} 2 \\ -1 \\ 4 \end{pmatrix}$ and

the plane $2x + y + z = 26$ is $(7, 0, 12)$.

② For each of the following, find the point of intersection of the line and the plane. Find also the angle between the line and the plane.

(i) $x + 2y + 3z = 11$ $\mathbf{r} = \begin{pmatrix} 1 \\ 2 \\ 4 \end{pmatrix} + \lambda \begin{pmatrix} 1 \\ 1 \\ 1 \end{pmatrix}$

(ii) $2x + 3y - 4z = 1$ $\dfrac{x + 2}{3} = \dfrac{y + 3}{4} = \dfrac{z + 4}{5}$

(iii) $3x - 2y - z = 14$ $\mathbf{r} = \begin{pmatrix} 8 \\ 4 \\ 2 \end{pmatrix} + \lambda \begin{pmatrix} 1 \\ 2 \\ 1 \end{pmatrix}$

(iv) $x + y + z = 0$ $\mathbf{r} = \lambda \begin{pmatrix} 1 \\ 1 \\ 2 \end{pmatrix}$

③ Find the distance from the point P to the plane π:

    (i)   P(5, 4, 0) π: $6x + 6y + 7z + 1 = 0$

    (ii)   P(7, 2, −2) π: $12x - 9y - 8z + 3 = 0$

    (iii) P(−4, −5, 3) π: $8x + 5y - 3z - 4 = 0$

④   (i)   Find the equation of the line L passing through A(4, 1, 3) and B(6, 4, 8).

    (ii)   Find the point of intersection of L with the plane $x + 2y - z + 3 = 0$.

    (iii) Find the angle between the line L and the plane.

⑤   (i)   Find the equation of the line through (13, 5, 0) parallel to the line

$$\mathbf{r} = \begin{pmatrix} 2 \\ -1 \\ 4 \end{pmatrix} + \lambda \begin{pmatrix} 3 \\ 1 \\ -2 \end{pmatrix}$$

    (ii)   Where does this line meet the plane $3x + y - 2z = 2$?

    (iii) How far is the point of intersection from (13, 5, 0)?

⑥ A plane passes through the points A(2, 3, −1), B(4, 0, 1) and C(−3, 5, −2).

    (i)   Find the vectors $\overrightarrow{AB}$ and $\overrightarrow{AC}$.

Hence confirm that the vector $\begin{pmatrix} 1 \\ 8 \\ 11 \end{pmatrix}$ is a normal to the plane and find the equation of the plane.

    (ii)   Find the points of intersection, P and Q, of the lines

$$L_1: \mathbf{r} = \begin{pmatrix} 4 \\ -1 \\ 3 \end{pmatrix} + \lambda \begin{pmatrix} 2 \\ 0 \\ 1 \end{pmatrix}$$

$$L_2: \mathbf{r} = \begin{pmatrix} 7 \\ -2 \\ 1 \end{pmatrix} + \mu \begin{pmatrix} -1 \\ 1 \\ 3 \end{pmatrix}$$

with the plane.

    (iii) Determine the point of intersection R of the lines L_1 and L_2.

    (iv) Find the angle between the vectors $\overrightarrow{PR}$ and $\overrightarrow{QR}$.

    (v)   Find the area of the triangle PQR.

⑦   (i)   Find the exact distance from the point A(2, 0, −5) to the plane π: $4x - 5y + 2z + 4 = 0$.

    (ii)   Write down the equation of the line l through the point A that is perpendicular to the plane π.

    (iii) Find the exact coordinates of the point M where the perpendicular from the point A meets the plane π.

⑧ A laser beam ABC is fired from the point A(1, 2, 4) and is reflected at B off the plane with equation $x + 2y - 3z = 0$, as shown in Figure 15.7.

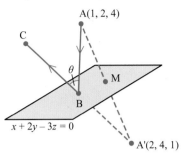

Figure 15.7

A′ is the point (2, 4, 1) and M is the midpoint of AA′.

(i) Show that AA′ is perpendicular to the plane $x + 2y - 3z = 0$ and that M lies in the plane.

The vector equation of the line AB is $\mathbf{r} = \begin{pmatrix} 1 \\ 2 \\ 4 \end{pmatrix} + \lambda \begin{pmatrix} 1 \\ -1 \\ 2 \end{pmatrix}$.

(ii) Find the coordinates of B and a vector equation of the line A′B.

(iii) Given that A′BC is a straight line, find the angle θ.

⑨ Figure 15.8 shows the tetrahedron ABCD. The coordinates of the vertices are A(−3, 0, 0), B(2, 0, −2), C(0, 4, 0) and D(0, 4, 5).

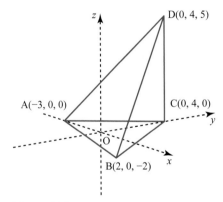

Figure 15.8

(i) Find the lengths of the edges AB and AC, and the size of the angle CAB. Hence calculate the area of the triangle ABC.

(ii) (a) Verify that $\begin{pmatrix} 4 \\ -3 \\ 10 \end{pmatrix}$ is normal to the plane ABC.

(b) Find the equation of this plane.

(iii) Write down the vector equation of the line through D that is perpendicular to the plane ABC. Find the point of intersection of this line with the plane ABC.

The area of a tetrahedron is given by $\frac{1}{3} \times$ base area $\times$ height.

(iv) Find the volume of the tetrahedron ABCD.

2 The vector product

You have already used the scalar product to find the angle between two vectors. It is particularly convenient if you want to test whether two vectors are perpendicular, as then the scalar product is zero.

The scalar product also enables you to write down the equation of a plane in the form $(\mathbf{r} - \mathbf{a}) \cdot \mathbf{n} = 0$ where $\mathbf{a}$ is the position vector of a specific point A in the plane, $\mathbf{r}$ is the position vector of a general point in the plane and $\mathbf{n}$ is a normal vector perpendicular to the plane (see Figure 15.10). This can be rearranged as $\mathbf{r} \cdot \mathbf{n} = d$, where $d = \mathbf{n} \cdot \mathbf{a}$; d is a scalar constant.

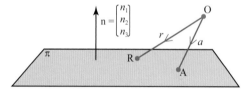

Figure 15.10

The **vector product** is a different method for 'multiplying' two vectors. As the name suggests, in this case the result is a vector rather than a scalar. The vector product of $\mathbf{a}$ and $\mathbf{b}$ is a vector perpendicular to both $\mathbf{a}$ and $\mathbf{b}$ and it is written $\mathbf{a} \times \mathbf{b}$.

It is given by

$$\mathbf{a} \times \mathbf{b} = |\mathbf{a}||\mathbf{b}|\sin\theta\,\hat{\mathbf{n}}$$

where θ is the angle between $\mathbf{a}$ and $\mathbf{b}$ and $\hat{\mathbf{n}}$ is a unit vector which is perpendicular to both $\mathbf{a}$ and $\mathbf{b}$.

> This is often described as having opposite senses.

There are two unit vectors perpendicular to both $\mathbf{a}$ and $\mathbf{b}$, but they point in opposite directions.

The vector $\hat{\mathbf{n}}$ is chosen such that $\mathbf{a}$, $\mathbf{b}$ and $\hat{\mathbf{n}}$ (in that order) form a **right-handed set** of vectors, as shown in Figure 15.11. If you point the thumb of your right hand in the direction of $\mathbf{a}$, and your index finger in the direction of $\mathbf{b}$, then your second finger coming up from your palm points in the direction $\mathbf{a} \times \mathbf{b}$ as shown below.

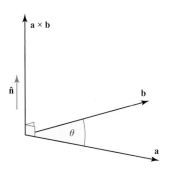
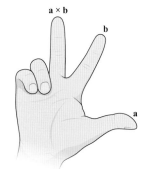

Figure 15.11

ACTIVITY 15.3

In this activity you might find it helpful to take the edges of a rectangular table to represent the unit vectors $\mathbf{i}$, $\mathbf{j}$ and $\mathbf{k}$, as shown in Figure 15.12.

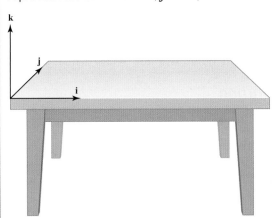

Figure 15.12

You could use pens to represent:

$\mathbf{i}$, the unit vector pointing to the right along the x-axis

$\mathbf{j}$, the unit vector pointing away from you along the y-axis

$\mathbf{k}$, the unit vector pointing upwards along the z-axis.

The vector product of $\mathbf{a}$ and $\mathbf{b}$ is defined as

$$\mathbf{a} \times \mathbf{b} = |\mathbf{a}||\mathbf{b}|\sin\theta\,\hat{\mathbf{n}}$$

where θ is the angle between $\mathbf{a}$ and $\mathbf{b}$ and $\hat{\mathbf{n}}$ is a unit vector which is perpendicular to both $\mathbf{a}$ and $\mathbf{b}$ such that $\mathbf{a}$, $\mathbf{b}$ and $\hat{\mathbf{n}}$ (in that order) form a right–handed set of vectors.

Using this definition, check the truth of each of the following results.

$$\mathbf{i} \times \mathbf{i} = 0 \qquad\qquad \mathbf{i} \times \mathbf{j} = \mathbf{k} \qquad\qquad \mathbf{i} \times \mathbf{k} = -\mathbf{j}$$

Give a further six results for vector products of pairs of $\mathbf{i}$, $\mathbf{j}$ and $\mathbf{k}$.

In component form, the vector product is expressed

> **Note**
>
> The vector product is sometimes referred to as the cross product.

$$\mathbf{a} \times \mathbf{b} = \begin{pmatrix} a_1 \\ a_2 \\ a_3 \end{pmatrix} \times \begin{pmatrix} b_1 \\ b_2 \\ b_3 \end{pmatrix} = \begin{pmatrix} a_2 b_3 - a_3 b_2 \\ a_3 b_1 - a_1 b_3 \\ a_1 b_2 - a_2 b_1 \end{pmatrix}$$

You will have the opportunity to prove this result in Exercise 15.2.

Notice that the first component of $\mathbf{a} \times \mathbf{b}$ is the value of the 2×2 determinant $\begin{vmatrix} a_2 & b_2 \\ a_3 & b_3 \end{vmatrix}$ obtained by covering up the top row of $\begin{pmatrix} a_1 \\ a_2 \\ a_3 \end{pmatrix} \times \begin{pmatrix} b_1 \\ b_2 \\ b_3 \end{pmatrix}$; the second component is the negative of the 2×2 determinant obtained by covering up the middle row; and the third component is the 2×2 determinant obtained by covering up the bottom row.

This means that the formula for the vector product can be expressed as a determinant:

$$\mathbf{a} \times \mathbf{b} = \begin{vmatrix} \mathbf{i} & a_1 & b_1 \\ \mathbf{j} & a_2 & b_2 \\ \mathbf{k} & a_3 & b_3 \end{vmatrix}$$

Expanding this determinant by the first column gives

$$\mathbf{a} \times \mathbf{b} = \begin{vmatrix} a_2 & b_2 \\ a_3 & b_3 \end{vmatrix} \mathbf{i} - \begin{vmatrix} a_1 & b_1 \\ a_3 & b_3 \end{vmatrix} \mathbf{j} + \begin{vmatrix} a_1 & b_1 \\ a_2 & b_2 \end{vmatrix} \mathbf{k}.$$

Note this sign.

Example 15.6

(i) Calculate $\mathbf{a} \times \mathbf{b}$ when $\mathbf{a} = 3\mathbf{i} + 2\mathbf{j} + 5\mathbf{k}$ and $\mathbf{b} = \mathbf{i} - 4\mathbf{j} + 2\mathbf{k}$.

(ii) Hence find $\hat{\mathbf{n}}$, a unit vector which is perpendicular to both $\mathbf{a}$ and $\mathbf{b}$.

Solution

(i) There are three possible methods:

Method 1
Expanding brackets and considering each component:

$$\mathbf{a} \times \mathbf{b} = (3\mathbf{i} + 2\mathbf{j} + 5\mathbf{k}) \times (\mathbf{i} - 4\mathbf{j} + 2\mathbf{k})$$
$$= (3\mathbf{i} \times \mathbf{i}) + (3\mathbf{i} \times -4\mathbf{j}) + (3\mathbf{i} \times 2\mathbf{k}) + (2\mathbf{j} \times \mathbf{i}) + (2\mathbf{j} \times -4\mathbf{j}) +$$
$$(2\mathbf{j} \times 2\mathbf{k}) + (5\mathbf{k} \times \mathbf{i}) + (5\mathbf{k} \times -4\mathbf{j}) + (5\mathbf{k} \times 2\mathbf{k})$$
$$= 0 - 12\mathbf{k} - 6\mathbf{j} - 2\mathbf{k} + 0 + 4\mathbf{i} + 5\mathbf{j} + 20\mathbf{i} + 0$$
$$= 24\mathbf{i} - \mathbf{j} - 14\mathbf{k}$$

Method 2
Using determinants:

$$\mathbf{a} \times \mathbf{b} = \begin{vmatrix} \mathbf{i} & 3 & 1 \\ \mathbf{j} & 2 & -4 \\ \mathbf{k} & 5 & 2 \end{vmatrix} \longleftarrow \qquad \mathbf{a} \times \mathbf{b} = \begin{vmatrix} \mathbf{i} & a_1 & b_1 \\ \mathbf{j} & a_2 & b_2 \\ \mathbf{k} & a_3 & b_3 \end{vmatrix}$$

$$= \mathbf{i} \begin{vmatrix} 2 & -4 \\ 5 & 2 \end{vmatrix} - \mathbf{j} \begin{vmatrix} 3 & 1 \\ 5 & 2 \end{vmatrix} + \mathbf{k} \begin{vmatrix} 3 & 1 \\ 2 & -4 \end{vmatrix}$$

$$= 24\mathbf{i} - \mathbf{j} - 14\mathbf{k}$$

Method 3
Using the result

$$\mathbf{a} \times \mathbf{b} = \begin{pmatrix} a_1 \\ a_2 \\ a_3 \end{pmatrix} \times \begin{pmatrix} b_1 \\ b_2 \\ b_3 \end{pmatrix} = \begin{pmatrix} a_2 b_3 - a_3 b_2 \\ a_3 b_1 - a_1 b_3 \\ a_1 b_2 - a_2 b_1 \end{pmatrix}$$

gives

$$\mathbf{a} \times \mathbf{b} = \begin{pmatrix} 3 \\ 2 \\ 5 \end{pmatrix} \times \begin{pmatrix} 1 \\ -4 \\ 2 \end{pmatrix} = \begin{pmatrix} 2 \times 2 - 5 \times (-4) \\ 5 \times 1 - 3 \times 2 \\ 3 \times (-4) - 2 \times 1 \end{pmatrix} = \begin{pmatrix} 24 \\ -1 \\ -14 \end{pmatrix}$$

(ii) So $\mathbf{a} \times \mathbf{b} = \begin{pmatrix} 24 \\ -1 \\ -14 \end{pmatrix}$, which is a vector perpendicular to the vectors $\mathbf{a}$ and $\mathbf{b}$.

$$\left| \mathbf{a} \times \mathbf{b} \right| = \sqrt{24^2 + (-1)^2 + (-14)^2} = \sqrt{773}$$

So a unit vector perpendicular to both $\mathbf{a}$ and $\mathbf{b}$ is $\hat{\mathbf{n}} = \dfrac{1}{\sqrt{773}} \begin{pmatrix} 24 \\ -1 \\ -14 \end{pmatrix}$.

Discussion point

→ How can you use the scalar product to check that the answer to Example 15.6 is correct?

TECHNOLOGY

Investigate whether your calculator will find the vector product of two vectors.

If so, use your calculator to check the vector product calculated in Example 15.6.

Properties of the vector product

1 **The vector product is anti-commutative**

The vector products $\mathbf{a} \times \mathbf{b}$ and $\mathbf{b} \times \mathbf{a}$ have the same magnitude but are in opposite directions, so $\mathbf{a} \times \mathbf{b} = -\mathbf{b} \times \mathbf{a}$. This is known as the **anti-commutative property**.

2 **The vector product of parallel vectors is zero**

This is because the angle θ between two parallel vectors is $0°$ or $180°$, so $\sin \theta = 0$.

In particular $\mathbf{i} \times \mathbf{i} = \mathbf{j} \times \mathbf{j} = \mathbf{k} \times \mathbf{k} = \mathbf{0}$.

3 **The vector product is compatible with scalar multiplication**

For scalars m and n,

$$(m\mathbf{a}) \times (n\mathbf{b}) = mn(\mathbf{a} \times \mathbf{b})$$

This is because the vector $m\mathbf{a}$ has magnitude $|m||\mathbf{a}|$; $m\mathbf{a}$ and $\mathbf{a}$ have the same direction if m is positive, but opposite directions if m is negative.

4 **The vector product is distributive over vector addition**

The result

$$\mathbf{a} \times (\mathbf{b} + \mathbf{c}) = \mathbf{a} \times \mathbf{b} + \mathbf{a} \times \mathbf{c}$$

enables you to change a product into the sum of two simpler products – in doing so the multiplication is 'distributed' over the two terms of the original sum.

Vector product form of the equation of a straight line

You are already familiar with the vector equation of a line in the form
$\mathbf{r} = \mathbf{a} + \lambda\mathbf{b}$. Note that this can be rewritten in the form $\mathbf{r} - \mathbf{a} = \lambda\mathbf{b}$.

Since $\mathbf{r} - \mathbf{a}$ is a multiple of $\mathbf{b}$, $\mathbf{r} - \mathbf{a}$ is parallel to $\mathbf{b}$, so it can be written in the form

$$(\mathbf{r} - \mathbf{a}) \times \mathbf{b} = 0.$$

This is called the vector product form of the equation of a line.

Example 15.7

Find the equation of the line joining points $(4, 2, 1)$ and $(3, -1, 2)$ in the form $(\mathbf{r} - \mathbf{a}) \times \mathbf{b} = 0$.

Solution

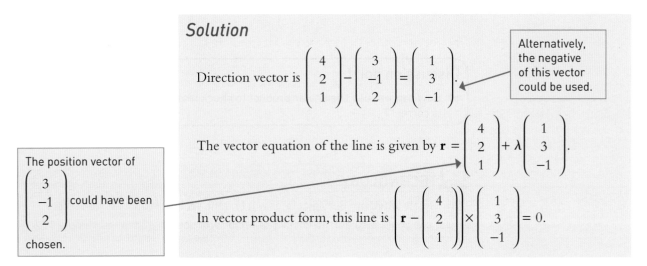

Direction vector is $\begin{pmatrix} 4 \\ 2 \\ 1 \end{pmatrix} - \begin{pmatrix} 3 \\ -1 \\ 2 \end{pmatrix} = \begin{pmatrix} 1 \\ 3 \\ -1 \end{pmatrix}$.

Alternatively, the negative of this vector could be used.

The vector equation of the line is given by $\mathbf{r} = \begin{pmatrix} 4 \\ 2 \\ 1 \end{pmatrix} + \lambda \begin{pmatrix} 1 \\ 3 \\ -1 \end{pmatrix}$.

The position vector of $\begin{pmatrix} 3 \\ -1 \\ 2 \end{pmatrix}$ could have been chosen.

In vector product form, this line is $\left(\mathbf{r} - \begin{pmatrix} 4 \\ 2 \\ 1 \end{pmatrix} \right) \times \begin{pmatrix} 1 \\ 3 \\ -1 \end{pmatrix} = 0$.

Using the vector product to find the equation of a plane

Since the equation of a plane involves a vector which is perpendicular to the plane, the vector product is very useful in finding the equation of a plane.

Example 15.8

Find the Cartesian equation of the plane which contains the points A$(3, 4, 2)$, B$(2, 0, 5)$ and C$(6, 7, 8)$.

Solution

Start by finding two vectors in the plane, for example $\overrightarrow{AB}$ and $\overrightarrow{BC}$.

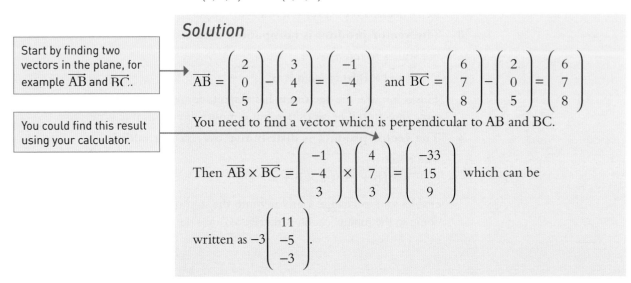

$\overrightarrow{AB} = \begin{pmatrix} 2 \\ 0 \\ 5 \end{pmatrix} - \begin{pmatrix} 3 \\ 4 \\ 2 \end{pmatrix} = \begin{pmatrix} -1 \\ -4 \\ 1 \end{pmatrix}$ and $\overrightarrow{BC} = \begin{pmatrix} 6 \\ 7 \\ 8 \end{pmatrix} - \begin{pmatrix} 2 \\ 0 \\ 5 \end{pmatrix} = \begin{pmatrix} 6 \\ 7 \\ 8 \end{pmatrix}$

You need to find a vector which is perpendicular to AB and BC.

You could find this result using your calculator.

Then $\overrightarrow{AB} \times \overrightarrow{BC} = \begin{pmatrix} -1 \\ -4 \\ 3 \end{pmatrix} \times \begin{pmatrix} 4 \\ 7 \\ 3 \end{pmatrix} = \begin{pmatrix} -33 \\ 15 \\ 9 \end{pmatrix}$ which can be

written as $-3 \begin{pmatrix} 11 \\ -5 \\ -3 \end{pmatrix}$.

Discussion point

➜ Another way of finding the equation through three given points is to form three simultaneous equations and solve them.

Compare these two methods.

So $\mathbf{n} = \begin{pmatrix} -11 \\ -5 \\ -3 \end{pmatrix}$ is a vector perpendicular to the plane containing A, B and C, and the equation of the plane is of the form $11x - 5y - 3z = d$.

Substituting the coordinates of one of the points, say A, allows you to find the value of the constant d:

$$(11 \times 3) - (5 \times 4) - (3 \times 2) = 7$$

The plane has equation $11x - 5y - 3z = 7$.

> Substituting for B and C provides a useful check of your answer.

Using the vector product to find the intersection of two planes

One application of the vector product is to find the intersection of two planes. This is an alternative approach to solving by using simultaneous equations.

Example 15.9

Find l, the line of intersection of the two planes $3x + 2y - 3z = -18$ and $x - 2y + z = 12$.

Solution

The planes have normals $\mathbf{n}_1 = \begin{pmatrix} 3 \\ 2 \\ -3 \end{pmatrix}$ and $\mathbf{n}_2 = \begin{pmatrix} 1 \\ -2 \\ 1 \end{pmatrix}$.

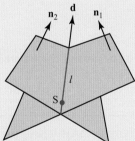

Figure 15.13

As line l is in each plane (see Figure 15.13) it must be perpendicular to both $\mathbf{n}_1$ and $\mathbf{n}_2$, so

$$\mathbf{n}_1 \times \mathbf{n}_2 = \begin{pmatrix} -4 \\ -6 \\ -8 \end{pmatrix} = -2 \begin{pmatrix} 2 \\ 3 \\ 4 \end{pmatrix}$$

> Removing factor -2 makes the arithmetic simpler.

gives a vector in the direction of l.

Use $\mathbf{d} = \begin{pmatrix} 2 \\ 3 \\ 4 \end{pmatrix}$ as the direction vector for l.

> Mentally check that $\mathbf{d} \cdot \mathbf{n}_1 = \mathbf{d} \cdot \mathbf{n}_2 = 0$.

Now you only need to find a point S on l.

Choosing to put $x = 1$ into the equation gives

$$\left. \begin{matrix} 2y - 3z = -21 \\ -2y + z = 11 \end{matrix} \right\} \quad \Leftrightarrow \quad \left\{ \begin{matrix} -2z = -10 \\ 2y = 3z - 21 \end{matrix} \right. \quad \Leftrightarrow \quad z = 5, y = -3$$

so S with coordinates $(1, -3, 5)$ is a point on line l.

The Cartesian equations of l are $\dfrac{x-1}{2} = \dfrac{y+3}{3} = \dfrac{z-5}{4}$, although you could give the parametric form

$$\left\{ \begin{matrix} x = 2t + 1 \\ y = 3t - 3 \\ z = 4t + 5 \end{matrix} \right. \text{ or the vector form } \mathbf{r} = \begin{pmatrix} 1 \\ -3 \\ 5 \end{pmatrix} + t \begin{pmatrix} 2 \\ 3 \\ 4 \end{pmatrix}.$$

Using the vector product to find areas

Given that $\mathbf{a} \times \mathbf{b} = |\mathbf{a}||\mathbf{b}| \sin\theta \hat{\mathbf{n}}$

The magnitude of this is given by

$$|\mathbf{a} \times \mathbf{b}| = |\mathbf{a}||\mathbf{b}| \sin\theta$$

By definition, this is the same as the area of the parallelogram spanned by $\mathbf{a}$ and $\mathbf{b}$, as shown in Figure 15.14.

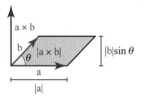

Figure 15.14

> Notice the similarity to
>
> Area of triangle $= \dfrac{1}{2} ab \sin\theta$.

So, it can be deduced that the area of the parallelogram $= |\mathbf{a} \times \mathbf{b}|$.

It follows from this that the area of the triangle spanned by $\mathbf{a}$ and $\mathbf{b} = \dfrac{1}{2}|\mathbf{a} \times \mathbf{b}|$.

Example 15.10

Find the area of the triangle OAB where two points A and B have position vectors $\mathbf{a} = 3\mathbf{i} + 5\mathbf{j} + 2\mathbf{k}$ and $\mathbf{b} = 2\mathbf{i} - \mathbf{j} + 4\mathbf{k}$.

Solution

$$\text{Area} = \frac{1}{2}|\mathbf{a} \times \mathbf{b}|$$

$$= \frac{1}{2} \left| \begin{pmatrix} 3 \\ 5 \\ 2 \end{pmatrix} \times \begin{pmatrix} 2 \\ -1 \\ 4 \end{pmatrix} \right|$$

$$= \frac{1}{2} \left| \begin{pmatrix} 22 \\ -8 \\ -13 \end{pmatrix} \right|$$

$$= \frac{\sqrt{717}}{2}$$

Using the vector product to find the distance between two parallel lines

The distance between two parallel lines l_1 and l_2 is measured along a line PQ, which is perpendicular to both l_1 and l_2, as shown in Figure 15.15.

You can find this distance by simply choosing a point P on l_1, say, and then finding the shortest distance from P to the line l_2.

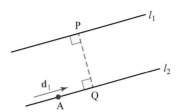

Figure 15.15

Two straight lines in three dimensions are given by the equations

$$l_1: \mathbf{r} = \begin{pmatrix} 4 \\ 2 \\ 1 \end{pmatrix} + \mu \begin{pmatrix} -2 \\ 6 \\ -4 \end{pmatrix} \text{ and } l_2: \mathbf{r} = \begin{pmatrix} 2 \\ -3 \\ 0 \end{pmatrix} + \lambda \begin{pmatrix} 1 \\ -3 \\ 2 \end{pmatrix}$$

(i) Show that the two lines are parallel.

(ii) Find the shortest distance between the two lines.

Solution

(i) The direction vectors of the two lines are $\mathbf{d}_1 = \begin{pmatrix} -2 \\ 6 \\ -4 \end{pmatrix}$ and

$\mathbf{d}_2 = \begin{pmatrix} 1 \\ -3 \\ 2 \end{pmatrix}$.

Since $\mathbf{d}_1 = -2\mathbf{d}_2$, the two lines are parallel.

> You could use any value for λ.

(ii) Choose a point P on l_1 by setting $\lambda = 0$ which gives $\mathbf{p} = \begin{pmatrix} 4 \\ 2 \\ 1 \end{pmatrix}$.

> **Note**
>
> You have already done this in question 1 of the Review exercise. Compare your two methods.

To find the shortest distance of P from l_2, use $\mathbf{a} = \begin{pmatrix} 2 \\ -3 \\ 0 \end{pmatrix}$ and

$\mathbf{d} = \begin{pmatrix} 1 \\ -3 \\ 2 \end{pmatrix}$.

$$\overrightarrow{AP} = \begin{pmatrix} 2 \\ 5 \\ 1 \end{pmatrix} \text{ and so } \overrightarrow{AP} \times \mathbf{d} = \begin{pmatrix} 2 \\ 5 \\ 1 \end{pmatrix} \times \begin{pmatrix} 1 \\ -3 \\ 2 \end{pmatrix} = \begin{pmatrix} 13 \\ -3 \\ -11 \end{pmatrix}.$$

$$\left| \overrightarrow{AP} \times \mathbf{d} \right| = \sqrt{13^2 + (-3)^2 + (-11)^2} = \sqrt{299}$$

$$\left| \mathbf{d} \right| = \sqrt{1^2 + (-3)^2 + 2^2} = \sqrt{14}$$

The shortest distance is $\dfrac{\left| \overrightarrow{AP} \times \mathbf{d} \right|}{\left| \mathbf{d} \right|} = \dfrac{\sqrt{299}}{\sqrt{14}} \approx 4.62.$

Example 15.12

Using the vector product to find the distance between skew lines

The distance between skew lines has already been shown earlier in this chapter. This section shows an alternative approach using the vector product.

Find the shortest distance between the skew lines $\quad l_1: \mathbf{r} = \begin{pmatrix} 2 \\ 1 \\ -2 \end{pmatrix} + \lambda \begin{pmatrix} 2 \\ 0 \\ 5 \end{pmatrix}$

and $l_2: \mathbf{r} = \begin{pmatrix} 3 \\ 5 \\ 0 \end{pmatrix} + \mu \begin{pmatrix} -5 \\ 3 \\ 4 \end{pmatrix}.$

Give your answer in simplified surd form.

Solution

The direction vector of l_1 is given by $\mathbf{d}_1 = \begin{pmatrix} 2 \\ 0 \\ 5 \end{pmatrix}.$

The direction vector of l_2 is given by $\mathbf{d}_2 = \begin{pmatrix} -5 \\ 3 \\ 4 \end{pmatrix}.$

$$\mathbf{n} = \mathbf{d}_1 \times \mathbf{d}_2 = \begin{pmatrix} 2 \\ 0 \\ 5 \end{pmatrix} \times \begin{pmatrix} -5 \\ 3 \\ 4 \end{pmatrix} = \begin{pmatrix} 5 \\ 11 \\ -2 \end{pmatrix}$$

> The vector product gives a vector that is perpendicular to both direction vectors so is normal to both planes.

The point P on l_1 and the point Q on l_2 are such that $\overrightarrow{PQ} = m\mathbf{n}$ for some scalar constant m.

$$\overrightarrow{PQ} = \begin{pmatrix} 3 - 5\mu \\ 5 + 3\mu \\ 0 + 4\mu \end{pmatrix} - \begin{pmatrix} 2 + 2\lambda \\ 1 + 0\lambda \\ -2 + 5\lambda \end{pmatrix} = \begin{pmatrix} 1 - 5\mu - 2\lambda \\ 4 + 3\mu \\ 2 + 4\mu - 5\lambda \end{pmatrix}$$

So

$$\begin{pmatrix} 1 - 5\mu - 2\lambda \\ 4 + 3\mu \\ 2 + 4\mu - 5\lambda \end{pmatrix} = m \begin{pmatrix} 5 \\ 11 \\ -2 \end{pmatrix}$$

Use simultaneous equations to eliminate λ and then solve to find m.

Discussion point

➜ Compare this method with the one in Example 15.2.

Solving gives $m = \dfrac{3}{10}$.

It can then be deduced that the distance between P and Q is

$$\frac{3}{10} \left| \begin{matrix} 5 \\ 11 \\ -2 \end{matrix} \right| = \frac{3\sqrt{6}}{2} \approx 3.67 \text{ units}$$

Exercise 15.2

In this exercise you should calculate the vector products by hand. You could check your answers using the vector product facility on a calculator.

① Calculate each of the following vector products.

(i) $\begin{pmatrix} 3 \\ 5 \\ 2 \end{pmatrix} \times \begin{pmatrix} 2 \\ 4 \\ -3 \end{pmatrix}$

(ii) $\begin{pmatrix} 7 \\ -4 \\ -5 \end{pmatrix} \times \begin{pmatrix} -4 \\ 5 \\ -3 \end{pmatrix}$

(iii) $(5\mathbf{i} - 2\mathbf{j} + 4\mathbf{k}) \times (\mathbf{i} + 5\mathbf{j} - 6\mathbf{k})$ (iv) $(3\mathbf{i} - 7\mathbf{k}) \times (2\mathbf{i} + 3\mathbf{j} + 5\mathbf{k})$

② Find a vector perpendicular to each of the following pairs of vectors.

(i) $\mathbf{a} = \begin{pmatrix} 2 \\ 0 \\ 5 \end{pmatrix}, \mathbf{b} = \begin{pmatrix} 3 \\ -1 \\ -2 \end{pmatrix}$

(ii) $\mathbf{a} = \begin{pmatrix} 12 \\ 3 \\ -2 \end{pmatrix}, \mathbf{b} = \begin{pmatrix} 7 \\ 1 \\ 4 \end{pmatrix}$

(iii) $\mathbf{a} = 2\mathbf{i} + 3\mathbf{j} + 4\mathbf{k}, \mathbf{b} = 3\mathbf{i} + 6\mathbf{j} + 7\mathbf{k}$

(iv) $\mathbf{a} = 3\mathbf{i} - 4\mathbf{j} + 6\mathbf{k}, \mathbf{b} = 8\mathbf{i} + 5\mathbf{j} - 3\mathbf{k}$

③ Three points A, B and C have coordinates $(1, 4, -2), (2, 0, 1)$ and $(5, 3, -2)$ respectively.

(i) Find the vectors $\overrightarrow{AB}$ and $\overrightarrow{AC}$.

(ii) Use the vector product to find a vector that is perpendicular to $\overrightarrow{AB}$ and $\overrightarrow{AC}$.

(iii) Hence find the equation of the plane containing points A, B and C.

(iv) Find the exact value of the area of triangle ABC.

④ Find, in vector product form, the equations of the line of intersections of these pairs of planes.

(i) $x + y - 6z = 4; 5x - 2y - 3z = 13$

(ii) $5x - y + z = 8; x + 3y + z = -4$

(iii) $3x + 2y - 6z = -4; x + 5y - 7z = 2$

(iv) $x + 2y + 3z = 6; 2x + 5y - 2z = -4$

⑤ Find a unit vector perpendicular to both $\mathbf{a} = \begin{pmatrix} 1 \\ 2 \\ 7 \end{pmatrix}$ and $\mathbf{b} = \begin{pmatrix} 3 \\ -1 \\ 6 \end{pmatrix}$.

⑥ Find the magnitude of $\begin{pmatrix} 3 \\ 1 \\ -4 \end{pmatrix} \times \begin{pmatrix} 1 \\ -1 \\ 1 \end{pmatrix}$.

⑦ Find the Cartesian equations of the planes containing the three points given.

(i) A(1, 4, 2), B(5, 1, 3) and C(1, 0, 0)

(ii) D(5, −3, 4), E(0, 1, 0) and F(6, 2, 5)

(iii) G(6, 2, −2), H(1, 4, 3) and L(−5, 7, 1)

(iv) M(4, 2, −1), N(8, 2, 4) and P(5, 8, −7)

⑧ Simplify the following.

(i) $4\mathbf{i} \times 2\mathbf{k}$

(ii) $2\mathbf{i} \times (5\mathbf{i} - 2\mathbf{j} - 3\mathbf{k})$

(iii) $(6\mathbf{i} + \mathbf{j} - \mathbf{k}) \times 2\mathbf{k}$

(iv) $(3\mathbf{i} - \mathbf{j} + 2\mathbf{k}) \times (\mathbf{i} - \mathbf{j} - 4\mathbf{k})$

⑨ Find the equation of the line, in vector form, which goes through (4, −2, −7) and which is parallel to both $2x - 5y -2z = 8$ and $x + 3y - 3z = 12$.

⑩ The four vertices of a parallelogram ABCD have coordinates A(1, 0, 5), B(6, 2, 6), C(8, 1, 10) and D(3, −1, 9)

(i) Find $\overrightarrow{AB} \times \overrightarrow{AD}$.

(ii) Show that the area of the parallelogram is $k\sqrt{6}$ where k is an integer to be found.

(iii) The diagonals AC and BD of the parallelogram meet at the point M. The line L passes through M and is perpendicular to the plane ABCD. Find an equation for the line L.

(iv) The plane π is parallel to the plane ABCD and passes through the point E(4, −1, 2). Find the equation of the plane.

(v) Find the coordinates of the point of intersection of the line L and the plane π.

⑪ Prove algebraically that, for two vectors $\mathbf{a} = a_1\mathbf{i} + a_2\mathbf{j} + a_3\mathbf{k}$ and $\mathbf{b} = b_1\mathbf{i} + b_2\mathbf{j} + b_3\mathbf{k}$,

$$\mathbf{a} \times \mathbf{b} = \begin{pmatrix} a_2b_3 - a_3b_2 \\ a_3b_1 - a_1b_3 \\ a_1b_2 - a_2b_1 \end{pmatrix}.$$

⑫ Four points have coordinates A(−2, −3, 2), B(−3, 1, 5), C(k, 5, −2) and D(0, 9, k).

(i) Find the vector product $\overrightarrow{AB} \times \overrightarrow{CD}$.

(ii) For the case when AB is parallel to CD:

(a) state the value of k

(b) find the shortest distance between the parallel lines AB and CD

(c) find, in the form $ax + by + cz + d = 0$, the equation of the plane containing AB and CD.

(iii) When AB is not parallel to CD, find the shortest distance between the lines AB and CD in terms of k.

(iv) Find the value of k for which the line AB intersects the line CD, and find the coordinates of the point of intersection in this case.

⑬ The point A(−1, 12, 5) lies on the plane P with equation $8x − 3y + 10z = 6$.
The point B(6, −2, 9) lies on the plane Q with equation $3x − 4y − 2z = 8$.
The planes P and Q intersect in the line L.

(i) (a) Show that the point $(0, −2, 0)$ lies on both planes.

(b) Use the vector product to find a vector perpendicular to both plane P and plane Q.

(c) Write down the equation of L, the line of intersection of planes P and Q.

(ii) Find the shortest distance between L and the line AB.

The lines M and N are both parallel to L, with M passing through A and N passing through B.

(iii) Find the distance between the parallel lines M and N.

The point C has coordinates $(k, 0, 2)$ and the line AC intersects the line N at the point D.

(iv) Find the value of k and the coordinates of D.

LEARNING OUTCOMES

Now you have finished this chapter, you should be able to:

➤ find the point of intersection of a line and a plane

➤ find the angle between a line and a plane

➤ use the vector product in component form to find a vector perpendicular to two given vectors

➤ know that $\mathbf{a} \times \mathbf{b} = |\mathbf{a}||\mathbf{b}| \sin\theta \hat{\mathbf{n}}$, where $\mathbf{a}$, $\mathbf{b}$ and $\hat{\mathbf{n}}$, in that order, form a right-handed triple

➤ find the vector product form of the equation of a straight line

➤ use the vector product to calculate the area of a triangle

➤ find the distance from a point to a line in two or three dimensions

➤ find the distance between two parallel lines

➤ find the shortest distance between two skew lines

➤ find the distance of a point from a plane.

KEY POINTS

1 The acute angle between a line and a plane is the acute angle between the line and its **orthogonal projection** onto the plane.

2 To find the point of intersection between a line, $\mathbf{r} = \mathbf{a} + \lambda\mathbf{b}$, and a plane, substitute the values for x, y and z into the equation of the plane and then solve for λ.

3 The vector product $\mathbf{a} \times \mathbf{b}$ of $\mathbf{a}$ and $\mathbf{b}$ is a vector perpendicular to both $\mathbf{a}$ and $\mathbf{b}$

$$\mathbf{a} \times \mathbf{b} = |\mathbf{a}||\mathbf{b}| \sin\theta \hat{\mathbf{n}}$$

where θ is the angle between $\mathbf{a}$ and $\mathbf{b}$ and $\hat{\mathbf{n}}$ is a unit vector which is perpendicular to both $\mathbf{a}$ and $\mathbf{b}$ such that $\mathbf{a}$, $\mathbf{b}$ and $\hat{\mathbf{n}}$ (in that order) form a right-handed set of vectors.

4 $\mathbf{a} \times \mathbf{b} = \begin{pmatrix} a_1 \\ a_2 \\ a_3 \end{pmatrix} \times \begin{pmatrix} b_1 \\ b_2 \\ b_3 \end{pmatrix} = \begin{pmatrix} a_2 b_3 - a_3 b_2 \\ a_3 b_1 - a_1 b_3 \\ a_1 b_2 - a_2 b_1 \end{pmatrix} = \begin{vmatrix} \mathbf{i} & a_1 & b_1 \\ \mathbf{j} & a_2 & b_2 \\ \mathbf{k} & a_3 & b_3 \end{vmatrix}$

5 The vector product has the following properties:
$$\mathbf{a} \times \mathbf{b} = -\mathbf{b} \times \mathbf{a}$$
$$\mathbf{a} \times (\mathbf{b} + \mathbf{c}) = \mathbf{a} \times \mathbf{b} + \mathbf{a} \times \mathbf{c}$$

6 The area of a triangle with sides $\mathbf{a}$ and $\mathbf{b}$ can be found using $\frac{1}{2}|\mathbf{a} \times \mathbf{b}|$.

7 The vector product form of the equation of a straight line is given as
$(\mathbf{r} - \mathbf{a}) \times \mathbf{b} = 0$ where $\mathbf{a}$ is the position vector of a point on the line and $\mathbf{b}$ is the direction vector of the line.

8 The shortest distance from a point P to a line l can be found by constructing a line through P perpendicular to l and finding the point M where this line intersects l. The distance PM can then be found.

9 The shortest distance from a point P to a plane π can be found by constructing a line through P perpendicular to π and finding the point M where this line intersects π. The distance PM can then be found.

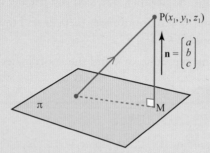

Figure 15.16

10 In three dimensions there are three possibilities for the arrangement of the lines. They are either parallel, intersecting or skew.

11 The shortest distance between two parallel lines can be found by choosing any point on one of the lines and finding the shortest distance from that point to the second line.

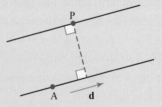

Figure 15.17

12 The shortest distance between two skew lines can be found by finding
$|(\mathbf{a}_2 - \mathbf{a}_1) \bullet \hat{\mathbf{n}}|$ where $\hat{\mathbf{n}}$ is a unit vector perpendicular to both planes. The vector product can be used to find $\mathbf{n}$.

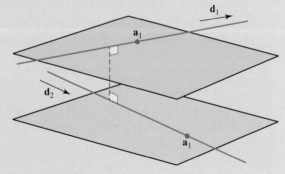

Figure 15.18

Second order differential equations

Discussion point

➜ The picture shows the Tacoma Narrows Bridge in the U.S. state of Washington, which collapsed on 7 November 1940. Find out what caused the collapse of the bridge.

1 Higher order differential equations

So far in this course you have mostly solved only first order differential equations. In this chapter you will extend these techniques to second (and higher) order differential equations. Second order equations are often needed in mechanics, particularly to model situations which involve acceleration.

A reasonable model of a parachutist (with parachute open) is provided by treating the parachutist as a particle of mass m, subject to two forces: weight, mg downwards, and resistance, kv, against the motion (i.e. upwards – where v is the velocity and k is a constant).

Applying Newton's second law with the downward direction as positive gives:
$mg - kv = ma$

Using some standard notation

s = distance fallen

$v = \dfrac{ds}{dt}$ = velocity

$a = \dfrac{dv}{dt} = \dfrac{d^2s}{dt^2}$ = acceleration

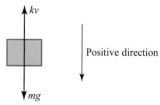

Figure 16.1

the equation can be written in several different ways:

$$1 \quad \frac{dv}{dt} = g - \frac{k}{m}v \qquad\qquad 2 \quad v\frac{dv}{ds} = g - \frac{k}{m}v \qquad\qquad 3 \quad \frac{d^2v}{dt^2} = g - \frac{k}{m}v$$

The first two equations are first order differential equations and can be solved using the method of separation of variables, covered in Chapter 12.

However, the third equation is of a type that is new to you. It is a second order differential equation, and this type of differential equation is the subject of this chapter.

Notation and vocabulary

As an example, the differential equation

$$\frac{d^3y}{dx^3} - 7\frac{d^2y}{dx^2} + 2\frac{dy}{dx} + 4y = 3\sin x$$

is described as:

- **third order** (because the highest derivative is a third derivative)
- **linear** (because where y and its derivatives appear they are to the power 1)
- having **constant coefficients** (because the coefficients of the terms involving y are constants – in this case 1, −7, 2 and 4)
- **non-homogeneous** (because the right-hand side, the part not containing y, is *not zero*). In cases where the right hand side *is zero* the differential equation is called **homogeneous**.

In this chapter you will meet **second order**, **linear differential equations**, with **constant coefficients**. In general, they can be written as

$$a\frac{d^2y}{dx^2} + b\frac{dy}{dx} + cy = f(x)$$

with a, b and c constant. This may, in fact be written without the a constant by dividing through by a, but the form given is helpful as it is very similar to the form of quadratic equations which you will be familiar with. The rest of this chapter concentrates on the method of solving this type of differential equation and its applications.

The auxiliary equation method

Before solving second and higher order differential equations you will find it helpful to look at the form of first order, linear, homogeneous differential equation with constant coefficients. In general they are of the form

| First order. | → | $\dfrac{dy}{dx} + ky = 0$ | ← | Homogeneous because the right-hand side is zero. |

| Constant coefficients because k is constant. |

You can solve this by separating the variables:

$$\frac{dy}{dx} = -ky$$

$$\int \frac{1}{y}\, dy = \int -k\, dx$$

$$\ln|y| = -kx + c$$

$$y = \pm e^{-kx+c}$$

$$= Ae^{-kx}$$

So, in general, the solution of *any* first order equation of this type will contain an *exponential function* and *one unknown constant*.

Knowing the form of the solution allows you to find it without doing any integration, as shown in the next example.

Example 16.1

Solve the differential equation

$$5\frac{dy}{dx} + y = 0$$

Solution

Since this is first order, linear, homogeneous and with constant coefficients, you know that the solution will be of the form

$$y = Ae^{\lambda x}$$

where λ and A are constants.

| Differentiating y with respect to x. |

| Substituting y and $\dfrac{dy}{dx}$ into the differential equation. |

If this is to be a solution, it must satisfy the original differential equation.

$$\frac{dy}{dx} = \lambda Ae^{\lambda x}$$

$$5\lambda Ae^{\lambda x} + Ae^{\lambda x} = 0$$

$$5\lambda + 1 = 0$$

| ← | Dividing by $Ae^{\lambda x}$ (since it is not zero) gives an equation just involving λ. |

$$\lambda = -\frac{1}{5}$$

So the general solution of the original differential equation is

$$y = Ae^{-\frac{1}{5}x}$$

Note

In Example 16.1 there cannot be any other solutions because this one already contains the one necessary arbitrary constant to give it the generality it needs.

This method of assuming the form of the solution is extremely powerful and is called the **auxiliary equation method**. The equation $5\lambda + 1 = 0$ is the **auxiliary equation**.

ACTIVITY 16.1

All of the following differential equations can be solved by at least one of the methods:

■ separation of variables

■ integrating factor

■ auxiliary equation.

(i) Which of the equations are linear? Which ones have constant coefficients?

(ii) For each equation, state which method (or methods) can be used and use it (or them) to solve the equation.

(a) $\dfrac{dy}{dx} - 17y = 0$

(b) $\dfrac{dy}{dx} - y^2 = 0$

(c) $\dfrac{dy}{dx} + xy = 0$

(d) $\dfrac{dy}{dx} - 3y = 0$

(e) $y\dfrac{dy}{dx} - y^2 = 0$

Second order homogenous differential equations

The ideas from the work above can be extended to cover second order equations.

Look at this differential equation. It is a second order, linear, homogeneous differential equation with constant coefficients.

$$\frac{d^2y}{dx^2} - 5\frac{dy}{dx} + 6y = 0$$

Suppose that you assume a solution of the form $y = Ae^{\lambda x}$ (just as before), where A and λ are constants. Then

$$\frac{dy}{dx} = A\lambda e^{\lambda x}$$

and

$$\frac{d^2y}{dx^2} = A\lambda^2 e^{\lambda x}$$

Substituting these into the differential equation gives

$$A\lambda^2 e^{\lambda x} - 5A\lambda e^{\lambda x} + 6Ae^{\lambda x} = 0$$

Dividing through by $Ae^{\lambda x}$ you obtain the auxiliary equation. It is a quadratic equation in λ:

$$\lambda^2 - 5\lambda + 6 = 0$$

Notice that the form of the auxiliary equation is very close to the form of the original differential equation, and with experience you will often write it down straight away, without the intermediate working shown above.

Factorising the auxiliary equation gives

$$(\lambda - 3)(\lambda - 2) = 0$$

You can see there are two different values for λ which satisfy the auxiliary equation: $\lambda = 3$ and $\lambda = 2$. So $y = Ae^{3x}$ and $y = Be^{2x}$ (with A and B constants) are two solutions of the differential equation.

Each of the two expressions Ae^{3x} and Be^{2x} is a **complementary function** of the differential equation. The sum of these expressions, $Ae^{3x} + Be^{2x}$, is usually called *the* complementary function, and is also a solution of the original differential equation. Since it has *two* arbitrary constants it is, in fact, the **general solution** of the original second order equation.

ACTIVITY 16.2

Verify that the complementary function $y = Ae^{3x} + Be^{2x}$ is a solution to the

original differential equation $\dfrac{d^2y}{dx^2} - 5\dfrac{dy}{dx} + 6y = 0$.

This method can be used on any linear differential equation with constant coefficients, whatever its order. If $\lambda_1, \lambda_2, \lambda_3, \ldots$ are the roots of the auxiliary equation, then, assuming there are no repeated roots, the general solution of the homogeneous differential equation is

$$y = Ae^{\lambda_1 x} + Be^{\lambda_2 x} + Ce^{\lambda_3 x} + \ldots$$

There are three important points to notice about the auxiliary equation method:

- this method does not involve any integration
- the number of terms in the complementary function is equal to the order of the differential equation
- the number of unknown constants in the solution is equal to the order of the differential equation.

In general, a second order, linear, homogenous differential equation with constant coefficients

$$a\frac{d^2y}{dx^2} + b\frac{dy}{dx} + cy = 0$$

has auxiliary equation

$$a\lambda^2 + b\lambda + c = 0.$$

There are three types of solution to this equation and they each lead to different types of solution:

- When $b^2 > 4ac$, there are real, distinct roots,
  e.g. $x^2 + 3x - 10 = 0$ has roots $x = 2, \ x = -5$.
- When $b^2 = 4ac$ there are real, repeated roots,
  e.g. $x^2 + 4x + 4 = 0$ has root $x = -2$ (twice).
- When $b^2 < 4ac$ there are complex, conjugate roots,
  e.g. $x^2 - 2x + 2 = 0$ has roots $x = 1 + i, \ x = 1 - i$.

In this section you will look at the first two cases, and you will also see how you can find values for the arbitrary constants to give a **particular solution**.

Auxiliary equation with real distinct roots, $b^2 > 4ac$

$$a\frac{d^2y}{dx^2} + b\frac{dy}{dx} + cy = 0$$

You have already seen an example of this situation at the start of this section. In general, if $b^2 > 4ac$ there will be two distinct roots of the auxiliary equation, λ_1 and λ_2, which lead to the complementary function:

$$y = Ae^{\lambda_1 x} + Be^{\lambda_2 x}$$

This complementary function is the general solution in this case, but to find a particular solution you need to eliminate the two unknown constants. This is only possible with two extra pieces of information, which may come in two different ways:

- Initial conditions:
  If the two conditions are given for the same value of the independent variable (often x), you say that you have two **initial conditions**.
  For example: $y = 0$ when $x = 0$, and $\frac{dy}{dx} = 1$ when $x = 0$. Since both conditions are for $x = 0$, these are initial conditions and the problem is called an **initial value problem**.

- Boundary conditions:
  If the two conditions are given for *different* values of the independent variable, you say that you have two **boundary conditions**.
  For example: $y = 0$ when $x = 0$, and $y = 1$ when $x = 1$. Since these use different x-values they are boundary conditions and the problem is called a **boundary value problem**. Often the solutions to boundary value problems are restricted to the region between the boundary points in the conditions.

The following examples demonstrate one of each type of problem.

Example 16.2

(i) Find the particular solution of $\frac{d^2y}{dx^2} + 4\frac{dy}{dx} + 3y = 0$

subject to the conditions $y = 0$ and $\frac{dy}{dx} = 1$ when $x = 0$.

(ii) Sketch the graph of the solution.

Solution

(i)   Auxiliary equation:

$$\lambda^2 + 4\lambda + 3 = 0$$

Factorise:

$$(\lambda + 3)(\lambda + 1) = 0$$

Solving this gives two distinct roots, $\lambda_1 = -1$ and $\lambda_2 = -3$.

Complementary function:

$$y = Ae^{-x} + Be^{-3x}$$

and this is also the general solution.

To find the values of A and B you need to use the conditions $y = 0$ and $\dfrac{dy}{dx} = 1$ when $x = 0$.

> You need to differentiate your general solution in order to use the $\dfrac{dy}{dx}$ condition.

$$\frac{dy}{dx} = -Ae^{-x} - 3Be^{-3x}$$

When $x = 0$, $y = 0$ $\Rightarrow A + B = 0$

When $x = 0$, $\dfrac{dy}{dx} = 1$ $\Rightarrow -A - 3B = 1$

Solving these two equations simultaneously gives

$$A = \frac{1}{2} \text{ and } B = -\frac{1}{2}$$

So the particular solution is

$$y = \frac{1}{2}e^{-x} - \frac{1}{2}e^{-3x}$$

(ii)

> You know that the curve passes through the origin, with positive gradient there.

> You also know that for positive x, e^{-x} and e^{-3x} are both positive, and $e^{-x} > e^{-3x}$. So the curve is above the x-axis for positive x.

> You also know that for large values of x, both e^{-x} and e^{-3x} tend to zero.

Note

You can check the shape of the graph using a graphical calculator or graphing software.

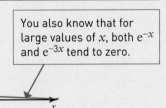

Figure 16.2

Example 16.3

The differential equation

$$\frac{d^2y}{dx^2} + 4\frac{dy}{dx} + 3y = 0$$

is used to model a situation in which the value of x lies between 0 and 2.

(i) Find the particular solution given the boundary conditions $y = 0$ at $x = 0$, and $y = 1$ when $x = 2$.

(ii) Sketch a graph of the solution for $0 \leqslant x \leqslant 2$.

Solution

(i) The general solution is the same as in the previous example:

$$y = Ae^{-x} + Be^{-3x}$$

Using the boundary conditions:

$y = 0, x = 0$ gives $0 = A + B$

$y = 1, x = 2$ gives $1 = Ae^{-2} + Be^{-6}$

Solving these simultaneously gives

> [!NOTE]
> Substituting for B from the first equation into the second.

$$\rightarrow 1 = Ae^{-2} - Ae^{-6}$$

$$A = \frac{e^2}{1 - e^{-4}}$$

and then

$$B = -\frac{e^2}{1 - e^{-4}}$$

The particular solution then is

$$y = \frac{e^2}{1 - e^{-4}}(e^{-x} - e^{-3x})$$

Note

Note that you have two points defined as the boundary conditions, but the same reasoning as before about the variable terms.

(ii)

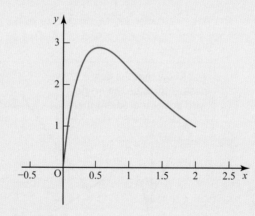

Figure 16.3

Discussion point

→ The last two examples started off exactly the same. However, when it came to finding the particular solutions, they required different approaches because of the different types of conditions involved. Which is the easier to work with?

Auxiliary equation with repeated roots, $b^2 = 4ac$

$$a\frac{d^2y}{dx^2} + b\frac{dy}{dx} + cy = 0$$

If $b^2 = 4ac$ then there is only one root from the auxiliary equation, and it is $\lambda = -\frac{b}{2a}$. (Check this with the quadratic formula).

This leads to a complementary function of $y = Ae^{\lambda x}$. But, since the original differential equation is second order, the general solution needs to have *two* arbitrary constants. So, you need to find another solution to the equation in order to provide this extra constant.

At this point, if someone suggested another possible solution of the form $y = Bxe^{\lambda x}$ (note the extra x in front), you could test to see if it satisfies the original differential equation:

Using the product rule

$$\frac{dy}{dx} = Be^{\lambda x} + B\lambda xe^{\lambda x}$$

and differentiating again

$$\frac{d^2y}{dx^2} = B\lambda e^{\lambda x} + B\lambda e^{\lambda x} + B\lambda^2 xe^{\lambda x}$$

$$= 2B\lambda e^{\lambda x} + B\lambda^2 xe^{\lambda x}.$$

If $y = Bxe^{\lambda x}$ is a solution, you should get 0 when you substitute these expressions into the left-hand side of the original equation. You get

$$a\left(2B\lambda e^{\lambda x} + B\lambda^2 xe^{\lambda x}\right) + b\left(Be^{\lambda x} + B\lambda xe^{\lambda x}\right) + c\left(Bxe^{\lambda x}\right)$$

$$= Be^{\lambda x}\left(2a\lambda + b + x\left(a\lambda^2 + b\lambda + c\right)\right).$$

But, in this case, $\lambda = -\frac{b}{2a}$, which rearranges to give $2a\lambda = -b$, meaning that the first two terms in the above bracket sum to zero. (Notice this only works in this case because of the value of λ).

The last bracket (the coefficient of x) is $a\lambda^2 + b\lambda + c$, which is the left-hand side of the auxiliary equation, and therefore equal to zero too.

The whole expression is therefore zero, which means the expression $y = Bxe^{\lambda x}$ *does* satisfy the original differential equation, and so is a complementary function – and it is different from your first solution $y = Ae^{\lambda x}$.

So, you can combine them to form the complementary function (which is also the general solution), with two arbitrary constants:

$$y = Ae^{\lambda x} + Bxe^{\lambda x}$$

or

$$y = (A + Bx)e^{\lambda x}$$

See question 11 in Exercise 16.1 for a way to derive this result.

Example 16.4

(i) Find the particular solution of the equation

$$\frac{d^2z}{dt^2} + 6\frac{dz}{dt} + 9z = 0$$

subject to the conditions $z = 0$ and $\frac{dz}{dt} = 5$ when $t = 0$.

(ii) Sketch the graph of the particular solution.

Solution

(i) Auxiliary equation:

$$\lambda^2 + 6\lambda + 9 = 0$$

$$(\lambda + 3)^2 = 0$$

$$\lambda = -3 \text{ (repeated)}$$

So the general solution is

$$z = (A + Bt)e^{-3t}$$

Differentiating this in order to use the condition gives

$$\frac{dz}{dt} = Be^{-3t} - 3(A + Bt)e^{-3t}$$

Using the conditions

$$0 = (A + B \times 0) \times 1$$

$$A = 0$$

and

$$5 = B - 3A$$

$$B = 5$$

So, the particular solution is

$$z = 5te^{-3t}$$

(ii)

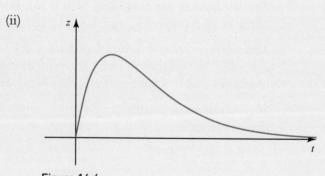

Figure 16.4

> **Note**
>
> For all graphs of the form $y = xe^{-kx}$, $y \to 0$ as $x \to \infty$. You are multiplying an increasing function, x, by a decreasing function, e^{-kx}, and in this case the decreasing one wins.

To summarise: if the auxiliary equation has a repeated real root λ, then the complementary function is

$$y = (A + Bx)e^{\lambda x}$$

① Use the auxiliary equation method to find the general solution of the following differential equations.

(i) $\dfrac{dy}{dx} - 3y = 0$ (ii) $\dfrac{dy}{dx} + 7y = 0$ (iii) $\dfrac{dx}{dt} + x = 0$

(iv) $\dfrac{dp}{dt} - 0.02p = 0$ (v) $5\dfrac{dz}{dt} - z = 0$

② Use the auxiliary equation method to find the particular solutions of the following differential equations.

(i) $\dfrac{dy}{dx} + 2y = 0$ $y = 3$ when $x = 0$

(ii) $2\dfrac{dy}{dx} - 5y = 0$ $y = 1$ when $x = 0$

(iii) $3\dfrac{dx}{dt} + x = 0$ $x = 2$ when $t = 1$

(iv) $\dfrac{dP}{dt} = kP$ $P = P_0$ when $t = 0$

(v) $\dfrac{dm}{dt} = -km$ $m = m_0$ when $t = 0$

③ Find the general solution of the following differential equations.

(i) $\dfrac{d^2y}{dx^2} - 16y = 0$ (ii) $\dfrac{d^2y}{dx^2} + \dfrac{dy}{dx} - 5y = 0$

(iii) $2\dfrac{d^2x}{dt^2} + 4\dfrac{dx}{dt} + x = 0$ (iv) $7\dfrac{d^2y}{dx^2} + 2\dfrac{dy}{dx} = 0$

(v) $9\dfrac{d^2y}{dx^2} - 12\dfrac{dy}{dx} + 4y = 0$

④ You are given the differential equation $\dfrac{d^2y}{dx^2} - 3\dfrac{dy}{dx} + 2y = 0$.

(i) Write down the auxiliary equation in terms of λ.

(ii) Solve the auxiliary equation to find its two real roots.

(iii) Write down the complementary function in the form $y(x) = \ldots$

(iv) Find the derivative of the complementary function.

(v) You are also told that $y = 1$ and $\dfrac{dy}{dx} = 0$ when $x = 0$. Use these facts, the complementary function and its derivative, to eliminate the arbitrary constants and find the particular solution that satisfies these constraints.

⑤ Given $\dfrac{d^2y}{dx^2} - 8\dfrac{dy}{dx} + 16y = 0$

(i) write down the auxiliary equation, and solve it

(ii) write down the complementary function (this is also the general solution)

(iii) find the particular solution which satisfies the initial conditions $y = 1$ and $\dfrac{dy}{dx} = -1$ when $x = 0$.

⑥ Find the particular solutions of the following differential equations that satisfy the conditions given. In each case sketch the graph of the solution (you may wish to use graphing software to help visualise this). Assuming each one models a real system, where x represents a variable changing over time, use the graph to describe the motion of each system in words.

(i) $\dfrac{d^2x}{dt^2} + 5\dfrac{dx}{dt} = 0$ $\qquad\qquad$ $x = 0,\ \dfrac{dx}{dt} = 4$, when $t = 0$

(ii) $4\dfrac{d^2x}{dt^2} + 4\dfrac{dx}{dt} + x = 0$ $\qquad\qquad$ $x = 1,\ t = 0$ and $x = 1,\ t = 1$

⑦ In an electrical circuit, the charge q coloumbs on a capacitor is modelled by the differential equation

$$0.2\dfrac{d^2q}{dt^2} + \dfrac{dq}{dt} + 1.25q = 0.$$

Initially, $q = 2$ and $\dfrac{dq}{dt}$ (the current in amperes) is 4.

(i) Find an equation for the charge as a function of time.

(ii) Sketch the graphs of charge and current against time. Describe how the charge and current change.

(iii) What is the charge on the capacitor and the current in the circuit after a long period of time?

⑧ The temperature of a chemical undergoing a reaction is modelled by the differential equation

$$2\dfrac{d^2T}{dt^2} + \dfrac{dT}{dt} = 0$$

where T is the temperature in °C and t is the time in minutes.

For a particular experiment, the temperature is initially 50 °C, and it is 45 °C one minute later.

(i) Find an expression for the temperature T at any time.

(ii) What will the temperature be after two minutes?

(iii) Sketch a graph of T against t.

(iv) What is the steady state temperature?

⑨ The metal fins on a motorcycle engine help to cool it. A model for the change of temperature along a fin is given by $\dfrac{d^2T}{dx^2} - 4T = 0$, where T °C is the temperature and x m is the distance from the hot end.

(i) Find the general solution of this differential equation.

When the engine has been running for some time the temperatures at the two ends of the fin are 100 °C and 80 °C. The fin is 5 cm long.

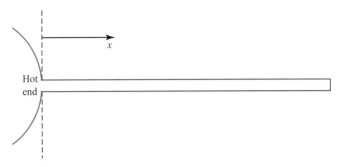

Figure 16.5

(ii) Find the particular solution.

(iii) Hence determine the temperature 3 cm from the hot end.

⑩ A car of mass 800 kg is travelling along a road at a constant velocity of 100 km h⁻¹. A catastrophic engine and brake failure renders the car unable to brake in any way other than natural slow-down from the friction from the air and road. The friction is modelled as a backwards force equal to 40 000 times the velocity (in km h⁻¹).

(i) Show that the equation of motion of the car as it slows (x in kilometres, t in hours), simplifies to $\dfrac{d^2x}{dt^2} + 50\dfrac{dx}{dt} = 0$.

(ii) Solve this differential equation, using appropriate initial conditions, to find a particular solution for x in terms of t as the distance (in km) travelled after the brake failure.

(iii) Find the time taken for the car to travel 1 km further, and the speed of the car at this moment.

(iv) State at least one problem with the solution given by this model.

⑪ In this question you will prove the result for the general solution of a second order differential equation for which the auxiliary equation has repeated roots.

(i) Find the general solution of the differential equation $\dfrac{d^2y}{dx^2} = 0$.

(ii) Assume that the differential equation

$$\frac{d^2y}{dx^2} + 4\frac{dy}{dx} + 4y = 0$$

has a solution of the form $y = f(x)e^{-2x}$. Differentiate this twice and substitute it into the original differential equation, and hence find the form for f(x).

(iii) Repeat for the differential equation $\dfrac{d^2y}{dx^2} + 2k\dfrac{dy}{dx} + k^2y = 0$.

2 Auxiliary equations with complex roots

$$a\frac{d^2y}{dx^2} + b\frac{dy}{dx} + cy = 0$$

If $b^2 < 4ac$ then the two roots will be complex (and conjugates of each other). For example, the differential equation

$$\frac{d^2y}{dx^2} + 4y = 0$$

has auxiliary equation

$$\lambda^2 + 4 = 0$$

The roots of this quadratic are $\lambda_1 = 2i$ and $\lambda_2 = -2i$ (where $i = \sqrt{-1}$).

The general solution of the differential equation is therefore

$$y = Ae^{2ix} + Be^{-2ix}.$$

The terms in the solution contain complex exponentials, and it is important to recognise that these can always be written in terms of sine and cosine functions (see Chapter 14) by using the relationships

$$e^{i\theta} = \cos\theta + i\sin\theta$$

$$e^{-i\theta} = \cos\theta - i\sin\theta.$$

So you can rewrite the general solution as

$$y = A(\cos 2x + i\sin 2x) + B(\cos 2x - i\sin 2x)$$
$$= (A + B)\cos 2x + i(A - B)\sin 2x.$$

Note

Notice that complex numbers are needed to find this form for the general solution, and yet this abstract result actually works as a model for oscillating systems in the real world.

Notice now that the coefficients of $\cos 2x$ and $\sin 2x$ are constants, which you can call P and Q, such that $P = A + B$ and $Q = i(A - B)$. This means the general solution can be written simply as

$$y = P\cos 2x + Q\sin 2x.$$

It is important to note that this solution has an oscillating nature, and in fact all differential equations with complex roots from their auxiliary equation will give oscillating solutions.

Simple harmonic motion

Many real-life situations can be modelled by a differential equation of the form

$$\frac{d^2x}{dt^2} = -\omega^2 x.$$

You will now be introduced to the mathematics of elastic strings and springs which will give you the key skills for modelling physical situations.

Hooke's law

In 1678, Robert Hooke formulated a *rule* or *law of nature in every springing body* which, for small extensions relative to the length of the string or spring, can be stated as follows.

> The tension in an elastic spring or string is proportional to the extension.
>
> If a spring is compressed the thrust is proportional to the decrease in length of the spring.

When a string or spring is described as elastic, it means that it is reasonable to apply the modelling assumption that it obeys Hooke's law. A further assumption that it is light (i.e. has zero mass) is usual and is made in this book.

There are three ways in which Hooke's law is commonly expressed for a string.

Which one you use depends on the extent to which you are interested in the string itself rather than just its overall properties. Denoting the natural length of the string by l_0 and its area of cross-section by A, the different forms are as follows.

- $T = \dfrac{EA}{l_0} x$ In this form, E is called **Young's modulus** and is a property of the material out of which the string is formed. This form is commonly used in physics and engineering, subjects in which properties of materials are studied. It is rarely used in mathematics. The S.I. unit for Young's modulus is $\mathrm{N\,m^{-2}}$.

- $T = \dfrac{\lambda}{l_0} x$ The constant λ is called the **modulus of elasticity** of the string and will be the same for any string of a given cross-section made out of the same material. Many situations require knowledge of the natural length of a string and this form may well be the most appropriate in such cases. The S.I. unit for the modulus of elasticity is N.

- $T = kx$ In this simplest form, k is called the **stiffness** of the string. It is a property of the string as a whole. You may choose to use this form if neither the natural length nor the cross-sectional area of the string is relevant to the situation. The S.I. unit for stiffness is $\mathrm{N\,m^{-1}}$.

Notice that $k = \dfrac{\lambda}{l_0} = \dfrac{EA}{l_0}$.

In this book, only the forms using stiffness $T = kx$ are used, and this can be applied to springs as well as strings.

The spring–mass oscillator

The spring–mass oscillator consists of an elastic spring attached to a fixed point, and at the other end is attached an object. The system is hung vertically as shown in Figure 16.6. If the object is displaced vertically from the equilibrium position and then released, the object oscillates along a straight vertical line.

Before you can analyse this motion you need to make some modelling assumptions:

- the effects of friction in the spring and air resistance on the object may be neglected

- the spring is light and perfectly elastic: this means that the tension in the spring is proportional to its extension.

The constant of proportionality between the tension and the extension of the spring is called its **stiffness**, and is denoted by k. The natural length of the spring is l_0, and the mass of the object is m. You can find the extension, e, of the spring when the object is in equilibrium by looking at the forces acting on the object in equilibrium (Figure 16.7).

Figure 16.6

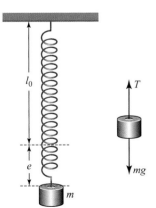

Figure 16.7

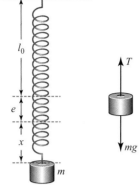

Figure 16.8

There is a downward force mg due to gravity, and a balancing upward tension T in the spring. The tension is given by $T = ke$.

Since the system is in equilibrium

$$ke = mg$$
$$\Rightarrow e = \frac{mg}{k}$$

Now look at the system when the spring has a further extension x below the equilibrium position, which means the total extension is $e + x$ (Figure 16.8). The tension T in the spring is now $T = k(e + x)$.

The object is not in equilibrium, and its acceleration in the downward direction (the direction of increasing x) is given by

$$\frac{\mathrm{d}^2 x}{\mathrm{d}t^2}$$

Applying Newton's second law gives the differential equation

$$
\begin{aligned}
m\frac{\mathrm{d}^2 x}{\mathrm{d}t^2} &= mg - T \\
&= mg - k(e + x) \\
&= mg - ke - kx
\end{aligned}
$$

But you have already seen that $mg - ke = 0$.

Substituting this into the differential equation and dividing by m gives the equation of motion

$$\frac{\mathrm{d}^2 x}{\mathrm{d}t^2} + \frac{k}{m}x = 0 \quad \text{or} \quad \frac{\mathrm{d}^2 x}{\mathrm{d}t^2} = -\frac{k}{m}x.$$

This differential equation models the motion of the idealised spring–mass oscillator.

For example:

- A mass fixed to the end of a spring is pulled down vertically and then released. The displacement x of the mass from its equilibrium position, at time t, can be modelled as

$$\frac{\mathrm{d}^2 x}{\mathrm{d}t^2} = -\frac{k}{m}x$$

where k is the stiffness of the spring and m is the mass.

- An alternative situation, that is also interesting, is where a mass is suspended from a fixed point, then is moved horizontally slightly from its equilibrium position and released, forming a simple pendulum. The angle θ which the string makes with the vertical at time t, can be modelled as

> The right-hand side of this equation should really involve $\sin \theta$ but the small angle approximation of $\sin \theta \approx \theta$ is very accurate and is used in this model.

$$\frac{\mathrm{d}^2 \theta}{\mathrm{d}t^2} = -\frac{g}{l}\theta$$

where l is the length of the string and g is the acceleration due to gravity.

Both these equations have the same form:

$$\frac{\mathrm{d}^2 x}{\mathrm{d}t^2} = -\omega^2 x$$

In the first case $\omega^2 = \frac{k}{m}$, in the second $\omega^2 = \frac{g}{l}$. In the second case the variable is θ rather than x.

In both cases the motion is called **simple harmonic motion** (SHM); it provides a model for many oscillations.

The general solution of the differential equation, as shown above, is

$$x = P\cos\omega t + Q\sin\omega t$$

The constants of integration, P and Q, are unknown at this stage, but if you know suitable initial or boundary conditions you can calculate their values.

ACTIVITY 16.3

Given that the general solution $x = P\cos\omega t + Q\sin\omega t$ can be written in the form $x = R\sin(\omega t + \varepsilon)$, use the compound angle formulae to prove that

$$R = \sqrt{P^2 + Q^2} \text{ and } \tan\varepsilon = \frac{Q}{P}.$$

Expressing the general solution $x = P\cos\omega t + Q\sin\omega t$ in the form $x = R\sin(\omega t + \varepsilon)$ tells you a lot about the solution.

- Since the sine function varies between $+1$ and -1, the solution varies between $-a$ and a.

- Since the sine function is periodic with period 2π, the solution is periodic with period $\frac{2\pi}{\omega}$.

ACTIVITY 16.4

Use graphing software to investigate the graph $x = R\sin(\omega t + \varepsilon)$ for different values of a, ω and ε.

In the activity above you should have noticed that the effect of ε is to translate the sine curve to the left by an amount $\frac{\varepsilon}{\omega}$, as shown in the diagram below. The quantity ε is called the **phase shift**.

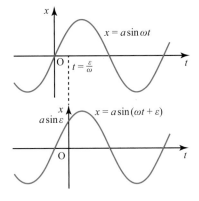

Figure 16.9

Example 16.5

(i) Find the general solution of the differential equation $\dfrac{d^2x}{dt^2} + 4x = 0$.

(ii) Find the particular solution in the case for which $x = 2$ and $\dfrac{dx}{dt} = -2$ when $t = 0$.

(iii) Find the period and amplitude of the oscillations, and sketch the graph of the particular solution.

Solution

(i) Auxiliary equation: $\qquad\qquad \lambda^2 + 4 = 0$

$\qquad\qquad\qquad\qquad\qquad\qquad \lambda = \pm 2i$

General solution: $\qquad\qquad x = A\cos 2t + B\sin 2t$

(ii) When $t = 0$, $x = 2$ $\qquad\qquad \Rightarrow 2 = A$

$$\frac{dx}{dt} = -2A\sin 2t + 2B\cos 2t$$

When $t = 0$, $\dfrac{dx}{dt} = -2$ $\qquad \Rightarrow -2 = 2B \quad \Rightarrow B = -1$

Particular solution: $\qquad\qquad x = 2\cos 2t - \sin 2t$

> To sketch the graph and find the amplitude, it is helpful to write the solution in the form $a\sin(2t + \varepsilon)$.

> Using the compound angle formula.

(iii) The period of the oscillations is $\dfrac{2\pi}{\omega} = \dfrac{2\pi}{2} = \pi$

$2\cos 2t - \sin 2t = a\sin(2t + \varepsilon)$

$\qquad\qquad\qquad\qquad = a\sin 2t\cos\varepsilon + a\cos 2t\sin\varepsilon$

Comparing coefficients of $\sin 2t$: $\qquad -1 = a\cos\varepsilon$

Comparing coefficients of $\cos 2t$: $\qquad 2 = a\sin\varepsilon$

So $a = \sqrt{2^2 + 1^2} = \sqrt5$.

The amplitude of the oscillations is $\sqrt5$.

$\tan\varepsilon = -2 \quad \Rightarrow \varepsilon = -1.107$

The particular solution can be written in the form $x = \sqrt5\sin(2t - 1.107)$.

> **Note**
> ----------
> The graph is obtained from the graph of $y = \sin t$ by
> - translation through 1.107 to the right
> - stretch scale factor 0.5 in the x-direction
> - stretch scale factor $\sqrt5$ in the y-direction.

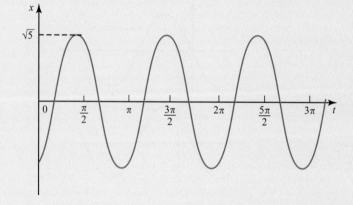

Figure 16.10

In the previous examples the roots of the auxiliary equation were purely imaginary. Often the roots also have a real part.

For example, suppose that the roots of the auxiliary equation are $2 \pm 3i$.

This gives the complementary function

$$y = Ae^{(2+3i)x} + Be^{(2+3i)x}$$

$$= e^{2x}\left(Ae^{3ix} + Be^{-3ix}\right).$$

> Notice that e^{2x} is a common factor.

But, as before, you can write $Ae^{3ix} + Be^{-3ix}$ as $P\cos 3x + Q\sin 3x$, so

$$y = e^{2x}\left(P\cos 3x + Q\sin 3x\right).$$

> Notice that the real part of the roots ends up in the exponential outside the bracket, and the imaginary part ends up as the coefficient inside the trigonometric terms.

This is the standard form of the general solution.

> **Note**
>
> The four forms of complementary function, for the four cases of roots of the auxiliary equation are:
> - Real distinct roots, λ_1 and λ_2: $y = Ae^{\lambda_1 x} + Be^{\lambda_2 x}$
> - Real, repeated root, λ: $y = (A + Bx)e^{\lambda x}$
> - Complex roots, $\alpha \pm \beta i$: $y = e^{\alpha x}(P\cos \beta x + Q\sin \beta x)$
> - Pure imaginary roots, $\pm \omega i$ $y = P\cos \omega x + Q\sin \omega x$

Example 16.6

(i) Find the particular solution of the differential equation

$$\frac{d^2 y}{dx^2} + 2\frac{dy}{dx} + 5y = 0$$

which satisfies the initial conditions $y = 0$ and $\dfrac{dy}{dx} = 1$ when $x = 0$.

(ii) Sketch the graph of this particular solution.

Solution

(i) The auxiliary equation is

$$\lambda^2 + 2\lambda + 5 = 0$$

Completing the square gives

$$(\lambda + 1)^2 - 1 + 5 = 0$$

$$\lambda + 1 = \pm\sqrt{-4}$$

$$\lambda = -1 \pm 2i$$

> Alternatively you could use the quadratic formula.

The general solution is $y = e^{-x}\left(P\cos 2x + Q\sin 2x\right)$.

To find the particular solution:

When $x = 0, y = 0 \implies P = 0$

So the general solution is now

$$y = Qe^{-x} \sin 2x$$

$$\Rightarrow \frac{dy}{dx} = -Qe^{-x} \sin 2x + 2Qe^{-x} \cos 2x$$

When $x = 0$, $\dfrac{dy}{dx} = 1 \quad \Rightarrow 1 = 2Q$

$$\Rightarrow Q = \frac{1}{2}$$

So the particular solution for these initial conditions is

$$y = \frac{1}{2}e^{-x} \sin 2x$$

(ii)   The graph of this particular solution is shown below. Notice that the oscillating solution moves between the curves $= \pm\dfrac{1}{2}e^{-x}$, which are indicated by the blue dotted lines.

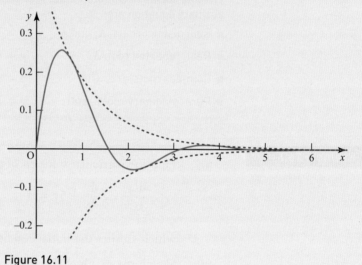

Figure 16.11

It is important to understand the relationship between the solution and its graph. In Example 16.6:

- the e^{-x} factor tells you that the solution will decay as x increases (because of the negative sign)

- the $\sin 2x$ factor tells you that there is an oscillation.

This form of solution can be described as **exponentially decaying oscillations** or **damped oscillations**, and it arises frequently when modelling real oscillating systems.

Damped oscillations

Simple harmonic motion has constant amplitude and goes on forever. For many real oscillating systems, SHM is not a very good model: usually the amplitude of the oscillations gradually decreases, and the motion dies away.

Example 16.7

A damped oscillating system is modelled by the differential equation

$$\frac{d^2x}{dt^2} + k\frac{dx}{dt} + 25x = 0$$

where k is a constant that can be varied.

When $t = 0$, $x = 0$ and $\frac{dx}{dt} = 1$.

Solve the differential equation for each of the following values of k, and sketch the graph of the solution in each case.

(i) $k = 26$

(ii) $k = 6$

(iii) $k = 10$

Solution

(i)   Auxiliary equation: $\lambda^2 + 26\lambda + 25 = 0$

$$(\lambda + 1)(\lambda + 25) = 0$$

$$\lambda = -1 \text{ or } -25$$

General solution: $x = Ae^{-t} + Be^{-25t}$

When $t = 0$, $x = 0$ $\Rightarrow A + B = 0$

$$\frac{dx}{dt} = -Ae^{-t} - 25Be^{-25t}$$

When $t = 0$, $\frac{dx}{dt} = 1$ $\Rightarrow -A - 25B = 1$

Solving these equations simultaneously gives $A = \frac{1}{24}$, $B = -\frac{1}{24}$.

Particular solution is $x = \frac{1}{24}e^{-t} - \frac{1}{24}e^{-25t}$.

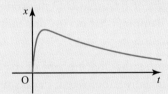

Figure 16.12

(ii)   Auxiliary equation: $\lambda^2 + 6\lambda + 25 = 0$

$$\lambda = \frac{-6 \pm \sqrt{36 - 100}}{2} = -3 \pm 4i$$

General solution: $x = e^{-3t}(A\sin 4t + B\cos 4t)$

When $t = 0$, $x = 0$ $\Rightarrow B = 0$

$$x = Ae^{-3t}\sin 4t$$

$$\frac{dx}{dt} = -3Ae^{-3t}\sin 4t + 4Ae^{-3t}\cos 4t$$

When $t = 0$, $\frac{dx}{dt} = 1$ $\Rightarrow 4A = 1 \Rightarrow A = \frac{1}{4}$

Particular solution is $x = \frac{1}{4}e^{-3t}\sin 4t$.

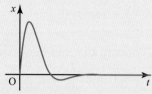

Figure 16.13

(iii) Auxiliary equation: $\lambda^2 + 10\lambda + 25 = 0$

$$(\lambda + 5) = 0$$

$$\lambda = -5$$

General solution: $x = (A + Bt)e^{-5t}$

When $t = 0$, $x = 0$ $\Rightarrow A = 0$ $x = Bte^{-5t}$

$$\frac{dx}{dt} = Be^{-5t} - 5Bte^{-5t}$$

When $t = 0$, $\frac{dx}{dt} = 1$ $\Rightarrow B = 1$

Particular solution is $x = te^{-5t}$.

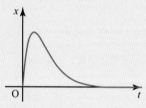

Figure 16.14

The three parts of Example 16.7 illustrate the three different types of damping. The type of damping that occurs depends on the nature of the roots of the auxiliary equation.

■ Overdamping: the discriminant of the auxiliary equation is positive and the roots are negative; the system decays without oscillating.

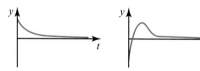

Figure 16.15

■ Underdamping: the discriminant of the auxiliary equation is negative and oscillations occur.

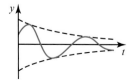

Figure 16.16

- Critical damping: the discriminant of the auxiliary equation is zero.

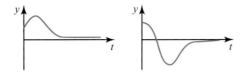

Figure 16.17

Critical damping is the borderline between overdamping and underdamping. It is not obvious in a physical situation when damping is critical, since the pattern of motion for critical damping can be very similar to that in the overdamped case.

Improving the physical model

When you find that a model is unsatisfactory, you need to look again at your assumptions. In this case, the assumption that you need to question is that the effects of air resistance and friction can be neglected. Real oscillating systems are almost always **damped**, that is they are affected to some degree by the resistive forces of friction and/or air resistance. They perform **damped oscillations**.

In many systems the damping force is proportional to the speed of the object. This is often represented on a diagram by a device called a linear dashpot, as shown in Figure 16.18.

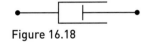

Figure 16.18

A dashpot exerts a force on the system that is proportional to the rate at which it is being extended or compressed, and which acts in the direction opposite to that of the motion. This is illustrated in Figure 16.19.

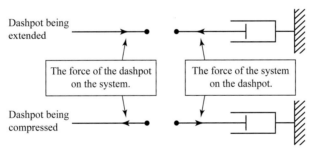

Figure 16.19

The force R that the dashpot exerts on the system at time t is given by

$$R = r \, \frac{\mathrm{d}L}{\mathrm{d}t}$$

where the constant of proportionality r is called the **dashpot constant** (or the **damping constant**). The amount of travel still left in the dashpot (see Figure 16.20) is denoted by L.

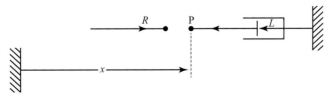

Figure 16.20

It is important to be clear in your mind about the direction of the force R and the signs involved. Look at the point P on the moving part of the dashpot.

- When P is moving from right to left, L is increasing; $\dfrac{dL}{dt}$ is positive and the force is in the same direction as that marked for R in Figure 16.20. The dashpot is opposing the right to left motion.

- When P is moving from left to right, L is decreasing; $\dfrac{dL}{dt}$ is negative and the force is in the opposite direction to that marked for R in Figure 16.20. The dashpot is now opposing the left to right motion.

Thus the sign of the dashpot force looks after itself as the motion changes.

However you will not usually be interested in the quantity L so much as the distance of the point P from some fixed point of the system. This distance is shown as x in Figure 16.20. All the systems that you will meet in this book are set up so that as x increases, L decreases, and vice versa, so

$$\frac{dx}{dt} = -\frac{dL}{dt}$$

Consequently, the force that the dashpot exerts on the system is given by

$$R = -r\,\frac{dx}{dt}$$

in the direction of increasing x.

> ### Note
>
> Here we are using a dashpot as a symbol to represent damping effects but a dashpot is actually a real device that is often used in man-made systems. It is used when a certain (predictable) amount of damping, over and above that provided by friction and air resistance, is desired. For example, a car's shock-absorber includes a linear dashpot, as shown in Figure 16.21.
>
> Such a dashpot consists of a cylinder containing a viscous liquid, often oil. When the cylinder is compressed or extended a disc moves along the cylinder and is opposed by a force that is proportional to the speed of compression or extension. The constant of proportionality depends on the viscosity of the liquid.
>
>
>
> **Figure 16.21**

Example 16.8

A simple oscillating system is being modelled as a damped spring–mass oscillator in which an object of mass $2\,\text{kg}$ is attached to fixed points by a spring of natural length $0.5\,\text{m}$, stiffness $20\,\text{N}\,\text{m}^{-1}$ and by a dashpot of constant $12\,\text{N}\,\text{m}^{-1}\,\text{s}$. The spring–mass–dashpot system lies on a smooth horizontal surface, as shown in Figure 16.22.

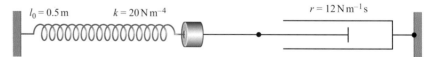

Figure 16.22

(i) Formulate the differential equation of motion for this system.

The system is released from rest when the spring length is 0.6 m.

(ii) Find the particular solution of the differential equation that models this situation.

(iii) State the initial amplitude and the period of the motion. What happens to the amplitude of the oscillations?

Solution

(i) Figure 16.23 shows the spring–mass–dashpot system at some general time t (seconds), when the extension of the spring is x. The horizontal forces are the tension in the spring, T, and the damping force R.

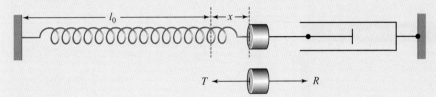

Figure 16.23

The tension in the spring is $T = kx = 20x$.

The dashpot force is $R = -r\dfrac{dx}{dt} = -12\dfrac{dx}{dt}$.

Applying Newton's second law $F = ma$ at any instant gives

> This side is ma.

$$2\frac{d^2x}{dt^2} = -12\frac{dx}{dt} - 20x$$

> This side is F, where $F = R - T$.

Dividing both sides by 2, and rearranging, you obtain the equation of motion of the spring–mass–dashpot system:

> Notice that you again have a linear homogeneous second order equation with constant coefficients.

$$\frac{d^2x}{dt^2} + 6\frac{dx}{dt} + 10x = 0$$

(ii) The auxiliary equation for the differential equation is

$$\lambda^2 + 6\lambda + 10 = 0$$

$$\Rightarrow \quad \lambda = \frac{-6 \pm \sqrt{6^2 - 40}}{2} = -3 \pm i$$

The general solution of the differential equation is $x = ae^{-3t}\sin(t + \varepsilon)$.

At the start of the motion, the length of the spring is 0.6 m and the object is at rest, so the initial conditions are $x = 0.1$ and $\dfrac{dx}{dt} = 0$ when $t = 0$.

When $t = 0$, $x = 0.1$ $\quad \Rightarrow \quad$ $a\sin\varepsilon = 0.1$ $\qquad\qquad$ ①

By differentiating the general solution you obtain

$$\frac{dx}{dt} = -3e^{-3t}\,a\sin(t + \varepsilon) + ae^{-3t}\cos(t + \varepsilon).$$

When $t = 0$, $\dfrac{dx}{dt} = 0$ $\quad \Rightarrow \quad$ $0 = -3a\sin\varepsilon + a\cos\varepsilon$

$$\Rightarrow \quad \tan\varepsilon = \frac{1}{3}$$

From ① you can see that $\sin \varepsilon = \dfrac{0.1}{a}$, which is positive, and so ε must be an angle in the first quadrant.

$$\Rightarrow \qquad \varepsilon = 0.322 \text{ (radians) and } a = \frac{0.1}{\sin 0.322} = 0.316.$$

The particular solution in this case is

$$x = 0.316\mathrm{e}^{-3t}\sin(t + 0.322)$$

(iii) The initial amplitude of the motion is 0.316 m (to 3 decimal places) and the period in seconds is $\dfrac{2\pi}{1} = 2\pi$.

The amplitude decays exponentially. In this case the oscillation decays very quickly.

Exercise 16.2

① For the differential equation

$$\frac{\mathrm{d}^2 x}{\mathrm{d}t^2} + 9x = 0$$

(i) Write down the general solution of the differential equation.

(ii) Given the initial conditions $x = 0$ and $\dfrac{\mathrm{d}x}{\mathrm{d}t} = 1$ when $t = 0$, find the particular solution.

(iii) Write down the period and amplitude of the oscillations, and sketch a graph of the solution.

② A spring–mass oscillator consists of a spring of stiffness $32\,\mathrm{N\,m^{-1}}$, natural length 0.5 m and an object of mass 0.5 kg. The oscillator lies on a smooth horizontal table with one end of the spring attached to a fixed point. The object is constrained to move in a straight line, and friction can be taken to be negligible.

Initially the object is at rest, and the spring is 0.6 m in length.

(i) Formulate the differential equation for the system and write down the initial conditions in this case.

(ii) Find the particular solution describing the motion.

(iii) Write down the period and amplitude of the motion.

(iv) Sketch a graph of the solution.

③ For the differential equation

$$4\frac{\mathrm{d}^2 x}{\mathrm{d}t^2} + x = 0$$

(i) Write down the general solution of the differential equation.

(ii) Given the initial conditions $x = 4$ and $\dfrac{\mathrm{d}x}{\mathrm{d}t} = 0$ when $t = 0$, find the particular solution.

(iii) Write down the period and amplitude of the oscillations, and sketch a graph of the solution.

④ A spring–mass oscillator consists of a spring of natural length $0.2\,\text{m}$ and stiffness $25\,\text{N m}^{-1}$ and an object of mass $0.4\,\text{kg}$. One end of the spring is attached to a fixed point and the system hangs vertically.

(i) Find the length of the spring when the system is in equilibrium.

The object is pulled down $10\,\text{cm}$ from its equilibrium position and released.

(ii) Formulate the differential equation for the system and write down the initial conditions in this case.

(iii) Find the particular solution describing the motion.

(iv) Write down the period and amplitude of the motion.

(v) Describe the motion of the object.

⑤ Find the general solution of each of the following differential equations.

(i) $\dfrac{d^2 y}{dx^2} - 4\dfrac{dy}{dx} + 5y = 0$

(ii) $\dfrac{d^2 y}{dx^2} - 2\dfrac{dy}{dx} + 5y = 0$

(iii) $\dfrac{d^2 x}{dt^2} + 2\dfrac{dx}{dt} + 4x = 0$

(iv) $4\dfrac{d^2 x}{dt^2} + 4\dfrac{dx}{dt} + 5x = 0$

⑥ Given $9\dfrac{d^2 x}{dt^2} + 4x = 0$

(i) Find the general solution.

(ii) Find the particular solution which satisfies the conditions $x = 4$ and $\dfrac{dx}{dt} = 2$ when $t = 0$. Express the solution in the form $x = a\sin(t + \varepsilon)$.

(iii) Sketch the graph of the particular solution.

⑦ Find the particular solutions of the following differential equations that satisfy the conditions given. In each case sketch the graph of the solution (you may wish to use graphing software to help visualise this). Assuming each one models a real system, where x represents a variable changing over time, use the graph to describe the motion of each system in words.

(i) $\dfrac{d^2 x}{dt^2} + 8x = 0$ $\qquad x = 1, \dfrac{dx}{dt} = 1$ when $t = 0$

(ii) $\dfrac{d^2 x}{dt^2} - \dfrac{dx}{dt} + x = 0$ $\qquad x = 1, \dfrac{dx}{dt} = 0$ when $t = 0$

(iii) $\dfrac{d^2 x}{dt^2} + 2\dfrac{dx}{dt} + 2x = 0$ $\qquad x = 0, \dfrac{dx}{dt} = 2$ when $t = 0$

(iv) $4\dfrac{d^2 x}{dt^2} - 8\dfrac{dx}{dt} + 5x = 0$ $\qquad x = 2, \dfrac{dx}{dt} = 0$ when $t = 0$

(v) $\dfrac{d^2 x}{dt^2} + 2\dfrac{dx}{dt} + 5x = 0$ $\qquad x = 0, t = 0$ and $x = 3, t = \dfrac{\pi}{4}$

⑧ The motion of a spring–mass oscillator is modelled by the differential equation

$$\dfrac{d^2 x}{dt^2} + 64x = 0$$

where x is the extension of the spring at time t.

(i) Find the general solution of the differential equation.

(ii) Initially $x = 0.1$ and $\dfrac{dx}{dt} = 0$. Find the particular solution corresponding to these initial conditions.

(iii) Write down the period and amplitude of the motion.

(iv) Sketch a graph of the solution.

⑨ A spring–mass oscillator consists of a spring with stiffness $20\,\text{N}\,\text{m}^{-1}$ and an object of mass $0.25\,\text{kg}$. The object is pulled down $10\,\text{cm}$ from its equilibrium position and released.

(i) Formulate the differential equation and write down the initial conditions for the system, stating any modelling assumptions you make.

(ii) Find the particular solution describing the motion.

(iii) Sketch a graph of the solution.

A linear damping device is now introduced to the system.

(iv) Calculate the value of the damping constant if the system is to be critically damped.

(v) If the damping constant does not change, describe the motion of the system if

(a) the mass of the object is increased to $0.3\,\text{kg}$

(b) the mass of the object is decreased to $0.2\,\text{kg}$.

⑩ A simple model of a delicate set of laboratory scales comprises a spring of stiffness $5\,\text{N}\,\text{m}^{-1}$ and a damping device with constant $7\,\text{N}\,\text{m}^{-1}\text{s}$.

An object of mass m is placed carefully on the scales when the spring is unstretched. Take g to be $10\,\text{m}\,\text{s}^{-2}$, and let $x\,$m be the displacement from the initial position.

(i) Formulate and solve a differential equation to model the motion of the object in each of the following cases:

(a) $m = 2$

(b) $m = 2.45$

(c) $m = 4$.

(ii) Sketch a graph of the solution for objects of each value of m. Describe the motion in each case.

⑪ The angular displacement from its equilibrium position of a swing door is modelled by the differential equation

$$\frac{d^2\theta}{dt^2} + 4\frac{d\theta}{dt} + 5\theta = 0$$

The door starts from rest at an angle of $\dfrac{\pi}{4}$ from its equilibrium position.

(i) Find the general solution of the differential equation.

(ii) Find the particular solution for the given initial conditions.

(iii) Sketch a graph of the particular solution, and hence describe the motion of the door.

(iv) What does your model predict as t becomes large?

⑫ A damped spring–mass oscillator consists of a spring and an object with mass m kg. The object is pulled down 10 cm from its equilibrium position and released. The motion of the system is modelled by the differential equation

$$m\frac{d^2x}{dt^2} + k\frac{dx}{dt} + 20x = 0$$

where x is the displacement from the equilibrium position at time t.

(i) In the case where $m = 0.25$, find the value of k if the system is to be critically damped.

(ii) If the value of k does not change, describe the motion of the system if:

(a) the mass of the object is increased to 0.3 kg

(b) the mass of the object is decreased to 0.2 kg.

⑬ Prove that, if a, b and c are positive constants, then all possible solutions of

$$a\frac{d^2x}{dt^2} + b\frac{dx}{dt} + cx = 0$$

approach zero as $t \to \infty$.

⑭ Find the general solution of the differential equation

$$\frac{d^4y}{dx^4} - 16y = 0.$$

⑮ Figure 16.24 shows a block of mass 10 kg being dragged across a horizontal surface by means of an elastic spring AB of unstretched length 0.5 m. The end A is being pulled with constant speed 5 m s⁻¹. The block is attached to the end B of the spring.

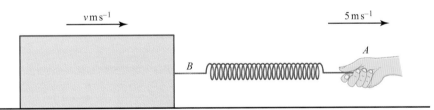

Figure 16.24

The block is subject to a resistance force given by $4v$ N, where v is the speed of the block in m s⁻¹. The only other horizontal force acting on the block is the tension in the spring given by $T = k(y - 0.5)$ N, where y is the length of the spring in metres and k is a constant (the stiffness of the spring).

(i) By considering the horizontal forces acting on the block, write down the equation of motion of the block in terms of $\frac{dv}{dt}$, v, y and k. Use the fact that $\frac{dy}{dt} = 5 - v$ to eliminate v, and show that

$$10\frac{d^2y}{dt^2} + 4\frac{dy}{dt} + ky = 0.5k + 20.$$

(ii) You are given that $k = 20$. Find the general solution of the differential equation.

At the beginning of the motion, the spring is unstretched and the block is at rest.

(iii) Find an expression for y at time $t > 0$.

(iv) What are the limiting values of v and y after a long period of time? Explain briefly how these values would be affected if a different spring were used with a lower value of k.

3 Simple harmonic motion: Further calculations

A particle P, of mass m kg is attached to a light, perfectly elastic spring which has stiffness k N m^{-1}. The other end, E, of the spring is attached to a fixed point and P moves horizontally.

The spring is at its natural length when P is at the fixed point O. The particle is pulled to a point A where OA = a m and released (see Figure 16.25).

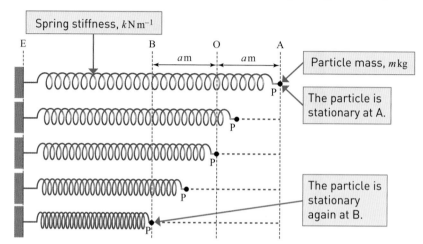

Figure 16.25

At the instant when the particle, P, is x m to the right of O, the forces acting on it are as shown in Figure 16.26.

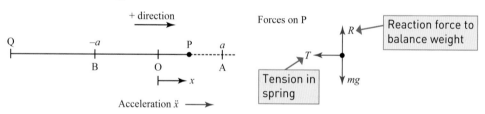

Figure 16.26

The horizontal equation of motion is

$$-T = m\ddot{x}$$

By Hooke's law, $T = kx$ so

$$-kx = m\ddot{x}$$
$$\ddot{x} = -\frac{k}{m}x \qquad ①$$

or

$$\frac{\mathrm{d}^2x}{\mathrm{d}t^2} = -\frac{k}{m}x$$

In the equation ①, k and m are both positive constants so $\frac{k}{m}$ can be written as one positive constant. By convention this is denoted by ω^2 so the equation may be written as

$$\ddot{x} = -\omega^2 x \qquad ②$$

The equation for the velocity that you found in the last section can then be written as

$$v^2 = \omega^2(a^2 - x^2) \qquad \text{or} \qquad \dot{x}^2 = \omega^2(a^2 - x^2) \quad \text{③}$$

To derive equation ③ from the differential equation ②, start by rewriting the acceleration $\ddot{x}$ in the form $v\dfrac{dv}{dx}$.

So equation ② becomes $v\dfrac{dv}{dx} = -\omega^2 x$.

Separating the variables gives

$$\int v\,dv = -\int \omega^2 x\,dx$$

$$\Rightarrow \qquad \frac{v^2}{2} = -\frac{\omega^2 x^2}{2} + C$$

Using the initial condition that $v = 0$ when $x = a$ gives

$$0 = -\frac{\omega^2 x^2}{2} + C$$

$$\Rightarrow \qquad C = \frac{\omega^2 a^2}{2}$$

$$\Rightarrow \qquad v^2 = \omega^2\left(a^2 - x^2\right)$$

There are many other similar systems which produce an equation of the same form as equation ②. Because the same equation of motion applies to all of them, they all have the same type of oscillating motion called **simple harmonic motion** (SHM).

Simple harmonic motion is defined by the equation $\ddot{x} = -\omega^2 x$. Remember that, in this equation, $\ddot{x}$ means acceleration and x means the displacement from the centre. It may be stated in words as:

> The acceleration is proportional to the magnitude of the displacement from the centre point of the motion and is directed towards the centre point.

The two equations $v^2 = \omega^2(a^2 - x^2)$ and $\ddot{x} = -\omega^2 x$ tell you a lot about the motion.

From the first equation you can see that:

- the motion is symmetrical about the point O where $x = 0$
- x must always lie between $-a$ and a, otherwise v^2 would be negative
- at the extreme points, when $x = \pm a$, the velocity is zero
- the maximum speed is when $x = 0$ and $v = \pm a\omega$.

The second equation tells you that:

- $\ddot{x}$ is always directed towards O, so the same must be true of the resultant force
- $\ddot{x}$ and hence the resultant force are zero when $x = 0$, so O is the equilibrium position
- $\ddot{x}$ has a maximum magnitude of $\omega^2 a$ when $x = \pm a$.

These results for $\ddot{x}$ and v in terms of x are shown in Figure 16.27.

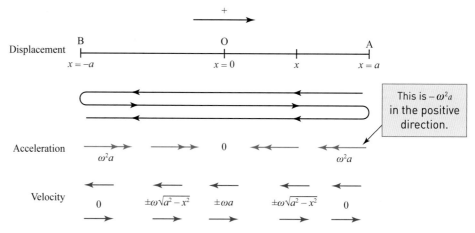

Figure 16.27

The frequency of the motion depends on the value of ω. In the next section, you will see how to write x, v and $\ddot{x}$ in terms of time. In the meantime, you can use the result

$$\omega = 2\pi \times \text{frequency}$$

Example 16.9

When a violin string is playing an 'A' with frequency $880\,\text{Hz}$, a particle on the string oscillates with amplitude of $0.5\,\text{mm}$.

(i) Calculate the time taken for one complete oscillation.

(ii) Express ω in terms of π and write down the equation of motion.

(iii) Find the maximum speed of the particle and its maximum acceleration.

(iv) Find the acceleration and velocity when the particle is $0.25\,\text{mm}$ from its central position.

Solution

(i) The particle does 880 oscillations per second, so the time for one oscillation is $\dfrac{1}{880} = 0.001\dot{1}3\dot{6}\,\text{s}$. This is the period of the motion.

(ii) $\omega = 2\pi \times \text{frequency} = 1760\pi$

The equation of motion is $\ddot{x} = -\omega^2 x$

$\Rightarrow \qquad \ddot{x} = -(1760\pi)^2\, x$

(iii) You can use the equation with mm along as your units are consistent.

$$
\begin{aligned}
\text{Maximum speed} \qquad a\omega &= 0.5 \times 1760\pi\,\text{mm s}^{-1} \\
&= 2764.60...\,\text{mm s}^{-1} \\
&= 2.76\,\text{m s}^{-1}\ (\text{3 s.f.})
\end{aligned}
$$

$$
\begin{aligned}
\text{Maximum acceleration} \qquad a\omega^2 &= 0.5 \times (1760\pi)^2\,\text{mm s}^{-2} \\
&= 1.528\,...\times 10^7\,\text{mm s}^{-2} \\
&= 15\,300\,\text{m s}^{-2}\ (\text{3 s.f.})
\end{aligned}
$$

(iv) When $x = 0.25$,

$$\ddot{x} = -(1760\pi)^2 \times 0.25\,\text{mm s}^{-2}$$

$\Rightarrow \qquad \text{acceleration} = -7640\,\text{m s}^{-2}\ (\text{3 s.f.})$

The velocity is given by

$$v^2 = \omega^2(0.5^2 - x^2)$$

$$\Rightarrow \text{velocity} = \pm 1760\pi \sqrt{0.25 - 0.0625}\,\text{mm s}^{-1}$$

$$= \pm 2.39\,\text{m s}^{-1}\,(3\text{ s.f.})$$

When $x = -0.25$

$$\ddot{x} = +(1760\pi)^2 \times 0.25\,\text{mm s}^{-2}$$

$$\Rightarrow \quad \text{acceleration} = +7640\,\text{m s}^{-2}$$

The velocity is $\pm 2.39\,\text{m s}^{-1}$ as before.

Exercise 16.3

In all these questions assume that time is measured in seconds.

① Write down the equation of motion in the following cases of simple harmonic motion. Calculate:

(i) the velocity and magnitude of the acceleration at the centre of the motion

(ii) the velocity and magnitude of the acceleration at the ends of the motion

(iii) the speed when x is half the amplitude.

    (a) $\omega = 3$, amplitude = 2 cm

    (b) $\omega = 10$, amplitude = 0.1 m

    (c) $\omega = \pi$, amplitude = a m

② A particle is moving with SHM with $\omega = 2$. Initially it is 10 cm from the centre of the motion and moving in the positive direction with a speed of 6 cm s^{-1}. Write down the equation of motion and calculate:

(i) the amplitude

(ii) the acceleration at the ends of the motion

(iii) the speed at the centre

(iv) the speed when it is 5 cm from the centre

(v) the distance from the centre when the speed is half its maximum.

③ Musical notes which are an octave apart have frequencies in the ratio 1 : 2. The note A above middle C has a frequency of 440 Hz. On a full size keyboard there are 4 As below it and 3 above it (the range on the keyboard is just over 7 octaves).

(i) Work out the values of ω corresponding to the frequencies of these seven As.

(ii) Find the maximum speed of points on piano strings which are vibrating with amplitude 1 mm to produce the highest and lowest of these notes.

④ A loudspeaker cone sounding a pure note of frequency 2000 Hz is modelled by SHM of amplitude 2 mm.

(i) Calculate ω and the maximum speed of the cone.

(ii) Write down the equation of motion and calculate the maximum acceleration.

⑤ A particle of dust lands at a point A on a surface which is vibrating vertically with SHM centre O and $\omega = 20$. The particle cannot stay in contact with the surface when the downwards acceleration is greater than g. How far (in cm) is the particle from O when this happens? Can you be certain whether it is above or below O?

⑥ A piston in an engine oscillates with a period of $0.03\,\text{s}$ and an amplitude of $0.35\,\text{m}$. Modelling the oscillations as SHM,

   (i)   draw a sketch graph to illustrate the oscillations

   (ii)   calculate the frequency of the motion and hence find ω

   (iii) calculate the maximum speed of the piston.

⑦ Air sickness might be caused by the rhythmic vibrations of an aircraft. It has been observed that about 50% of the passengers of an aircraft suffer air sickness when it bounces up and down with a frequency of about $0.3\,\text{Hz}$ and a maximum acceleration of $4\,\text{m}\,\text{s}^{-2}$. Assuming SHM, find:

   (i)   the value of ω and hence the amplitude of the motion

   (ii)   the greatest vertical speed during this motion.

⑧ A jig-saw operates at 3000 strokes per minute with the tip of the blade moving 17 mm from the top to the bottom of the stroke. (One stroke is a complete cycle.) Assuming that the motion is simple harmonic, find:

   (i)   the maximum speed of the blade

   (ii)   the maximum acceleration of the blade

   (iii) the speed of the blade when it is 6 mm from the central position.

⑨ The waveforms of two musical notes, P and Q, are shown in Figure 16.28.

Figure 16.28

Which of the following statements are true?

   (i)   P is lower than Q.

   (ii)   P is louder than Q.

   (iii) The frequency of P is twice that of Q.

   (iv) The period of P is twice that of Q.

   (v) The amplitude of Q is larger than that of P.

4 Non-homogeneous differential equations

So far you have seen that equations of the form

$$a\frac{d^2y}{dx^2} + b\frac{dy}{dx} + cy = 0$$

(linear, homogeneous, second order differential equations, with constant coefficients) have general solutions of the form

$$y = Au(x) + Bv(x)$$

where u(x) and v(x) may involve exponential and/or trigonometric functions (and possibly a factor x). In modelling real situations, equations like this often arise in which the right-hand side is non-zero. These equations are called **non-homogeneous** linear equations:

$$a\frac{d^2y}{dx^2} + b\frac{dy}{dx} + cy = f(x)$$

There are many situations where such equations arise, such as the equation modelling the motion of the parachutist near the beginning of this chapter, $\frac{d^2x}{dt^2} + \frac{k}{m}\frac{dx}{dt} = g$, or when an external force causes a structure to vibrate (forced oscillations).

In order to learn how to deal with this type of second order differential equation, it is useful to look at similar first order equations, e.g.

Linear, first order.

Constant coefficients.

$$\frac{dy}{dx} + 2y = 3x - 1$$

Non-homogeneous because the right-hand side is non-zero.

You should remember how to solve this using an integrating factor method, but it is instructive to compare the solutions to several similar looking equations.

ACTIVITY 16.5

Verify that the following solutions satisfy their respective first order linear differential equations.

(i) $\dfrac{dy}{dx} + 2y = 0$ $y = Ae^{-2x}$

(ii) $\dfrac{dy}{dx} + 2y = 3x - 1$ $y = Ae^{-2x} + \frac{3}{2}x - \frac{5}{4}$

(iii) $\dfrac{dy}{dx} + 2y = e^{3x}$ $y = Ae^{-2x} + \frac{1}{5}e^{3x}$

(iv) $\dfrac{dy}{dx} + 2y = \sin x$ $y = Ae^{-2x} - \frac{1}{5}\cos x + \frac{3}{5}\sin x$

Compare the different solutions: what is the same and what is different?

You should notice that all the solutions have strong similarities, and in particular they consist of two distinct parts.

- The first part Ae^{-2x} contains an arbitrary constant, and is the general solution for the homogeneous version (part (i) of the activity). It appears to depend only on the left-hand side of the differential equation. This part of the solution is the **complementary function**.

- The second part only exists in the non-homogeneous cases, contains no arbitrary constants, appears to depend only on the right-hand side of the differential equation – and is of a similar form to the original right-hand side (e.g. linear in part (ii), exponential in part (iii), trigonometric in part (iv)). This part of the solution is called the **particular integral**.

Since you already know how to find the complementary function, all that remains in order for you to use this method of constructing a solution is a way to find a particular integral. The following first order example demonstrates a way to approach this, using a **trial function**.

Example 16.10

Solve the differential equation

$$\frac{\mathrm{d}y}{\mathrm{d}x} + 2y = \mathrm{e}^{3x}.$$

> This is one of the examples from the previous activity.

Solution

The auxiliary equation is $\lambda + 2 = 0$

$$\Rightarrow \quad \lambda = -2$$

The complementary function is $y = A\mathrm{e}^{-2x}$.

To find the particular integral use a trial function of $y = a\mathrm{e}^{3x}$, since the right-hand side is of the form $k\mathrm{e}^{3x}$ where k is a constant.

$$y = a\mathrm{e}^{3x}$$

$$\frac{\mathrm{d}y}{\mathrm{d}x} = 3a\mathrm{e}^{3x}$$

> Differentiate.

$$3a\mathrm{e}^{3x} + 2a\mathrm{e}^{3x} = \mathrm{e}^{3x}$$

> Substitute into the differential equation.

Comparing coefficients of e^{3x} gives

$$5a = 1$$

so $a = \frac{1}{5}$.

> Notice that you still have one arbitrary constant, as you would expect for a first order equation.

$$y = A\mathrm{e}^{-2x} + \frac{1}{5}\mathrm{e}^{3x}$$

> The general solution is constructed by adding the complementary function and the particular integral.

The previous examples could also have been solved using first order methods, such as an integrating factor, but this method of complementary functions and particular integrals is more powerful since it can be used for higher order equations. The next example shows how this method is used for a second order differential equation, and also demonstrates a different form of particular integral.

Example 16.11

Find the general solution of the differential equation

$$\frac{\mathrm{d}^2 y}{\mathrm{d}x^2} - 2\frac{\mathrm{d}y}{\mathrm{d}x} - 3y = 6x - 2.$$

Solution

The auxiliary equation is $\lambda^2 - 2\lambda - 3 = 0$

$$(\lambda - 3)(\lambda + 1) = 0$$

$$\lambda = 3 \text{ or } -1$$

> The right-hand side is a linear function, so your trial function should be a general linear function.

So the complementary function is $y = A\mathrm{e}^{3x} + B\mathrm{e}^{-x}$.

To find a particular integral, use a trial function $y = ax + b$.

$$\frac{\mathrm{d}y}{\mathrm{d}x} = a \text{ and } \frac{\mathrm{d}^2 y}{\mathrm{d}x^2} = 0$$

> Differentiating twice.

> Substituting into the original differential equation.

$$0 - 2a - 3(ax + b) = 6x - 2$$

Comparing coefficients of x gives

$$-3a = 6$$
$$a = -2$$

and comparing constant terms

$$-2a - 3b = -2$$
$$4 - 3b = -2$$
$$-3b = -6$$
$$b = 2$$

The particular integral is $y = -2x + 2$.

So the general solution is

$$y = Ae^{3x} + Be^{-x} - 2x + 2$$

Notice that the general solution has the properties you need:

- it satisfies the full differential equation (you may wish to check this)
- it has two arbitrary constants, which is consistent with it being the general solution to a second order differential equation.

Particular integrals

In order to find a suitable particular integral, the trial function needs to match the right-hand side of the differential equation.

Table 16.1 is a guide to what trial function to use for the differential equation

$$a\frac{d^2y}{dx^2} + b\frac{dy}{dx} + cy = f(x)$$

Right-hand side: $f(x)$	Trial function
linear function	$ax + b$
polynomial of order n	$a_n x^n + a_{n-1} x^{n-1} + \ldots + a_1 x + a_0$
trigonometric function involving $\cos px$ and/or $\sin px$	$a \cos px + b \sin px$
exponential function involving e^{px}	ae^{px}
sum of different functions	sum of matching functions

Table 16.1

You should note that this method finds a simple particular integral, but that there are infinitely many possible particular integrals, which can be constructed by adding on any term from the complementary function.

The following examples demonstrate this trial function approach, for the cases you have not yet seen.

Example 16.12

Find a particular integral of the differential equation

$$\frac{d^2z}{dt^2} - 2\frac{dz}{dt} - 3z = 6\cos 3t.$$

Solution

Use the trial function: $z = a\cos 3t + b\sin 3t$

Notice that both the $\cos 3t$ and $\sin 3t$ functions are used, since they differentiate to become each other.

Differentiating the trial function twice:

$$\frac{dz}{dt} = -3a\sin 3t + 3b\cos 3t$$

$$\frac{d^2z}{dt^2} = -9a\cos 3t - 9b\sin 3t$$

Substituting these into the differential equation gives

$$(-9a\cos 3t - 9b\sin 3t) - 2(-3a\sin 3t + 3b\cos 3t) - 3(a\cos 3t + b\sin 3t)$$

$$= 6\cos 3t$$

$$(-12a - 6b)\cos 3t + (-12b + 6a)\sin 3t = 6\cos(3t)$$

Equating coefficients of $\cos 3t$ gives $\quad -12a - 6b = 6$

Equating coefficients of $\sin 3t$ gives $\quad -12b + 6a = 0$

Solving for a and b gives $\qquad a = -\frac{2}{5}$ and $b = -\frac{1}{5}$

The particular integral is $-\frac{2}{5}\cos 3t - \frac{1}{5}\sin 3t$.

Special cases

In some differential equations the function on the right-hand side has the same form as one of the complementary functions. For example, the complementary function of the differential equation

$$\frac{d^2y}{dx^2} - 5\frac{dy}{dx} + 6y = 4e^{3x}$$

is $Ae^{2x} + Be^{3x}$, and e^{3x} occurs on the right-hand side. In this situation it is no good using the trial function ae^{3x}, since upon substituting $y = ae^{3x}$, $\frac{dy}{dx} = 3ae^{3x}$ and $\frac{d^2y}{dx^2} = 9ae^{3x}$ into the differential equation, you obtain

$$9ae^{3x} - 5(3ae^{3x}) + 6(ae^{3x}) = 4e^{3x}$$

$$\Rightarrow 0 = 4e^{3x}$$

and so clearly this trial function does not work.

Instead $y = axe^{3x}$ is used as a trial function.

This gives $\qquad \dfrac{dy}{dx} = ae^{3x} + 3axe^{3x}$

and $\qquad\qquad \dfrac{d^2y}{dx^2} = 6ae^{3x} + 9axe^{3x}$

Substituting these in the differential equation gives

$$(6ae^{3x} + 9axe^{3x}) - 5(ae^{3x} + 3axe^{3x}) + 6axe^{3x} = 4e^{3x}$$

$$\Rightarrow ae^{3x} = 4e^{3x}$$

$$\Rightarrow a = 4$$

A particular integral is $4xe^{3x}$.

This illustrates a general rule. If the function on the right-hand side of the differential equation has exactly the same form as one of the complementary functions, you multiply the usual trial function by the independent variable to give a new trial function. In order to recognise these special cases when they arise, it is worth getting into the habit of finding the complementary function before the particular integral.

Exercise 16.4

① Find a particular integral for each of the following differential equations.

(i) $\dfrac{d^2y}{dx^2} - 4\dfrac{dy}{dx} + y = -2x + 3$ (ii) $\dfrac{d^2x}{dt^2} + 4x = t + 2$

(iii) $\dfrac{d^2y}{dx^2} + 2\dfrac{dy}{dx} + y = \cos 3x$ (iv) $\dfrac{d^2x}{dt^2} + \dfrac{dx}{dt} + 2x = 3e^{-2t}$

② Find the general solutions of the following differential equations.

(i) $\dfrac{d^2y}{dx^2} + 4y = e^{2x}$ (ii) $\dfrac{d^2x}{dt^2} - 2\dfrac{dx}{dt} - 3x = 5e^{-2t}$

(iii) $\dfrac{d^2y}{dx^2} + 2\dfrac{dy}{dx} + 5y = \cos x$ (iv) $\dfrac{d^2y}{dx^2} + 2\dfrac{dy}{dx} + 5y = 3x + 2$

③ Find the general solutions of the following differential equations.

(i) $\dfrac{d^2x}{dt^2} + x = \sin t$ (ii) $\dfrac{d^2y}{dx^2} + 3\dfrac{dy}{dx} - 4y = e^x$

(iii) $\dfrac{d^2x}{dt^2} - 4\dfrac{dx}{dt} + 3x = e^{3t}$ (iv) $\dfrac{d^2x}{dt^2} - 6\dfrac{dx}{dt} + 9x = 4e^{3t}$

④ Find the particular solution of each of the following differential equations with the given initial conditions.

(i) $\dfrac{d^2y}{dx^2} - 5\dfrac{dy}{dx} + 6y = 36x$ $y = 0$ and $\dfrac{dy}{dx} = -10$ when $x = 0$

(ii) $\dfrac{d^2x}{dt^2} + 9x = 20e^{-t}$ $x = 0$ and $\dfrac{dx}{dt} = 1$ when $t = 0$

(iii) $\dfrac{d^2y}{dx^2} + 2\dfrac{dy}{dx} + 5y = 4e^{-x}$ $y = 0$ and $\dfrac{dy}{dx} = 0$ when $x = 0$

(iv) $\dfrac{d^2x}{dt^2} + \dfrac{dx}{dt} - 2x = 20\sin 2t$ $x = 2$ and $\dfrac{dx}{dt} = 0$ when $t = 0$

(v) $\dfrac{d^2x}{dt^2} - 3\dfrac{dx}{dt} + 2x = 1 - e^t$ $x = 0$ and $\dfrac{dx}{dt} = 1$ when $t = 0$

(vi) $\dfrac{d^2y}{dx^2} + 4y = 12\sin 2x$ $y = 0$ and $\dfrac{dy}{dx} = 1$ when $x = 0$

(vii) $\dfrac{d^2y}{dx^2} + 4\dfrac{dy}{dx} + 5y = 8\sin x$ $y = 1$ and $\dfrac{dy}{dx} = 0$ when $x = 0$

(viii) $\dfrac{d^2y}{dx^2} + 4\dfrac{dy}{dx} + 4y = 8e^{2x} + 4x$ $y = 0$ and $\dfrac{dy}{dx} = 1$ when $x = 0$

⑤ A biological population of size P at time t is growing in an environment which can support a maximum population which is subject to seasonal variation. The growth of the population is described by the first order linear differential equation

$$\frac{dP}{dt} + P = 100 + 50\sin t$$

Find:

(i) the complementary function of this differential equation

(ii) the particular integral

(iii) the complete solution given that initially $P = 20$

(iv) the mean size of the population after a long time has elapsed

(v) the amplitude of the oscillations of the population.

⑥ The pointer on a set of kitchen scales oscillates before settling down at its final reading.

If x is the reading at time t then the oscillation of the pointer is modelled by the differential equation

$$\frac{d^2x}{dt^2} + 3\frac{dx}{dt} + 10x = 0.5$$

(i) Find the general solution of the equation for x.

(ii) Given that $x = 0.1$ and $\frac{dx}{dt} = 0$ when $t = 0$, find the particular solution. At what reading will the pointer settle?

(iii) What length of time will elapse before the amplitude of the oscillations of the pointer is less than 20% of the final value of x?

⑦ In an electrical circuit, the charge q coulombs stored in a capacitor at time t is given by the differential equation

$$\frac{d^2q}{dt^2} + 100\frac{dq}{dt} + 10000q = 1000\sin 100t$$

where t is the time in seconds.

(i) Find the complementary function of the differential equation.

(ii) Decide if the circuit is overdamped, critically damped or underdamped.

(iii) Find the particular integral and hence the general solution.

(iv) Initially both q and $\frac{dq}{dt}$, the current in the circuit, are zero. Find the particular solution.

(v) Use a graphical calculator or graphing software to sketch the solution and describe how the charge in the circuit changes with time.

⑧ A simple model for the motion of a delicate set of laboratory scales, when an object of mass m is placed gently on the scales, is given by the differential equation

$$m\frac{d^2x}{dt^2} + 7\frac{dx}{dt} + 5x = 10m$$

where x is the displacement from the initial position at time t.

(i) Find the particular solution that models the motion of the object in each of the cases:

    (a) $m = 2$ (b) $m = 2.45$ (c) $m = 4$.

(ii) Sketch a graph of the solution for each value of m. Describe the motion in each case.

⑨ (i) Solve the equation $\dfrac{dy}{dx} - ky = e^{pt}$, where $k \neq p$,

    (a) using the integrating factor method

    (b) by finding the complementary function and particular integral.

(ii) Repeat part (i) for the equation $\dfrac{dy}{dx} - ky = e^{kt}$.

⑩ *You are advised to have a graphical calculator or graphing software to hand when doing this question.*

In normal running, a machine vibrates slightly and this is monitored by a marker on it. The marker moves along a straight line. Its displacement in mm from an origin is denoted by x.

The motion of the marker is subject to the differential equation

$$\frac{d^2x}{dt^2} + 9x = 0$$

At the start of the motion, when $t = 0$, the marker is stationary and $x = 2$.

(i) Find the particular solution.

    Draw a graph of x against t.

One day the machine's operator attaches another mechanism to it with the result that a forced oscillation is applied to the machine and the differential equation governing x becomes

$$\frac{d^2x}{dt^2} + 9x = 5\cos 2t$$

The same initial conditions apply.

(ii) Verify that the solution is $x = \cos 3t + \cos 2t$.

    Draw a graph of x against t for $0 \leqslant t \leqslant 2\pi$.

    Describe what has happened to the motion of the marker.

Another day the operator changes the frequency of the forced oscillation so that the differential equation becomes

$$\frac{d^2x}{dt^2} + 9x = 5\cos 3t$$

Again the same initial conditions apply.

(iii) Solve the equation.

    Draw a sketch graph of x against t.

    Predict what happens to the machine when this forced oscillation is applied.

5 Systems of differential equations

Many situations that are modelled by differential equations involve a number of variables which may depend on one another, a **system**. Sometimes this means that the situation may be modelled by two or more differential equations.

Example 16.13

Figure 16.29 shows how lead from pollution can build up in the body. Biological research shows that it is reasonable to assume that the rate of transfer of lead from one part of the body to another is proportional to the amount of lead present. The symbols p, q, r and s are the constants of proportionality for the routes shown in the diagram.

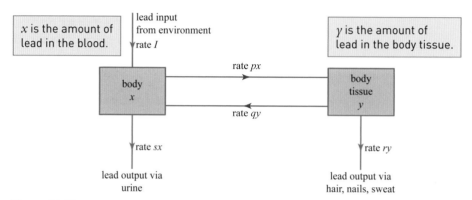

Figure 16.29

Formulate a system of differential equations to model this situation.

Solution

For the blood, at any time t:

Rate of increase of lead in the bloodstream = rate of input − rate of output

$$\frac{dx}{dt} = (I + qy) - (px + sx)$$
$$= I - (p + s)x + qy$$

Similarly for the body tissue:

Rate of increase of lead in the body tissues = rate of input − rate of output

$$\frac{dy}{dt} = px - qy - ry$$
$$= px - (q + r)y$$

The example above resulted in two differential equations, both involving the dependent variables x and y and the independent value t. When two or more differential equations involve the same combination of variables they are called a **system of differential equations**. In this case, since both of the equations are linear, it is a **linear system**.

Solving linear simultaneous differential equations

To solve a pair of linear simultaneous differential equations, you need to eliminate one of the variables so that you end up with a differential equation involving just one dependent variable. This is shown in the following example.

Example 16.14	Solve the simultaneous equations

$$\frac{dx}{dt} = 2x + 4y \qquad ①$$

$$\frac{dy}{dt} = x - y \qquad ②$$

with initial conditions $x = 2$ and $y = -2$ when $t = 0$.

Solution

$$x = \frac{dy}{dt} + y \qquad ③$$

Make x the subject in equation ②.

$$\frac{dx}{dt} = \frac{d^2y}{dy^2} + \frac{dy}{dt}$$

Differentiate with respect to t.

$$\frac{d^2y}{dy^2} + \frac{dy}{dt} = 2\left(\frac{dy}{dt} + y\right) + 4y$$

Substitute for x and $\frac{dx}{dt}$ in equation ①.

$$\frac{d^2y}{dy^2} - \frac{dy}{dt} - 6y = 0$$

Simplify to give a second order differential equation which involves only y and not x.

Auxiliary equation: $\qquad \lambda^2 - \lambda - 6 = 0$

$$(\lambda - 3)(\lambda + 2) = 0$$

$$\lambda = 3 \text{ or } -2$$

General solution for y: $\quad y = Ae^{3t} + Be^{-2t}$

$$x = \frac{dy}{dt} + y$$

Substituting into equation ③ to give the general solution for x.

$$= \left(3Ae^{3t} - 2Be^{-2t}\right) + \left(Ae^{3t} + Be^{-2t}\right)$$

$$= 4Ae^{3t} - Be^{-2t}$$

When $t = 0, x = 2 \implies 2 = 4A - B$

When $t = 0, y = -2 \implies -2 = A + B$

Solving for A and B gives $A = 0$ and $B = -2$.

The particular solution is $x = 2e^{-2t}$

$$y = -2e^{-2t}$$

In the example above, both x and y tend to zero as t tends to infinity. Figure 16.30 shows the behaviour of the solution graphically.

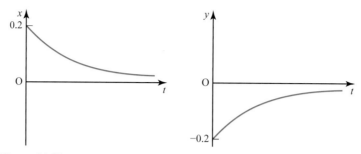

Figure 16.30

Sometimes it is important to understand the relationship between x and y, in which case the solution can be thought of as a pair of parametric equations with parameter t. In the example above, t can be eliminated simply by adding the two solutions:

$$x = 2e^{-2t}$$
$$\underline{y = -2e^{-2t}}$$
$$x + y = 0$$

This is illustrated in Figure 16.31. Notice that only part of the line $x + y = 0$ is required, since the starting point is $x = 2$, $y = -2$, and as $t \to \infty$, $(x, y) \to (0, 0)$.

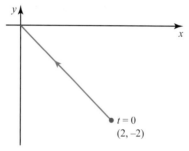

Figure 16.31

A graph like this is called a solution curve. It shows the relationship between the dependent variables (in this case x and y) as the independent variable (in this case t) increases through its permitted range.

Example 16.15

A system of differential equations is given by

$$\frac{dx}{dt} = -x + y - 1 \quad ①$$

$$\frac{dy}{dt} = -x - y + 3 \quad ②$$

When $t = 0, x = 0$ and $y = 3$.

(i)   Find expressions for x and y in terms of t.

(ii) Draw the graph of y against x for values of $t \geqslant 0$. Describe what happens as $t \to \infty$.

Solution

(i)   Equation ① gives $y = x + \dfrac{dx}{dt} + 1$ ③

Differentiating with respect to t: $\dfrac{dy}{dt} = \dfrac{dx}{dt} + \dfrac{d^2x}{dt^2}$

Substituting for y and $\dfrac{dy}{dt}$ into equation ②:

$$\frac{dx}{dt} + \frac{d^2x}{dt^2} = -x - \left(x + \frac{dx}{dt} + 1 \right) + 3$$

Rearranging: $\dfrac{d^2x}{dt^2} + 2\dfrac{dx}{dt} + 2x = 2$

> **Note**
> Verify this for yourself.

The general solution of this equation is $x = e^{-t}(A\sin t + B\cos t) + 1$.

Substituting for x and $\dfrac{dx}{dt}$ in equation ③ gives the general solution for y:

$$y = e^{-t}(A\cos t - B\sin t) + 2$$

The initial conditions give the values of A and B:

When $t = 0, x = 0$ $\Rightarrow 0 = B + 1 \Rightarrow B = -1$

When $t = 0, y = 3$ $\Rightarrow 3 = A + 2 \Rightarrow A = 1$

The particular solution that satisfies the initial conditions is therefore

$$x = e^{-t}(\sin t - \cos t) + 1$$

$$y = e^{-t}(\cos t + \sin t) + 2$$

(ii) Figure 16.32 shows the graph of y against x.

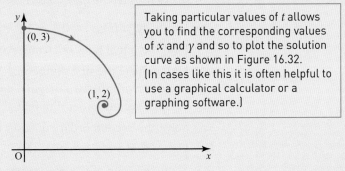

Taking particular values of t allows you to find the corresponding values of x and y and so to plot the solution curve as shown in Figure 16.32. (In cases like this it is often helpful to use a graphical calculator or a graphing software.)

Figure 16.32

As t increases, $e^{-t} \to 0$, so $x \to 1$ and $y \to 2$. In the long term the system approaches the point $(1, 2)$.

In each of questions 1–4:

(i)   solve the equations to find expressions for x and y in terms of t

(ii) find the particular solutions for which $x = 1$ and $y = 2$ at $t = 0$

(iii) describe the long-term behaviour of the system.

① $\dfrac{dx}{dt} = 3x - y$

     $\dfrac{dy}{dt} = 2x$

② $\dfrac{dx}{dt} = 2x + 3y$

     $\dfrac{dy}{dt} = 3x + 2y$

③ $\dfrac{dx}{dt} = x + 5y$

     $\dfrac{dy}{dt} = -x - 3y$

④ $\dfrac{dx}{dt} = 2x - y - 1$

     $\dfrac{dy}{dt} = 2y - 6$

⑤   A system of differential equations is given by

$$\frac{dx}{dt} = x + y$$

$$\frac{dy}{dt} = x - y$$

When $t = 0$, $x = 0$ and $y = 1$.

(i)   Find expressions for x and y in terms of t.

(ii) Describe the long-term behaviour of the system.

⑥   A system of differential equations is given by

$$\frac{dx}{dt} = x + 2y - 3$$

$$\frac{dy}{dt} = -3x + y + 2$$

When $t = 0$, $x = 0$ and $y = 1$.

(i)   Find expressions for x and y in terms of t.

(ii) Describe the long-term behaviour of the system.

⑦   Each of two competing species of insect reproduces at a rate proportional to its own number and is adversely affected by the other species at a rate proportional to the number of that other species. At time t, measured in centuries, the populations of these two species are x million and y million. The situation is modelled by the pair of simultaneous equations

$$\frac{dx}{dt} = 2x - 3y \text{ and } \frac{dy}{dt} = y - 2x$$

where, initially, at time $t = 0$, $x = 15$ and $y = 10$.

(i)   Find the initial rates of changes of both x and y.

(ii)   Differentiate the first differential equation with respect to t and use this together with the two original differential equations to eliminate y and obtain a second order differential equation for x.

(iii) Solve this second order equation to find x as a function of t.

(iv) Hence find y as a function of t.

(v) Show that one of the species becomes extinct and determine the time at which this occurs.

⑧ In a chemical process, compound P reacts with an abundant supply of a gas to form compound Q. The process is governed by two reaction rates, one for the forward reaction, and a lesser one for the reverse reaction, where Q decomposes into P and the gas. P is introduced into a reaction chamber at a constant rate of 21 kg per hour and Q is extracted at a rate proportional to the quantity of Q in the chamber. The equations which describe the process are

$$\frac{dp}{dt} = -5p + q + 21$$

$$\frac{dq}{dt} = -8q + 10p$$

where p and q are the quantities (in kg) of P and Q respectively and t is the time in hours.

(i) Calculate the values of p and q for which $\frac{dp}{dt}$ and $\frac{dq}{dt}$ are both zero.

(ii) Eliminate q to show that

$$\frac{d^2p}{dt^2} + 13\frac{dp}{dt} + 30p = 168$$

(iii) Find the general solution of the differential equation in part (ii). Hence find the particular solutions for p and q for which $p = q = 0$ when $t = 0$.

(iv) Sketch the graphs of the solutions, showing the significance of your answers to part (i).

⑨ In a predator-prey environment the rate of growth of the predator population is found to be proportional to the size of the prey proportion. The rate of change of the prey proportion, however, is found to depend upon the sizes of both the predator and prey populations. The population dynamics are modelled by assuming that both populations vary continuously. The differential equations governing the relationships between the two populations are

$$100\frac{dx}{dt} = y \quad \text{and} \quad 100\frac{dy}{dt} = 2y - x$$

where x and y are the numbers of predator and prey respectively, and t is the time in years.

Initially the predator population is 10 thousand and the prey population 5 million.

(i) By eliminating y between the two equations show that the predator population, x, satisfies the second order differential equation

$$10\,000\frac{d^2x}{dt^2} - 200\frac{dx}{dt} + x = 0$$

(ii) Solve this equation to find the predator population as a function of time.

(iii) Find the prey population, y, as a function of time.

(iv) Determine the size of each population after five years.

⑩ In a chemical decomposition a compound X produces a compound Y which in turn gives a compound Z. These decompositions are governed by the system of differential equations

$$\frac{dx}{dt} = -4x, \quad \frac{dy}{dt} = 4x - 2y, \quad \frac{dz}{dt} = 2y$$

where x, y and z are the masses in grams of X, Y and Z respectively, and time t is measured in hours.

Initially $x = 8$, $y = 0$ and $z = 0$.

(i)   Find x, y and z in terms of t.

(ii) Determine the maximum value of y.

(iii) Determine the final value of z.

⑪ A population of cells consisting of a mixture of two-chromosome and four-chromosome cells is described approximately by the equations

$$\frac{dT}{dt} = (a - b)T, \quad \frac{dF}{dt} = bT + cF$$

where T is the number of two-chromosome cells and F is the number of four-chromosome cells. The variables T and F clearly cannot be negative, and a, b and c are constants with $a \neq b$ and $c \neq 0$.

(i)   Show that the proportion of two-chromosone cells in the population can be written in the form $\dfrac{p}{q + re^{-(a+b+c)t}}$, where p, q are r are constants which depend on the values of a, b and c and the initial conditions.

(ii) Hence show that whatever the values of a, b and c, the proportion of two-chromosome cells in the population tends to a constant value in the long term, independent of the initial conditions.

(iii) Find conditions on the values of a, b and c which ensure that this limiting value of the proportion is non-zero, and find an expression for the limit when this is the case.

LEARNING OUTCOMES

Now you have finished this chapter, you should be able to:

➤ solve differential equations of the form $\dfrac{d^2y}{dx^2} + a\dfrac{dy}{dx} + by = 0$ using the auxiliary equation method

➤ understand and use the relationship between different cases of the solution and the nature of the roots of the auxiliary equation

➤ solve differential equations of the form $\dfrac{d^2y}{dx^2} + a\dfrac{dy}{dx} + by = f(x)$ by solving the homogeneous case and adding a particular integral to the complementary function

➤ find particular integrals for cases where $f(x)$ is a polynomial, trigonometry or exponential function, including cases where the form of the complementary function affects the form required for the particular integral

➤ solve the equation for simple harmonic motion, $\dfrac{d^2x}{dt^2} = -\omega^2 x$, and be able to relate the solution to the motion

➤ model damped oscillations using second order differential equations

➤ interpret the solutions of equations modelling damped oscillations in words and graphically

➤ solve coupled first order simultaneous linear differential equations involving one independent variable and two dependent variables.

KEY POINTS

1 A second order linear differential equation with constant coefficients can be written in the form $\dfrac{d^2y}{dx^2} + a\dfrac{dy}{dx} + by = f(x)$, where a and b are constants.

2 The equation is homogeneous if $f(x) = 0$. Otherwise it is non-homogeneous.

3 When you are given a differential equation of the form above, you can immediately write down the auxiliary equation $\lambda^2 + a\lambda + b = 0$.

4 Each root of the auxiliary equation determines the form of one of the complementary functions.

5 If the auxiliary equation has two real, distinct roots λ_1 and λ_2, then the complementary function of the differential equation is:
$$y = Ae^{\lambda_1 x} + Be^{\lambda_2 x}$$

6 If the auxiliary equation has a repeated root α, then the complementary function of the differential equation is:
$$y = e^{\alpha x}(A + Bx)$$

7 If the auxiliary function has complex roots $\lambda = \alpha \pm \beta i$, then the complementary function of the differential equation is:
$$y = e^{\alpha x}(A\sin\beta x + B\cos\beta x)$$

8 Motion for which the differential equation is $\dfrac{d^2x}{dt^2} + \omega^2 x = 0$ is called simple harmonic motion (SHM). The solution of this differential equation is of the form:
$$x = A\sin\omega t + B\cos\omega t \text{ or } x = R\sin(\omega t + \varepsilon)$$
The period of this motion is $\dfrac{2\pi}{\omega}$ and the amplitude is $\sqrt{A^2 + B^2}$.

The velocity can be given by the equation $v^2 = \omega^2 (a^2 - x^2)$, where a is the amplitude of oscillation.

9 Motion for which the differential equation is $\dfrac{d^2x}{dt^2} + \alpha \dfrac{dx}{dt} + \omega^2 x = 0$, where $\alpha > 0$, is called damped harmonic motion. The following table shows the features of damped harmonic motion.

$\alpha^2 - 4\omega^2$	Type of damping
$\alpha = 0$	no damping
$\alpha^2 - 4\omega^2 > 0$	overdamping
$\alpha^2 - 4\omega^2 = 0$	critical damping
$\alpha^2 - 4\omega^2 < 0$	underdamping

Table 16.2

10 The general solution of the non-homogeneous linear differential equation with constant coefficients

$$\frac{d^2y}{dx^2} + a\frac{dy}{dx} + by = f(x)$$

is the sum of the complementary function and a particular integral.

11 The number of unknown constants is the same as the order of the equation.

12 A particular integral is any function that satisfies the full equation; it does not contain any arbitrary constants.

13 To find the particular integral, use the trial function shown in the following table:

Function	Trial function
linear function	$ax + b$
polynomial of order n	$a_n x^n + a_{n-1} x^{-1} + \ldots + a_1 x + a_0$
trigonometric function involving $\sin px$ and/or $\cos px$	$a \sin px + b \cos px$
exponential function involving e^{px}	ae^{px}
sum of different functions	sum of matching functions

Table 16.3

14 If the trial function for a particular integral is the same as one of the complementary functions, you multiply the trial function by x.

15 A system of differential equations involves two or more dependent variables and one independent variable.

16 In a linear system

$$\frac{dx}{dt} = a_1 x + b_1 y + f_1(t)$$

$$\frac{dy}{dt} = a_2 x + b_2 y + f_2(t)$$

x and y are the dependent variables, and t is the independent variable.

17 Solving such a system of equations involves finding x in terms of t and y in terms of t. This is done by differentiating one equation and substituting into the other, and then solving the resulting second order linear differential equation.

① It is required to find a numerical solution to the differential equation
$\frac{dy}{dx} = \sqrt{x^2 + y^2}$, with the initial conditions $y = 2$ when $x = 0$.

(i) Jordan constructs a spreadsheet to find a solution to the equation using Euler's method. Part of his spreadsheet is shown in Table 1.

	A	B	C	D $\frac{dy}{dx}$	E δy
	x	y	h		
2	0	2	0.1	2	0.2
3	0.1	2.2	0.1	2.202272	0.220227
4	0.2	2.420227	0.1		
5	0.3				

Table 1

Find the numbers in cells D4, E4 and B5. [3 marks]

(ii) Jana uses the modified Euler method to find a solution to the equation. She uses numerical values from Table 1. Find the value she obtains for y when $x = 0.1$. [3 marks]

② The variables x and y satisfy $\frac{dx}{dt} = x - 0.09y$ and $\frac{dy}{dt} = y - 0.16x$.

(i) Find a differential equation satisfied by x. [4 marks]

(ii) Show that the general solution of the equation in part (i) can be written as
$$x = Ae^{1.12t} + Be^{0.88t}.$$ [2 marks]

(iii) Given that $A = 6$ and $B = 9$, find y in terms of t. [3 marks]

③ (i) Find, in the form $a + ib$, the complex number with modulus 1 and argument $\frac{2\pi}{3}$. [2 marks]

The point P represents the complex number $2 + i\sqrt{3}$ in an Argand diagram. P is one vertex of an equilateral triangle centred on the origin.

(ii) Find, in the form $a + ib$, the complex numbers represented by the other two vertices. [3 marks]

(iii) The midpoint of the side of the equilateral triangle opposite P represents the complex number w. Find w in the form $a + ib$. [3 marks]

MP ④ (i) Prove that $e^{2i\theta} - 1 = 2ie^{i\theta}\sin\theta$. [3 marks]

(ii) Given that $\theta \neq n\pi$ for any integer n, find the sum of the series
$$e^{i\theta} + e^{3i\theta} + \ldots + e^{(2n-1)i\theta}.$$

Hence prove that $\cos\theta + \cos 3\theta + \ldots \cos(2n-1)\theta = \frac{\sin n\theta \cos n\theta}{\sin\theta}$ provided $\theta \neq n\pi$. [7 marks]

(iii) Find the corresponding expression for $\sin\theta + \sin 3\theta + \ldots + \sin(2n-1)\theta$. [2 marks]

⑤ The current flowing through a particular component in an electrical circuit is x amps at time t seconds, where x and t are modelled by the differential equation

$$\frac{d^2x}{dt^2} + 6\frac{dx}{dt} + 34x = 25\sin 4t.$$

(i) Find the general solution of this differential equation. **[8 marks]**

(ii) Write down an expression, in terms of t, for the current in the component when the circuit has been operating for a long time. Explain why this expression does not depend on the initial current in the component. **[2 marks]**

(iii) Discuss briefly whether or not the model will break down if the coefficient of $\frac{dx}{dt}$ in the differential equation is -6 instead of 6. **[2 marks]**

⑥ An area of moorland is modelled as a flat (but not horizontal) plane π. Two paths on the moorland are AB and AC, where the coordinates of A, B, C, relative to a convenient origin, are as follows: A(0, 3, 1), B(10, 12, 2), C(4, −5, 3). The units are metres.

(i) Find the vector product of $\overrightarrow{AB}$ and $\overrightarrow{AC}$. Hence show that the equation of the plane π may be written as $13x - 8y - 58z = -82$. **[5 marks]**

A drone is hovering above the moorland at the point P with coordinates (5, 2, 51). Charlie is standing on the path AB at the point Q that is closest to the drone.

(ii) Find the coordinates of Q and hence the distance from Charlie's position to the drone. **[6 marks]**

In order to get a better look, Charlie walks to the point on the moorland that is closest to the drone.

(iii) Find the distance from Charlie's new position to the drone. Hence determine how far Charlie has walked. **[4 marks]**

Chapter 1

Review exercise (Page 11)

1 (i)

    (a) 17   (b) −10   (c) −5   (d) 0

    (e) 19   (f) 14   (g) 30   (h) 0

    (ii) (d) and (h)

2 $p = 5, q = 3$ or $p = −5, q = −7$

In questions 3 to 7, alternative answers may be valid.

3 (i) $\mathbf{r} = \begin{pmatrix} 5 \\ 6 \end{pmatrix} + \lambda \begin{pmatrix} 2 \\ 1 \end{pmatrix}$

   (ii) $\mathbf{r} = \begin{pmatrix} 3 \\ -2 \end{pmatrix} + \lambda \begin{pmatrix} 1 \\ -4 \end{pmatrix}$

   (iii) $\mathbf{r} = \begin{pmatrix} 2 \\ 3 \end{pmatrix} + \lambda \begin{pmatrix} 5 \\ 2 \end{pmatrix}$

   (iv) $\mathbf{r} = \begin{pmatrix} -3 \\ 3 \end{pmatrix} + \lambda \begin{pmatrix} 1 \\ -8 \end{pmatrix}$

4 (i) $\mathbf{r} = \begin{pmatrix} 1 \\ 5 \\ 2 \end{pmatrix} + \lambda \begin{pmatrix} 1 \\ 2 \\ 1 \end{pmatrix}$

   (ii) $\mathbf{r} = \begin{pmatrix} -2 \\ -2 \\ 2 \end{pmatrix} + \lambda \begin{pmatrix} -3 \\ 0 \\ 4 \end{pmatrix}$

   (iii) $\mathbf{r} = \begin{pmatrix} 4 \\ 5 \\ 2 \end{pmatrix} + \lambda \begin{pmatrix} 3 \\ 1 \\ 4 \end{pmatrix}$

   (iv) $\mathbf{r} = \begin{pmatrix} -3 \\ 0 \\ -2 \end{pmatrix} + \lambda \begin{pmatrix} 2 \\ -5 \\ -3 \end{pmatrix}$

5 (i) $\dfrac{x-4}{1} = \dfrac{y-3}{1} = \dfrac{z-4}{4}$

   (ii) $\dfrac{x+1}{-2} = \dfrac{y}{2} = \dfrac{z-2}{7}$

   (iii) $\dfrac{x-2}{3} = \dfrac{z+6}{-2}, y = 3$

   (iv) $y = 1, z = 1, x = 1$

6 (i) $\mathbf{r} = \begin{pmatrix} 3 \\ 2 \\ 8 \end{pmatrix} + \lambda \begin{pmatrix} 5 \\ 4 \\ 1 \end{pmatrix}$

   (ii) $\mathbf{r} = \begin{pmatrix} -1 \\ 3 \\ 0 \end{pmatrix} + \lambda \begin{pmatrix} 2 \\ -3 \\ 3 \end{pmatrix}$

   (iii) $\mathbf{r} = \begin{pmatrix} \frac{1}{2} \\ 0 \\ 0 \end{pmatrix} + \lambda \begin{pmatrix} 6 \\ 5 \\ 3 \end{pmatrix}$

   (iv) $\mathbf{r} = \begin{pmatrix} \frac{3}{2} \\ \frac{5}{2} \\ -2 \end{pmatrix} + \lambda \begin{pmatrix} 1 \\ -8 \\ 0 \end{pmatrix}$

7 $\mathbf{r} = \begin{pmatrix} 3 \\ -2 \\ -5 \end{pmatrix} + \lambda \begin{pmatrix} 0 \\ 0 \\ 1 \end{pmatrix}, x = 3, y = -2, z = -5$

8 (i) Yes, at $(6, -4)$ (ii) No, parallel

   (iii) Yes, at $(4, 3, -1)$ (iv) No, skew

   (v) No, parallel

9 (i) 71.6°   (ii) 0   (iii) 82.4°

   (iv) 72.3°   (v) 0

Discussion points (Page 13)

The pencil is at right angles to any line in the plane. It would not alter.

Discussion point (Page 16)

One method would start by calculating the vectors $\overrightarrow{AB}$ and $\overrightarrow{AC}$. Use the scalar product to find a vector perpendicular to these two vectors, which can be used as the normal $\begin{pmatrix} n_1 \\ n_2 \\ n_3 \end{pmatrix}$ to the plane.

Substitute one of the points A, B or C into the equation $n_1x + n_2y + n_3z + d = 0$ to find the value of d. Alternatively, substitute the three points into the equation $ax + by + cz + d = 0$ to form three simultaneous equations and use a matrix method to solve these equations and hence find the equation of the plane.

Activity 1.1 (Page 17)

$$\begin{cases} a + b + c + d = 0 \\ a - b + d = 0 \\ -a + 2c + d = 0 \end{cases}$$

$a = 1$ gives $b = -\dfrac{2}{3}, c = \dfrac{4}{3}, d = -\dfrac{5}{3}$

So the equation of the plane is

$$x - \frac{2}{3}y + \frac{4}{3}z - \frac{5}{3} = 0$$

or

$$3x - 2y + 4z - 5 = 0$$

Exercise 1.1 (Page 17)

1 (i) $\begin{pmatrix} 5 \\ -3 \\ 2 \end{pmatrix}$

2 (i) $\mathbf{r} \cdot \begin{pmatrix} 1 \\ 1 \\ 1 \end{pmatrix} = 6$

(ii) $\mathbf{r} \cdot \begin{pmatrix} 1 \\ 1 \\ 1 \end{pmatrix} = 0$

(iii) $\mathbf{r} \cdot \begin{pmatrix} -1 \\ -1 \\ -1 \end{pmatrix} = -6$

(iv) $\mathbf{r} \cdot \begin{pmatrix} 1 \\ 1 \\ 1 \end{pmatrix} = 8$

3 (i) $x + y + z = 6$ (ii) $x + y + z = 0$
   (iii) $x - y - z = -6$ (iv) $x + y + z = 8$
   The planes are all parallel.

4 $\mathbf{r} \cdot \begin{pmatrix} -1 \\ 3 \\ -2 \end{pmatrix} = 5$

5 $4x - 5y + 6z = -29$

6 (i) $\begin{pmatrix} 2 \\ 2 \\ -2 \end{pmatrix}, \begin{pmatrix} 5 \\ 2 \\ -1 \end{pmatrix}$

   (iii) $x - 4y - 3z = -2$

7 (iii) B

8 (i) $\begin{pmatrix} 2 \\ -3 \\ 2 \end{pmatrix}, 10$

(ii) e.g. $(-5, 0, 0)$

(iii) $\left(\mathbf{r} - \begin{pmatrix} -5 \\ 0 \\ 0 \end{pmatrix} \right) \cdot \begin{pmatrix} 2 \\ -3 \\ 2 \end{pmatrix} = 0$

9 (i) $\begin{pmatrix} 1 \\ 4 \\ 1 \end{pmatrix}, \begin{pmatrix} 3 \\ 8 \\ 2 \end{pmatrix}$

(ii) $\overrightarrow{AB} \neq k\overrightarrow{AC}$

(iii) $\begin{pmatrix} 0 \\ 1 \\ -4 \end{pmatrix}$

(iv) $\left(\mathbf{r} - \begin{pmatrix} 2 \\ -3 \\ 2 \end{pmatrix} \right) \cdot \begin{pmatrix} 0 \\ 1 \\ -4 \end{pmatrix} = 0$

(vi) Perpendicular to the y-z plane

Discussion points (Page 19)

Yes, provided the angle between two parallel or identical planes is taken to be 0.

No, if the planes are perpendicular then there is only one angle, 90°.

Discussion points (Page 20)

$\mathbf{n} = 0$ or $\mathbf{d} = 0$ are impossible because then the plane or the direction of the line would not be defined. So the only possible cases are when the cosine of the angle between the normal and the direction vector is zero. This means that the line is either in the plane or parallel to it.

The line is perpendicular to the plane if the direction vector of the line is a scalar multiple of the normal vector to the plane.

Discussion point (Page 21)

(i) -8 (ii) any other value of d

Discussion points (Page 22)

No, provided you have one equation in two unknowns.

Provided the planes are not identical or parallel, the method will always work, but it may need modification. See Exercise 1.2 question 3 for examples.

Exercise 1.2 (Page 22)

*(Other answers are possible, particularly for the vector **a**.)*

1 (i) 80.4°   (ii) 68.4°   (iii) 69.9°

2 (i) $\mathbf{r} = \begin{pmatrix} -1 \\ 6 \\ 0 \end{pmatrix} + \lambda \begin{pmatrix} 1 \\ -2 \\ 1 \end{pmatrix}$

   (ii) $\mathbf{r} = \begin{pmatrix} 1.2 \\ 2.4 \\ 0 \end{pmatrix} + \lambda \begin{pmatrix} 0.8 \\ -1.4 \\ 1 \end{pmatrix}$

3 $\lambda = \dfrac{(x - 2)}{1} = \dfrac{y}{1} = \dfrac{z}{1}$

4 $-1, 4$

5 $\dfrac{16 + 9\sqrt{6}}{5}$

6 $x - 4y + 7z = 27$

7 (i) $\mathbf{r} = \begin{pmatrix} 0 \\ 4 \\ 1 \end{pmatrix} + \lambda \begin{pmatrix} 1 \\ -1 \\ 0 \end{pmatrix}$

   (ii) $\mathbf{r} = \begin{pmatrix} 1 \\ 0 \\ 2 \end{pmatrix} + \lambda \begin{pmatrix} 0 \\ 1 \\ -1 \end{pmatrix}$

   (iii) $\mathbf{r} = \begin{pmatrix} 1 \\ 0 \\ 0 \end{pmatrix} + \lambda \begin{pmatrix} 0 \\ 0 \\ 1 \end{pmatrix}$

8 (i) $\mathbf{r} = \begin{pmatrix} 0 \\ 8 \\ 9 \end{pmatrix} + \lambda \begin{pmatrix} 1 \\ -5 \\ -7 \end{pmatrix}$

   (ii) $(1, 3, 2)$

   (iii) $x = 1, y = 3, z = 2$

9 (i) $\lambda = \dfrac{x + 1}{12} = \dfrac{1 - y}{7} = \dfrac{z + 1}{11}$

   (ii) $\lambda = \dfrac{x + 1}{12} = \dfrac{1 - y}{7} = \dfrac{z + 1}{11}$;

   $\lambda = \dfrac{x + 6}{12} = \dfrac{4 - y}{7} = \dfrac{z + 6}{11}$;

   $\lambda = \dfrac{x + 3}{12} = \dfrac{2 - y}{7} = \dfrac{z + 3}{11}$

11 π_1 and π_3 are parallel.
Pairs π_1 and π_2, and π_2 and π_3 are perpendicular.

Review: Matrices and transformations

Exercise R.1 (Page 29)

1 (i) **A**: 2×2 **B**: 1×3 **C**: 2×1 **D**: 2×3
   E: 3×2 **F**: 3×3 **G**: 1×1 **H**: 1×5

   (ii) (a) $\begin{pmatrix} -10 \\ -28 \end{pmatrix}$ (b) $\begin{pmatrix} 3 \\ 27 \\ -5 \end{pmatrix}$

   (c) $(19 \quad 14 \quad 3)$
   (d) non-conformable
   (e) $(-4 \quad 8 \quad -12 \quad 16 \quad -20)$
   (f) non-conformable

   (g) $\begin{pmatrix} -5 & 11 & -3 \\ 6 & 1 & 2 \end{pmatrix}$

   (iii) (a) $\begin{pmatrix} -4 & 3 \\ 1 & 10 \end{pmatrix}$ (b) $\begin{pmatrix} 8 & 3 \\ -1 & 4 \end{pmatrix}$

   (c) $\begin{pmatrix} 8 \\ -17 \end{pmatrix}$

   (d) non-conformable

   (e) $\begin{pmatrix} 1 & -12 & 7 \\ 3 & -4 & -1 \\ -4 & 5 & 2 \end{pmatrix}$

   (f) non-conformable
   (g) non-conformable

2 (i) $\mathbf{MN} = \begin{pmatrix} 6 & 0 \\ -3 & 3 \end{pmatrix}$, $\mathbf{NM} = \begin{pmatrix} 5 & 1 \\ 2 & 4 \end{pmatrix}$

   (ii) Matrix multiplication is not commutative

3 (i) $\mathbf{PQ} = \begin{pmatrix} 3 & 0 & -3 \\ -3 & 2 & -1 \end{pmatrix}$

   $(\mathbf{PQ})\mathbf{R} = \begin{pmatrix} 3 & 3 & 3 \\ -7 & -9 & 3 \end{pmatrix}$

   (ii) $\mathbf{QR} = \begin{pmatrix} -2 & -3 & 3 \\ 7 & 9 & -3 \end{pmatrix}$

   $\mathbf{P}(\mathbf{QR}) = \begin{pmatrix} 3 & 3 & 3 \\ -7 & -9 & 3 \end{pmatrix}$

   (iii) Matrix multiplication is associative

4 $a = -\dfrac{1}{5}$ or 3, $b = \pm 4$

5 (i) $\begin{pmatrix} 25 & 6 \\ 0 & 1 \end{pmatrix}$ (ii) $\begin{pmatrix} 125 & 31 \\ 0 & 1 \end{pmatrix}$

   (iii) $\begin{pmatrix} 625 & 156 \\ 0 & 1 \end{pmatrix}$

(iv) Top right entry is the sum of powers of

5 from 0 to $n-1$, so $A^n = \begin{pmatrix} 5^n & \sum\limits_{r=0}^{n-1} 5^r \\ 0 & 1 \end{pmatrix}$

(v) $\begin{pmatrix} 15\,625 & 3906 \\ 0 & 1 \end{pmatrix}$

6 (i) $\begin{pmatrix} 10+3x & -15+x^2 \\ -5+3x & 18-x \end{pmatrix}$

(ii) $x = -2$ or 5

(iii) $\begin{pmatrix} 4 & -11 \\ -11 & 20 \end{pmatrix}$ or $\begin{pmatrix} 25 & 10 \\ 10 & 13 \end{pmatrix}$

7 $a = -7, b = 2$

Exercise R.2 (Page 36)

1 (i) $\begin{pmatrix} 0 & 1 \\ 1 & 0 \end{pmatrix}$

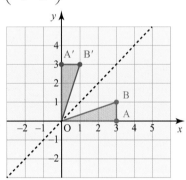

(ii) $\begin{pmatrix} -1 & 0 \\ 0 & -1 \end{pmatrix}$

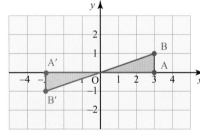

(iii) $\begin{pmatrix} 4 & 0 \\ 0 & 4 \end{pmatrix}$

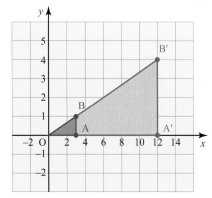

(iv) $\begin{pmatrix} 1 & 2 \\ 0 & 1 \end{pmatrix}$

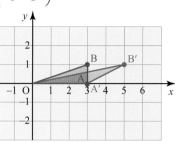

2 (i) Reflection in $y = -x$

(ii) Stretch, scale factor 4, parallel to the x-axis

(iii) Enlargement scale factor 4, centre the origin

(iv) Reflection in the x-axis

(v) Rotation of 90° clockwise about the origin

3 (i)

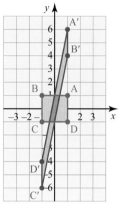

(ii) Shear with the y-axis fixed; A is mapped from $(1, 1)$ to $(1, 6)$

4 (i) $\begin{pmatrix} 1 & 0 & 0 \\ 0 & 0 & 1 \\ 0 & -1 & 0 \end{pmatrix}$ (ii) $\begin{pmatrix} 1 & 0 & 0 \\ 0 & 1 & 0 \\ 0 & 0 & -1 \end{pmatrix}$

(iii) $\begin{pmatrix} -1 & 0 & 0 \\ 0 & 1 & 0 \\ 0 & 0 & -1 \end{pmatrix}$

5 (i) Enlargement scale factor 2, centre $(0, 0)$

(ii) Reflection in the plane $z = 0$

(iii) Rotation of 90° clockwise about the x-axis

(iv) Three way stretch of factor 4 in the x-direction, factor 0.5 in the y-direction and factor 3 in the z-direction

6 (i) $\mathbf{P} = \begin{pmatrix} 0 & -1 \\ 1 & 0 \end{pmatrix}, \mathbf{Q} = \begin{pmatrix} 0 & -1 \\ -1 & 0 \end{pmatrix}$

(ii) $\mathbf{PQ} = \begin{pmatrix} 1 & 0 \\ 0 & -1 \end{pmatrix}$ Reflection in the x-axis

(iii) $\mathbf{QP} = \begin{pmatrix} -1 & 0 \\ 0 & 1 \end{pmatrix}$ Reflection in the y-axis

(iv) $\begin{pmatrix} 1 & 0 \\ 0 & -1 \end{pmatrix}\begin{pmatrix} x \\ y \end{pmatrix} = \begin{pmatrix} x \\ -y \end{pmatrix}$ and

$\begin{pmatrix} -1 & 0 \\ 0 & 1 \end{pmatrix}\begin{pmatrix} x \\ y \end{pmatrix} = \begin{pmatrix} -x \\ y \end{pmatrix}$

so if the points have the same image,

$\begin{pmatrix} x \\ -y \end{pmatrix} = \begin{pmatrix} -x \\ y \end{pmatrix}$ which is only true

when $x = y = 0$. The point that has the same image under both transformations is the origin $(0, 0)$.

7　(i)　$\mathbf{A} = \begin{pmatrix} -\dfrac{1}{2} & -\dfrac{\sqrt{3}}{2} \\ \dfrac{\sqrt{3}}{2} & -\dfrac{1}{2} \end{pmatrix}$

(ii)　$\mathbf{B} = \begin{pmatrix} \dfrac{\sqrt{3}}{2} & -\dfrac{1}{2} \\ \dfrac{1}{2} & \dfrac{\sqrt{3}}{2} \end{pmatrix}$, $\mathbf{C} = \begin{pmatrix} 0 & -1 \\ 1 & 0 \end{pmatrix}$

$\mathbf{BC} = \begin{pmatrix} \dfrac{\sqrt{3}}{2} & -\dfrac{1}{2} \\ \dfrac{1}{2} & \dfrac{\sqrt{3}}{2} \end{pmatrix}\begin{pmatrix} 0 & -1 \\ 1 & 0 \end{pmatrix}$

$= \begin{pmatrix} -\dfrac{1}{2} & -\dfrac{\sqrt{3}}{2} \\ \dfrac{\sqrt{3}}{2} & -\dfrac{1}{2} \end{pmatrix} = \mathbf{A}$

(iii)　$\mathbf{B}^3 = \begin{pmatrix} \dfrac{\sqrt{3}}{2} & -\dfrac{1}{2} \\ \dfrac{1}{2} & \dfrac{\sqrt{3}}{2} \end{pmatrix}\begin{pmatrix} \dfrac{\sqrt{3}}{2} & -\dfrac{1}{2} \\ \dfrac{1}{2} & \dfrac{\sqrt{3}}{2} \end{pmatrix}$

$\begin{pmatrix} \dfrac{\sqrt{3}}{2} & -\dfrac{1}{2} \\ \dfrac{1}{2} & \dfrac{\sqrt{3}}{2} \end{pmatrix} = \begin{pmatrix} 0 & -1 \\ 1 & 0 \end{pmatrix}$

This verifies that three successive anticlockwise rotations of 30° about the origin is equivalent to a single anticlockwise rotation of 90° about the origin.

8　(i)　$\mathbf{J} = \begin{pmatrix} 1 & 0 & 0 \\ 0 & -1 & 0 \\ 0 & 0 & 1 \end{pmatrix}$ $\mathbf{K} = \begin{pmatrix} 0 & -1 & 0 \\ 1 & 0 & 0 \\ 0 & 0 & 1 \end{pmatrix}$

$\mathbf{L} = \begin{pmatrix} -1 & 0 & 0 \\ 0 & 1 & 0 \\ 0 & 0 & 1 \end{pmatrix}$ $\mathbf{M} = \begin{pmatrix} 1 & 0 & 0 \\ 0 & -1 & 0 \\ 0 & 0 & -1 \end{pmatrix}$

(ii)　(a)　$\mathbf{LJ} = \begin{pmatrix} -1 & 0 & 0 \\ 0 & -1 & 0 \\ 0 & 0 & 1 \end{pmatrix}$

(b)　$\mathbf{KJ} = \begin{pmatrix} 0 & 1 & 0 \\ 1 & 0 & 0 \\ 0 & 0 & 1 \end{pmatrix}$

(c)　$\mathbf{K}^2 = \begin{pmatrix} -1 & 0 & 0 \\ 0 & -1 & 0 \\ 0 & 0 & 1 \end{pmatrix}$

(d)　$\mathbf{JLM} = \begin{pmatrix} -1 & 0 & 0 \\ 0 & 1 & 0 \\ 0 & 0 & -1 \end{pmatrix}$

9　(i)　$\begin{pmatrix} 1 & 0 \\ 0 & 3 \end{pmatrix}$

(ii)　A reflection in the x-axis and a stretch of scale factor 2 parallel to the x-axis

(iii)　$\begin{pmatrix} 2 & 0 \\ 0 & -3 \end{pmatrix}$ reflection in the x-axis; stretch scale factor 2 parallel to the x-axis; stretch factor 3 parallel to the y-axis. The outcome of these three transformations would be the same regardless of the order in which they are applied. There are 6 different possible orders.

(iv)　$\begin{pmatrix} \dfrac{1}{2} & 0 \\ 0 & -\dfrac{1}{3} \end{pmatrix}$

10　(i)　$\mathbf{A} = \begin{pmatrix} \cos\theta & -\sin\theta \\ \sin\theta & \cos\theta \end{pmatrix}$

$\mathbf{B} = \begin{pmatrix} \cos\phi & -\sin\phi \\ \sin\phi & \cos\phi \end{pmatrix}$

(ii)

$\mathbf{BA} = \begin{pmatrix} \cos\theta\cos\phi - \sin\theta\sin\phi & -\sin\theta\cos\phi - \cos\theta\sin\phi \\ \sin\theta\cos\phi + \cos\theta\sin\phi & -\sin\theta\sin\phi + \cos\theta\cos\phi \end{pmatrix}$

(iii)　$\mathbf{C} = \begin{pmatrix} \cos(\theta + \phi) & -\sin(\theta + \phi) \\ \sin(\theta + \phi) & \cos(\theta + \phi) \end{pmatrix}$

(iv)　$\sin(\theta + \phi) = \sin\theta\cos\phi + \cos\theta\sin\phi$
$\cos(\theta + \phi) = \cos\theta\cos\phi - \sin\theta\sin\phi$

(v)　A rotation through angle θ followed by rotation through angle ϕ has the same effect as a rotation through angle ϕ followed by angle θ.

11　A reflection in a line followed by a second reflection in the same line returns a point to its original position.

12 (i) **A** represents an anticlockwise rotation through 90° about the origin; $\mathbf{A}^4$ represents four rotations each of 90°, totalling 360° which leaves an object unchanged – this is equivalent to the identity matrix **I**.

(ii) $\mathbf{B} = \begin{pmatrix} 0 & 1 \\ -1 & 0 \end{pmatrix}$ which represents a rotation of 90° clockwise about the origin.

(iii) $\mathbf{C} = \begin{pmatrix} \dfrac{1}{2} & -\dfrac{\sqrt{3}}{2} \\ \dfrac{\sqrt{3}}{2} & \dfrac{1}{2} \end{pmatrix}$

(iv) $m = 2$, $n = 3$. $\mathbf{A}^2 = \mathbf{C}^3$ because both represent a rotation through 180°.

(v) $\mathbf{AC} = \begin{pmatrix} -\dfrac{\sqrt{3}}{2} & -\dfrac{1}{2} \\ \dfrac{1}{2} & -\dfrac{\sqrt{3}}{2} \end{pmatrix}$ Rotations of 60° and 90° can be carried out in either order, both result in a rotation of 150°.

Exercise R.3 (Page 41)

1 (i) $(0, 0)$ is the only invariant point
  (ii) $(0, 0)$ is the only invariant point
  (iii) Invariant points have the form $(\lambda, -\lambda)$
  (iv) Invariant points have the form $(2\lambda, 3\lambda)$

2 (i) x-axis, y-axis, lines of the form $y = mx$
  (ii) x-axis, y-axis, lines of the form $y = mx$
  (iii) No invariant lines
  (iv) $y = x$, lines of the form $y = -x + c$
  (v) $y = -x$, lines of the form $y = x + c$
  (vi) x-axis

3 (ii) $y = \pm x$

4 (i) $y = \dfrac{3}{2}x$ (ii) $y = \dfrac{3}{2}x$

(iii)

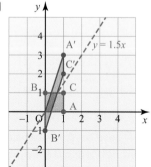

5 (i) $(3, 6)$; $y = 2x$
  (ii) $(-2, 3)$; Rotation 90° anticlockwise about the origin

(iii) $\begin{pmatrix} -0.8 & -0.6 \\ -0.6 & 0.8 \end{pmatrix}$

(iv) $3x + y = 0$

Chapter 2

Review exercise (Page 50)

1 $x = -7$, $\dfrac{3}{2}$

2 (i) $\mathbf{A} = \begin{pmatrix} -1 & 0 \\ 0 & 1 \end{pmatrix}$ $\mathbf{B} = \begin{pmatrix} 0 & -1 \\ -1 & 0 \end{pmatrix}$

(iii)

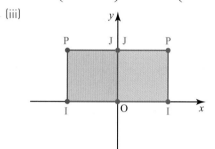

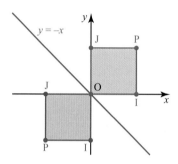

3 (i) $\begin{pmatrix} 1 & 0 \\ 4 & 1 \end{pmatrix}$

(ii) determinant = 1 so area is preserved

4 (i) $(11, -3)$

(ii) $\dfrac{1}{20}\begin{pmatrix} 5 & 5 \\ -1 & 3 \end{pmatrix}$

(iii) $(1, 0)$

5 $k = 1, 6$

6 (i) $\det(\mathbf{M}) = -9$, $\det(\mathbf{N}) = -67$

(ii) $\mathbf{MN} = \begin{pmatrix} 19 & -17 \\ 50 & -13 \end{pmatrix}$
  $\det(\mathbf{MN}) = 603$ and $603 = -9 \times -67$

7 $\begin{pmatrix} 3 & -1 & 3 & 6 \\ -1 & 0 & 0 & 0 \end{pmatrix}$

8 (i)

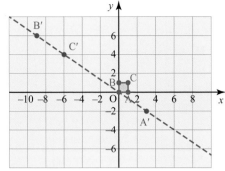

(ii) The image of all points lie on the line $y = -\frac{2}{3}x$. The determinant of the matrix is zero which shows that the image will have zero area.

9 (i) $x = 2, y = -1$ (ii) $x = -3, y = 4$

10 $k = \pm 6$; $k = 6$ gives the same line so an infinite number of solutions

$k = -6$ gives parallel lines so there are no solutions

11 (ii) $\mathbf{M}^n = (a + d)^{n-1}\mathbf{M}$

12 $\begin{pmatrix} x' \\ y' \end{pmatrix} = \begin{pmatrix} a & b \\ c & d \end{pmatrix}\begin{pmatrix} x \\ y \end{pmatrix} \Rightarrow$

$\begin{pmatrix} x' \\ y' \end{pmatrix} = \begin{pmatrix} ax + by \\ cx + dy \end{pmatrix} \Rightarrow \begin{matrix} x' = ax + by \\ y' = cx + dy \end{matrix}$

Solving simultaneously and using the fact $ad - bc = 0$ gives the result.

13 (i) $(4, 1), (2, 2)$ and $(-12, -3)$

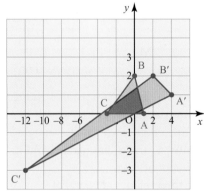

(ii) Area of T $= 4$; Area of T$' = 12$
Ratio 12:4 or 3:1. 3 is the determinant of $\mathbf{M}$.

(iii) $\mathbf{M}^{-1} = \frac{1}{3}\begin{pmatrix} 1 & -1 \\ -1 & 4 \end{pmatrix}$

14 (i) (b) $\mathbf{S}^{-1} = \frac{1}{2}\begin{pmatrix} 4 & -2 \\ 3 & -1 \end{pmatrix}$

(ii) $\mathbf{T}\begin{pmatrix} x \\ y \end{pmatrix} = \begin{pmatrix} x \\ y \end{pmatrix}$

$\Rightarrow \mathbf{T}^{-1}\mathbf{T}\begin{pmatrix} x \\ y \end{pmatrix} = \mathbf{T}^{-1}\begin{pmatrix} x \\ y \end{pmatrix}$

$\Rightarrow \begin{pmatrix} x \\ y \end{pmatrix} = \mathbf{T}^{-1}\begin{pmatrix} x \\ y \end{pmatrix}$ so (x, y)

is invariant under $\mathbf{T}^{-1}$

15 (i) $\mathbf{M} = \begin{pmatrix} 2 & -1 \\ 3 & k \end{pmatrix}$

(ii) $k = -\frac{3}{2}$; $x = 2, y = 3$

(iii) There are no unique solutions.

(iv) (a) Lines intersect at a unique point
(b) Lines are parallel
(c) Lines are coincident

16 (i) $\mathbf{M}^{-1} = \frac{1}{3}\begin{pmatrix} 1 & -2 \\ 0 & 3 \end{pmatrix}$, $\mathbf{N}^{-1} = \frac{1}{7}\begin{pmatrix} 4 & 3 \\ -1 & 1 \end{pmatrix}$

(ii) $\mathbf{MN} = \begin{pmatrix} 5 & -1 \\ 1 & 4 \end{pmatrix}$

$(\mathbf{MN})^{-1} = \frac{1}{21}\begin{pmatrix} 4 & 1 \\ -1 & 5 \end{pmatrix} = \mathbf{N}^{-1}\mathbf{M}^{-1}$

Activity 2.1 (Page 53)

$\det(\mathbf{A}) = -17$ $\mathbf{A}^{-1} = \begin{pmatrix} -\frac{1}{17} & -\frac{2}{17} & \frac{5}{17} \\ -\frac{8}{17} & \frac{1}{17} & \frac{6}{17} \\ \frac{4}{17} & \frac{8}{17} & -\frac{3}{17} \end{pmatrix}$

$\mathbf{B}$ is singular.

$\mathbf{C}^{-1} = \begin{pmatrix} -\frac{1}{15} & -\frac{2}{3} & \frac{1}{5} \\ \frac{2}{5} & 0 & -\frac{1}{5} \\ -\frac{1}{15} & \frac{1}{3} & \frac{1}{5} \end{pmatrix}$

$\mathbf{D}^{-1} = \begin{pmatrix} -1 & -3 & \frac{4}{3} \\ 1 & 2 & -\frac{2}{3} \\ 1 & 3 & -1 \end{pmatrix}$

Exercise 2.1 (Page 57)

1 (i) (a) 5   (b) 5

(ii) (a) −5   (b) −5

Interchanging the rows and columns has not changed the determinant.

(iii) (a) 0   (b) 0

If a matrix has a repeated row or column the determinant will be zero.

2 (i) $\dfrac{1}{3}\begin{pmatrix} 3 & 0 & -6 \\ -4 & 2 & 3 \\ 2 & -1 & 0 \end{pmatrix}$

(ii) Matrix is singular

(iii) $\begin{pmatrix} -0.06 & -0.1 & -0.1 \\ 0.92 & 0.2 & 0.7 \\ 0.66 & 0.1 & 0.6 \end{pmatrix}$

(iv) $\dfrac{1}{21}\begin{pmatrix} 34 & 11 & 32 \\ 9 & 6 & 6 \\ -38 & -16 & -37 \end{pmatrix}$

3 $\dfrac{1}{7}\begin{pmatrix} 2 & 18 & -11 \\ 2 & 39 & -25 \\ 3 & 41 & -27 \end{pmatrix}$; $x = 8$, $y = 4$, $z = -3$

4 $\mathbf{M}^{-1} = \dfrac{1}{28 - 10k}\begin{pmatrix} 4 & -10 & 12 \\ -(4k - 8) & 8 & -(4 + 2k) \\ -k & 7 & -3k \end{pmatrix}$;

$k = 2.8$

5 (i) The columns of the matrix have been moved one place to the right, with the final column moving to replace the first. This is called **cyclical interchange** of the columns.

(ii) $\det(\mathbf{A}) = \det(\mathbf{B}) = \det(\mathbf{C}) = -26$
Cyclical interchange of the columns leaves the determinant unchanged.

6 $x = \dfrac{-1 \pm \sqrt{41}}{2}$

7 $x = 1$, $x = 4$

8 $1 < k < 5$

9 (i) Let $\mathbf{X} = (\mathbf{PQ})^{-1}$ so $\mathbf{X}(\mathbf{PQ}) = \mathbf{I}$.
$\Rightarrow \mathbf{X}(\mathbf{PQ})\mathbf{Q}^{-1} = \mathbf{IQ}^{-1} = \mathbf{Q}^{-1}$
$\Rightarrow \mathbf{XP}(\mathbf{QQ}^{-1}) = \mathbf{XP} = \mathbf{Q}^{-1}$
$\Rightarrow \mathbf{XPP}^{-1} = \mathbf{Q}^{-1}\,\mathbf{P}^{-1}$
$\Rightarrow \mathbf{X} = \mathbf{Q}^{-1}\,\mathbf{P}^{-1}$

(ii) $\mathbf{P}^{-1} = \begin{pmatrix} -\dfrac{1}{9} & \dfrac{1}{6} & -\dfrac{4}{9} \\ \dfrac{2}{9} & \dfrac{1}{6} & -\dfrac{1}{9} \\ -\dfrac{1}{3} & \dfrac{1}{2} & -\dfrac{1}{3} \end{pmatrix}$

$\mathbf{Q}^{-1} = \begin{pmatrix} \dfrac{3}{2} & -4 & \dfrac{1}{2} \\ 1 & -2 & 0 \\ -\dfrac{3}{2} & 5 & -\dfrac{1}{2} \end{pmatrix}$

$(\mathbf{PQ})^{-1} = \mathbf{Q}^{-1}\mathbf{P}^{-1} = \begin{pmatrix} -\dfrac{11}{9} & -\dfrac{1}{6} & -\dfrac{7}{18} \\ -\dfrac{5}{9} & -\dfrac{1}{6} & -\dfrac{2}{9} \\ \dfrac{13}{9} & \dfrac{1}{3} & \dfrac{5}{18} \end{pmatrix}$

10 (ii) Multiplying only the first column by k equates to a stretch of scale factor k in one direction, so only multiplies the volume by k.

(iii) Multiplying any column by k multiplies the determinant by k.

11 (i) $10 \times 43 = 430$

(ii) $4 \times 5 \times -7 \times 43 = -6020$

(iii) $x \times 2 \times y \times 43 = 86xy$

(iv) $x^4 \times \dfrac{1}{2x} \times 4y \times 43 = 86x^3 y$

Exercise 2.2 (Page 62)

1 (i)
$\det \mathbf{M} = 20$, $\mathbf{M}^{-1} = \begin{pmatrix} 0.2 & 0.4 & -0.6 \\ -0.25 & 0.25 & -0.25 \\ -0.15 & -0.05 & 0.45 \end{pmatrix}$

2 (i) Planes π_1 and π_3 are parallel and the second is not parallel to either, so will cross through both to form two parallel straight lines.

(ii) Planes π_1 and π_3 are parallel and π_2 is coincident to π_1.

(iii) All three planes are parallel.

3 (i) $\begin{pmatrix} 5 & 3 & -2 \\ 6 & 2 & 3 \\ 7 & 1 & 8 \end{pmatrix}\begin{pmatrix} x \\ y \\ z \end{pmatrix} = \begin{pmatrix} 6 \\ 11 \\ 12 \end{pmatrix}$

(ii) Eliminating y gives the equations $-8x - 13z = -13$ and $-8x - 13z = -15$ which are inconsistent so the planes form a prism.

4 (i) Planes meet at the unique point $(3, -14, 8)$

(ii) Inconsistent, the planes form a prism

(iii) Consistent, then planes form a sheaf

(iv) Planes meet at the unique point $(-15, 24, -1)$

(v) Three coincident planes

5 (i) The planes intersect in the unique point $(-0.8, 0.6, 1.5)$

(ii) Inconsistent, the planes form a prism

(iii) Consistent, the planes form a sheaf

6 (i) $\dfrac{1}{13k - 65} \begin{pmatrix} 13 & -26 & -13 \\ 7 & -2k-4 & -3k+8 \\ -4 & 3k-7 & -2k+14 \end{pmatrix}$;

$k = 5$

(ii) Unique point of intersection at

$$\left(\frac{52 - 13p}{13}, \frac{20 - 7p}{13}, \frac{4p - 17}{13} \right)$$

(iii) Form a sheaf of planes when $p = 4$

Discussion point (Page 70)

Check that substituting $x = 1$ or 2 or 4 into the determinant gives zero.

Exercise 2.3 (Page 71)

1 1900

2 (i) -13 (ii) -13 (iii) 26 (iv) -39

(v) 13 (vi) -78 (vii) 13 (viii) 13

3 (i) $r_1 + r_3 = r_2$

(ii) Interchange c_2 and c_3

(iii) Interchange c_1 and c_2, then interchange c_2 and c_3

(iv) $x = 3 \Rightarrow c_1 = c_3$

4 (i) 430 (ii) -6020 (iii) $86x^3y$

5 (i) 43 (ii) 7 (iii) -1

7 $(6x - 1)(2x - 5)(3x + 7)$

9 (ii) $5, -8$

11 $a = 0$ or $b = 0$ or $c = 0$ or $a = b$ or $b = c$ or $c = a$

12 (i) $(a - b)(b - c)(c - a)$

(ii) $(x - y)(y - z)(z - x)$

(iii) $(x - y)(y - z)(z - x)(x + y + z)$

(iv) $(x - y)(y - z)(z - x)(xy + yz + zx)$

13 $x(x - 1)^3(x + 1)$

Chapter 3

Discussion points (Page 81)

The equation $-\dfrac{y^2}{b^2} = 1$ has no real solutions.

Instead of the arms of the hyperbola being to the left and right of the asymptotes, they are above and below. There are y-intercepts but no x-intercepts.

Discussion point (Page 81)

The equation has no finite solutions.

Discussion points (Page 81)

One possible systematic method is to consider planes that are (i) horizontal, (ii) vertical, (iii) neither horizontal nor vertical but shallower than the slope of the cone, (iv) neither horizontal nor vertical but steeper than the slope of the cone, and in each case then to consider planes that pass through the common vertex of the two cones separately from those that do not.

One 'degenerate case' is that the plane cuts the double cone at the common vertex of the two cones, and nowhere else. This produces a circle of zero radius, that is, a point.

Another is that the plane could pass through the common vertex and be steeper than the slope of the cones. This produces a pair of straight lines. A third possibility is that the plane passes through the common vertex and has the same slope as the cone, so that the plane touches the cone. This produces a single straight line

Review exercise (Page 82)

1 $x^2 + y^2 = 100$

2 $\dfrac{x^2}{25} + 9y^2 = 1$

3 (i)

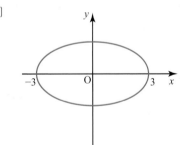

(ii)

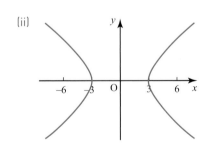

(iii)

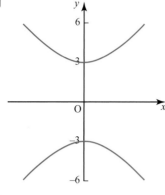

(iv)

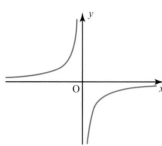

4 $5\sqrt{2}$

5 $\dfrac{y^2}{25} - \dfrac{x^2}{100} = 1$

6 (i) $xy = -13$

(ii) $xy = -13$

(iii) Because of the symmetry of the graph, its image is the same under either transformation.

7 (i) $\dfrac{(x-3)^2}{3} + \dfrac{y^2}{4} = 1$

(ii) $(3 \pm \sqrt{3}, 0); (0, \pm\sqrt{2})$

8 (i) $\dfrac{x^2}{2916} - \dfrac{y^2}{16} = 1$

(ii) $C = (\pm 54, 0), C' = (\pm 6, 0)$

9 (i)

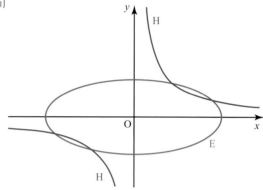

(ii) $\pm(2\sqrt{5}, \tfrac{2}{5}\sqrt{5}), \pm(\sqrt{5}, \tfrac{4}{5}\sqrt{5})$

(iii) Parallelogram

Discussion point (Page 86)

The revised equation would be $(x+1)(y-4) = -16$.

Exercise 3.1 (Page 87)

1

	Equation of original curve	Equation of transformed curve	Description of transformation
(i)	$x^2 + y^2 = 25$	$\dfrac{x^2}{9} + \dfrac{y^2}{9} = 25$	Enlargement centre O scale factor 3
(ii)	$\dfrac{x^2}{3} - \dfrac{y^2}{5} = 1$	$\dfrac{y^2}{3} - \dfrac{x^2}{5} = 1$	Reflection in the line $y = -x$
(iii)	$\dfrac{x^2}{36} + \dfrac{y^2}{36} = 1$	$\dfrac{x^2}{36} + \dfrac{y^2}{324} = 1$	Stretch of scale factor 3 in the y-direction.
(iv)	$xy = 4$	$(x-3)(y-2) = 4$	Translation by vector $\begin{pmatrix} 3 \\ 2 \end{pmatrix}$
(v)	$xy = 9$	$xy = -9$	Reflection in the x-axis
(vi)	$xy = 9$	$xy = -9$	Reflection in the y-axis
(vii)	$a = 1, b = 2\left[x^2 + \dfrac{y^2}{4} = 1\right];$ $p = 2, q = 1$	$4x^2 + y^2 + 16 = 2y + 16x$	Translation by vector $\begin{pmatrix} p \\ q \end{pmatrix}$ where p and q are to be found.
(viii)	$\dfrac{x^2}{10} - \dfrac{y^2}{5} = 1$	$\dfrac{y^2}{10} - \dfrac{x^2}{5} = 1$	Rotation by 90° anticlockwise about O

3 y-direction, scale factor $\frac{b}{a}$

4 (i) $\dfrac{(x-4)^2}{3} + \dfrac{(y+2)^2}{4} = 1$

(ii) $\dfrac{x^2}{3} + \dfrac{y^2}{324} = 1$

(iii) $\dfrac{(x-4)^2}{3} + \dfrac{(y+2)^2}{36} = 1$

(iv) $\dfrac{(x-4)^2}{3} + \dfrac{(y+6)^2}{36} = 1$

5 (i) $\dfrac{y^2}{25} - \dfrac{x^2}{169} = 1$

(ii) $\left(\pm\dfrac{65}{12}, \pm\dfrac{65}{12}\right)$

(iii) Square

6 (i) Translation by the vector $\begin{pmatrix} -\frac{1}{2} \\ \frac{3}{4} \end{pmatrix}$

(ii) Rotate 90° clockwise

(iii) $a = \dfrac{3}{4}$

7 (i) Translation by the vector $\begin{pmatrix} 2 \\ -11 \end{pmatrix}$

(ii) Reflect in $y = -x$

(iii) $a = \dfrac{1}{8}$

8 (ii) Rotate 90° anticlockwise

(iii) $c^2 = \dfrac{a^2}{2}$

(iv) $2\sqrt{2}c$

9 (i) Replacing either x by $-x$ or y by $-y$ leaves x^2 and y^2 unchanged.

(ii) Both simply interchange x^2 and y^2.

Chapter 4

Review exercise (Page 93)

1 (i) Converge, decrease

(ii) Converge, oscillate

(iii) Oscillate (this series actually tends to an oscillation between two specific numbers, 0.513 044… and 0.799 455, but this behaviour is not usually called convergent)

(iv) Divergent (decreases after the second term)

2 $2n^2 + n$

3 $n(n+1)^2$

4 $1 - \dfrac{1}{(n+1)^2}$

5 (i) $4r^3$

6 (i) $\dfrac{1}{r} - \dfrac{1}{r+2}$

(ii) $\dfrac{3}{2} - \dfrac{1}{n+1} - \dfrac{1}{n+2}$

7 (i) $\frac{1}{4}n(n+1)(n+2)(n+3)$ (ii) 26 527 650

8 (ii) $\frac{1}{3}(n+1)(n+2)(n+3) - 2$

(iii) $\frac{1}{3}n(n^2 + 6n + 11)$

9 (i) $\frac{1}{3}S = \frac{1}{3} - 2\left(\frac{1}{3}\right)^2 + 3\left(\frac{1}{3}\right)^3 - 4\left(\frac{1}{3}\right)^4 + \ldots$

(ii) $\frac{4}{3}S = 1 - \frac{1}{3} + \left(\frac{1}{3}\right)^2 - \left(\frac{1}{3}\right)^3 + \left(\frac{1}{3}\right)^4 + \ldots$

This is a geometric series with common ratio $-\frac{1}{3}$. Sum to infinity is $\frac{3}{4}$ and so $S = \frac{9}{16}$.

Discussion point (Page 95)

The numerator of one fraction is the negative of the numerator of the other, so that fractions cancel out in pairs.

Exercise 4.1 (Page 96)

1 (i) $\dfrac{1}{2(2r-1)} - \dfrac{1}{2(2r+1)}$

(ii) $\dfrac{1}{2}\left[1 - \dfrac{1}{2r+1}\right]$

(iii) $\dfrac{1}{2}$

2 (iii) $1 - \dfrac{1}{(n+1)^2}$

(iv) As n increases, $\dfrac{1}{(n+1)^2} \to 0$

3 $\dfrac{1}{4} - \dfrac{1}{2(n+1)(n+2)}, \dfrac{1}{4}$

4 (i) $\dfrac{1}{r} - \dfrac{1}{r+2}$

(ii) $\dfrac{3}{2} - \dfrac{2n+3}{(n+1)(n+2)}, \dfrac{3}{2}$

5 (i) $\dfrac{1}{3(r-1)} - \dfrac{1}{3(r+2)}$

(ii) $\dfrac{1}{3}\left[\dfrac{11}{6} - \dfrac{3n^2 + 6n + 2}{n(n+1)(n+2)}\right]$

(iii) $\dfrac{11}{18}$

6 $\dfrac{1}{4} - \dfrac{2n+3}{2(n+2)(n+3)}$

7 (ii) $\dfrac{5}{4} - \dfrac{4n+5}{2(n+1)(n+2)}$

Review exercise (Page 100)

1 (ii) $\mathbf{A}^{+1} = \begin{pmatrix} 1 - 3(k+1) & 9(k+1) \\ -(k+1) & 1 + 3(k+1) \end{pmatrix}$

(iii) $\mathbf{A}^k = \begin{pmatrix} -2 - 3k & 9k + 9 \\ -k - 1 & 3k + 4 \end{pmatrix}$

2 (ii) $\frac{1}{2}(k+1)(3(k+1)+1)$

(iii) $\frac{1}{2}(k+1)(3k+4)$

3 (ii) $u_{k+1} = 2^{k+2} - 1$

(iii) $u_{k+1} = 2(2^{k+1} - 1) + 1$

14 (i) $\mathbf{M}^2 = \begin{pmatrix} 9 & 6 & -3 \\ 0 & 9 & 0 \\ 0 & 18 & 0 \end{pmatrix}$

$\mathbf{M}^3 = \begin{pmatrix} 27 & 18 & -9 \\ 0 & 27 & 0 \\ 0 & 54 & 0 \end{pmatrix}$

$\mathbf{M}^4 = \begin{pmatrix} 81 & 54 & -27 \\ 0 & 81 & 0 \\ 0 & 162 & 0 \end{pmatrix}$

(ii) $\mathbf{M}^n = \begin{pmatrix} 3^n & 2 \times 3^{n-1} & -3^{n-1} \\ 0 & 3^n & 0 \\ 0 & 2 \times 3^n & 0 \end{pmatrix}$

Practice questions 1 (Page 103)

1 $\begin{pmatrix} \frac{1}{2} & -\frac{\sqrt{3}}{2} \\ \frac{\sqrt{3}}{2} & \frac{1}{2} \end{pmatrix}$ or $\begin{pmatrix} \frac{1}{2} & \frac{\sqrt{3}}{2} \\ -\frac{\sqrt{3}}{2} & \frac{1}{2} \end{pmatrix}$ [4]

2 c_n is $\dfrac{1 + 3 + 5 + \cdots + (2n-1)}{(2n+1) + \cdots + (4n-1)}$ in some form. [1]

The sum of the n terms of the numerator is n^2. [2]

The sum of the $2n$ combined terms of the numerator and denominator is $4n^2$. [1]

Hence c_n is $\dfrac{n^2}{4n^2 - n^2} = \dfrac{n^2}{3n^2} = \dfrac{1}{3}$. [2]

3 Base case: $\dfrac{\mathrm{d}y}{\mathrm{d}x} = \mathrm{e}^{2x} + 2x\mathrm{e}^{2x}$ [1]

$(2^1 x + 1 \times 2^0)\mathrm{e}^{2x} = (2x+1)\mathrm{e}^{2x} = \dfrac{\mathrm{d}y}{\mathrm{d}x}$ [1]

Assume true for $n = k$, so that

$\dfrac{\mathrm{d}^k y}{\mathrm{d}x^k} = (2^k x + k \times 2^{k-1})\mathrm{e}^{2x}$. [1]

Then $\dfrac{\mathrm{d}^{k+1} y}{\mathrm{d}x^{k+1}} = \dfrac{\mathrm{d}}{\mathrm{d}x}\left[(2^k x + k \times 2^{k-1})\mathrm{e}^{2x}\right]$ [1]

Therefore $\dfrac{\mathrm{d}^{k+1} y}{\mathrm{d}x^{k+1}} = \left[2^k + 2(2^k x + k \times 2^{k-1})\right]\mathrm{e}^{2x}$ [1]

$= \left[2^k + 2^{k+1} x + k \times 2^k\right]\mathrm{e}^{2x}$

$= \left[2^k + 2^{k+1} x + k \times 2^k\right]\mathrm{e}^{2x}$

$= \left[2^{k+1} x + (k+1) \times 2^k\right]\mathrm{e}^{2x}$ [1]

Hence, true for $n = k \Rightarrow$ true for $n = k + 1$.
Hence true for all positive integer n by mathematical induction. [1]

4 (i) $\overrightarrow{AB} = \begin{pmatrix} 2 \\ -2 \\ 0 \end{pmatrix}$, $\overrightarrow{AC} = \begin{pmatrix} 1 \\ 1 \\ 4 \end{pmatrix}$ [1]

Show that $\overrightarrow{AB} \cdot \begin{pmatrix} 2 \\ 2 \\ -1 \end{pmatrix} = \overrightarrow{AC} \cdot \begin{pmatrix} 2 \\ 2 \\ -1 \end{pmatrix} = 0$ [1]

(ii) $2x + 2y - z = 5$ [2]

(iii) Calculate $\begin{pmatrix} 2 \\ 2 \\ -1 \end{pmatrix} \cdot \begin{pmatrix} 3 \\ -4 \\ 12 \end{pmatrix}$ [1]

$\cos^{-1}\left(\dfrac{-14}{13 \times 3}\right)$ [$= 111.0°$] [1]

$180° - 111.0 = 69.0°$ [1]

5 (i) $\dfrac{1}{(2r+3)(2r+5)} \equiv \dfrac{1}{2(2r+3)} - \dfrac{1}{2(2r+5)}$ [3]

(ii) $\displaystyle\sum_{r=1}^{n} \dfrac{1}{(2r+3)(2r+5)} = \dfrac{1}{2} \cdot \dfrac{1}{5} - \dfrac{1}{2} \cdot \dfrac{1}{2n+5}$ [2]

$= \dfrac{n}{5(2n+5)}$ [2]

(iii) $\dfrac{1}{10}$ [1]

6 (i) C_2 is $9(x-3)^2 + (y+5)^2 = 36$ [1]

or $\dfrac{(2x-6)^2}{16} - \dfrac{(\frac{1}{2}(y+5))^2}{9} = 1$ [1]

Stretch, scale factor $\frac{1}{2}$ in x-direction, [1]
2 in y-direction, [1]

then translation $\begin{pmatrix} 3 \\ -5 \end{pmatrix}$ [1]

[or: translation $\begin{pmatrix} 6 \\ -2.5 \end{pmatrix}$, then stretch, scale factor $\frac{1}{2}$ in x-direction, 2 in y-direction]

(ii) $\dfrac{(2x-6)^2}{16} = \dfrac{(\frac{1}{2}(y+5))^2}{9}$ [1]

$y = 3x - 14$ and $y = -3x + 4$ [2]

7 (i) $\Delta = \begin{vmatrix} 1 & 1 \\ k & 0 \end{vmatrix} + k \begin{vmatrix} k & 1 \\ 1 & k \end{vmatrix}$ [1]

$\Delta = -k + k(k^2 - 1) = k^3 - 2k$ [1]

No inverse if $\Delta = 0$. [1]

No inverse for $k = 0$ or $k = \pm\sqrt{2}$. [1]

Inverse $= \dfrac{1}{k^3 - 2k} \begin{pmatrix} k^2 & -k & -k \\ -k & k^2 - 1 & 1 \\ -k & 1 & k^2 - 1 \end{pmatrix}^{\mathrm{T}}$ [1]

Inverse $= \dfrac{1}{k^3 - 2k} \begin{pmatrix} k^2 & -k & -k \\ -k & k^2 - 1 & 1 \\ -k & 1 & k^2 - 1 \end{pmatrix}$ [1]

(ii) Two planes are parallel. [1]

The other plane ($y + z = 5$) intersects each of them in a line. [1]

(iii) If there is a point of intersection, coordinates are

$\dfrac{1}{k^3 - 2k} \begin{pmatrix} k^2 & -k & -k \\ -k & k^2 - 1 & 1 \\ -k & 1 & k^2 - 1 \end{pmatrix} \begin{pmatrix} k \\ 1 \\ 1 \end{pmatrix}$ [1]

$= \dfrac{1}{k^3 - 2k} \begin{pmatrix} k^3 - 2k \\ 0 \\ 0 \end{pmatrix}$ [1]

The point of intersection is $(1, 0, 0)$ and the coordinates are independent of k. [1]

(iv) (a) $\mathrm{I} = \left(\dfrac{1}{\sqrt{2}}\right)(\mathrm{II} + \mathrm{III})$ but

$4 \neq \left(\dfrac{1}{\sqrt{2}}\right)(3 + 5)$ [1]

Therefore planes form a triangular prism. [1]

(b) $\mathrm{I} = \left(\dfrac{1}{\sqrt{2}}\right)(\mathrm{II} + \mathrm{III})$ and

$4\sqrt{2} = \left(\dfrac{1}{\sqrt{2}}\right)(3 + 5)$ [1]

Planes meet in a sheaf [1]

8 $3(x - y)(y - z)(z - x)$

E.g.: $c_1' = c_1 - c_2$: $\begin{vmatrix} 3 & x - 1 & (x - 1)^2 \\ 3 & y - 1 & (y - 1)^2 \\ 3 & z - 1 & (z - 1)^2 \end{vmatrix}$

$r_2' = r_2 - r_1, r_3' = r_3 - r_1$:

$\begin{vmatrix} 3 & x - 1 & (x - 1)^2 \\ 0 & y - x & (y - x)(x + y - 2) \\ 0 & z - x & (z - x)(x + z - 2) \end{vmatrix}$

Factorise:

$3(y - x)(z - x) \begin{vmatrix} 1 & x - 1 & (x - 1)^2 \\ 0 & 1 & x + y - 2 \\ 0 & 1 & x + z - 2 \end{vmatrix}$

$r_3' = r_3 - r_1$:

$3(y - x)(z - x) \begin{vmatrix} 1 & x - 1 & (x - 1)^2 \\ 0 & 1 & x + y - 2 \\ 0 & 0 & z - y \end{vmatrix}$

Evaluate: $3(x - y)(y - z)(z - x)$

(Completely correct answer, obtained by any method, 5 marks. If not completely correct, up to 4 marks available for each correct row or column operation performed.)

Chapter 5

Discussion points (Page 105)

Many different types of equation are possible, but a quadratic equation seems to produce a good fit. Different parabolas could be superimposed on photographs taken from sufficiently far away to minimise perspective distortion.

More advanced techniques such as *non-linear regression* can be used for fitting curves to data; these are beyond the scope of this book.

Discussion point (Page 107)

(i) The function equals a constant for all other values of x.

(ii) This depends on whether the corresponding factor is repeated or not. Suppose the value of x that makes $N(X)$ and $D(X)$ equal to 0 is $x = x_0$. Then both $N(x)$ and $D(x)$ have a factor of $(x - x_0)$. If both of them have the factor $(x - x_0)$ to the same power, then this can be divided out and therefore ignored (at least for all other values of x). If the powers are different, then it depends on which is larger.

For example, $\dfrac{(x-1)^2(x-2)}{(x-1)^2(x-3)}$ is identically equal

to $\dfrac{(x-2)}{(x-3)}$ provided $x \neq 1$. $\dfrac{(x-1)^2}{(x-1)(x-3)}$ has

a zero at $x = 1$; $\dfrac{(x-1)(x-2)}{(x-1)^2}$ has a vertical

asymptote at $x = 1$.

Discussion points (Page 107)

When $ax + b$ is a constant multiple k of $cx + d$, the

expression $y = \dfrac{ax + b}{cx + d}$ reduces to $y = k$ (except for

$x = -\dfrac{d}{c}$), and then there is no asymptote.

Review exercise (Page 111)

1. (a) 3; $x + 4 = 0$ (b) $5, -3$; $x = 4$ or $x = -4$
   (c) 0 or 4; none

2. (i) 0 (ii) 1 (iii) -3

3. (i) (a) $1 + \dfrac{7}{x-2}$ (b) $3 - \dfrac{13}{x+4}$

   (c) $\dfrac{1}{2} + \dfrac{\frac{1}{2}}{2x+1}$ (d) $-1 + \dfrac{2}{1+2x}$

   (ii) (a) Translate $\begin{pmatrix} 2 \\ 0 \end{pmatrix}$; stretch, scale factor

   7, in y-direction; translate $\begin{pmatrix} 0 \\ 1 \end{pmatrix}$

   (b) Translate $\begin{pmatrix} -4 \\ 0 \end{pmatrix}$; stretch, scale factor

   -13, in y-direction; translate $\begin{pmatrix} 0 \\ 3 \end{pmatrix}$

   (c) Translate $\begin{pmatrix} -1 \\ 0 \end{pmatrix}$; stretch, scale factor $\dfrac{1}{2}$,

   in x-direction; stretch, scale factor $\dfrac{1}{2}$, in

   y-direction; translate $\begin{pmatrix} 0 \\ \frac{1}{2} \end{pmatrix}$

   (d) Translate $\begin{pmatrix} -1 \\ 0 \end{pmatrix}$; stretch, scale factor $\dfrac{1}{2}$,

   in x-direction; stretch, scale factor 2, in

   y-direction; translate $\begin{pmatrix} 0 \\ -1 \end{pmatrix}$

   (iii) (a) $x = 2$, $y = 1$ (b) $x = -4$, $y = 3$

   (c) $x = -\dfrac{1}{2}$, $y = \dfrac{1}{2}$ (d) $x = -\dfrac{1}{2}$, $y = -1$

4. (i)

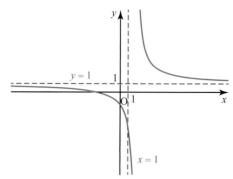

   (ii) $x \leqslant -2$, $x > 1$

5. (i)

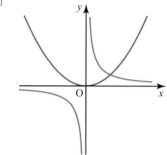

   (ii) $x < 0$ and $x \geqslant 2$

6. $c = 1$ or -7

7. (i) $y \leqslant -1$ or $y \geqslant 3$

   (ii) $y \leqslant -\dfrac{10}{3}$ or $y \geqslant 1$

8. (i) $y = 1$

   (i) (a) above   (b) below

9. (i)

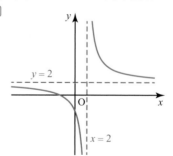

   (ii)

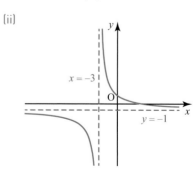

10 (i)

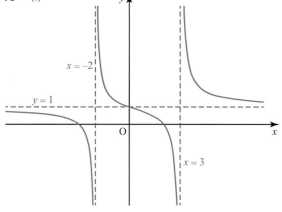

(ii)

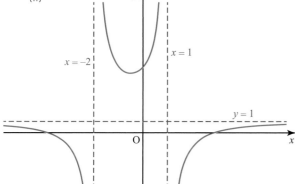

11 $0 < x < \dfrac{1}{3}, x > \dfrac{5}{4}$

12 (i) $x = \dfrac{1 \pm \sqrt{5}}{2}$

(ii) $\dfrac{1 - \sqrt{5}}{2} < x < \dfrac{1 + \sqrt{5}}{2}$

13 (ii)

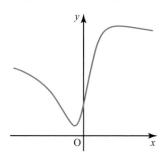

Turning points are at $(-0.5, 1)$ and $(2, 11)$

14 (i) $\dfrac{1}{2} < x \le \dfrac{5}{3}$ (ii) $-3 < x \le \dfrac{5}{3}$

15 (i) Line $y = -2$ with a gap at the point $x = \dfrac{2}{3}$

(ii) (a) undefined (b) equal to $\dfrac{x - 1}{x - 4}$

(iii) For $x \ne a$, the expression equals $1 + x + \ldots + x^{n-1}$, which is a polynomial and does not tend to $\pm\infty$ except when x tends to $\pm\infty$.

16

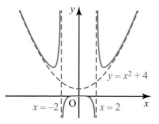

The graph is asymptotic to $y = x^2 + 4$ as $x \to \pm\infty$.

Discussion points (Page 113)

The graph of $y = \dfrac{x^2 + 1}{x + 1}$ has neither a horizontal asymptote or a vertical asymptote other than $x = 1$. If x is very large in magnitude then x^2 and x dominate, so that $y \approx x$.

Write y in the form $y = x - 1 + \dfrac{2}{x + 1}$. You can see that the second term tends to 0 as x tends to infinity, so y approaches $x - 1$.

Exercise 5.1 (Page 115)

1 (i) $x + 5 + \dfrac{15}{x - 2}$

(ii) $x = 2, y = x + 5$

2 (i) $7 - x - \dfrac{20}{x + 3}$

(ii) $x = -3, y = 7 - x$

3 (i) $x + 3 + \dfrac{10x - 1}{x^2 - 3x + 2}$

(ii) $x = 1, x = 2, y = x + 3$

4 (i) $7 - x + \dfrac{71 - 31x}{x^2 + 3x - 10}$

(ii) $x = 2, x = -5, y = 7 - x$

5 (iii) only

6 (i)

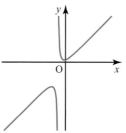

$x = -2, y = x - 1$

(ii)

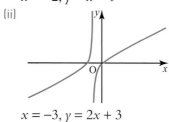

$x = -3, y = 2x + 3$

(iii)

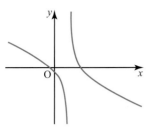

$x = 1, y = 1 - 2x$

7 (i)

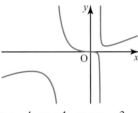

$x = -4, x = 1, y = x - 3$

(ii)

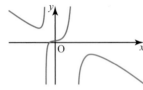

$x = -1, x = 2, y = -2x - 2$

8

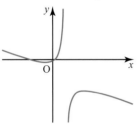

$x = 3, y = -2x - 15; (3 + 2\sqrt{5}, -21 + 8\sqrt{5})$ and $(3 - 2\sqrt{5}, -21 - 8\sqrt{5})$

9 (i) $a = 0, 1$

(ii) $b = -5$

10 (i) $y = x + 1$

(ii) Once (at $x = 2$)

(iii) The graph cuts the asymptote at $x = 2$. As it is above the asymptote to the right of $x = 2$, it is below the asymptote to the left. Therefore as x tends to $-\infty$ the graph approaches the asymptote from below.

Discussion points (Page 118)

$y = f(x)$ needs to have zero gradient at the relevant point. For instance, $y = |x^3|$.

The gradient at a cusp is not defined; the gradient tends to different limits on the two sides of the cusp. There is no derivative of the function at a cusp; it is not defined. The gradient of $y = |f(x)|$ changes from the gradient of $y = f(x)$ on one side of the cusp to the gradient of $y = -f(x)$ at the other side.

Exercise 5.2 (Page 121)

1

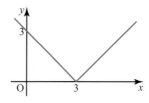

2

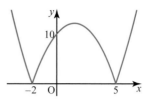

3

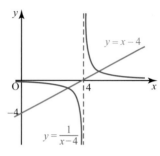

4

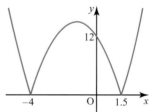

5

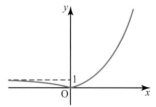

6

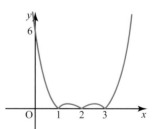

7

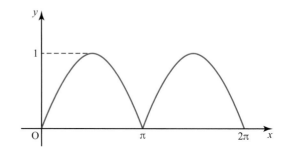

8

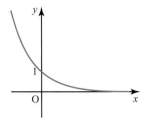

This is the same as the graph of $y = e^x$, reflected in the y-axis. This is because it is the same as $y = e^{-x}$.

9 (i)

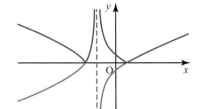

(ii)

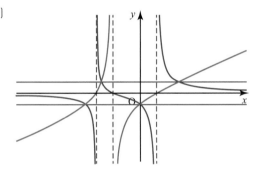

10

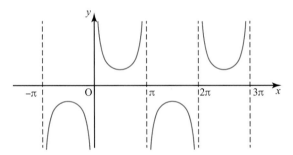

11

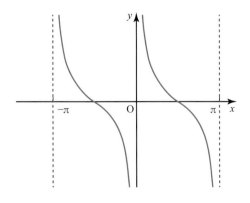

12

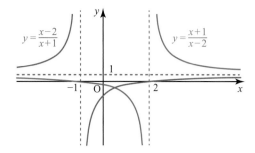

13 (i)

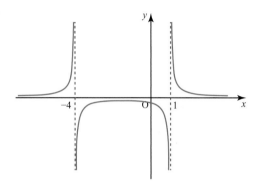

(ii)

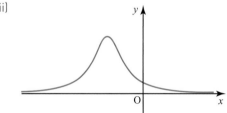

14

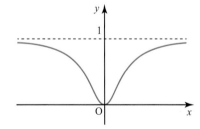

15 $a = 2, b = 5$

16

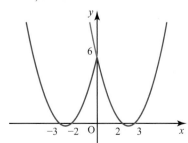

The graphs of $y = f(x)$ and $y = f(|x|)$ are identical for $x \geqslant 0$. For $x \leqslant 0$, the graph of $y = f(|x|)$ is the reflection of the graph of $y = f(x)$ (for $x \geqslant 0$) in the y-axis.

Exercise 5.3 (Page 125)

1   $x = e^2$ or e^{-2}

2   $x = \pm\dfrac{\pi}{4}, \pm\dfrac{3\pi}{4}$

3   0.305 or −1.30

4   $\pm\dfrac{\pi}{4}$

5   $x = -\dfrac{10}{3}, 0$

6   $x = \dfrac{1}{2}$

7   $-5 < x < 4$

8   $-5 \leqslant x \leqslant 3$

9   (i)

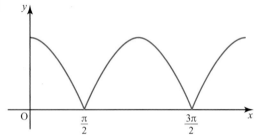

    (ii) $\dfrac{\pi}{3} < x < \dfrac{2\pi}{3}, \dfrac{4\pi}{3} < x < \dfrac{5\pi}{3}$

10   (i)

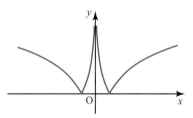

    (ii) $x = \pm2, \pm\dfrac{1}{2}$

    (iii) $x < -2, -\dfrac{1}{2} < x < \dfrac{1}{2}, x > 2$

11   $\dfrac{\pi}{24}, \dfrac{5\pi}{24}, \dfrac{13\pi}{24}, \dfrac{17\pi}{24}$

12   $x = \pm3, \pm1$

13   (i) $a = \dfrac{1}{2}$

    (ii) $x = 1, 4, \dfrac{1}{2}(5 \pm \sqrt{89})$

14   $x = \pm1, \pm3, \pm(2 + \sqrt{7}), \pm(2 - \sqrt{7})$
    For $x \geqslant 0$ the graphs of $y = f(x)$ and $y = f(|x|)$ are identical.
    For $x < 0$, the graph of $y = f(|x|)$ is the reflection in the y-axis of the graph of $y = f(x)$ for $x > 0$.

Chapter 6

Discussion point (Page 128)

The curve may be discontinuous, or it may approach the x-axis but never actually reach it.

Activity 6.1 (Page 131)

The graph of $y = \dfrac{1}{(x-2)^2}$ is above the x-axis for all values of x, so the value of the integral cannot be negative.

Karen has integrated from 1 to 3, but there is a discontinuity at $x = 2$.

Exercise 6.1 (Page 132)

1   (i)

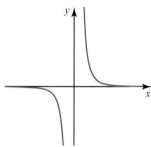

    (ii) $\dfrac{1}{8} - \dfrac{1}{2a^2}$

    (iii) $\dfrac{1}{8}$

2   (i) $3(b-1)^{\frac{1}{3}} + 3,\ 3\sqrt[3]{2} - 3(c-1)^{\frac{1}{3}}$

    (ii) $3\sqrt[3]{2} + 3$

    (iii)

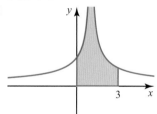

3   (i)

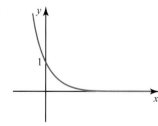

    (ii) $1 - e^{-d}$

    (iii)   1

4   One of the limits is infinity, Convergent, 0.5

5   One of the limits is infinity, Divergent

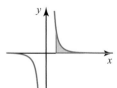
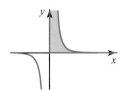

6 The integral approaches infinity at the start of the interval, Divergent

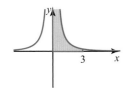

7 One of the limits is infinity (notice zero is not in the interval so the integrand not being defined there is not pertinent). Convergent, 0.5

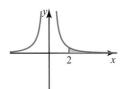

8 One of the limits is negative infinity, Convergent, 1

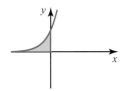

9 One of the limits is infinity, Divergent

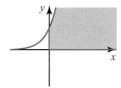

10 The integrand is not defined at $x = 2$ which is within the interval, Convergent, 9.25 to 3 s.f.

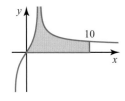

11 One of the limits is infinity, Convergent, -0.5

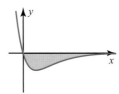

12 1

13 $-\ln 2$

14 1.5

Discussion point (Page 135)

If you replace x with $x\sqrt{3}$ you are doing integration by substitution, so you need to change the variable, e.g. let $u = x\sqrt{3}$ and then use $\mathrm{d}u = \sqrt{3}\,\mathrm{d}x$.

Exercise 6.2 (Page 136)

1 arcsin has domain $[-1, 1]$ and range $\left[-\frac{\pi}{2}, \frac{\pi}{2}\right]$

arccos has domain $[-1, 1]$ and range $[0, \pi]$

arctan has domain $\mathbb{R}$ and range $\left[-\frac{\pi}{2}, \frac{\pi}{2}\right]$

2 (i) $\arcsin\frac{x}{5} + c$ (ii) $\frac{1}{4}\arctan\frac{t}{4} + c$

3 (i) 0.615 (3 s.f.)   (ii) 0.464 (3 s.f.)

4 (i) $\dfrac{3}{\sqrt{1 - 9x^2}}$ (ii) $-\dfrac{1}{2\sqrt{4 - x^2}}$

(iii) $\dfrac{5}{25x^2 + 1}$ (iv) $\dfrac{6x}{\sqrt{1 - 9x^4}}$

(v) $\dfrac{e^x}{e^{2x} + 1}$ (vi) $-\dfrac{6x}{x^4 - 2x^2 + 2}$

5 (i) $\frac{1}{2}\arctan\frac{x}{2} + c$ (ii) $\frac{1}{2}\arctan 2x + c$

(iii) $\arcsin\frac{x}{2} + c$ (iv) $\frac{1}{2}\arcsin 2x + c$

6 (i) $\frac{5}{6}\arctan\frac{x}{6} + c$ (ii) $\frac{2}{5}\arctan\frac{2x}{5} + c$

(iii) $\frac{1}{2}\arcsin\frac{2x}{3} + c$ (iv) $\frac{7}{\sqrt{3}}\arcsin\left(x\sqrt{\frac{3}{5}}\right) + c$

7 (i) $\dfrac{\pi}{12}$ (ii) $\dfrac{\pi}{4}$

(iii) $\dfrac{7\pi}{36}$ (iv) $\dfrac{\pi}{12}$

10 $\arcsin(x^2) + \dfrac{2x^2}{\sqrt{1 - x^4}}$

11 π

12 $\dfrac{\pi}{12\sqrt{10}}$

Exercise 6.3 (Page 140)

1 (ii) $2\ln|x + 1| + \ln|5x - 1| + c$

2 (ii) $-\ln|2x + 3| + \ln|x + 1| - \dfrac{1}{x + 1} + c$

3 (ii) $\ln|x - 3| + \frac{1}{2}\ln(x^2 + 5) + c$

4 (i) $\dfrac{1}{x + 7} + \dfrac{2 - x}{x^2 + 3}$

(ii) $\ln|x + 7| - \frac{1}{2}\ln\left(\frac{1}{3}x^2 + 1\right) + \frac{2}{\sqrt{3}}\arctan\frac{x}{\sqrt{3}} + c$

5 (i) $2\ln|x + 1| - \ln(x^2 + 1) + 2\arctan x + c$

(ii) $\ln|x - 2| - \frac{1}{2}\ln(x^2 + 9) - \frac{1}{3}\arctan\frac{x}{3} + c$

(iii) $\frac{1}{8}\ln(4x^2 + 1) + \ln|x + 2| + \frac{1}{2}\arctan 2x + c$

6 (i) $\frac{1}{2}(\pi + \ln 2)$

(ii) 2.23

(iii) $-\dfrac{\pi}{3\sqrt{3}} - \ln\dfrac{3}{2}$

7 $\ln\dfrac{\sqrt{13}}{3}$

8 $\dfrac{1}{x + 1} - \dfrac{3}{x - 1} + \dfrac{2}{(x - 1)^2} + \dfrac{2x}{x^2 + 1}$

9 $1 - \ln 2$

10 (i) 0.6319 (iv) 9.3%

(iii) 0.579

Activity 6.4 (Page 142)

The first one can be written as $\dfrac{1}{(x+1)^2 - 3}$ and the second can be written as $\dfrac{1}{\sqrt{(x-1)^2 + 2}}$. Neither of these are the correct form for using the standard integrals.

Exercise 6.4 (Page 144)

1   $\dfrac{1}{4\sqrt{15}}$

2   $\dfrac{9\pi}{4}$

3   $\dfrac{1}{4\sqrt{2}}$

4   $\dfrac{1}{2}\arctan\dfrac{3x+1}{2} + c$

5   $\arcsin\dfrac{x-1}{2} + c$

6   (i) $x\arcsin x + \sqrt{1 - x^2} + c$

    (ii)   (a) $x\arccos x - \sqrt{1 - x^2} + c$

           (b) $x\arctan x - \dfrac{1}{2}\ln(x^2 + 1) + c$

           (c) $x\operatorname{arccot} x + \dfrac{1}{2}\ln(x^2 + 1) + c$

7   (i) $\dfrac{1}{16\sqrt{3}}$ (ii) $\dfrac{1}{\sqrt{2}}$ (iii) $\dfrac{\pi}{5}$

8   (i) $\dfrac{1}{2}b\sqrt{a^2 - b^2} + \dfrac{1}{2}a^2\arcsin\dfrac{b}{a}$

    (ii)

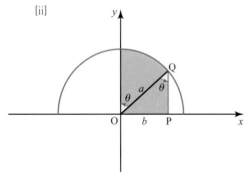

    $\sin\theta = \dfrac{b}{a}$, so $\dfrac{1}{2}a^2\arcsin\dfrac{b}{a}$ is the area of the sector OPQ and $\dfrac{1}{2}b\sqrt{a^2 - b^2}$ is the area of the triangle.

10   (i) $\dfrac{1}{2}\ln(x^2 + 1) + \arctan x + c$

     (ii) $\dfrac{1}{2}\ln\left|x^2 + 2x + 3\right| + c$

     (iii) $\dfrac{1}{\sqrt{2}}\arctan\dfrac{(x+1)}{\sqrt{2}} + c$

11   (i) $\sqrt{4 - x^2} + 2\arcsin\dfrac{x}{2} + c$

     (ii) $\arcsin\dfrac{x-2}{2} + c$

     (iii) $\sqrt{4x - x^2} + c$

12   $\dfrac{1}{|x|\sqrt{x^2 - 1}}$, $\operatorname{arcsec}\dfrac{x}{a} + c$

14   $\dfrac{1}{2}\left(\sqrt{2} + \ln(1 + \sqrt{2})\right)$

Chapter 7

Review exercise (Page 152)

1   $\left(2, \dfrac{8\pi}{3}\right)$ $\left(-2, \dfrac{5\pi}{3}\right)$ $\left(2, \dfrac{14\pi}{3}\right)$

2   (i) $\left(-3\sqrt{3}, -3\right)$

    (ii) $\left(5\sqrt{2}, \dfrac{7\pi}{4}\right)$

3   $r = 10\sin\theta$

4   $\theta = \arctan\left(\dfrac{3}{4}\right)$

5

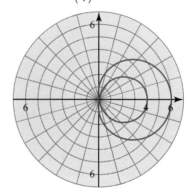

6   (ii) $\left(2, -\dfrac{\pi}{6}\right)$

    (iii)   They are all $\dfrac{\pi}{3}$

    (iv)   equilateral (or equivalent)

7   (i)

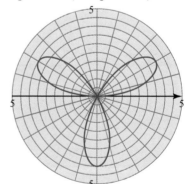

    (ii) $\theta = \dfrac{\pi}{3}$

    (iii) $r\sin\theta = -4$

    (iv) $r = 4$

8   (i) $r = 4\sin\theta - 2\cos\theta$

(iii)

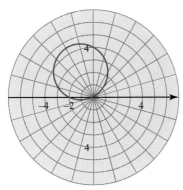

(iv) $\left(2\sqrt{5}, \arctan\dfrac{1}{2}\right)$

(v) $r = \dfrac{10}{2\sin\theta - \cos\theta}$

Exercise 7.1 (Page 154)

1 (i) Area is 25π
   (ii) Area is 50π (i.e. two loops of the circle)

2 (i)

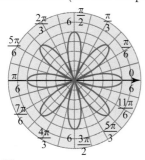

   (ii) $\dfrac{25\pi}{16}$

3 (i)

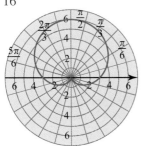

   (ii) $\dfrac{27\pi}{2}$

4 $\dfrac{64\pi}{3}$

5 (i) $2\pi + \dfrac{3\sqrt{3}}{2}$ (ii) $\pi + 3\sqrt{3}$

6

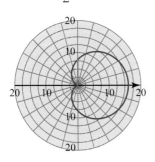

Larger part $= 24\pi + 64$ Smaller part $= 24\pi - 64$

7

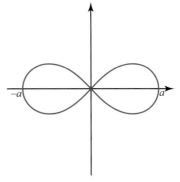

One loop has area $\dfrac{1}{2}a^2$.

8 $A = \dfrac{a^2}{4k}\left(e^{\frac{k\pi}{2}} - 1\right), B = \dfrac{a^2}{4k}e^{4k\pi}\left(e^{\frac{k\pi}{2}} - 1\right),$

   $C = \dfrac{a^2}{4k}e^{8k\pi}\left(e^{\frac{k\pi}{2}} - 1\right)$

   Common ratio $e^{4k\pi}$

9 1.15

Chapter 8

Activity 8.1 (Page 157)

2   $p(x) = 1 + x + \dfrac{1}{2}x^2 + \dfrac{1}{6}x^3 + \dfrac{1}{24}x^4 + \dfrac{1}{120}x^5$

Activity 8.2 (Page 158)

$\left(1 + x + \dfrac{1}{2}x^2 + \dfrac{1}{6}x^3\right)\left(1 - x + \dfrac{1}{2}x^2 - \dfrac{1}{6}x^3\right)$
$\qquad = 1 - \dfrac{1}{12}x^4 - \dfrac{1}{36}x^6$

If more terms were used, the terms in x^4 and x^6 and higher terms would cancel out, leaving 1.

Discussion point (Page 158)

It is often the case that if the terms of a series converge, the sum of the series also converges. However, it is not always the case: for example, the terms of the series $u_n = \dfrac{1}{n}$ converge to zero but the sum of the terms does not converge.

Activity 8.3 (Page 159)

Order = 7

Maclaurin approximation up to term in x^{10} is 7.388994709, percentage error = -0.00083%.

Activity 8.5 (Page 164)

$1 + \frac{1}{2}x^2 + \frac{5}{24}x^4$

This works because

$$\sec x = \frac{1}{\cos x} = \frac{1}{1 - \frac{x^2}{2} + \frac{x^4}{24} + \cdots} = \frac{1}{1 + y},$$

where $y = -\frac{x^2}{2} + \frac{x^4}{24}$

Exercise 8.1 (Page 164)

1 (ii) 0.9800666667

   (iii) 0.9800665778, 0.00000907%

2 $e^{-\frac{1}{2}x^2} = 1 - \frac{1}{2}x^2 + \frac{1}{8}x^4 - \frac{1}{48}x^6 + \frac{1}{384}x^8;$

   valid for all x; $(-1)^r \dfrac{x^{2r}}{2^r \times r!}$

3 (i) (a) $(1 + x)e^x = 1 + 2x + \frac{3}{2}x^2 + \frac{2}{3}x^3 + \frac{5}{24}x^4$

     (b) $\sin x + \cos x = 1 + x - \frac{1}{2}x^2 - \frac{1}{6}x^3 + \frac{1}{24}x^4$

     (c) $e^x - \sin x = 1 + \frac{1}{2}x^2 + \frac{1}{3}x^3 + \frac{1}{24}x^4$

   (ii) (a) all x (b) all x (c) all x

4 (i) $k = \frac{9}{8}$

   (ii) $-1 < x < 1$

5 (i) 0.7074

   (ii) A fourth term is needed giving 0.7071 which is correct to 4 decimal places.

6 (i) (a) $\cos^2 x = 1 - x^2 + \frac{1}{3}x^4$

     (b) $\cos 2x = 1 - 2x^2 + \frac{2}{3}x^4$

7 $f'(0) = 0,\ f''(0) = -3,\ f^{(3)}(0) = 15$

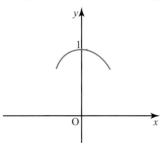

8 (i) $\frac{1}{2}\arcsin 2x + c$

   (ii) $2x + \frac{4}{3}x^3 + \frac{12}{5}x^5 + \frac{40}{7}x^7 + \cdots$

   (iii) Using $x = 0.25$ gives $0.523525856\ldots$, percentage error $= -0.0139\%$

9 (i) $1 + x + x^2 + x^3 + \cdots$

   (ii) $\dfrac{1}{(1 - x)^2}$

10 (iv) 3.14159

11 (ii) $a_1 = 1,\ a_2 = 0$

(iv) $x + \dfrac{x^3}{3!} + \dfrac{9x^5}{5!} + \cdots$

$$+ \frac{(2r - 1)^2 (2r - 3)^2 \ldots 5^2 \times 3^2}{(2r + 1)!} x^{2r+1} + \cdots$$

Activity 8.6 (Page 167)

The first term, $\dfrac{1}{x}$ is undefined.

Activity 8.7 (Page 170)

Using Maclaurin:

$$\lim_{x \to 0} \frac{\sin x^2}{1 - \cos 2x} = \lim_{x \to 0} \frac{x^2 - \dfrac{x^6}{3!} + \cdots}{1 - \left(1 - \dfrac{(2x)^2}{2} + \dfrac{(2x)^4}{4!} + \cdots\right)}$$

$$= \lim_{x \to 0} \frac{x^2 - \dfrac{x^6}{6} + \cdots}{2x^2 - \dfrac{2}{3}x^4 + \cdots} = \lim_{x \to 0} \frac{1 - \dfrac{x^4}{6} + \cdots}{2 - \dfrac{2}{3}x^2 + \cdots} = \frac{1}{2}$$

Using L'Hôpital's rule:

$$\lim_{x \to 0} \frac{\sin x^2}{1 - \cos 2x} = \lim_{x \to 0} \frac{2x \cos x^2}{2 \sin 2x}$$

$$= \lim_{x \to 0} \frac{2\cos x^2 - 4x^2 \sin x^2}{4 \cos 2x} = \frac{2}{4}$$

In this case Maclaurin's seems easier as the techniques used are easier but there is one more step in the working as shown.

Exercise 8.2 (Page 170)

1 (i) 1   (ii) 1   (iii) $-\dfrac{1}{2}$ (iv) $\dfrac{1}{6}$

2 (i) -1 (ii) $-\dfrac{2}{3}$ (iii) $-\dfrac{\pi}{12\sqrt{e}}$ (iv) $\dfrac{16}{\pi}$

3 The $\lim\limits_{x \to \infty} \dfrac{f'(x)}{g'(x)}$ does not exist and that means the conditions for applying L'Hôpital's rule are not satisfied.

4 (i) ∞ (ii) 1   (iii) 0

5 (ii) $\dfrac{1}{4}$

6 $-\dfrac{1}{4}$

7 (i) -1 (ii) -1

   (iii) Maclaurin because known series can be used and the expressions simplify to a quotient that can be evaluated, whereas L'Hôpital's rule requires 2 iterations.

   (iv) It depends on whether the functions use standard Maclaurin's series and how easy they are to differentiate.

8 (iii) 0.6931471806 and 0.9830598741

(iv) Large percentage error suggests slow convergence and so more terms needed to improve the approximation.

9 (ii) $f^{(3)}(x) = 4f''(x) - 13f'(x)$
$f^{(4)}(x) = 4f^{(3)}(x) - 13f''(x)$

(iii) $f(x) = 3x + 6x^2 + \frac{3}{2}x^3 - 5x^4 + \ldots$

(iv) 3

10 Given $0 < r < 1$, the area of the square
$= \left(\frac{1}{1-r}\right)^2 = 1 + 2r + 3r^2 + 4r^3 + 5r^4 + \ldots$

Also the side of the square
$= (1-r)\left(1 + 2r + 3r^2 + 4r^3 + 5r^4 + \ldots\right)$
$= (1-r)\left(\frac{1}{1-r}\right)^2 = \frac{1}{1-r}$

Practice questions 2 (Page 174)

1 $x(5x-6) > x^3(x-2)$ $\qquad$ [1]

$x(x^3 - 2x^2 - 5x + 6) > 0$

$x(x-1)(x+2)(x-3) > 0$ $\qquad$ [2]

$x < -2, 0 < x < 1, x > 3$ $\qquad$ [2]

2 (a) (i) $A = \int_1^a \frac{1}{x}\,dx$ $\qquad$ [1]

$A = [\ln x]_1^a = \ln a$ $\qquad$ [1]

(ii) As $a \to \infty$, $\ln a \to \infty$ $\qquad$ [1]

(b) $V = \pi \int_1^\infty \frac{1}{x^2}\,dx = \pi\left[-\frac{1}{x}\right]_1^\infty$ $\qquad$ [2]

$V = \pi \lim_{N\to\infty}\left(-\frac{1}{N} + 1\right) = \pi$ $\qquad$ [1]

3 Require $\int_0^2 \left(\frac{2x(6-x)}{(3x+2)(x^2+4)}\right)dx$ $\qquad$ [1]

Using partial fractions gives

$\frac{2x(6-x)}{(3x+2)(x^2+4)} \equiv \frac{-2}{3x+2} + \frac{4}{x^2+4}$ $\qquad$ [3]

$\int_0^2 \left(\frac{-2}{3x+2} + \frac{4}{x^2+4}\right)dx$

$= \left[-\frac{2}{3}\ln(3x+2) + 2\tan^{-1}\frac{x}{2}\right]_0^2$ $\qquad$ [4]

$= \left(-\frac{2}{3}\ln(8) + \frac{2\pi}{4}\right) - \left(-\frac{2}{3}\ln(2) + 0\right)$

$= -\frac{2}{3}\ln 4 + \frac{\pi}{2}$ $\qquad$ [2]

4 (i) $f(x) \equiv x - 3 + \frac{-2x+7}{x^2 + 3x + 2}$ $\qquad$ [1]

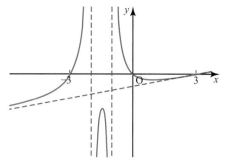

$\qquad$ [4]

Asymptotes $x = -2, x = -1, y = x - 3$ $\qquad$ [2]

(ii) $\qquad$ [4]

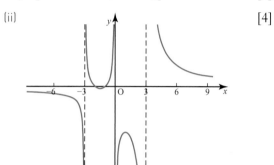

5 (i) $y_1 = -\tan x; \; y_2 = -\sec^2 x;$
$y_3 = -2\sec^2 x \tan x$ $\qquad$ [2]
so $y_3 + 2y_2 y_1 = 0$. $\qquad$ [1]

(ii) $y(0) = 0; \; y_1(0) = 0; \; y_2(0) = -1; \; y_3(0) = 0$. $\qquad$ [2]
Find y_4 and obtain $y_4(0) = -2$. $\qquad$ [1]
Establish $\ln(\cos x) \approx -\frac{x^2}{2} - \frac{x^4}{12} + \ldots$ $\qquad$ [2]

(iii) $\ln\left(\cos\frac{\pi}{3}\right) = \ln\frac{1}{2} = -\ln(2)$ $\qquad$ [1]

so $\ln(2) \approx \frac{\left(\frac{\pi}{3}\right)^2}{2} + \frac{\left(\frac{\pi}{3}\right)^4}{12} + \ldots$

$= \frac{\pi^2}{18}\left(1 + \frac{\pi^2}{54}\right)$ $\qquad$ [2]

(iv) $\ln\left(\cos\frac{\pi}{4}\right) = \ln\frac{1}{\sqrt{2}} = -\frac{1}{2}\ln(2)$,

so $\ln(2) \approx \frac{2\left(\frac{\pi}{4}\right)^2}{2} + \frac{2\left(\frac{\pi}{4}\right)^4}{12} + \ldots$

$= \frac{\pi^2}{16}\left(1 + \frac{\pi^2}{96}\right)$ $\qquad$ [2]

$\ln(2) \approx 0.69315\ldots$; using $\frac{\pi}{4}$ in the expansion gives $0.68027\ldots$ relative error $\approx 2\%$; using $\frac{\pi}{3}$ in the expansion gives $0.64853\ldots$ relative error $\approx 6\%$ $\qquad$ [1]
Using $\frac{\pi}{4}$ is better as it is smaller. $\qquad$ [1]

6 (i) $\frac{1}{2}\int_0^{\frac{\pi}{6}} \theta^{2n}\cos(a\theta)\,d\theta$ $\qquad$ [2]

(ii) Use $\cos(a\theta) \approx 1 - \dfrac{a^2\theta^2}{2}$ [2]

$$\frac{1}{2}\int_0^{\frac{\pi}{6}} \theta^{2n}\left(1 - \frac{a^2\theta^2}{2}\right)d\theta$$

$$= \frac{1}{2}\left[\frac{\theta^{2n+1}}{2n+1} - \frac{\theta^{2n+3}}{2(2n+3)}\right]_0^{\frac{\pi}{6}}$$ [2]

$$= \frac{1}{2}\left(\frac{\pi}{6}\right)^{2n+1}\left[\frac{1}{2n+1} - \frac{a^2\left(\frac{\pi}{6}\right)^2}{2(2n+3)}\right]$$ [1]

$$= \left(\frac{\pi}{6}\right)^{2n+1}\left[\frac{72(2n+3) - a^2\pi^2(2n+1)}{144(2n+1)(2n+3)}\right]$$ [1]

(iii) $\dfrac{\pi}{12}$ [1]

(iv) It is the sector of a circle of radius 1, angle $\dfrac{\pi}{6}$ [1]

(v) $r = \theta\sqrt{\cos\theta}$

$$\sqrt{x^2+y^2} = \frac{\sqrt{x}}{\sqrt[4]{x^2-y^2}}\tan^{-1}\frac{y}{x}$$ [2]

$$y = x\tan\left(\frac{(x^2+y^2)^{3/4}}{\sqrt{x}}\right)$$ [2]

Chapter 9

Discussion points (Page 176)

In a reflection, points on the mirror line will map to themselves.

In a rotation/enlargement, the centre of rotation/enlargement is the only point that maps to itself.

In a reflection, lines perpendicular to the mirror line will map to themselves.

In a rotation, there are no lines that map to themselves (accept trivial cases of rotation of 360°).

In an enlargement, all ray lines through the centre of enlargement will map to themselves.

Activity 9.1 (Page 178)

(i) (a) $\begin{pmatrix} 4 \\ 1 \end{pmatrix}$ (b) $\begin{pmatrix} 10 \\ 5 \end{pmatrix}$ (c) $\begin{pmatrix} 2 \\ 3 \end{pmatrix}$

(d) $\begin{pmatrix} 0 \\ 5 \end{pmatrix}$ (e) $\begin{pmatrix} -2 \\ 2 \end{pmatrix}$ (f) $\begin{pmatrix} -6 \\ 1 \end{pmatrix}$

(ii) (a) $y = \dfrac{1}{4}x$ (b) $y = \dfrac{1}{2}x$ (c) $y = \dfrac{3}{2}x$

(d) $x = 0$ (e) $y = -x$ (f) $y = -\dfrac{1}{6}x$

Activity 9.2 (Page 181)

Since $\mathbf{u}_1$ and $\mathbf{u}_2$ are non-zero and non-parallel you can express any position vector $\mathbf{p}$ as $\alpha\mathbf{u}_1 + \beta\mathbf{u}_2$.

Then:

$$\mathbf{Mp} = \mathbf{M}(\alpha\mathbf{u}_1 + \beta\mathbf{u}_2)$$
$$= \alpha\mathbf{Mu}_1 + \beta\mathbf{Mu}_2$$
$$= 2\alpha\mathbf{u}_1 + 5\beta\mathbf{u}_2$$

Showing that the image of $\mathbf{p}$ is attracted towards the eigenvector with the numerically larger eigenvalue.

Activity 9.3 (Page 183)

$\det(\mathbf{M} - \lambda\mathbf{I}) \equiv (\lambda_1 - \lambda)(\lambda_2 - \lambda)(\lambda_3 - \lambda)$ ①

Putting $\lambda = 0$ gives $\det\mathbf{M} = \lambda_1\lambda_2\lambda_3$

Coefficient of λ^2 on RHS of ① $= \lambda_1 + \lambda_2 + \lambda_3$.

When expanding $\det(\mathbf{M} - \lambda\mathbf{I})$ the terms in λ^2 arise only from the product of the elements on the leading diagonal, $(a_1 - \lambda)(b_2 - \lambda)(c_3 - \lambda)$ since all other terms are linear in λ. The coefficient of λ is $a_1 + b_2 + c_3$, and the result follows.

Exercise 9.1 (Page 183)

1 (i) $\det\mathbf{M} = (2 \times (-2)) - ((-1) \times 3) = -1$

The area is preserved under the transformation but has been 'flipped' over.

(ii) $\det\begin{pmatrix} 2-\lambda & 3 \\ -1 & -2-\lambda \end{pmatrix} = 0 \Leftrightarrow$

$\Leftrightarrow (2-\lambda)(-2-\lambda) - (-3) = 0 \Leftrightarrow \lambda^2 - 1 = 0$

$(\lambda - 1)(\lambda + 1) = 0$. So $\lambda = 1, -1$.

(iii) When $\lambda = 1$ eigenvector is $\begin{pmatrix} 3 \\ -1 \end{pmatrix}$.

When $\lambda = -1$ eigenvector is $\begin{pmatrix} 1 \\ -1 \end{pmatrix}$.

(iv) $y = -\dfrac{1}{3}x$, $y = -x$

2 (i) $7, k(3\mathbf{i} + 2\mathbf{j}); 2, k(\mathbf{i} - \mathbf{j})$ where $k \neq 0$

(ii) $4, k(2\mathbf{i} - 3\mathbf{j}); -1, k(\mathbf{i} - 4\mathbf{j})$ where $k \neq 0$

(iii) $1 + \sqrt{2}, k(\sqrt{2}\mathbf{i} + \mathbf{j}); 1 - \sqrt{2}, k(\sqrt{2}\mathbf{i} - \mathbf{j})$ where $k \neq 0$

(iv) 2 (repeated), $k(\mathbf{i} - \mathbf{j})$ where $k \neq 0$

(v) $1, k(4\mathbf{i} + \mathbf{j}); 0.3, k(\mathbf{i} + 2\mathbf{j})$ where $k \neq 0$

(vi) $p, k\mathbf{i}; q, k\mathbf{j}$ where $k \neq 0$

3 (i) $3, c\mathbf{i}; 2, c\mathbf{j}; -1, c(\mathbf{j} - 3\mathbf{k})$ where $c \neq 0$

(ii) $4, c(\mathbf{i} - \mathbf{j} + 2\mathbf{k}); 3, c(\mathbf{i} - 2\mathbf{j} + 2\mathbf{k}); -1,$
$c(9\mathbf{i} - 14\mathbf{j} - 2\mathbf{k})$ where $c \neq 0$

(iii) $4, c(\mathbf{i} + \mathbf{j} + \mathbf{k}); -3, c(\mathbf{i} - 6\mathbf{j} + \mathbf{k}); 0,$
$c(5\mathbf{i} + 9\mathbf{j} - 7\mathbf{k})$ where $c \neq 0$

(iv) $3, c\mathbf{j}; 2, c(\mathbf{i} + \mathbf{k}); -2, c(\mathbf{i} - \mathbf{k})$ where $c \neq 0$

(v) $9, c(\mathbf{j} - \mathbf{k}); 4, c(\mathbf{i} + \mathbf{j} - 2\mathbf{k}); 1, c(\mathbf{i} + \mathbf{j} - \mathbf{k})$
where $c \neq 0$

4 (i) Characteristic equation is
$(\lambda - 1)(\lambda - 2)^2 = 0$ so $\lambda = 2, 2, 1$

(ii) When $\lambda = 1$ eigenvector is $\begin{pmatrix} 1 \\ 5 \\ 1 \end{pmatrix}$ (or any multiple).

When $\lambda = 2$ (repeated) eigenvectors are
$\begin{pmatrix} p \\ 2p + 4q \\ q \end{pmatrix}$ (or any multiple), where p

and q are not both 0.

(iii) The line through the origin with direction
$\begin{pmatrix} 1 \\ 5 \\ 1 \end{pmatrix}$ is a line of invariant points whereas

lines through the origin with direction

$\begin{pmatrix} p \\ 2p + 4q \\ q \end{pmatrix}$ are invariant lines of the

transformation T.

5 (i) $1, k(\mathbf{i} + \tan \theta \mathbf{j}); -1, k(\tan \theta \mathbf{i} - \mathbf{j})$

(ii) $\theta \neq \pi n \Rightarrow$ no real eigenvalues
$\theta = n\pi \Rightarrow$ eigenvalues are $(-1)^n$, and all non-zero vectors are eigenvectors.

6 (i) $\alpha + \beta$

(ii) $\alpha\beta$

10 (i) (a) $2, 3$

(b) $4, 9$

(c) $32, 243$

(d) $\dfrac{1}{2}, \dfrac{1}{3}$

(ii) (a) $1, 2, 3$

(b) $1, 4, 9$

(c) $1, 32, 243$

(d) $1, \dfrac{1}{2}, \dfrac{1}{3}$

11 (i) $\mathbf{M}^n\mathbf{v}$ converges to 0

(ii) If $\lambda_1 = 1$, $\mathbf{M}^n\mathbf{v}$ converges to $\mathbf{u}_1$; if $\lambda_1 = -1$
$\mathbf{M}^n\mathbf{v}$ eventually alternates between $\pm\mathbf{u}_1$.

(iii) The magnitude of $\mathbf{M}^n\mathbf{v}$ increases without limit; the direction of $\mathbf{M}^n\mathbf{v}$ becomes parallel to $\mathbf{u}_1$.

13 (i) $\mathbf{M} = \begin{pmatrix} 0.5 & 0.3 \\ 0.5 & 0.7 \end{pmatrix}$

(ii) $\mathbf{M}^2 \begin{pmatrix} 100 \\ 100 \end{pmatrix} = \begin{pmatrix} 76 \\ 124 \end{pmatrix}$

So 76 at Calgary, 124 at Vancouver

(iii) $\mathbf{x} = \begin{pmatrix} 75 \\ 125 \end{pmatrix}$; $\mathbf{Mx} = \lambda\mathbf{x}$ with $\lambda = 1$

(iv) 75 at Calgary, 125 at Vancouver

14 (iv) $4, c(3\mathbf{i} + \mathbf{j}); 2, c(\mathbf{i} + \mathbf{j})$

$\begin{pmatrix} r \\ w \end{pmatrix} = 475e^{4t} \begin{pmatrix} 3 \\ 1 \end{pmatrix} - 425e^{2t} \begin{pmatrix} 1 \\ 1 \end{pmatrix}$

Activity 9.4 (Page 186)

An eigenvalue 2 with the eigenvector of $\begin{pmatrix} -1 \\ 1 \end{pmatrix}$

means that $\begin{pmatrix} 4 & 2 \\ 1 & 3 \end{pmatrix}\begin{pmatrix} -1 \\ 1 \end{pmatrix} = 2\begin{pmatrix} -1 \\ 1 \end{pmatrix}$

An eigenvalue of 5 with eigenvector of $\begin{pmatrix} 2 \\ 1 \end{pmatrix}$ means

that $\begin{pmatrix} 4 & 2 \\ 1 & 3 \end{pmatrix}\begin{pmatrix} 2 \\ 1 \end{pmatrix} = 5\begin{pmatrix} 2 \\ 1 \end{pmatrix}$

These can be combined to give

$\begin{pmatrix} 4 & 2 \\ 1 & 3 \end{pmatrix}\begin{pmatrix} -1 & 2 \\ 1 & 1 \end{pmatrix} = \begin{pmatrix} -1 & 2 \\ 1 & 1 \end{pmatrix}\begin{pmatrix} 2 & 0 \\ 0 & 5 \end{pmatrix}$

It then follows that

$\begin{pmatrix} 4 & 2 \\ 1 & 3 \end{pmatrix}\begin{pmatrix} -1 & 2 \\ 1 & 1 \end{pmatrix}\begin{pmatrix} -1 & 2 \\ 1 & 1 \end{pmatrix}^{-1} = \begin{pmatrix} -1 & 2 \\ 1 & 1 \end{pmatrix}\begin{pmatrix} 2 & 0 \\ 0 & 5 \end{pmatrix}\begin{pmatrix} -1 & 2 \\ 1 & 1 \end{pmatrix}^{-1}$

$\begin{pmatrix} 4 & 2 \\ 1 & 3 \end{pmatrix}\begin{pmatrix} 1 & 0 \\ 0 & 1 \end{pmatrix} = \begin{pmatrix} -1 & 2 \\ 1 & 1 \end{pmatrix}\begin{pmatrix} 2 & 0 \\ 0 & 5 \end{pmatrix}\begin{pmatrix} -1 & 2 \\ 1 & 1 \end{pmatrix}^{-1}$

Discussion points (Page 187)

If there is a repeated eigenvalue, the corresponding 'vectors' in the matrix $\mathbf{U}$ will be scalar multiples of each other. Using properties established with determinants, row operations could be carried out to give a column of zeros hence there would be a determinant of 0 and no inverse can be calculated.

Activity 9.5 (Page 188)

You get the zero matrix.

General characteristic equation is given by

$$\begin{vmatrix} a - \lambda & c \\ b & d - \lambda \end{vmatrix} = 0$$

$$(a - \lambda)(d - \lambda) - bc = 0$$

$$\lambda^2 - (a + d)\lambda + ad - bc = 0$$

It needs to be shown that

$$\mathbf{M}^2 - (a + d)\mathbf{M} + (ad - bc)\mathbf{I} = 0$$

$$\begin{pmatrix} a & c \\ b & d \end{pmatrix}^2 - (a + d)\begin{pmatrix} a & c \\ b & d \end{pmatrix} + (ad - bc)\begin{pmatrix} 1 & 0 \\ 0 & 1 \end{pmatrix}$$

$$= \begin{pmatrix} a^2 + bc & ac + cd \\ ab + bd & bc + d^2 \end{pmatrix} - \begin{pmatrix} a^2 + ad & ac + cd \\ ab + bd & ad + d^2 \end{pmatrix}$$

$$+ \begin{pmatrix} ad - bc & 0 \\ 0 & ad - bc \end{pmatrix}$$

$$= \begin{pmatrix} 0 & 0 \\ 0 & 0 \end{pmatrix}$$

Exercise 9.2 (Page 189)

1   Note: the columns of **U** may be reversed provided the eigenvalues are also reversed. Each column of **U** may (independently) be multiplied by a non-zero constant.

(i) $\mathbf{U} = \begin{pmatrix} 4 & 1 \\ -3 & 1 \end{pmatrix}$, $\mathbf{D} = \begin{pmatrix} 2 & 0 \\ 0 & 9 \end{pmatrix}$

(ii) $\mathbf{U} = \begin{pmatrix} 4 & 5 \\ 2 & 3 \end{pmatrix}$, $\mathbf{D} = \begin{pmatrix} 2 & 0 \\ 0 & 1 \end{pmatrix}$

(iii) $\mathbf{U} = \begin{pmatrix} 1 & 5 \\ 1 & -3 \end{pmatrix}$ $\mathbf{D} = \begin{pmatrix} 1 & 0 \\ 0 & 0.2 \end{pmatrix}$

2   $\begin{pmatrix} 5 & 3 \\ 3 & 2 \end{pmatrix}\begin{pmatrix} 1 & 0 \\ 0 & 0.9 \end{pmatrix}\begin{pmatrix} 2 & -3 \\ -3 & 5 \end{pmatrix}$,

$\begin{pmatrix} 4.0951 & -5.1585 \\ 2.0634 & -2.4390 \end{pmatrix}$;

approximates to $\begin{pmatrix} 10 & -15 \\ 6 & -9 \end{pmatrix}$

3   (i) $\begin{pmatrix} 876 & -1266 \\ 422 & -601 \end{pmatrix}$

(ii) $\begin{pmatrix} 524800 & -523776 \\ -523776 & 524800 \end{pmatrix}$

(iii) $\begin{pmatrix} 0.6667 & 0.3333 \\ 0.6666 & 0.3334 \end{pmatrix}$

4   $5; \begin{pmatrix} 1 & 1 & 1 \\ 1 & 1 & 0 \\ 0 & 1 & -1 \end{pmatrix}\begin{pmatrix} 5 & 0 & 0 \\ 0 & 4 & 0 \\ 0 & 0 & 2 \end{pmatrix}\begin{pmatrix} 1 & 1 & 1 \\ 1 & 1 & 0 \\ 0 & 1 & -1 \end{pmatrix}^{-1}$

5   For example,

(i) $\begin{pmatrix} 3 & 1 \\ 0 & 3 \end{pmatrix}$

(ii) $\begin{pmatrix} 1 & 0 \\ 0 & 1 \end{pmatrix}$

(iii) $\begin{pmatrix} 0 & 1 \\ 0 & 0 \end{pmatrix}$

(iv) $\begin{pmatrix} 1 & 3 \\ 0 & 0 \end{pmatrix} =$

$\begin{pmatrix} 3 & 1 \\ -1 & 0 \end{pmatrix}\begin{pmatrix} 0 & 0 \\ 0 & 1 \end{pmatrix}\begin{pmatrix} 0 & -1 \\ 1 & 3 \end{pmatrix}$ or

$\begin{pmatrix} 6 & 4 \\ 3 & 2 \end{pmatrix} =$

$\begin{pmatrix} 2 & 2 \\ -3 & 1 \end{pmatrix}\begin{pmatrix} 0 & 0 \\ 0 & 8 \end{pmatrix}\begin{pmatrix} \frac{1}{8} & -\frac{1}{4} \\ \frac{3}{8} & \frac{1}{4} \end{pmatrix}$

6   (i) (a) $-2k - 8$

(b) $\dfrac{1}{2k + 8}\begin{pmatrix} -2 - 2k & 1 & 2 + k \\ 6 & 1 & -6 - k \\ 12 & 2 & -4 \end{pmatrix}$

(ii) $k = -3$

(iii) $p = 2, q = 1, r = -2$

7   (i) 1

(ii) $\begin{pmatrix} 1 \\ -1 \end{pmatrix}, \begin{pmatrix} 5 \\ -3 \end{pmatrix}$

(iii) $\begin{pmatrix} 1 & 5 \\ -1 & -3 \end{pmatrix}$

8 Consider the case when $n = 1$

LHS $\mathbf{M}^1 = \mathbf{M}$

RHS $\mathbf{UD}^1\mathbf{U}^{-1} = \mathbf{UDU}^{-1} = \mathbf{M}$

So true for $n = 1$

Assume true for some value of $n = k$

$\mathbf{M}^k = \mathbf{UD}^k\mathbf{U}^{-1}$

Consider the case when $n = k + 1$

$$\mathbf{M}^{k+1} = \mathbf{MM}^k$$
$$= \left(\mathbf{UDU}^{-1}\right)\left(\mathbf{UD}^k\mathbf{U}^{-1}\right)$$
$$= \mathbf{UD}(\mathbf{U}^{-1}\mathbf{U})\mathbf{D}^k\mathbf{U}^{-1}$$
$$= \mathbf{UDID}^k\mathbf{U}^{-1}$$
$$= \mathbf{UDD}^k\mathbf{U}^{-1}$$
$$= \mathbf{UD}^{k+1}\mathbf{U}^{-1}$$

So, true for all $n \geq 1$

9 (i) There are four such products, generally distinct, but **AB** has at most two distinct eigenvalues which may or may not be the product of an eigenvalue of **A** and an eigenvalue of **B**.

(ii) 'Proof' assumes that eigenvector of **A** is eigenvector of **B**.

Review: Complex numbers

Review Exercise R.1 (Page 193)

1 (i) (a) -1 (b) $-i$ (c) 1 (d) i (e) -1

(ii) The powers of i form a cycle:

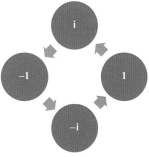

All numbers of the form i^{4n} are equal to 1.
All numbers of the form i^{4n+1} are equal to i.
All numbers of the form i^{4n+2} are equal to -1.
All numbers of the form i^{4n+3} are equal to $-i$.

2 (i) $9 + 12i$ (ii) $-21 - 20i$

(iii) $53 - 89i$

3 (i) $-1 + 2i$ (ii) $\dfrac{9}{5} + \dfrac{7}{5}i$

(iii) $\dfrac{9}{5} - \dfrac{7}{5}i$

4 $\dfrac{2}{29} - \dfrac{179}{29}i$

5 $a = 1, -2$ and $b = 2, 3$

Possible complex numbers are
$1 + 15i, 1 + 10i, 4 + 15i$ and $4 + 10i$

6 $\dfrac{4}{29}, \dfrac{4}{29}$

7 $z + z^* = 2x$ which is real

$zz^* = x^2 + y^2$ which is real

8 $a = \dfrac{5}{3}, b = -\dfrac{5}{3}$

Review Exercise R.2 (Page 196)

1 (i)

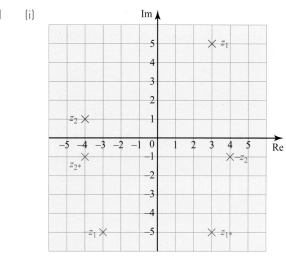

(ii) $-z$ represents a rotation of z about the origin, through an angle of $180°$.

(iii) z^* represents a reflection of z in the real axis.

2

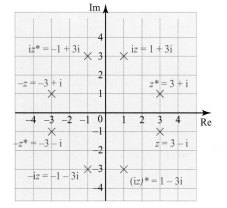

3

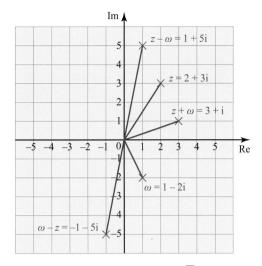

4 (i) (a) $z^0 = 1$, $z^1 = -\frac{1}{2} - \frac{\sqrt{3}}{2}$ i,

$z^2 = -\frac{1}{2} + \frac{\sqrt{3}}{2}$ i

(b)

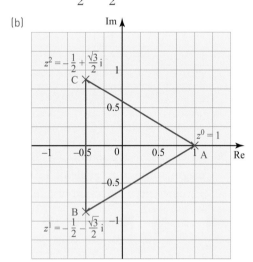

(ii) Using coordinate geometry:

Length AC $= \sqrt{\left(\frac{3}{2}\right)^2 + \left(\frac{\sqrt{3}}{2}\right)^2} = \sqrt{3}$

Length AB $= \sqrt{\left(\frac{3}{2}\right)^2 + \left(\frac{\sqrt{3}}{2}\right)^2} = \sqrt{3}$

Length BC $= 2 \times \frac{\sqrt{3}}{2} = \sqrt{3}$

So the triangle is equilateral.

5 (i) $z_1 = -\frac{7}{29} + \frac{3}{29}$ i $z_2 = -\frac{3}{29} - \frac{7}{29}$ i

(ii)

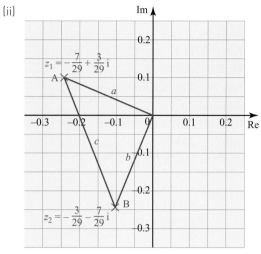

(iii) z_2 is a rotation of z_1 through an angle of 90° anticlockwise.

(iv) Length OA $= \sqrt{\left(\frac{7}{29}\right)^2 + \left(\frac{3}{29}\right)^2} = \frac{\sqrt{58}}{29}$

Length OB $= \sqrt{\left(\frac{3}{29}\right)^2 + \left(\frac{7}{29}\right)^2} = \frac{\sqrt{58}}{29}$

Length AB $= \sqrt{\left(\frac{4}{29}\right)^2 + \left(\frac{10}{29}\right)^2} = \frac{2\sqrt{29}}{29}$

So the triangle is isosceles.

Chapter 10

Activity 10.1 (Page 199)

0 for $k < 0$, 1 for $k = 0$, 2 for $k > 0$

Activity 10.2 (Page 200)

$\dfrac{2}{e^x - e^{-x}}$

Exercise 10.1 (Page 201)

1 (a) (i) $\frac{1}{2}\left(e^{2x} - e^{-2x}\right)$

(ii)

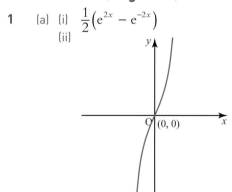

(iii) domain $x \in \mathbb{R}$, range $y \in \mathbb{R}$

(b) (i) $\frac{3}{2}\left(e^x + e^{-x}\right)$

(ii)

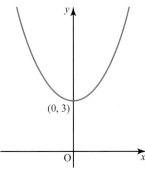

(0, 3)

O   x

(iii) domain $x \in \mathbb{R}$, range $y \geqslant 3$

(c) (i) $\frac{1}{2}\left(e^x + e^{-x}\right) + 1$

(ii)

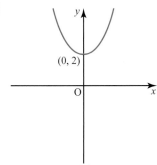

(0, 2)

O   x

(iii) domain $x \in \mathbb{R}$, range $y \geqslant 2$

(d) (i) $\frac{1}{2}\left(e^{x-3} - e^{-x+3}\right)$

(ii)

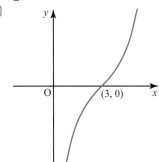

O   (3, 0)   x

(iii) domain $x \in \mathbb{R}$, range $y \in \mathbb{R}$

2 (i)

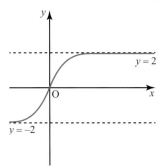

y = 2

O   x

y = −2

(ii) Stretch, scale factor 2, parallel to the y-axis

(iii) domain $x \in \mathbb{R}$, range $-2 < y < 2$

(iv) $\dfrac{2\left(e^x - e^{-x}\right)}{e^x + e^{-x}}$

3 $\ln(\frac{1}{7}(5 + 4\sqrt{2}))$

4 $\ln(\frac{1}{2}(1 + \sqrt{13}))$

5 $\ln(-\frac{3}{2} + \frac{1}{2}\sqrt{21})$

7 (i)

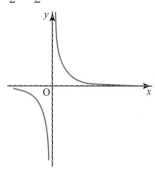

O   x

(ii) domain $x \in \mathbb{R}$, $x \neq 0$
range $y \in \mathbb{R}$, $y \neq 0$

(iii)

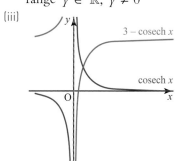

3 − cosech x

cosech x

O   x

(iv) a reflection in the x-axis followed by a
translation of $\begin{pmatrix} 0 \\ 3 \end{pmatrix}$

(v) domain $x \in \mathbb{R}$, $x \neq 0$
range $y \in \mathbb{R}$, $y \neq 3$

8 (i)

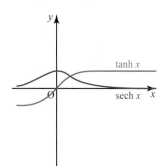

tanh x

O   sech x   x

(ii) $\ln(1 + \sqrt{2})$

9 (i) $\dfrac{2e^{4x} - 4}{e^{4x} - 1}$ (ii) $\frac{1}{4}\ln 3$

10 (i) $p = \frac{1}{2}\ln\left(2 + \sqrt{5}\right)$, $q = \ln\left(1 + \sqrt{2}\right)$

(ii) (a) tanh $x <$ sinh $x <$ sech $x <$ cosh $x <$
cosech $x <$ coth x

(b) tanh $x <$ sech $x <$ sinh $x <$ cosech $x <$
cosh $x <$ coth x

Activity 10.4 (Page 208)

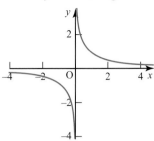

Domain $x \in \mathbb{R}$, $x \neq 0$, range $y \in \mathbb{R}$, $y \neq 0$

Exercise 10.3 (Page 209)

1　(a)　$\ln(5 + 2\sqrt{6})$　　(b)　$\frac{1}{2}\ln\frac{3}{2}$

　　(c)　$\ln\left(\sqrt{10} - 3\right)$　　(d)　$\frac{1}{2}\ln\frac{1}{3}$

　　(e)　$\ln 2$　　(f)　$\ln\left(2 + \sqrt{3}\right)$

2　(a)　$\ln\left(2 + \sqrt{3}\right)$　　(b)　$\frac{1}{2}\ln\frac{3}{2}$

　　(c)　$\ln(\sqrt{10} - 3)$

3　One solution

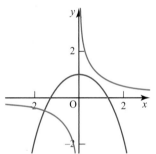

4　$\frac{1}{2}\ln 3$,　$\frac{1}{2}\ln 5$

5　(i)

(ii)　$x \in \mathbb{R}$, $y \in \mathbb{R}$

6　$\ln 5, \ln\frac{1}{2}\left(\sqrt{5} - 1\right)$

7　$\frac{1}{2}\ln\left(\frac{1 + x}{x - 1}\right)$

8　(i)　$\pm\frac{1}{2}\ln 3$　　(ii)　$0, \ln 7$　　(iii)　$0, \frac{1}{2}\ln 2$

9　(i)

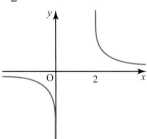

(ii)　Domain $x < 0$, $x > 2$, range $y \in \mathbb{R}$, $y \neq 0$

10　$\pm\dfrac{\sqrt{5}}{2}$

11　$\ln\left(x - 1 + \sqrt{2x^2 - 2x + 1}\right) - \ln x$

Activity 10.5 (Page 210)

$-\tanh x \operatorname{sech} x$

Activity 10.6 (Page 210)

$\ln\cosh x + c$

Activity 10.8 (Page 211)

$\dfrac{1}{1 - x^2}$

Exercise 10.4 (Page 212)

1　(i)　$4\cosh 4x$　　(ii)　$2x\sinh x^2$

　　(iii)　$2\cosh x \sinh x$　　(iv)　$\cosh^2 x + \sinh^2 x$

2　(i)　$-\coth x \operatorname{cosech} x$　　(ii)　$\dfrac{1}{1 - x^2}$

3　(i)　$\dfrac{1}{3}\cosh 3x + c$　　(i)　$x\cosh x - \sinh x + c$

　　(iii)　$\dfrac{1}{2}\sinh\left(1 + x^2\right) + c$

4　(i)　$\dfrac{3}{\sqrt{1 + 9x^2}}$　　(ii)　$\dfrac{4}{\sqrt{16x^2 - 1}}$

　　(iii)　$\dfrac{4x}{\sqrt{1 + 4x^4}}$　　(iv)　$\dfrac{1}{\sqrt{x^2 + x}}$

6　(i)　$x\operatorname{arcosh} x - \sqrt{x^2 - 1} + c$

　　(ii)　$x\operatorname{arsinh} x - \sqrt{x^2 + 1} + c$

　　(iii)　$x\operatorname{artanh} x - \frac{1}{2}\ln\left(1 - x^2\right) + c$

7　(i)　$\operatorname{arcosh}\dfrac{x}{2} + c$　　(ii)　$\operatorname{arsinh}\dfrac{x}{2} + c$

　　(iii)　$\dfrac{1}{3}\operatorname{arcosh} 3x + c$　　(iv)　$\dfrac{1}{3}\operatorname{arsinh} 3x + c$

　　(v)　$\dfrac{1}{2}\operatorname{arsinh}\dfrac{2x}{3} + c$　　(vi)　$\dfrac{1}{2}\operatorname{arcosh}\dfrac{2x}{3} + c$

8　(i)　$\ln\left(2 + \sqrt{3}\right)$　　(ii)　$\ln\left(\dfrac{2 + \sqrt{5}}{1 + \sqrt{2}}\right)$

　　(iii)　$\dfrac{1}{5}\ln\left(\dfrac{1}{4}\left(5 + \sqrt{21}\right)\right)$　　(iv)　$\dfrac{1}{3}\ln\left(3 + \sqrt{10}\right)$

9　(i)　$\dfrac{1 + x\left(x^2 - 1\right)^{-\frac{1}{2}}}{x + \sqrt{x^2 - 1}}$

10　(ii)　$\operatorname{artanh} x = \dfrac{1}{2}\ln\dfrac{(1 + x)}{(1 - x)}$

11　(ii)　$\dfrac{1}{2}x\sqrt{x^2 - 9} - \dfrac{9}{2}\operatorname{arcosh}\dfrac{x}{3} + c$

12　$\left(-\dfrac{1}{2}\ln 3, \ -\dfrac{1}{3\sqrt{3}}\right)$

13　$10\ln 2 - 6$

14 (i) $4\left(x + \dfrac{3}{2}\right)^2 - 49$ (ii) 0.322

15 (i) $\dfrac{1}{2}\cosh(x^2) + c$

    (ii) $\dfrac{1}{2}x^2\cosh(x^2) - \dfrac{1}{2}\sinh(x^2) + c$

16 $1 - \dfrac{1}{2}\sqrt{3}$

Chapter 11

Review Exercise (Page 220)

1 $\dfrac{\pi^2}{2}$

2 $\dfrac{96\pi}{5}$

3 $\dfrac{\pi}{6}$

4 $a = 2$

5 0.1537

6 $\dfrac{32\pi}{3}$, it is a sphere of radius 2 so use the formula for the volume of a sphere

7 $\dfrac{1}{e}$

8 (ii) 32π

9 (i) $\dfrac{1}{8}(25\ln 5 - 12)$ (ii) $\dfrac{1}{8}(25\ln 5 - 12) + a$

10 $\dfrac{25\pi}{2}$

Exercise 11.1 (Page 224)

1 (i) $\arcsin\dfrac{x}{3} + c$ (ii) $\operatorname{arcosh}\dfrac{x}{3} + c$

    (iii) $\operatorname{arsinh}\dfrac{x}{3} + c$ (iv) $\dfrac{1}{3}\arctan\dfrac{x}{3} + c$

    (v) $\dfrac{1}{6}\ln\left|\dfrac{3+x}{3-x}\right| + c$

2 (i) $\dfrac{\pi}{20}$ (ii) $\dfrac{1}{5}\ln 3$

    (iii) $\ln(1 + \sqrt{2})$ (iv) $\ln(2 + \sqrt{3})$

    (v) $\dfrac{\pi}{2}$

3 (i) $\dfrac{1}{3}\arcsin\dfrac{3x}{2} + c$ (ii) $\dfrac{1}{3}\operatorname{arsinh}\dfrac{3x}{2} + c$

    (iii) $\dfrac{1}{3}\operatorname{arcosh}\dfrac{3x}{2} + c$

4 (i) $\dfrac{\pi}{3\sqrt{3}}$ (ii) $\dfrac{1}{\sqrt{3}}\log\dfrac{3+\sqrt{5}}{2}$

    (iii) $\dfrac{\pi}{3\sqrt{3}}$

6 $9\sqrt{3} - 12 = 3.59$ (3 s.f.)

7 $\dfrac{3}{320}$

8 $\dfrac{3}{\sqrt{2}}\arctan\dfrac{x}{\sqrt{2}} - \dfrac{1}{x-1} + c$

9 $\dfrac{3}{2}\ln 2 - \dfrac{1}{8}\pi\sqrt{2}$

10 (i) $\dfrac{\pi}{4} + \ln\dfrac{32\sqrt{2}}{9} \approx 2.40$

    (ii) There are discontinuities at $x = -1$ and $x = -2$

    (iii) Only (d), area $\dfrac{\pi}{2} + 2\ln 2$

11 (i) $\dfrac{4}{3}$ (ii) $\dfrac{2}{3}$ (iii) $\dfrac{128}{105}\pi$ (iv) 0.781 (3 s.f.)

12 $2\ln\left(\dfrac{4 + \sqrt{17}}{3 + \sqrt{10}}\right)$

Exercise 11.2 (Page 228)

1 0

2 0

3 0

5 (i) $e^{x\ln x}$

7 (i) One of its limits is infinite (ii) $\dfrac{e^{-12}}{9}$

8 (i) One of its limits is undefined (ii) $4\ln 2 - 1$

Discussion point (Page 230)

Answer the first one is easier to keep track of.

Discussion point (Page 230)

Successively differentiating either sine or cosine alternates between the two functions so repeats after two iterations.

Discussion point (Page 231)

That will be an individual matter.

Activity 11.1 (Page 231)

$\dfrac{1}{2}e^x(\sin x + \cos x)$

Exercise 11.3 (Page 231)

1 (ii) $I_0 = 0$ (iii) $I_1 = -2$ (iv) $I_2 = -2\pi$

2 (ii) $I_3 = x(\ln x)^3 - 3x(\ln x)^2 + 6x\ln x - 6x$

3 (ii) $4\pi^4 - 192(\pi^2 - 8)$

4 $\dfrac{1}{2}e^x(\cos x + \sin x)$

6 $I_n = \dfrac{1}{3}\left(x^n e^{3x} - nI_{n-1}\right)$,

    $I_5 = \dfrac{e^{3x}}{81}\left(27x^5 - 45x^4 + 60x^3 - 60x^2 + 120x - 40\right)$

7 $\dfrac{128}{945}$

8 (ii) 0.2762

Exercise 11.4 (Page 237)

1   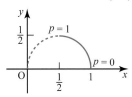 $\dfrac{8\left(10^{\frac{3}{2}} - 1\right)}{27} \approx 9.07$

2   (i) 45π

    (ii) $4\pi a^2$

    (iii) $2\pi ac + \pi c^2 \sinh\left(\dfrac{2a}{c}\right)$

3

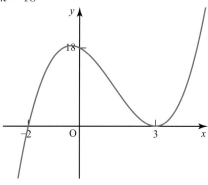

$\dfrac{\pi}{4}$

4   $c \sinh\left(\dfrac{X}{c}\right)$

5   (i) $\dfrac{8\pi a^2 \left(5^{\frac{3}{2}} - 2^{\frac{3}{2}}\right)}{3}$

    (ii) $\dfrac{64\pi a^2}{3}$

    (iii) $\dfrac{12\pi a^2}{5}$

8   $24a$

Review: Roots of polynomials

Exercise R.1 (Page 242)

1   (i) $\dfrac{3}{2}$ (ii) $-\dfrac{5}{2}$

2   (i) $-\dfrac{1}{3}$ (ii) $\dfrac{4}{3}$ (iii) $\dfrac{7}{3}$

3   (i)   2   (ii) -5 (iii)   0   (iv) -3

4   $z^3 - 3z^2 - 18z + 40 = 0$

5   (i) $3w^2 + 8w + 9 = 0$

    (ii) $3w^2 + 4w + 16 = 0$

    (iii) $3w^2 - 14w + 31 = 0$

6   (i) $w^3 - 9w^2 + 23w - 17 = 0$

    (ii) $8w^3 - 12w^2 - 2w + 1 = 0$

    (iii) $w^3 + 6w^2 - 23w - 10 = 0$

7   (i)   Roots are $3, 3, -2$ and $k = 18$

      or roots are $-\dfrac{1}{3}, -\dfrac{1}{3}, \dfrac{14}{3}$ and $k = -\dfrac{14}{27}$

    (ii) $k = 18$

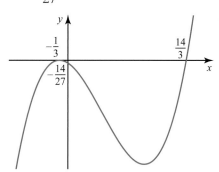

$k = -\dfrac{14}{27}$

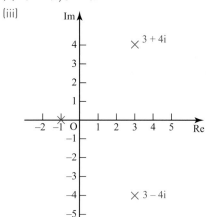

9   $z = 3, 6$ or $-2, k = 36$

10   $p = -3, q = -6$, roots are $1, 2, \pm\sqrt{2}$

      or $p = 3, q = 6$, roots are $-1, -2, \pm\sqrt{2}$

Exercise R.2 (Page 245)

1   (i) $2 + 5i$ (ii) $p = -4, q = 29$

2   (ii) $3 + 4i, 3 - 4i$

    (iii)

3   $z^4 - 13z^3 + 65z^2 - 151z + 130 = 0$

4   $p = 0$, roots are $x = -2, 1 + i$ and $1 - i$

5   $z = \dfrac{2}{3}, 2 + 3i, 2 - 3i$

6   $z^4 - 2z^3 + 2z^2 - 10z + 25 = 0$

7 $z = -2, -1 + 3i, -1 - 3i$

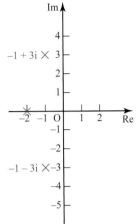

8 (i) $a = -9, b = -5$ (ii) $z = 2 - i, 2 + i, \frac{1}{2}$

9 $z = 1 + i, 1 - i, \frac{1}{2}i, -\frac{1}{2}i$

10 $p = 3, q = -10, z = -1 + 2i, -1 - 2i, 3 + i, 3 - i$

Chapter 12

Discussion point (Page 247)

The first is the aeroplane landing and the second is the aeroplane taking off. When landing, $\dfrac{dh}{dt}$ is decreasing, reaching almost zero at the point of landing. When taking off, $\dfrac{dh}{dt}$ increases rapidly.

Exercise 12.1 (Page 253)

1 $\dfrac{dm}{dt} = -km$

2 $\dfrac{dP}{dt} = \dfrac{P}{34}$

3 $\dfrac{dV}{dt} = -2\sqrt{h}$

5 (iii) (i) is the coffee, (ii) is the juice

6 (ii) 69.3 hours   (iii) 20 mg

    (iv) $m = 50e^{\frac{-t}{100}}$

7 $\dfrac{dr}{dt} = -\dfrac{1}{100\pi}$

8 $\dfrac{dp}{dh} = \begin{cases} 9800(1 + 0.001h) & 0 \leqslant h \leqslant 100 \\ 10780 & h > 100 \end{cases}$

9 $\dfrac{dV}{dt} = -\sqrt{20h}, \dfrac{dh}{dt} = -\dfrac{\sqrt{5h}}{2}$

11 $y = x^3 - \frac{1}{2}x^2 + x + c, \ y = x^3 - \frac{1}{2}x^2 + x + 2.5$

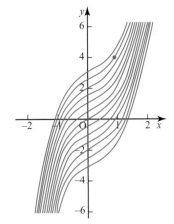

12 (i) and (v)

13 (ii) $\alpha + A, \alpha$ (iii) $\alpha = 25, A = 65$

    (iv)

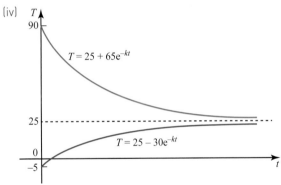

14 $\dfrac{dy}{dx} = \dfrac{b - y}{a + vt - x}$

15 $\dfrac{dh}{dt} = \dfrac{4.8}{\pi h^2}$

Exercise 12.2 (Page 261)

1 (i) $y = Ae^x$ (ii) $y = Ae^{\frac{1}{2}x^2}$

   (iii) $y = Ae^{\frac{1}{4}x^4}$ (iv) $y = \ln(x^3 + c)$

   (v) $\ln|1 + y| = \dfrac{x^2}{2} + c$ or $y = A^{\frac{1}{2}x^2}$

   (vi) Not possible   (vii) $e^x + e^{-y} = c$

   (viii) $y = \dfrac{A(x - 1)}{x}$ (ix) Not possible

   (x) $y^3 = K - 3\cos x$

   (xi) $y = A(x - 2) - 2$

   (xii) $\ln\left|\dfrac{x - 8}{x}\right| = 8t + c$ or $x = \dfrac{8}{1 - Ae^{8t}}$

2 (i) $y = \dfrac{10}{1 - 10\ln x}$ (ii) $z = \dfrac{2}{1 - 2\ln t}$

   (iii) $y^2 = \dfrac{2x^3 + 298}{3}$ (iv) $p = \ln\left(\dfrac{3 - \cos 2s}{2}\right)$

(v) $y = \ln\left(e^{10} + \dfrac{x^3}{3}\right)$

(vi) $e^{-y}(1 + y) = 3e^{-2} - t$

3 (i) $m = Ae^{-5t}$ (ii) $m = 10e^{-5t}$

4 (i) $P = Ae^{0.7t}$ (ii) $P = 100e^{0.7t}$

  (iii) 0.99 minutes

5 (i) $v = \dfrac{20}{1 + 2t}$ (ii) 4.5 seconds

6 (i) $h = \dfrac{4}{\pi t + k}$ (ii) $t = \dfrac{4}{9\pi H}$

7 (i) $h = \left(2 - \dfrac{\sqrt{5}}{4}t\right)^2$

  (ii) $t = \dfrac{8}{\sqrt{5}} \approx 3.58$ minutes

8 (i) $|10 - 0.2v| = Ae^{-0.2t}$

  (ii) (a) $v = 50(1 - e^{-0.2t})$

    (b) $v = 50 + 30e^{-0.2t}$

9 (i) $T = 20 + 80e^{-0.5t}$ (ii) $t = 1.96$ minutes

10 $v = \dfrac{40e^{-0.2t}}{41 - 40e^{-0.2t}}$

11 (i) $I = 0.2 - Ae^{-2500t}$ (ii) $I = 0.2(1 - e^{-2500t})$

  (iii) $I = \dfrac{V}{R} - Ae^{\frac{-Rt}{L}}$

12 6 minutes before midnight

13 (i) In running out at 8 litres per minute, salt is removed at a rate of $8C\,$kg per minute and as the salt inflow is zero $\Rightarrow \dfrac{\mathrm{d}M}{\mathrm{d}t} = -8C$.

$C\,$kg of salt per litre $\Rightarrow 2000C\,$kg of salt in the tank $\Rightarrow M = 2000C$

  (ii) $\dfrac{\mathrm{d}C}{\mathrm{d}t} = -0.004C$ (iii) 101 minutes

Activity 12.1 (Page 264)

(i) (a) (ii) (a), (b), (c)

Exercise 12.3 (Page 270)

1 (i) $e^{\frac{1}{3}x^3}$ (ii) $e^{-\cos x}$ (iii) $x^{-\frac{1}{4}}$

  (iv) x (v) e^{7x} (vi) $\sec x$

2 (i) $\dfrac{\mathrm{d}y}{\mathrm{d}x} + \dfrac{1}{x}y = \dfrac{1}{x^2}$ (ii) x

  (iii) $x\dfrac{\mathrm{d}y}{\mathrm{d}x} + y = \dfrac{1}{x}$ (iv) $\dfrac{\mathrm{d}}{\mathrm{d}x}(xy) = \dfrac{1}{x}$

(v) $xy = \ln x + c$ (vi) $y = \dfrac{1}{x}\ln x + \dfrac{c}{x}$

(vi) $c = 0$ (vii) $y = \dfrac{1}{x}\ln x$

3 (i) $y^2 = 4x^2 \ln|x| + 4Cx^2$

  (ii) $y = \pm\sqrt{x^2 - \dfrac{x}{k}}$

  (iii) $y^2 = \dfrac{x^2\left(kx^2 - 3\right)}{2}$

5 (i) $y = \dfrac{x^2}{4} - \dfrac{1}{4x^2}$ (ii) $y = 4 - 2e^{-\frac{1}{2}x^2}$

  (iii) $y = 3e^{-3(x^2 - 1)}$ (iv) $x = \dfrac{1}{2}(3e^{t^2} - 1)$

  (v) $y = x^3 - x^2$ (vi) $v = \dfrac{1}{3}(1 - 4e^{-t^3})$

6 (i) $v = 25 + Ae^{-0.4t}$

  (ii)–(iii) $v = 25(1 - e^{-0.4t})$

  (iv) The method of separation of variables is usually preferred as it involves less work.

7 (i) $k = \dfrac{1}{3}$ (ii) $\dfrac{\mathrm{d}v}{\mathrm{d}t} = 10 - \dfrac{1}{3}v$

  (iii) $v = 30 + Ae^{-\frac{1}{3}t}$ (iv) $v = 30\left(1 + e^{-\frac{1}{3}t}\right)$

8 (i) $y = \sin x + \dfrac{\cos x}{x} + \dfrac{A}{x}$

  (ii) $y = \sin x + \dfrac{1 + \cos x}{x}$ (iii) $y = \sin x$

  (iv) As $x \to 0, y \to \infty$

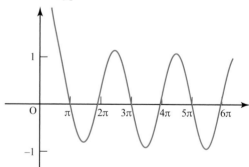

9 (i) $y = \dfrac{2}{2Cx^2 - x^4}$

  (ii) $y = \dfrac{1}{\sin 2x + C\sin x}$

  (iii) $y = \dfrac{4e^{x^2}}{4C - e^{2x^2}}$

  (iv) $y = \dfrac{10x}{\left(1 + x^2\right)^5 + 10C}$

10 (i) $e^{k_2 t},\ y = Ae^{-k_2 t} + \dfrac{k_1 a}{k_2 - k_1}e^{-k_1 t}$

(ii) $y = \dfrac{k_1 a}{k_2 - k_1}\left(e^{-k_1 t} - e^{-k_2 t}\right)$

(iii) $\dfrac{dx}{dt} = -k_1 x$ (iv) $x = ae^{-k_1 t}$

The amount y of Th-234 is affected by both its own decay (the $k_2 y$ part of the differential equation) and the rate at which U-238 decays into it – given by $k_1 x$, i.e. $ak_1 e^{-k_1 t}$.

(v) $y = kate^{-kt}$

(vi) $\dfrac{dz}{dt} = -ky,\ z = a(kt + 1)e^{-kt} - a$

11 (i) $y = \dfrac{1-x}{1+x}\left(A + x + \dfrac{x^2}{2}\right)$

(ii) (a) $y = \dfrac{1-x}{1+x}\left(x + \dfrac{x^2}{2}\right)$

As $x \to -1,\ y \to -\infty$

(b) $y = \dfrac{1-x}{1+x}\left(1 + x + \dfrac{x^2}{2}\right)$

As $x \to -1,\ y \to \infty$

(iii) $y = \dfrac{1}{2}(1 - x^2)$

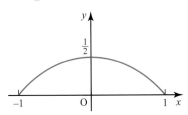

12 (i) $x^2 y = A + \dfrac{1}{3}(1 + x^2)^{\frac{3}{2}}$

(ii) $x^2 y = \left(1 - \dfrac{2\sqrt{2}}{3}\right) + \dfrac{1}{3}(1 + x^2)^{\frac{3}{2}}$

$y \to \infty$ as $x \to 0$

(iii) $1 + \dfrac{3x^2}{2} + \dfrac{3x^4}{8}$

(iv) $y \approx \dfrac{A}{x^2} + \dfrac{1}{3x^2}\left(1 + \dfrac{3x^2}{2} + \dfrac{3x^4}{8} + \ldots\right)$

Take $A = -\dfrac{1}{3} \Rightarrow y \approx -\dfrac{1}{3x^2} + \dfrac{1}{3x^2}\left(1 + x^2\right)^{\frac{3}{2}}$

13 (i) $x\dfrac{du}{dx} + u = 3x^2$

(ii) $\dfrac{dy}{dx} = x^2 + \dfrac{C}{x}$

(iii) $y = \dfrac{x^3}{3} + C\ln|x| + D$

14 $y = \dfrac{x^4}{8} + \dfrac{2x^3}{3} + C\ln|x| + D$

Practice questions 3 (Page 275)

1 $\operatorname{Re}(z) \geqslant -2,\ \operatorname{Im}(z) \geqslant 1,\ \arg(z) \geqslant \dfrac{\pi}{4},\ |z| \leqslant 4.$

2 $y = \dfrac{1}{2}e^{x^2}(e^{2x} + 1)$

3 $1 + i$ is a root so $(1 + i)^2 = 2i,\ (1 + i)^3 = -2 + 2i$ [1]

$(-2 + 2i)(m + 2) + 2i(m^2 - 8)$

$+ (1 + i)(m + 3) - 2 = 0$ [1]

Real parts: $-2(m + 2) + (m + 3) - 2 = 0$

so $m = -3$ [2]

Equation is $-x^3 + x^2 - 2 = 0$

Second root is $1 - i$ [1]

Sum of roots is 1 so third root is -1.   [1]

4 (i) $\begin{vmatrix} \dfrac{1}{2} - \lambda & \dfrac{1}{4} & 0 \\[2mm] \dfrac{1}{2} & \dfrac{1}{2} - \lambda & \dfrac{1}{2} \\[2mm] 0 & \dfrac{1}{4} & \dfrac{1}{2} - \lambda \end{vmatrix} = 0$ [1]

$\left(\dfrac{1}{2} - \lambda\right)\left[\left(\dfrac{1}{2} - \lambda\right)^2 - \dfrac{1}{8}\right] - \dfrac{1}{8}\left(\dfrac{1}{2} - \lambda\right) = 0$ [1]

$\left(\dfrac{1}{2} - \lambda\right)\left[\left(\dfrac{1}{2} - \lambda\right)^2 - \dfrac{1}{4}\right] = 0$

$\lambda^3 - \dfrac{3}{2}\lambda^2 + \dfrac{1}{2}\lambda = 0$ [1]

$\lambda(\lambda - 1)\left(\lambda - \dfrac{1}{2}\right) = 0$

$\lambda = 0,\ 1,\ \dfrac{1}{2}$ [2]

Corresponding eigenvectors are

$\begin{pmatrix} 1 \\ -2 \\ 1 \end{pmatrix},\ \begin{pmatrix} 1 \\ 2 \\ 1 \end{pmatrix},\ \begin{pmatrix} 1 \\ 0 \\ -1 \end{pmatrix}$ [3]

(ii) $\mathbf{M} = \mathbf{U\Lambda U^{-1}}$, where $\mathbf{\Lambda} = \begin{pmatrix} 0 & 0 & 0 \\ 0 & 1 & 0 \\ 0 & 0 & \dfrac{1}{2} \end{pmatrix}$,

$\mathbf{U} = \begin{pmatrix} 1 & 1 & 1 \\ -2 & 2 & 0 \\ 1 & 1 & -1 \end{pmatrix},$ **U**

$\mathbf{U^{-1}} = \begin{pmatrix} \dfrac{1}{4} & -\dfrac{1}{4} & \dfrac{1}{4} \\[2mm] \dfrac{1}{4} & \dfrac{1}{4} & \dfrac{1}{4} \\[2mm] -\dfrac{1}{2} & 0 & -\dfrac{1}{2} \end{pmatrix}$ [3]

5 (i) $I_n = \int_0^1 \frac{x}{\sqrt{1+x^2}} x^{n-1}\,dx = \left[x^{n-1}\sqrt{1+x^2}\right]_0^1$

$\qquad - (n-1)\int_0^1 x^{n-2}\sqrt{1+x^2}\,dx$ [2]

$\qquad = \sqrt{2} - (n-1)\int_0^1 \frac{x^{n-2}(1+x^2)}{\sqrt{1+x^2}}\,dx$ [3]

$\qquad = \sqrt{2} - (n-1)(I_{n-2} + I_n)$

$\therefore nI_n = \sqrt{2} - (n-1)I_{n-2}$ [1]

(ii) $I_0 = \int_0^1 \frac{1}{\sqrt{1+x^2}}\,dx = \left[\sinh^{-1}(x)\right]_0^1 = s$ [1]

$2I_2 = \sqrt{2} - I_0$ [1]

$\therefore I_2 = \frac{1}{2}\sqrt{2} - \frac{1}{2}s$

$4I_4 = \sqrt{2} - 3I_2$

$\therefore I_4 = \frac{1}{4}\sqrt{2} - \frac{3}{4}\left(\frac{1}{2}\sqrt{2} - \frac{1}{2}s\right)$ [1]

$\qquad = \frac{3}{8}s - \frac{1}{8}\sqrt{2}$ [1]

6 (i) $\frac{dx}{dt} = 6t$, $\frac{dy}{dt} = 3t^2 - 3$ so arc length

$\qquad = \int \sqrt{36t^2 + 9t^4 - 18t^2 + 9}\,dt$ [2]

Limits are $(0, \sqrt{3})$ [1]

$\int_0^{\sqrt{3}} 3(t^2 + 1)\,dt$ [1]

$= 6\sqrt{3}$ [2]

(ii) $2\pi\int_0^{\sqrt{3}} (t^3 - 3t)(t^2 + 1)\,dt$ [2]

$= 2\pi\left[-\frac{9}{2}\right]$ [1]

so area $= 9\pi$ [1]

7 (i) $2\sinh^2 x + 1 = 2\left(\frac{e^x - e^{-x}}{2}\right)^2 + 1$ [1]

$\qquad = 2\left(\frac{e^{2x} + e^{-2x} - 2}{4}\right) + 1$ [1]

$\qquad = \frac{e^{2x} + e^{-2x}}{2} = \cosh 2x$ [1]

(ii) $2\sinh^2 x - 3\cosh x = 0$

$2(\cosh^2 x - 1) - 3\cosh x = 0$ [1]

$2\cosh^2 x - 3\cosh x - 2 = 0$

$(2\cosh x + 1)(\cosh x - 2) = 0$

$\cosh x = 2$ [1]

$x = \pm\ln(2 + \sqrt{3})$ [1]

(iii) $\int_{-\ln(2+\sqrt{3})}^{\ln(2+\sqrt{3})} (2\sinh^2 x - 3\cosh x)\,dx$ [1]

$= \int_{-\ln(2+\sqrt{3})}^{\ln(2+\sqrt{3})} (\cosh 2x - 1 - 3\cosh x)\,dx$ [1]

$= \left[\frac{1}{2}\sinh 2x - x - 3\sinh x\right]_{-\ln(2+\sqrt{3})}^{\ln(2+\sqrt{3})}$ [1]

$= \sinh(2\ln(2 + \sqrt{3})) - 2\ln(2 + \sqrt{3})$

$\quad - 6\sinh(\ln(2 + \sqrt{3}))$ [1]

$= \left(\frac{e^{\ln(2+\sqrt{3})^2} - e^{-\ln(2+\sqrt{3})^2}}{2}\right) - 2\ln(2 + \sqrt{3})$

$\quad - 6\left(\frac{e^{\ln(2+\sqrt{3})} - e^{-\ln(2+\sqrt{3})}}{2}\right)$ [1]

$= \frac{1}{2}\left((2 + \sqrt{3})^2 - \frac{1}{(2+\sqrt{3})^2}\right) - 2\ln(2 + \sqrt{3})$

$\quad - 3\left((2 + \sqrt{3}) - \frac{1}{(2+\sqrt{3})}\right)$ [1]

$= -2\sqrt{3} - 2\ln(2 + \sqrt{3})$ (simplification of surds by calculator) [1]
Area positive, so area is
$2\sqrt{3} + 2\ln(2 + \sqrt{3})$. [1]

Chapter 13

Discussion point (Page 277)

The area is the integral of the line between values 0 and 5. You need the equation of the line. You might calculate an approximation by breaking up the area under the curve into small rectangles, or by using a series of straight lines or curves to approximate the curve.

Discussion point (Page 279)

The width of each strip, h, is the total width divided by n.

Activity 13.1 (Page 282)

(i) (a) underestimate (b) overestimate
(c) overestimate

(ii) See page 279

Activity 13.2 (Page 284)

(i) $\displaystyle\int_{-h}^{h} r(x)\,dx = \int_{a}^{b} q(x)\,dx$ because, thinking of these as areas, one is a translation of the other.

$r(-h) = f_0$, $r(0) = f_1$ and $r(h) = f_2$

(ii) $\displaystyle\int_{-h}^{h} r(x)\,dx = \frac{2ah^3}{3} + 2ch$ by direct integration.

$f_0 = r(-h) = ah^2 - ah + c$,

$f_1 = r(0) = c$ $f_2 = r(h) = ah^2 + bh + c$.

The results follow.

(iii) Substituting the above into $\dfrac{h}{3}\big(f_0 + 4f_1 + f_2\big)$

gives $\dfrac{2ah^3}{3} + 2ch$. Hence the result.

(iv) If two quadratics were used across the interval, the area under the second one would be $\dfrac{h}{3}\big(f_2 + 4f_3 + f_4\big)$. Adding gives

$\dfrac{h}{3}\big(f_0 + 4f_1 + f_2\big) + \dfrac{h}{3}\big(f_2 + 4f_3 + f_4\big)$

$= \dfrac{h}{3}\big(f_0 + 4f_1 + 2f_2 + 4f_3 + f_4\big)$.

This can be extended to more quadratics.

Activity 13.3 (Page 287)

1.024 019 162 656 to 12 d.p.

Exercise 13.1 (Page 287)

1 1.109

2 (i) 1.08   (ii) underestimate

(iii) the curve is concave in the interval

3 (i) 2.936 452   (ii) 2.936 053

(iii) 2.936 048 accuracy increases with more strips

4 (i) 2

(ii) If f(0.3) becomes available then it is possible to calculate with 4 strips as well.

5 (a) (i) 1.0986   (ii) 1.0898   (iii) 1.0963
(iv) part (iii) is closer than part
(ii) to the value of the integral

(b) (i) 4.000   (ii) 3.875   (iii) 3.9688
(iv) part (iii) is closer than part
(ii) to the value of the integral

(c) (i) 0.6366   (ii) 0.6533   (iii) 0.6407
(iv) part (iii) is closer than part
(ii) to the value of the integral

(d) (i) −6.2832   (ii) −6.9789   (iii) −6.4476
(iv) part (iii) is closer than part
(ii) to the value of the integral

6 (a) underestimate, (concave)

(b) underestimate, (concave)

(c) overestimate, (convex)

(d) difficult to tell as both concave and convex in the interval.

7 (i) 0.541 401; 0.540 137; 0.540 056;
0.5401; use more strips

(ii) 4.366 236; 4.372 954; 4.373 262;
4.373; use more strips

Activity 13.4 (Page 288)

(b) and (d)

Discussion point (Page 290)

B3 is the value of y when $x = 1.1$. This is the previous value of y (B2) plus the y increment $= h \times$ gradient at $(1, 1) = 0.1 \star C2$.

Activity 13.5 (Page 290)

(i)

x	y	y'	$y = \frac{1}{3}x^3 + \frac{2}{3}$
1	1	1	1
1.1	1.1	1.21	1.110 333 333
1.2	1.221	1.44	1.212 666 667
1.3	1.365	1.69	1.399
1.4	1.534	1.96	1.581 333 333
1.5	1.73	2.25	1.791 666 667
1.6	1.955	2.56	2.032
1.7	2.211	2.89	2.304 333 333
1.8	2.5	3.24	2.610 666 667
1.9	2.824	3.61	2.953
2	3.185	4	3.333 333 333

(ii) The error is greater for larger step lengths.

Activity 13.6 (Page 292)

1.34466

Exercise 13.2 (Page 293)

1 1.6319 with $h = 0.2$

2 2.7387 with $h = 0.1$

3 2.2666 with $h = 0.2$

By using smaller steps

4 1.7680

5 0.712 884

6 (i) 97.6   (ii) $1280\,e^{-0.05(x-5)} + 80x - 1600$

    (iii) 97.5737   (iv) using shorter step lengths

7 (i) 1.1952   (ii) $y = \dfrac{2}{5 - x^2}$ $y(2) = 2$

    (iii) using shorter step lengths

8 (i) 1.25   (ii) 1.6267

9 (i) 0.22

    (ii)

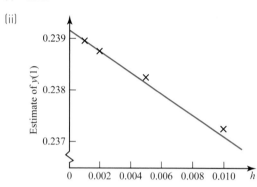

The approximation improves as h gets smaller.

From the graph, $y(1.2) \approx 0.23915$ but only $y(1.2) = 0.239$ can be justified (i.e. 3 d.p.).

10 (i) 0.1

    (ii)

x_n	y_n	y'_n	y_{n+1}
0.5			0.4926
0.6	0.4926	0.6427	0.5569
0.7	0.5569	0.6101	0.6179
0.8	0.6179		

    (iii)

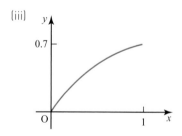

    (iv) The gradient is decreasing. For $x > 1$ the curve either turns upwards, in which case $y'' > 0$ or the curve intersects $y^2 = x - 1$ in which case $y' = 0$ (a maximum) with y'' continuing to be negative.

    (v) Less accurate as the error increases as h increases.

    (vi) Less, because $y'' < 0$, for all h by this method.

Chapter 14

Discussion point (Page 298)

Similarity – both use r and θ

Difference – Modulus argument form is used to represent a complex number on the Argand diagram, polar coordinates are used to represent a real number on the coordinate plane

Activity 14.1 (Page 301)

1 (i)

$$i = \cos\frac{\pi}{2} + i\sin\frac{\pi}{2}, \; -2 = 2\big(\cos(-\pi) + i\sin(-\pi)\big)$$

    (ii) (a) Anticlockwise rotation of the vector z through $\dfrac{\pi}{2}$ radians

       (b) Anticlockwise rotation of the vector z through π radians (which is equivalent to two successive rotations through $\dfrac{\pi}{2}$ radians) and enlargement by scale factor 2

Review exercise (Page 305)

1 (i) $3\sqrt{2} - 3\sqrt{2}i$ (ii) $\dfrac{3\sqrt{3}}{2} - \dfrac{3}{2}i$

    (iii) $-\sqrt{2} + \sqrt{2}i$ (iv) $-\dfrac{7\sqrt{3}}{2} - \dfrac{7}{2}i$

2 (i) 4 $\dfrac{\pi}{6}$ $4\left(\cos\dfrac{\pi}{6} + i\sin\dfrac{\pi}{6}\right)$

    (ii) 4 $\dfrac{5\pi}{6}$ $4\left(\cos\left(\dfrac{5\pi}{6}\right) + i\sin\left(\dfrac{5\pi}{6}\right)\right)$

    (iii) 4 $-\dfrac{\pi}{6}$ $4\left(\cos\left(-\dfrac{\pi}{6}\right) + i\sin\left(-\dfrac{\pi}{6}\right)\right)$

    (iv) 4 $-\dfrac{5\pi}{6}$ $4\left(\cos\left(-\dfrac{5\pi}{6}\right) + i\sin\left(-\dfrac{5\pi}{6}\right)\right)$

3 (i) $\dfrac{3\sqrt{2}}{2} + \dfrac{3\sqrt{2}}{2}i$

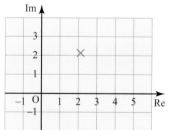

(ii) 5i

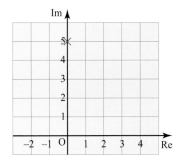

(iii) $-2\sqrt{3} - 2i$

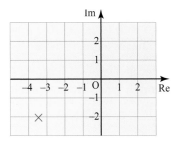

(iv) $3\sqrt{2} - 3\sqrt{2}i$

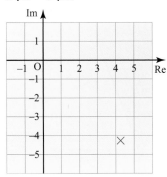

4 (i) $15\left(\cos\frac{\pi}{2} + i\sin\frac{\pi}{2}\right)$

(ii) $\frac{3}{5}\left(\cos\left(-\frac{5\pi}{6}\right) + i\sin\left(-\frac{5\pi}{6}\right)\right)$

(iii) $\frac{5}{3}\left(\cos\frac{5\pi}{6} + i\sin\frac{5\pi}{6}\right)$

(iv) $\frac{1}{5}\left(\cos\frac{\pi}{3} + i\sin\frac{\pi}{3}\right)$

5 (i) Enlargement scale factor $5\sqrt{2}$ and a clockwise rotation through $\frac{\pi}{4}$ radians

(ii) Enlargement scale factor $\frac{1}{2}$ and a clockwise rotation through $\frac{2\pi}{3}$ radians

6 (i) $1 + \sqrt{3}i$

(ii) $-\frac{1}{6} - \frac{1}{6}\sqrt{3}i$

7 (i)

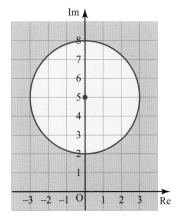

(ii)

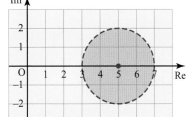

(iii)

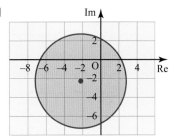

8 (i)

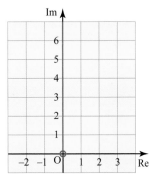

(ii)

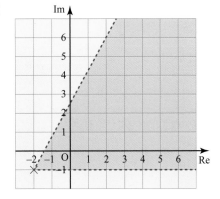

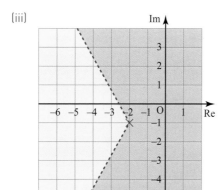

(iii)

9 (i)

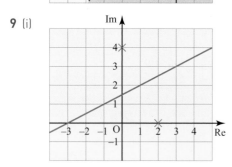

(ii)

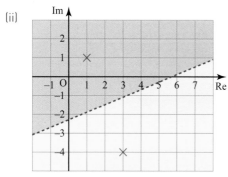

(iii)

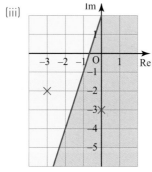

10 (i) $\left|z-(1+2i)\right| < 3$

(ii) $\left|z-3i\right| \geqslant \left|z-(6+i)\right|$

(iii) $-\dfrac{\pi}{4} < \arg\left(z-(-2+2i)\right) < 0$

11

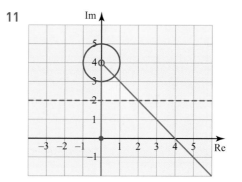

12 $a = \pm2,\ b = \pm3\quad$ or $\quad a = \pm3,\ b = \pm2$

13 (i) (a) Enlarge z scale factor 3, centre O

(b) Enlarge z scale factor 2, centre O, and rotate through $\dfrac{\pi}{2}$ radians anticlockwise

(c) Enlarge z scale factor $\sqrt{13}$, centre O, and rotate through 0.588 radians anticlockwise

(ii) For example, if $z = 1 + 2i$

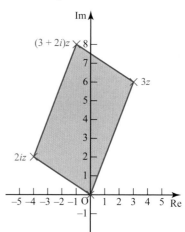

The points form a rectangle.

Activity 14.2 (Page 307)

$z^2 = 4(\cos0.2 + i\sin0.2)$

$z^3 = 8(\cos0.3 + i\sin0.3)$

$z^4 = 16(\cos0.4 + i\sin0.4)$

$z^5 = 32(\cos0.5 + i\sin0.5)$

$z^n = 2^n(\cos0.1n + i\sin0.1n)$

$z^n = r^n(\cos n\theta + i\sin n\theta)$

Discussion point (Page 308)

The lines will continue all the way round. What happens then depends on whether or not $\dfrac{\theta}{\pi}$ is a rational number. If it is rational, a limited number of lines get repeated over and over again as $n \to \infty$. If it is not rational the number of lines $\to \infty$ but no line ever gets repeated.

Activity 14.3 (Page 310)

$$\cos(-\phi) + i\sin(-\phi) = \cos\phi - i\sin\phi \longleftarrow$$

Using the two results given.

$$\Rightarrow \left[\cos(-\phi) + i\sin(-\phi)\right]^n = \left[\cos\phi - i\sin\phi\right]^n$$

$$\Rightarrow \cos(-n\phi) + i\sin(-n\phi) = \left[\cos\phi - i\sin\phi\right]^n$$

Using de Moivre's theorem on the left-hand side.

$$\Rightarrow \cos(n\phi) - i\sin(n\phi) = \left[\cos\phi - i\sin\phi\right]^n \longleftarrow$$

Using the two results given in the question.

Exercise 14.1 (Page 310)

1 (a) (i) $\cos\dfrac{2\pi}{3} + i\sin\dfrac{2\pi}{3}$

    (ii) $\cos\left(-\dfrac{2\pi}{3}\right) + i\sin\left(-\dfrac{2\pi}{3}\right)$

    (iii) $\cos\left(-\dfrac{5\pi}{6}\right) + i\sin\left(-\dfrac{5\pi}{6}\right)$

  (b) (i) $-\dfrac{1}{2} + \dfrac{\sqrt{3}}{2}i$ (ii) $-\dfrac{1}{2} - \dfrac{\sqrt{3}}{2}i$

    (iii) $-\dfrac{\sqrt{3}}{2} - \dfrac{1}{2}i$

2 (i) $z_1 = w^3$ (ii) $z_2 = w^2$ (iii) $z_3 = w^4$

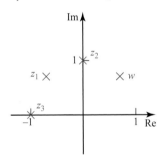

3 (i) $\cos(-3\theta) + i\sin(-3\theta)$

  (ii) $\cos\left(\dfrac{5\pi}{12}\right) + i\sin\left(\dfrac{5\pi}{12}\right)$ (iii) -1

4 (i) $z_1 = w^{-1}$

  (ii) $z_2 = w^{-3}$

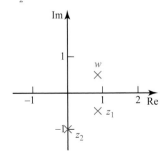

5 (i) $\left[2\left(\cos\left(-\dfrac{\pi}{3}\right) + i\sin\left(-\dfrac{\pi}{3}\right)\right)\right]^4$

$$= 16\left(-\dfrac{1}{2} + \dfrac{\sqrt{3}}{2}i\right) = -8 + 8\sqrt{3}i$$

  (ii) $\left[2\sqrt{2}\left(\cos\left(\dfrac{3\pi}{4}\right) + i\sin\left(\dfrac{3\pi}{4}\right)\right)\right]^7$

$$= 1024\sqrt{2}\left(-\dfrac{\sqrt{2}}{2} - \dfrac{\sqrt{2}}{2}i\right) = -1024 - 1024i$$

  (iii) $\left[6\left(\cos\dfrac{\pi}{6} + i\sin\dfrac{\pi}{6}\right)\right]^6$

$$= 46656(-1 + 0i) = -46656$$

6   $64\left(\sqrt{3} + i\right)$

7 (i) $81\left(\cos 8\theta + i\sin 8\theta\right)$

  (ii) $i\cos 15\theta - \sin 15\theta$

  (iii) $\dfrac{1}{8}\left(i\cos(-21\theta) - \sin(-21\theta)\right)$

8   $k = 41472$

9 (i) $z_1 = \cos 0 + i\sin 0,\ z_2 = \cos\dfrac{2\pi}{3} + i\sin\dfrac{2\pi}{3},$

     $z_3 = \cos\left(-\dfrac{2\pi}{3}\right) + i\sin\left(-\dfrac{2\pi}{3}\right)$

  (iii)

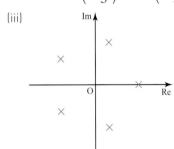

$$\cos 0 + i\sin 0,\ \cos\dfrac{2\pi}{5} + i\sin\dfrac{2\pi}{5},$$

$$\cos\dfrac{4\pi}{5} + i\sin\dfrac{4\pi}{5},\ \cos\left(-\dfrac{2\pi}{5}\right) + i\sin\left(-\dfrac{2\pi}{5}\right),$$

$$\cos\left(-\dfrac{4\pi}{5}\right) + i\sin\left(-\dfrac{4\pi}{5}\right)$$

Activity 14.4 (Page 314)

$n = 2$ $1 + (-1) = 0$

$n = 3$ $1 + \dfrac{-1 + \sqrt{3}i}{2} + \dfrac{-1 - \sqrt{3}i}{2} = 0$

$n = 4$ $1 + (-1) + i + (-i) = 0$

Exercise 14.2 (Page 318)

1   $\omega^0 = 1, \quad \omega^1 = 0.309 + 0.951i,$

    $\omega^2 = -0.809 + 0.588i, \quad \omega^3 = -0.809 - 0.588i,$

$\omega^4 = 0.309 - 0.951\mathrm{i}$

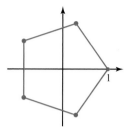

The points form a regular pentagon with one vertex at the point 1.

2 $\omega^0 = 1, \quad \omega = \dfrac{\sqrt{2}}{2} + \dfrac{\sqrt{2}}{2}\mathrm{i}, \quad \omega^2 = \mathrm{i},$

$\omega^3 = -\dfrac{\sqrt{2}}{2} + \dfrac{\sqrt{2}}{2}\mathrm{i}, \quad \omega^4 = -1,$

$\omega^5 = -\dfrac{\sqrt{2}}{2} - \dfrac{\sqrt{2}}{2}\mathrm{i} \quad \omega^6 = -\mathrm{i}, \quad \omega^7 = \dfrac{\sqrt{2}}{2} - \dfrac{\sqrt{2}}{2}\mathrm{i}$

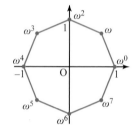

The points form a regular octagon with one vertex at the point 1.

3 $0.90 + 2.79\mathrm{i}, \ -0.90 - 2.79\mathrm{i}$

4 $1 + \mathrm{i}, \ 1 - \mathrm{i}, \ -1 + \mathrm{i}, \ -1 - \mathrm{i}$

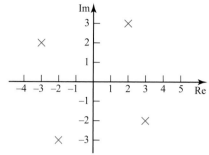

5 (i) $\left(2^{\frac{1}{6}}, -\dfrac{\pi}{12}\right) \left(2^{\frac{1}{6}}, \dfrac{7\pi}{12}\right) \left(2^{\frac{1}{6}}, -\dfrac{3\pi}{4}\right)$

(ii) $(1.38, 0.25) \quad (1.38, 1.82)$

$(1.38, -2.90) \quad (1.38, -1.39)$

(iii) $(1.38, 0.44) \quad (1.38, 1.70) \quad (1.38, 2.96)$

$(1.38, -2.07) \quad (1.38, -0.81)$

6 The fifth roots give alternate tenth roots and their negatives (given by a half turn about the origin) fill the gaps.

7 $w = -119 - 120\mathrm{i}$

$-3 + 2\mathrm{i}, \ -2 - 3\mathrm{i}, \ 3 - 2\mathrm{i}$

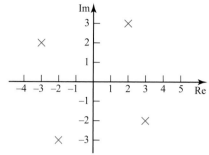

8 $2\left(\cos\dfrac{\pi}{18} + \mathrm{i}\sin\dfrac{\pi}{18}\right), \ 2\left(\cos\dfrac{13\pi}{18} + \mathrm{i}\sin\dfrac{13\pi}{18}\right),$

$2\left(\cos\left(-\dfrac{11\pi}{18}\right) + \mathrm{i}\sin\left(-\dfrac{11\pi}{18}\right)\right)$

9 $(z + 1 - 3\mathrm{i})^7 = 2187$

10 (i) $\sqrt{3}\left(\cos\dfrac{3\pi}{8} + \mathrm{i}\sin\dfrac{3\pi}{8}\right), \ \sqrt{3}\left(\cos\dfrac{7\pi}{8} + \mathrm{i}\sin\dfrac{7\pi}{8}\right),$

$\sqrt{3}\left(\cos\dfrac{11\pi}{8} + \mathrm{i}\sin\dfrac{11\pi}{8}\right), \ \sqrt{3}\left(\cos\dfrac{15\pi}{8} + \mathrm{i}\sin\dfrac{15\pi}{8}\right)$

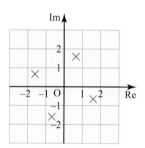

(ii) Midpoint of PS has argument $\dfrac{\pi}{8}$ and modulus $\sqrt{\dfrac{3}{2}}$.

So $\arg(w) = 4 \times \dfrac{\pi}{8} = \dfrac{\pi}{2}$ and

$|w| = \left(\sqrt{\dfrac{3}{2}}\right)^4 = \dfrac{9}{4}$

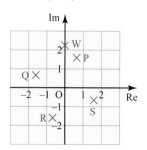

Activity 14.5 (Page 319)

$\sin 5\theta = 5c^4 s - 10c^2 s^3 + s^5$

$= 5s\left(1 - s^2\right)^2 - 10s^3\left(1 - s^2\right) + s^5$

$= 5s - 10s^3 + 5s^5 - 10s^3 + 10s^5 + s^5$

$\sin 5\theta = 16\sin^5\theta - 20\sin^3\theta + 5\sin\theta$

Activity 14.6 (Page 320)

$2i\sin\theta = z - z^{-1}$

$\Rightarrow \quad (2i\sin\theta)^5 = \left(z - z^{-1}\right)^5$

$\Rightarrow \quad 32i\sin^5\theta = z^5 - 5z^3 + 10z - 10z^{-1}$
$\qquad\qquad\qquad + 5z^{-3} - z^{-5}$

$\qquad\qquad = \left(z^5 - z^{-5}\right) - 5\left(z^3 - z^{-3}\right)$
$\qquad\qquad\quad + 10\left(z - z^{-1}\right)$

$\qquad\qquad = 2i\sin 5\theta - 10i\sin 3\theta + 20i\sin\theta$

$\Rightarrow \qquad \sin^5\theta = \dfrac{\sin 5\theta - 5\sin 3\theta + 10\sin\theta}{16}$

Exercise 14.3 (Page 320)

1 (ii) $\tan 3\theta = \dfrac{3\tan\theta - \tan^3\theta}{1 - 3\tan^2\theta}$

2 (i) $z^3 = \cos 3\theta + i\sin 3\theta$

$z^{-3} = \cos 3\theta - i\sin 3\theta$

3 (i) $z^{-1} = \cos\theta - i\sin\theta$

(ii) (b) $\dfrac{\cos 4\theta + 4\cos 2\theta + 3}{8}$

(iii) (b) $\dfrac{\sin 5\theta - 5\sin 3\theta + 10\sin\theta}{16}$

4 $\cos 6\theta = 32\cos^6\theta - 48\cos^4\theta + 18\cos^2\theta - 1$

$\dfrac{\sin 6\theta}{\sin\theta} = 32\cos^5\theta - 32\cos^3\theta + 6\cos\theta$

5 $\sin^6\theta = \dfrac{-\cos 6\theta + 6\cos 4\theta - 15\cos 2\theta + 10}{32}$

$\displaystyle\int \sin^6\theta\, d\theta = -\dfrac{1}{192}\sin 6\theta + \dfrac{3}{64}\sin 4\theta$
$\qquad\qquad\qquad - \dfrac{15}{64}\sin 2\theta + \dfrac{5}{16}\theta + c$

6 $\cos^4\theta\sin^3\theta = \dfrac{-\sin 7\theta - \sin 5\theta + 3\sin 3\theta + 3\sin\theta}{64}$

$\displaystyle\int_0^\pi \cos^4\theta\sin^3\theta\, d\theta = \dfrac{4}{35}$

7 $\dfrac{203}{480}$

8 (ii) $\cos^2\theta = \dfrac{5 \pm \sqrt{5}}{8}$

(iii) $\theta = 18° \Rightarrow \cos 5\theta = 0$

$\cos\theta = \pm\left(\dfrac{5 + \sqrt{5}}{8}\right)^{\frac{1}{2}}$ or $\pm\left(\dfrac{5 - \sqrt{5}}{8}\right)^{\frac{1}{2}}$

$\cos 18°$ is close to 1 so $\cos 18° = \left(\dfrac{5 + \sqrt{5}}{8}\right)^{\frac{1}{2}}$

and using $\cos^2\theta + \sin^2\theta \equiv 1$ gives

$\sin 18° = \left(\dfrac{3 - \sqrt{5}}{8}\right)^{\frac{1}{2}}$

9 $a = \dfrac{1}{2}z^{-(2n-1)},\ r = z^2,\ 2n$ terms

10 (i) $z^n = \cos n\theta + i\sin n\theta$

$\dfrac{1}{z^n} = \cos n\theta - i\sin n\theta$

$z^n + \dfrac{1}{z^n} = 2\cos n\theta$

$z^n - \dfrac{1}{z^n} = 2i\sin n\theta$

(ii) $p = \dfrac{1}{16},\ q = -\dfrac{1}{32},\ r = -\dfrac{1}{16},\ s = \dfrac{1}{32}$

Discussion point (Page 322)

$\cos(-\theta) = \cos\theta$ and $\sin(-\theta) = -\sin\theta$

so

$r(\cos\theta - i\sin\theta) = r(\cos(-\theta) + i\sin(-\theta))$

Therefore

$r(\cos\theta - i\sin\theta) = re^{-i\theta}$

Activity 14.7 (Page 323)

(i) $e^{x+yi} = e^x e^{yi} = e^x(\cos y + i\sin y)$

(ii) $e^{z+2\pi ni} = e^z e^{2\pi ni} = e^z\left(\cos 2\pi n + i\sin 2\pi n\right)$
$\qquad\qquad\qquad\qquad = e^z \times 1 = e^z$

(iii) Using (i) with $x = 0$, $y = \pi$ gives
$\qquad e^0(\cos\pi + i\sin\pi) = -1$

Discussion point (Page 323)

The earlier results $\cos\theta = \dfrac{z + z^{-1}}{2}$ and
$\sin\theta = \dfrac{z - z^{-1}}{2i}$, where $z = \cos\theta + i\sin\theta$, can
be rewritten with $z = e^{i\theta}$, so $z^{-1} = e^{-i\theta}$, giving
$\cos\theta = \dfrac{e^{i\theta} + e^{-i\theta}}{2}$ and $\sin\theta = \dfrac{e^{i\theta} - e^{-i\theta}}{2i}$.

Activity 14.8 (Page 325)

$S = 2^n\cos^n\dfrac{\theta}{2}\sin\dfrac{n\theta}{2}$

Exercise 14.4 (Page 325)

1 (i) $4e^{\frac{\pi}{3}i}$ (ii) $\sqrt{3}e^{-\frac{5\pi}{6}i}$ (iii) $5e^{-\frac{\pi}{2}i}$

    (iv) $3\sqrt{2}e^{-\frac{3\pi}{4}i}$ (v) $2e^{-\frac{\pi}{6}i}$

2 (i) -5 (ii) $-1+i$

    (iii) $-1-i$ (iv) $\dfrac{5\sqrt{2}}{2} - \dfrac{5\sqrt{2}}{2}i$

3 $zw = 6e^{-\frac{11}{12}\pi i}$

   $\dfrac{z}{w} = \dfrac{2}{3}e^{\frac{5\pi}{12}i}$

4 (i) $w = 32e^{\frac{\pi}{2}i}$

    (ii) $2e^{\frac{\pi}{10}i},\ 2e^{\frac{\pi}{2}i},\ 2e^{\frac{9\pi}{10}i},\ 2e^{-\frac{3\pi}{10}i},\ 2e^{-\frac{7\pi}{10}i}$

5 (iii) $\dfrac{\sin\theta + \sin(n-1)\theta - \sin n\theta}{2 - 2\cos\theta}$

6 (ii) $S = 2^n \cos^n\theta \sin n\theta$

7 (i) $a = 16,\ b = -20,\ c = 5$

8 $z^* = a - bi$ so

   $e^{z^*} = e^{a-bi} = e^a\left(\cos(-b) + i\sin(-b)\right)$

        $= e^a(\cos b - i\sin b)$

   $e^z = e^{a+bi} = e^a(\cos b + i\sin b)$ so

   $\left(e^z\right)^* = e^a(\cos b - i\sin b)$

9 (i) $C = \dfrac{e^{3x}(3\cos 2x + 2\sin 2x)}{13} + c_1$

     $S = \dfrac{e^{3x}(-2\cos 2x + 3\sin 2x)}{13} + c_2$

10 $C = \dfrac{2\cos\theta + 1}{5 + 4\cos\theta}$ $S = \dfrac{2\sin\theta}{5 + 4\cos\theta}$

Chapter 15

Review exercise (Page 333)

1 (i) Since $d_2 = -2d_1$ the lines are parallel

    (ii) $\mathbf{p} = \begin{pmatrix} 4 \\ 2 \\ 1 \end{pmatrix}$

    (iii) $\overrightarrow{PQ} = \begin{pmatrix} 2+\lambda \\ -3-3\lambda \\ 2\lambda \end{pmatrix} - \begin{pmatrix} 4 \\ 2 \\ 1 \end{pmatrix} = \begin{pmatrix} -2+\lambda \\ -5-3\lambda \\ 2\lambda-1 \end{pmatrix}$

    (iv) $\lambda = -\dfrac{11}{14}$

    (v) $\dfrac{\sqrt{299}}{\sqrt{14}} = 4.62$

2 (i) $\mathbf{d}_1 = \begin{pmatrix} 1 \\ 3 \\ 2 \end{pmatrix}$ and $\mathbf{d}_2 = \begin{pmatrix} 2 \\ 4 \\ 5 \end{pmatrix}$

    (ii) $\begin{pmatrix} n_1 \\ n_2 \\ n_3 \end{pmatrix} \bullet \begin{pmatrix} 1 \\ 3 \\ 2 \end{pmatrix} = 0$ and $\begin{pmatrix} n_1 \\ n_2 \\ n_3 \end{pmatrix} \bullet \begin{pmatrix} 2 \\ 4 \\ 5 \end{pmatrix} = 0$

    so $n_1 + 3n_2 + 2n_3 = 0$ and

      $2n_1 + 4n_2 + 5n_3 = 0$

      $-2n_2 + n_3 = 0 \Rightarrow n_3 = 2n_2$

    $\mathbf{n} = \begin{pmatrix} -7 \\ 1 \\ 2 \end{pmatrix}$

    (iii) Shortest distance is

    $\left|(a_1 - a_2) \bullet \hat{\mathbf{n}}\right| = \dfrac{1}{\sqrt{54}}\left|\begin{pmatrix} -2 \\ 4 \\ -1 \end{pmatrix} \bullet \begin{pmatrix} -7 \\ 1 \\ 2 \end{pmatrix}\right|$

      $= \dfrac{16}{\sqrt{54}} = \dfrac{8\sqrt{6}}{9} = 2.18$

3 (i) $2\sqrt{2}$ (ii) $\dfrac{1}{11}\sqrt{110}$ (iii) $\dfrac{1}{10}\sqrt{10}$ (iv) 2

4 (i) 2, skew   (ii) 0, intersect

    (iii) $\dfrac{4}{7}\sqrt{21}$, parallel

Activity 15.1 (Page 335)

(i) The line $\mathbf{r} = \begin{pmatrix} 2 \\ 3 \\ 4 \end{pmatrix} + \lambda \begin{pmatrix} 1 \\ 2 \\ 7 \end{pmatrix}$ and the

plane $5x + y - z = 1$

   $5(2+\lambda) + (3+2\lambda) - (4+7\lambda) = 1$

     $10 + 5\lambda + 3 + 2\lambda - 4 - 7\lambda = 1$

              $8 + 0\lambda = 0$

In this case there are no possible solutions so the line and the plane do not intersect.

(ii) The line $\mathbf{r} = \begin{pmatrix} 1 \\ 1 \\ 5 \end{pmatrix} + \lambda \begin{pmatrix} 1 \\ 2 \\ 7 \end{pmatrix}$ and the

plane $5x + y - z = 1$

   $5(1+\lambda) + (1+2\lambda) - (5+7\lambda) = 1$

     $5 + 5\lambda + 1 + 2\lambda - 5 - 7\lambda = 1$

     $5 + 5\lambda + 1 + 2\lambda - 5 - 7\lambda = 1$

                $0\lambda = 0$

In this case there are an infinite number of values for λ so the line is contained within the plane.

The direction vector will be perpendicular to the normal vector if there is not a unique solution.

Activity 15.2 (Page 337)

The line through P and M has equation $r = \mathbf{p} + \lambda\mathbf{n}$.

Substituting this into the equation of the plane $r \cdot \mathbf{n} + d = 0$ gives $(\mathbf{p} + \lambda\mathbf{n}) \cdot \mathbf{n} + d = 0$.

$\mathbf{p} \cdot \mathbf{n} + \lambda\mathbf{n} \cdot \mathbf{n} + d = 0$

$\mathbf{p} \cdot \mathbf{n} + \lambda|\mathbf{n}|^2 + d = 0$

$\lambda|\mathbf{n}|^2 = -\mathbf{p} \cdot \mathbf{n} - d$

$\lambda = -\dfrac{\mathbf{p} \cdot \mathbf{n} + d}{|\mathbf{n}|^2}$

The distance $PM = \left|\overrightarrow{PM}\right| = \left|\mathbf{m} - \mathbf{p}\right| = \left|\mathbf{p} + \lambda\mathbf{n} - \mathbf{p}\right| = \left|\lambda\mathbf{n}\right|$

So, it follows that $\left|\overrightarrow{PM}\right| = \dfrac{\left|\mathbf{p} \cdot \mathbf{n} + d\right|}{|\mathbf{n}|}$.

Exercise 15.1 (Page 334)

2 (i) $(0, 1, 3)$, $67.8°$ (ii) $(1, 1, 1)$, $3.01°$
  (iii) $(8, 4, 2)$, $12.6°$ (iv) $(0, 0, 0)$, $70.5°$

3 (i) 5   (ii) 5   (iii) $5\sqrt{2}$

4 (i) $\mathbf{r} = \begin{pmatrix} 4 \\ 1 \\ 3 \end{pmatrix} + \lambda\begin{pmatrix} 2 \\ 3 \\ 5 \end{pmatrix}$ or

$\mathbf{r} = \begin{pmatrix} 6 \\ 4 \\ 8 \end{pmatrix} + \mu\begin{pmatrix} 2 \\ 3 \\ 5 \end{pmatrix}$ or equivalent

(ii) Intersect at $(0, -5, -7)$
(iii) $11.5°$

5 (i) $\mathbf{r} = \begin{pmatrix} 13 \\ 5 \\ 0 \end{pmatrix} + \lambda\begin{pmatrix} 3 \\ 1 \\ -2 \end{pmatrix}$

(ii) $(4, 2, 6)$
(iii) $\sqrt{126}$

6 (i) $\overrightarrow{AB} = \begin{pmatrix} 2 \\ -3 \\ 2 \end{pmatrix}$

$\overrightarrow{AC} = \begin{pmatrix} -5 \\ 2 \\ -1 \end{pmatrix}$ $x + 8y + 11z = 15$

(ii) $P\left(\dfrac{24}{13}, -1, \dfrac{25}{13}\right)$ $Q\left(\dfrac{267}{40}, \dfrac{-67}{40}, \dfrac{79}{40}\right)$

(iii) $R(6, -1, 4)$ (iv) $22.4°$ (v) 0.89

7 (i) $\dfrac{2\sqrt{5}}{15}$

(ii) $\mathbf{r} = \begin{pmatrix} 2 \\ 0 \\ -5 \end{pmatrix} + \lambda\begin{pmatrix} 4 \\ -5 \\ 2 \end{pmatrix}$

(iii) $M\left(\dfrac{82}{45}, \dfrac{10}{45}, -\dfrac{229}{45}\right)$

8 (i) $\overrightarrow{AA'} = \begin{pmatrix} 1 \\ 2 \\ -3 \end{pmatrix}$ so AA' is perpendicular to

$x + 2y - 3z = 0$

M has coordinates $(1.5, 3, 2.5)$ and $1.5 + (2 \times 3) - (3 \times 2.5) = 0$ so lies in the plane.

(ii) $(0, 3, 2)$; $\mathbf{r} = \begin{pmatrix} 2 \\ 4 \\ 1 \end{pmatrix} + \lambda\begin{pmatrix} -2 \\ -1 \\ 1 \end{pmatrix}$

(iii) $80.4°$

9 (i) $AB = \sqrt{29}$ $AC = 5$; $56.1°$; 11.2
(ii) (b) $4x - 3y + 10z = -12$

(iii) $\mathbf{r} = \begin{pmatrix} 0 \\ 4 \\ 5 \end{pmatrix} + \lambda\begin{pmatrix} 4 \\ -3 \\ 10 \end{pmatrix}$;

Intersect at $(-1.6, 5.2, 1)$
(iv) 16.7

Activity 15.3 (Page 341)

$\mathbf{j} \times \mathbf{i} = -\mathbf{k}$ $\mathbf{j} \times \mathbf{j} = 0$ $\mathbf{j} \times \mathbf{k} = \mathbf{i}$

$\mathbf{k} \times \mathbf{i} = \mathbf{j}$ $\mathbf{k} \times \mathbf{k} = 0$ $\mathbf{k} \times \mathbf{j} = -\mathbf{i}$

Discussion point (Page 343)

Use the scalar product to check that the vector $\mathbf{a} \times \mathbf{b}$ is perpendicular to both $\mathbf{a}$ and $\mathbf{b}$:

$(\mathbf{a} \times \mathbf{b}) \cdot \mathbf{a} = \begin{pmatrix} 24 \\ -1 \\ -14 \end{pmatrix} \cdot \begin{pmatrix} 3 \\ 2 \\ 5 \end{pmatrix}$

$= (24 \times 3) + (-1 \times 2) + (-14 \times 5) = 0$

$(\mathbf{a} \times \mathbf{b}) \cdot \mathbf{b} = \begin{pmatrix} 24 \\ -1 \\ -14 \end{pmatrix} \cdot \begin{pmatrix} 1 \\ -4 \\ 2 \end{pmatrix}$

$= (24 \times 1) + (-1 \times -4) + (-14 \times 2) = 0$

Exercise 15.2 (Page 349)

1 (i) $\begin{pmatrix} -23 \\ 13 \\ 2 \end{pmatrix}$ (ii) $\begin{pmatrix} 37 \\ 41 \\ 19 \end{pmatrix}$

(iii) $\begin{pmatrix} -8 \\ 34 \\ 27 \end{pmatrix}$ (iv) $\begin{pmatrix} 21 \\ -29 \\ 9 \end{pmatrix}$

2 (i) $\begin{pmatrix} 5 \\ 19 \\ -2 \end{pmatrix}$ (ii) $\begin{pmatrix} 14 \\ -62 \\ -9 \end{pmatrix}$

(iii) $\begin{pmatrix} -3 \\ -2 \\ 3 \end{pmatrix}$ (iv) $\begin{pmatrix} -18 \\ 57 \\ 47 \end{pmatrix}$

3 (i) $\overrightarrow{AB} = \begin{pmatrix} 1 \\ -4 \\ 3 \end{pmatrix}, \overrightarrow{AC} = \begin{pmatrix} 4 \\ -1 \\ 0 \end{pmatrix}$

(ii) $\begin{pmatrix} 3 \\ 12 \\ 15 \end{pmatrix}$ or $\begin{pmatrix} 1 \\ 4 \\ 5 \end{pmatrix}$

(iii) $x + 4y + 5z = 7$

(iv) $\dfrac{3\sqrt{42}}{2}$

4 (i) $\left(r - \begin{pmatrix} 3 \\ 1 \\ 0 \end{pmatrix} \right) \times \begin{pmatrix} 15 \\ 27 \\ 7 \end{pmatrix} = 0$

(ii) $\left(r - \begin{pmatrix} 0 \\ -3 \\ 5 \end{pmatrix} \right) \times \begin{pmatrix} 1 \\ 1 \\ -4 \end{pmatrix} = 0$

(iii) $\left(r - \begin{pmatrix} 0 \\ 2.5 \\ 1.5 \end{pmatrix} \right) \times \begin{pmatrix} 16 \\ 15 \\ 13 \end{pmatrix} = 0$

(iv) $\left(r - \begin{pmatrix} 0 \\ 0 \\ 2 \end{pmatrix} \right) \times \begin{pmatrix} -19 \\ 8 \\ 1 \end{pmatrix} = 0$

5 $\dfrac{1}{\sqrt{635}} \begin{pmatrix} 19 \\ 15 \\ -7 \end{pmatrix}$

6 $\sqrt{74}$

7 (i) $5x + 4y - 8z = 5$

(ii) $24x + y - 29z = 1$

(iii) $19x + 40y + 3z = 188$

(iv) $30x - 29y - 24z = 86$

8 (i) $-8\mathbf{j}$ (ii) $6\mathbf{j} - 4\mathbf{k}$

(iii) $2\mathbf{i} - 12\mathbf{j}$ (iv) $6\mathbf{i} + 14\mathbf{j} - 2\mathbf{k}$

9 $\left(r - \begin{pmatrix} 4 \\ -2 \\ -7 \end{pmatrix} \right) \times \begin{pmatrix} 21 \\ 4 \\ 11 \end{pmatrix} = 0$

10 (i) $\overrightarrow{AB} \times \overrightarrow{AD} = \begin{pmatrix} 5 \\ 2 \\ 1 \end{pmatrix} \times \begin{pmatrix} 2 \\ -1 \\ 4 \end{pmatrix} = \begin{pmatrix} 9 \\ -18 \\ -9 \end{pmatrix}$

(ii) Area of the parallelogram is
$\left| \overrightarrow{AB} \times \overrightarrow{AD} \right| = \sqrt{486} = 9\sqrt{6}$ so $k = 9$

(iii) $\left| \overrightarrow{OM} = \overrightarrow{OA} \right| + \dfrac{1}{2}\overrightarrow{AC} = \begin{pmatrix} 1 \\ 0 \\ 5 \end{pmatrix} + \dfrac{1}{2}\begin{pmatrix} 7 \\ 1 \\ 5 \end{pmatrix}$

$= \begin{pmatrix} 4.5 \\ 0.5 \\ 7.5 \end{pmatrix}, L = \begin{pmatrix} 4.5 \\ 0.5 \\ 7.5 \end{pmatrix} + \lambda \begin{pmatrix} 9 \\ -18 \\ -9 \end{pmatrix}$

or $L = \begin{pmatrix} 4.5 \\ 0.5 \\ 7.5 \end{pmatrix} + \lambda \begin{pmatrix} -1 \\ 2 \\ 1 \end{pmatrix}$

(iv) Equation of parallel plane will have same normal

$r \cdot \mathbf{n} = a \cdot \mathbf{n}, r \cdot \begin{pmatrix} -1 \\ 2 \\ 1 \end{pmatrix} = \begin{pmatrix} 4 \\ -1 \\ 2 \end{pmatrix} \cdot \begin{pmatrix} -1 \\ 2 \\ 1 \end{pmatrix}$

$= -4$ or $-x + 2y + z = -4$

(v) Line L and the plane π intersect when
$-(4.5 - \lambda) + 2(0.5 + 2\lambda) + (7.5 + \lambda) = -4$ giving
intersection coordinates as $\left(\dfrac{35}{6}, \dfrac{-13}{6}, \dfrac{37}{6} \right)$.

12 (i) $\begin{pmatrix} 4k - 4 \\ 2 - 2k \\ 4k - 4 \end{pmatrix}$

(ii) (a) $k = 1$ (b) $\dfrac{45}{\sqrt{26}}$

(c) $8x - y + 4z + 5 = 0$

(iii) $\dfrac{1}{3}\left| 12 - 2k \right|$

(iv) $(-6, 13, 14), k = 6$

13 (i) (b) $\begin{pmatrix} 2 \\ 2 \\ -1 \end{pmatrix}$

(c) $\mathbf{r} = \begin{pmatrix} 0 \\ -2 \\ 0 \end{pmatrix} + \lambda \begin{pmatrix} 2 \\ 2 \\ -1 \end{pmatrix}$

(ii) $\dfrac{46}{5}$ (iii) 15

(iv) $k = 50; (16, 8, 4)$

Chapter 16

Activity 16.1 (Page 356)

(i) (a) and (d) are linear, (a), (b), (d) and (e) have constant coefficients

(ii) (a) Can be solved by any of the three methods, $y = Ae^{17x}$

(b) Separation of variables, $y = \dfrac{1}{c - x}$

(c) Separation of variables or integrating factor, $y = Ae^{-\frac{1}{2}x^2}$

(d) Can be solved by any of the three methods, $y = Ae^{3x}$

(e) After division by y, any of the three methods can be used, $y = Ae^x$

Exercise 16.1 (Page 363)

1 (i) $y = Ae^{3x}$ (ii) $y = Ae^{-7x}$

(iii) $x = Ae^{-t}$ (iv) $p = Ae^{0.02t}$

(v) $z = Ae^{0.2t}$

2 (i) $y = 3e^{-2x}$ (ii) $y = e^{\frac{5}{2}x}$

(iii) $x = 2e^{\frac{1}{3}(1-t)}$ (iv) $P = P_0 e^{kt}$

(v) $m = m_0 e^{-kt}$

3 (i) $y = Ae^{4x} + Be^{-4x}$

(ii) $y = Ae^{1.79x} + Be^{-2.79x}$

(iii) $x = Ae^{-1.71t} + Be^{-0.29t}$

(iv) $y = A + Be^{-\frac{2}{7}x}$

(v) $y = e^{\frac{2}{3}x}(Ax + B)$

4 (i) $\lambda^2 - 3\lambda + 2 = 0$

(ii) $\lambda = 1, 2$ (iii) $y = Ae^x + Be^{2x}$

(iv) $\dfrac{dy}{dx} = Ae^x + 2Be^{2x}$

(v) $y = 2e^x - e^{2x}$

5 (i) $\lambda^2 - 8\lambda + 16 = 0, \lambda = 4$

(ii) $y = (A + Bx)e^{4t}$ (iii) $y = (1 - 5t)e^{4t}$

6 (i) $x = \dfrac{4}{5}\left(1 - e^{-5t}\right)$

Initially moves quickly towards its limiting position $x = 0.8$

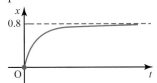

(ii) $x = \left(1 + t\left(\sqrt{e} - 1\right)\right)e^{-\frac{1}{2}t}$

Initially increases briefly then decays to zero

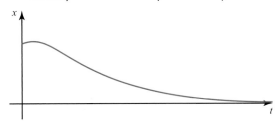

7 (i) $q = (2 + 9t)e^{-2.5t}$

(ii)

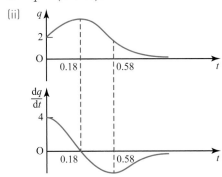

(iii) $q \to 0$ and $\dfrac{dq}{dt} \to 0$ as $t \to \infty$

8 (i) $T = 37.3 + 12.7e^{-0.5t}$

(ii) $T = 42.0\,°C$

(iii)

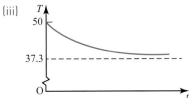

(iv) $37.3\,°C$

9 (i) $T = Ae^{2x} + Be^{-2x}$

(ii) $T = -52.3e^{2x} + 152.3e^{-2x}$

(iii) $87.9\,°C$

10 (ii) $x = 2 - 2e^{-50t}$

(iii) 49.9 seconds, $50\,km\,h^{-1}$

(iv) The car takes an infinite amount of time to stop.

Exercise 16.2 (Page 378)

1 (i) $x = A\sin 3t + B\cos 3t$

(ii) $x = \frac{1}{3}\sin 3t$

(iii) Period $= \frac{2\pi}{3}$, amplitude $= \frac{1}{3}$

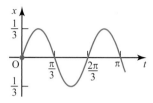

2 (i) $\frac{d^2x}{dt^2} + 64x = 0$ $\frac{dx}{dt} = 0$, $x = 0.1$ when $t = 0$

(ii) $x = 0.1\cos 8t$

(iii) period $= \frac{\pi}{4}$ s, amplitude $= 0.1$ m

(iv)

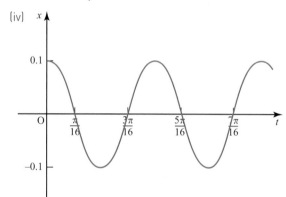

3 (i) $x = A\sin\frac{1}{2}t + B\cos\frac{1}{2}t$

(ii) $x = 4\cos\frac{1}{2}t$

(iii) Period $= 4\pi$, amplitude $= 4$

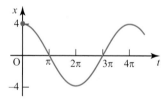

4 (i) 0.357 m

(ii) $\frac{d^2x}{dt^2} + 62.5x = 0$

$\frac{dx}{dt} = 0$, $x = 0.1$ when $t = 0$

(iii) $x = 0.1\cos\sqrt{62.5}\,t$

(iv) period $= \frac{2\pi}{\sqrt{62.5}} = 0.79$ s,

amplitude $= 0.1$ m (10 cm)

5 (i) $y = e^{2x}(A\sin x + B\cos x)$

(ii) $y = e^{x}(A\sin 2x + B\cos 2x)$

(iii) $x = e^{-t}\left(A\sin\sqrt{3}t + B\cos\sqrt{3}t\right)$

(iv) $x = e^{-0.5t}(A\sin t + B\cos t)$

6 (i) $x = A\sin\frac{2}{3}t + B\cos\frac{2}{3}t$

(ii) $x = 5\sin\left(\frac{2}{3}t + 0.927\right)$

(iii)

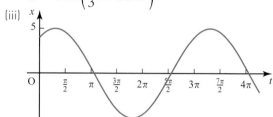

7 (i) $x = \frac{\sqrt{2}}{4}\sin 2\sqrt{2}t + \cos 2\sqrt{2}t$

$= \frac{3\sqrt{2}}{4}\sin\left(2\sqrt{2}t + 1.23\right)$

Oscillations with constant amplitude

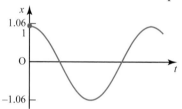

(ii) $x = e^{\frac{t}{2}}\left(\cos\frac{\sqrt{3}}{2}t - \frac{1}{\sqrt{3}}\sin\frac{\sqrt{3}}{2}t\right)$

$= \frac{2}{\sqrt{3}}e^{\frac{t}{2}}\cos\left(\frac{\sqrt{3}}{2}t + 0.52\right)$

Exponentially increasing oscillations

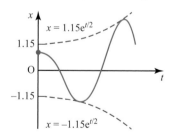

(iii) $x = 2e^{-t}\sin t$

Decaying oscillations

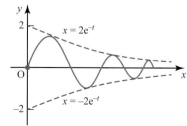

(iv) $x = e^t\left(2\cos\frac{1}{2}t - 4\sin\frac{1}{2}t\right)$

$= \sqrt{20}e^t\cos\left(\frac{1}{2}t + 1.11\right)$

Increasing oscillations

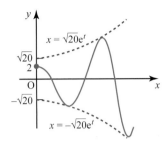

(v) $x = 3e^{-t+\frac{\pi}{4}}\sin 2t = 6.58e^{-t}\sin 2t$

Decaying oscillations

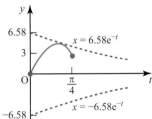

8 (i) $x = A\sin 8t + B\cos 8t$

(ii) $x = 0.1\cos 8t$

(iii) Period $= \frac{\pi}{4}$, amplitude $= 0.1$

(iv)

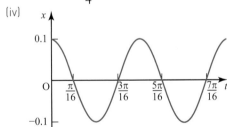

9 (i) $\dfrac{d^2x}{dt^2} + 80x = 0,\ \dfrac{dx}{dt} = 0\text{ and } x = 0.1$

when $t = 0$

(ii) $x = 0.1\cos\sqrt{80}\,t$

(iii)

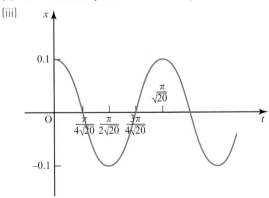

(iv) For critical damping,
damping constant $= \sqrt{20}(\approx 4.47)$

(v) For $m = 0.3$, underdamped system; for
$m = 0.2$, overdamped system

10 (i) (a) $x = 4 - \dfrac{20}{3}e^{-t} + \dfrac{8}{3}e^{-2.5t}$

(b) $x = 4.9 - (7.01t + 4.9)e^{-1.43t}$

(c)
$x = 8 - e^{-0.875t}(10.0\sin 0.70t + 8\cos 0.70t)$

$= 8 - 12.8e^{-0.875t}\cos(0.70t - 0.90)$

(ii)

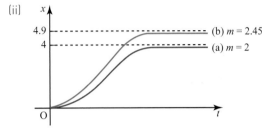

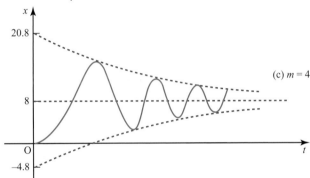

11 (i) $\theta = e^{-2t}(A\sin t + B\cos t)$ or
$\theta = e^{-2t}A\cos(t + \varepsilon)$

(ii) $\theta = \dfrac{\pi}{4}e^{-2t}(2\sin t + \cos t)$ or

$\theta = 1.76e^{-2t}\cos(t - 1.11)$

(iii) The motion effectively ends at about
$t = 2.5$. Although θ subsequently does
become negative, the amplitude of the
oscillation is very small and decreases rapidly.

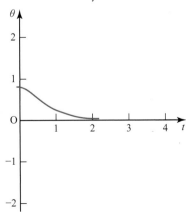

(iv) $\theta \to 0$ as $t \to \infty$

12 (i) $k = \sqrt{20}$

   (ii) (a) Underdamped   (b) Overdamped

14 $y = Ae^{2t} + Be^{-2t} + C\sin 2t + D\cos 2t$

15 (i) $10\dfrac{dv}{dt} = k(y - 0.5) - 4v$

   (ii) $y = \dfrac{3}{2} + e^{\frac{-t}{5}}\left(A\sin\dfrac{7t}{5} + B\cos\dfrac{7t}{5}\right)$

   (iii) $y = \dfrac{3}{2} + e^{\frac{-t}{5}}\left(\dfrac{24}{7}\sin\dfrac{7t}{5} - \cos\dfrac{7t}{5}\right)$

   (iv) $y \to \dfrac{3}{2}$, $v \to 5v$ is unaffected, but y increases as the PI is now $0.5 + \dfrac{20}{k}$, since for large $t, y \to 0.5 + \dfrac{20}{k}$ which is $> \dfrac{3}{2}$ for $k < 20$.

Exercise 16.3 (Page 385)

1 (a) $\ddot{x} = -9x$

   (i) $\pm 6\,\text{cm s}^{-1}, 0$

   (ii) $0, 18\,\text{cm s}^{-2}$

   (iii) $5.2\,\text{cm s}^{-1}$

  (b) $\ddot{x} = -100x$

   (i) $\pm 1\,\text{ms}^{-1}, 0$

   (ii) $0, 10\,\text{ms}^{-2}$

   (iii) $0.87\,\text{ms}^{-1}$

  (c) $\ddot{x} = -\pi^2 x$

   (i) $\pm a\,\pi\,\text{ms}^{-1}, 0$

   (ii) $0, a\,\pi^2\,\text{ms}^{-2}$

   (iii) $\dfrac{\sqrt{3}}{2}a\,\pi\,\text{ms}^{-1}$

2 (i) $10.4\,\text{cm}$ (ii) $\pm 41.8\,\text{cm s}^{-2}$

   (iii) $20.9\,\text{cm s}^{-1}$ (iv) $18.3\,\text{cm s}^{-1}$

   (v) $9.04\,\text{cm}$

3 (i) $55\pi, 110\pi, 220\pi, 440\pi$, below; $1760\pi, 3520\pi, 7040\pi$, above

   (ii) $22.1\,\text{ms}^{-1}, 0.173\,\text{ms}^{-1}$

4 (i) $4000\pi, 25.1\,\text{ms}^{-1}$

   (ii) $\ddot{x} = -16\,000\,000\,\pi^2 x$, $316\,000\,\text{ms}^{-2}$

5 $2.45\,\text{cm}$ above O

6 (i)

   (ii) $33.3\,\text{Hz}, 209$

   (iii) $73.3\,\text{ms}^{-1}$

7 (i) $1.885, 1.13\,\text{m}$

   (ii) $2.12\,\text{ms}^{-1}$

8 (i) $2.67\,\text{ms}^{-1}$

   (ii) $839\,\text{ms}^{-2}$

   (iii) $1.89\,\text{ms}^{-1}$

9 (i), (iv) and (v) are true

Exercise 16.4 (Page 391)

1 (i) $y = -2x - 5$

   (ii) $x = \dfrac{1}{4}t + \dfrac{1}{2}$

   (iii) $y = -0.08\cos 3x + 0.06\sin 3x$

   (iv) $x = \dfrac{3}{4}e^{-2t}$

2 (i) $y = A\cos 2x + B\sin 2x + \dfrac{1}{8}e^{2x}$

   (ii) $x = Ae^{3t} + Be^{-t} + e^{-2t}$

   (iii) $y = e^{-x}(A\cos 2x + B\sin 2x)$ $+ 0.1\sin(x) + 0.2\cos(x)$

   (iv) $y = e^{-x}(A\cos 2x + B\sin 2x) + 0.6x + 0.16$

3 (i) $x = A\cos t + B\sin t - \dfrac{1}{2}t\cos t$

   (ii) $y = Ae^{-4x} + Be^{x} + \dfrac{1}{5}xe^{x}$

   (iii) $x = Ae^{t} + Be^{3t} + \dfrac{1}{2}te^{3t}$

   (iv) $x = (At + B)e^{3t} + 2t^2e^{3t}$

4 (i) $y = -6e^{3x} + e^{2x} + 6x + 5$

   (ii) $x = -2\cos 3t + \sin 3t + 2e^{-t}$

   (iii) $y = e^{-x}(-\cos 2x + 1)$

   (iv) $y = -e^{-2t} + 4e^{t} - \cos 2t - 3\sin 2t$

   (v) $x = \dfrac{1}{2}e^{2t} - e^{t} + \dfrac{1}{2} + te^{t}$

   (vi) $y = 2\sin 2x - 3x\cos 2x$

   (vii) $y = e^{-2x}(2\cos x + 3\sin x) - \cos x + \sin x$

   (viii) $y = \dfrac{1}{2}e^{-2x} + \dfrac{1}{2}e^{2x} + x - 1$

5 (i) $P = Ae^{-t}$

   (ii) $P = 100 - 25(\cos t - \sin t)$

   (iii) $P = 100 - 25(\cos t - \sin t) - 55e^{-t}$

   (iv) 100

   (v) $25\sqrt{2}$

6 (i) $x = e^{-1.5t}\left(A\cos\dfrac{\sqrt{31}}{2}t + B\sin\dfrac{\sqrt{31}}{2}t\right) + 0.05$

   (ii)

$x = e^{-1.5t}\left(0.05\cos\dfrac{\sqrt{31}}{2}t + \dfrac{0.15}{\sqrt{31}}\sin\dfrac{\sqrt{31}}{2}t\right) + 0.05$

    $x \to 0.05$ as $t \to \infty$

   (iii) 1.16

7 (i) $e^{-50t}\left(A\sin 50\sqrt{3}t + B\cos 50\sqrt{3}t\right)$

   (ii) Underdamped

(iii) $-0.1\cos 100t$

$q = e^{-50t}\left(A\sin 50\sqrt{3}t + B\cos 50\sqrt{3}t\right)$
$\quad - 0.1\cos 100t$

(iv) $q = e^{-50t}\left(\dfrac{1}{10\sqrt{3}}\sin 50\sqrt{3}t + \dfrac{1}{10}\cos 50\sqrt{3}t\right)$
$\quad - 0.1\cos 100t$

(v)

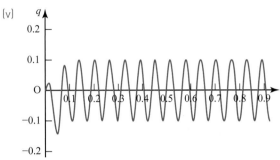

The oscillations quickly settle to almost constant amplitude.

8 (i) (a) $x = 4 - \dfrac{20}{3}e^{-t} + \dfrac{8}{3}e^{-2.5t}$

(b) $x = 4.9 - (7.01t + 4.9)e^{-1.43t}$

(c) $x = 8 - e^{-0.875t}(10.0\sin 0.70t + 8\cos 0.70t)$

(ii) (a) (b)

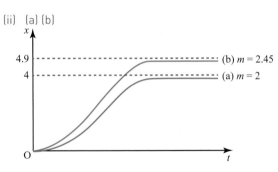

(c)

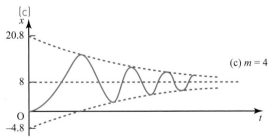

9 (i) (a) (b) $y = Ae^{kt} + \dfrac{1}{p-k}e^{pt}$

(ii) (a) (b) $y = Ae^{kt} + te^{kt}$

10 (i) $x = 2\cos 3t$

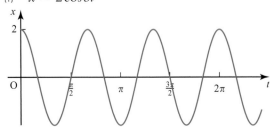

(ii)

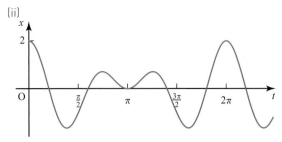

(iii) $x = 2\cos 3t + \dfrac{5}{6}t\sin 3t$

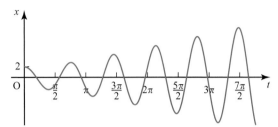

The machine will shake itself to bits (this situation is described as **resonance**).

Exercise 16.5 (Page 398)

1 (i) $x = Ae^{2t} + Be^{t}$, $y = Ae^{2t} + 2Be^{t}$

(ii) $x = e^{t}$, $y = 2e^{t}$

(iii) $(x, y) \to \infty$ as $t \to \infty$ along the line $y = 2x$

2 (i) $x = Ae^{5t} + Be^{-t}$, $y = Ae^{5t} - Be^{-t}$

(ii) $x = \dfrac{1}{2}\left(3e^{5t} - e^{-t}\right)$, $y = \dfrac{1}{2}\left(3e^{5t} + e^{-t}\right)$

(iii) $(x, y) \to \infty$ as $t \to \infty$ along the line $y = x$

3 (i) $x = e^{-t}(A\sin t + B\cos t)$,

$y = \dfrac{e^{-t}}{5}((A - 2B)\cos t - (2A + B)\sin t)$

(ii) $x = e^{-t}(12\sin t + \cos t)$,

$y = e^{-t}(2\cos t - 5\sin t)$

(iii) $(x, y) \to (0, 0)$ as $t \to \infty$

4 (i) $x = e^{2t}(A + Bt)$, $y = -Be^{2t} + 3$

(ii) $x = 2 + (t - 1)e^{2t}$, $y = 3 - e^{2t}$

(iii) $x \to \infty$, $y \to -\infty$ as $t \to \infty$

5 (i) $x = \dfrac{\sqrt{2}}{4}\left(e^{\sqrt{2}t} - e^{-\sqrt{2}t}\right)$,

$y = \dfrac{1}{4}\left(2 - \sqrt{2}\right)e^{\sqrt{2}t} + \dfrac{1}{4}\left(2 + \sqrt{2}\right)e^{-\sqrt{2}t}$

(ii) $(x, y) \to \infty$ as $t \to \infty$ along the line $y = \left(\sqrt{2} - 1\right)x$

6 (i) $x = 1 - e^{t}\cos\sqrt{6}t$, $y = \dfrac{\sqrt{6}e^{t}}{2}\sin\sqrt{6}t + 1$

(ii) Spirals away from $(0, 1)$ to ∞

7 (i) $0, -20$

(ii) $\dfrac{d^2x}{dt^2} - 3\dfrac{dx}{dt} - 4x = 0$

(iii) $x = 3e^{4t} + 12e^{-t}$

(iv) $y = 12e^{-t} - 2e^{4t}$

(v) The second species becomes extinct after nearly 35.8 years.

8 (i) $p = \dfrac{28}{5}, q = 7$

(iii) $p = Ae^{-10t} + Be^{-3t} + \dfrac{28}{5}$

$p = \dfrac{28}{5} - \dfrac{3}{5}e^{-10t} - 5e^{-3t}$

$q = 7 + 3e^{-10t} - 10e^{-3t}$

(iv)

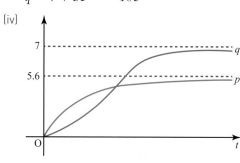

$p \rightarrow \dfrac{28}{5} = 5.6, q \rightarrow 7$

9 (ii) $x = e^{0.01t}(49\,900t + 10\,000)$

(iii) $y = e^{0.01t}(49\,900t + 5\,000\,000)$

(iv) $x = 273\,000, y = 5.52$ million

10 (i) $x = 8e^{-4t}, y = 16e^{-2t} - 16e^{-4t},$
$z = 8e^{-4t} - 16e^{-2t} + 8$

(ii) 4

(iii) $z \rightarrow 8$ as $t \rightarrow \infty$

11 (iii) $a > b + c$, limit is $\dfrac{a - b - c}{a - c}$

Practice questions 4 (Page 403)

1 (i) 2.428477 [1]

0.242848 [1]

2.663075 [1]

(ii) $k_1 = 0.2, f(x_1 + h, y_1 + k_1) = 2.202272$ [1]

$k_2 = 0.220227$

$y_2 = y_1 + \dfrac{1}{2}(k_1 + k_2) = 2.210144$ [2]

2 (i) $\ddot{x} = \dot{x} - 0.09\dot{y}$ [1]

$= \dot{x} - 0.09(y - 0.16x)$ [1]

$= \dot{x} + 0.0144x - 0.09y$

$= \dot{x} + 0.0144x - (x - \dot{x})$ [1]

$\ddot{x} = 2\dot{x} - 0.9856x$ [1]

(ii) $m^2 - 2m + 0.9856 = 0$ has solutions

$m = 1.12, 0.88$ [1]

Hence $x = Ae^{1.12t} + Be^{0.88t}$ is the general solution. [1]

(iii) $\dot{x} = 6.72e^{1.12t} + 7.92e^{0.88t}$ [1]

$0.09y = x - \dot{x}$ [1]

$y = 8e^{1.12t} + 12e^{0.88t}$ [1]

3 (i) $-\dfrac{1}{2} + i\dfrac{\sqrt{3}}{2}$ [3]

(ii) $-\dfrac{5}{2} + i\dfrac{\sqrt{3}}{2}, \quad \dfrac{1}{2} - i\dfrac{3\sqrt{3}}{2}$ [3]

(iii) $-1 - \dfrac{i\sqrt{3}}{2}$ [3]

4 (ii) $\dfrac{e^{i\theta}\left(e^{2ni\theta} - 1\right)}{e^{2i\theta} - 1}$ [7]

(iii) $\dfrac{\sin^2 n\theta}{\sin\theta}$ [2]

5 (i) $x = e^{-3t}\left(A\cos 5t + B\sin 5t\right) - \dfrac{2}{3}\cos 4t + \dfrac{1}{2}\sin 4t$

(ii) $x = -\dfrac{2}{3}\cos 4t + \dfrac{1}{2}\sin 4t$ as the initial conditions affect only A and B, and these are in terms tending to zero because of the factor e^{-3t}.

(iii) The auxiliary equation has roots $3 \pm 5i$, and so the exponential factor is e^{3t} which tends to infinity. This leads to the model breaking down (infinite current is not possible) unless the initial conditions give $A = B = 0$.

6 (i) $\begin{pmatrix} 10 \\ 9 \\ 1 \end{pmatrix} \times \begin{pmatrix} 4 \\ -8 \\ 2 \end{pmatrix} = \begin{pmatrix} 26 \\ -16 \\ -116 \end{pmatrix}$

(ii) $Q\,(5, 7.5, 1.5)$
Distance is $49.80\,\text{m}$

(iii) New distance is $47.14\,\text{m}$
Charlie has walked $16.08\,\text{m}$

Index